Regolith Geology and Geomorphology

Regolith Geology and Geomorphology

G. Taylor
University of Canberra

R. A. Eggleton
Australian National University

JOHN WILEY & SONS, LTD

Chichester • New York • Weinheim • Brisbane • Singapore • Toronto

Baffins Lane, Chichester,
West Sussex PO19 1UD, England

National 01243 779777
International (+44) 1243 779777
e-mail (for orders and customer service enquiries): cs-books@wiley.co.uk
Visit our Home Page on http://www.wiley.co.uk
or http://www.wiley.com

Other Wiley Editorial Offices

John Wiley & Sons, Inc., 605 Third Avenue,
New York, NY 10158-0012, USA

WILEY-VCH Verlag GmbH, Pappelallee 3,
D-69469 Weinheim, Germany

John Wiley & Sons Australia, Ltd, 33 Park Road, Milton,
Queensland 4064, Australia

John Wiley & Sons (Asia) Pte Ltd, 2 Clementi Loop #02-01,
Jin Xing Distripark, Singapore 129809

John Wiley & Sons (Canada) Ltd, 22 Worcester Road,
Rexdale, Ontario M9W 1L1, Canada

Library of Congress Cataloging-in-Publication Data

Taylor, G. (Graham)
Reglith geology & geomorphology / G. Taylor, R.A. Eggleton.
p. cm.
Includes bibligraphical references and index.
ISBN 0-471-97454-4
1. Regolith. 2. Weathering. 3. Geomorphology. I. Title: Regolith geology and geomorphology. II. Eggleton, R.A.

QE511 .T285 2001
551.48—dc21 2001024879

British Library Cataloguing in Publication Data

A catalogue record for this book is available from the British Library

ISBN 0-471-97454-4

Typeset in 9/11pt Plantin by Mayhew Typesetting, Rhayader, Powys
Printed and bound in Great Britain by Bookcraft (Bath) Ltd, Midsomer Norton
This book is printed on acid-free paper responsibly manufactured from sustainable forestry, in which at least two trees are planted for each one used for paper production.

Contents

Preface

The decision to write this book dates back about 10 years, but it only came together during the last few. In writing it we were conscious of the fact that both of us are Australian and that the majority of our work has been in Australia. We had the view that perhaps our experience being so geographically restricted our efforts may have been too localized for an international readership. The final impetus to get on with writing it came when we joined other regolith researchers and teachers in the Cooperative Research Centre for Landscape Evolution and Mineral Exploration. Joining this, and reading the literature showed us that there was a need for such a book, as a summary, as a source, and as a general reference book for working professionals in many fields including mineral exploration, environmental sciences and in geomorphology.

Few syntheses of the type we present here have been published, and certainly not with the slant this book takes. There are many books that approach various aspects of the subject from a particular view, for example climate, weathering, geomorphology, exploration geochemistry and others. We wanted to try to bring many aspects of the subject together to look at the general principles underlying an understanding of the regolith, its description and genesis.

In doing this from an Australian perspective we bring some new insights developed away from the yoke of traditional Northern Hemisphere ideas and concepts. In Australia our geological and geomorphological science has developed under this yoke and because of this many of the precepts of the science that were taken for granted have been challenged as research has led to new understandings. Australia, while not unique, is a continent that experienced its last major glaciation during the Permian. It has regolith and landscapes that can be shown to date back to the mid-Palaeozoic. It is tectonically less active than many others, but this is not to suggest that tectonism is not affecting post-Palaeozoic landscapes and regolith. But, overall it is sufficiently different to have provoked a new thinking on regolith evolution. For these reasons we are not now concerned about any parochialism the book may show and hope that the work sourced in Australia will serve to open others' minds as they pursue their study and research.

The book is organized rather differently from many others on aspects of the subject because we believe it brings a new approach to the problems of understanding regolith geology and landscapes. Because regolith forms in a landscape context we start, after an introduction, with some views on landscapes and their evolution. Within landscapes minerals and water interact to produce regolith, so the first chapters address the nature of minerals, the basic controls and factors involved in rock weathering and, since water is so important in weathering, the behaviour of water at a landscape scale. Chapters 6–10 focus on weathering, its reactions, chemistry and products.

Rocks having weathered are available for erosion and deposition as transported regolith, so Chapter 12 considers the physically transported debris from weathering and its character in deposits. Complementing 12, Chapter 13 addresses the fate of solutes that remain in the regolith after weathering reactions have changed the water composition, as well as exploring water behaviour at various scales in regolith materials.

Chapter 14 looks at methods of conveying information about regolith via maps and in GIS packages, while the last chapter briefly looks at some applications for our understanding of regolith geology.

Although we have borrowed extensively in preparing this book, we are solely responsible for its content. Having said this, we cannot possibly list all the help and influences that have led to our being in a position to write the book. To all our colleagues over our careers we thank you for either stimulating us or for annoying us sufficiently that we got on and worked on the problem. Both of us would like to acknowledge with sincere thanks our students, undergraduate and research, who have conspired to keep us on our toes and thinking. To our families who have put up with what it takes to acquire geological experience, thanks for your understanding and support. We do not know what it says, but of the four children we have between us none are scientists, much less geologists. They have obviously learned well from their dads.

GT wishes to acknowledge three people in particular. Firstly, Griffith Taylor my great uncle. It was Griff who handed down his iconoclasm and his quest for knowledge to me as well as his love of geology. Secondly, Geoff Sykes, my high school geology teacher and Keith Crook my PhD adviser without whose support I never would have had the opportunities I did early in my career. Geoff's and Keith's enthusiasm and tolerance of off-beat ideas were critical in my geological development. Like Graham, RAE's life in science that led to this book has been greatly influenced by others; my parents who engendered the curiosity essential for sustained research, Grade 7 Primary School Teacher Mr R. Hole who gave focus to never-ending questions and Bull Bailey at the University of Wisconsin, the greatest teacher of all.

GT and RAE have been colleagues and friends for about 15 years and the trust and friendship that have evolved over that time has enabled writing this book to be a pleasurable experience. Thanks Tony, thanks Graham.

The Visual Resources Unit of the Cooperative Research Centre for Landscape Evolution and Mineral Exploration was responsible for drafting all the figures, and what a great group they proved to be, thanks to you all for your support and the quality product you have produced. Judy Papps provided those keyboard and presentation skills that writing a book demands; a wizard with the software, coping with problematic handwriting, providing style to the text, and all done with unfailing good humour. Andrew McPherson sought and found innumerable references and sources; his efforts were invaluable to us. Thanks also to Bernadette Kovacs, particularly for her help in proof-reading.

The cover is a painting done at Weipa in north Queensland by the late Fred Williams, who is widely regarded as Australia's premier landscape artist. For us, he captures the essence of that landscape. We feel very privileged to be able to use one of his many paintings for our book and it was his wife Mrs Lyn Williams who cleared the way for its use. Our sincere thanks Mrs Williams

We thank the Cooperative Research Centre for Landscape Evolution and Mineral Exploration (CRC LEME) to which we belong. The support of the Australian Government's Cooperative Research Centres Scheme, and of our colleagues in the Centre is acknowledged; their work has enhanced what we offer in this book. Many diagrams come from CRC LEME publications, and where they are used here we acknowledge the author.

Graham Taylor and Tony Eggleton
Canberra
May 2001

1 Introduction: of rotten rocks, soils and solutions

Of all the material making up the solid Earth, the most important to humans is its surface mantle, the regolith. Most continents are covered by it to varying degrees and it is the regolith that records much of the Earth's more recent history. The regolith is host to many of the resources that maintain our society and living standards: aluminium from bauxite, diamonds, clays, nickel, many iron ores, and opals. It is also in the upper part of the regolith that our soils are formed; the regolith is thus the host for agriculture.

By contrast, the regolith can be a barrier to the exploration for economic mineral resources in the rocks beneath. If the exploration industry is to find new mineral resources then it must learn how to read the signals in the regolith that indicate the presence of concealed mineral deposits. Very finally, most of us find our ultimate resting place within the regolith.

Many features that now form part of the rock record, and many ore deposits, had a regolithic origin. The type bauxite at Le Baux, in France, occurs in a sequence of dipping Mesozoic strata, revealing that for a time prior to their tilting, these rocks were exposed and underwent prolonged chemical weathering. There have even been suggestions some massive sulfide ores such as those at Broken Hill in Australia have a regolithic origin. The study of regolith has been used to interpret past geological and climatic conditions. For example, it is widely believed that a 'laterite' or bauxite indicates that the climate was tropical at the time and place of its formation. Thus the study of the regolith is able to provide insights to the evolution of the Earth and provides one of the few windows to our past which are so important in predicting how shifts in climate are likely to affect us.

It is the regolith that is the natural environment for humankind and as such it is prudent for us to learn as much about it as we can. This book is dedicated to this end, but we will touch on more specific reasons noted above throughout the book.

What is the regolith?

The term 'regolith' was first defined by Merrill (1897), who derived the word from the Greek *ρηγοσ* (regos) meaning blanket or cover and *λιθοσ* (lithos) meaning rock or stone. Extracts from his work read:

> Everywhere, with the exception of comparatively limited portions laid bare by ice or stream erosion, or on the steepest mountain slopes, the underlying rocks are covered by an incoherent mass of varying thickness composed of materials essentially the same as those which make up the rocks themselves, but in greatly varying conditions of mechanical aggradation and chemical combination.
>
> In places this covering is made up of material originating through rock weathering or plant growth *in situ*. In other instances it is of fragmental and more or less decomposed material drifted by wind, water or ice from other sources. This entire mantle of unconsolidated material, whatever its nature or origin it is proposed to call the regolith . . .

Merrill also provided a tabulation of some regolith materials and their genesis (Table 1.1).

The American Geological Institute's *Glossary of Geology* (Jackson 1997) defines regolith as:

> A general term for the layer or mantle of fragmental and unconsolidated rock material, whether residual or transported and of highly varied character, that nearly everywhere forms the surface of the land and overlies or covers the bedrock. It includes rock debris of all kinds, volcanic ash, glacial drift, alluvium, loess and aeolian deposits, vegetal accumulations and soil.

Table 1.1 *Merrill's (1897) summary of his view of what he defined as regolith*

The regolith	Sedentary	Residual deposits	Residuary gravels, sands, and clays, wacke, laterite, terra rosa, etc.
		Cumulose deposits	Peat, muck and swamp soils, in part
	Transported	Colluvial deposits	Talus and cliff debris, material of avalanches
		Alluvial deposits (including aqueo-glacial)	Modern alluvium, marsh and swamp (paludal) deposits, the Champlain clays, loess and adobe, in part
		Aeolian deposits	Wind-blown material, sand dunes, adobe and loess, in part
		Glacial deposits	Morainal material, drumlins, eskers, osars, etc.

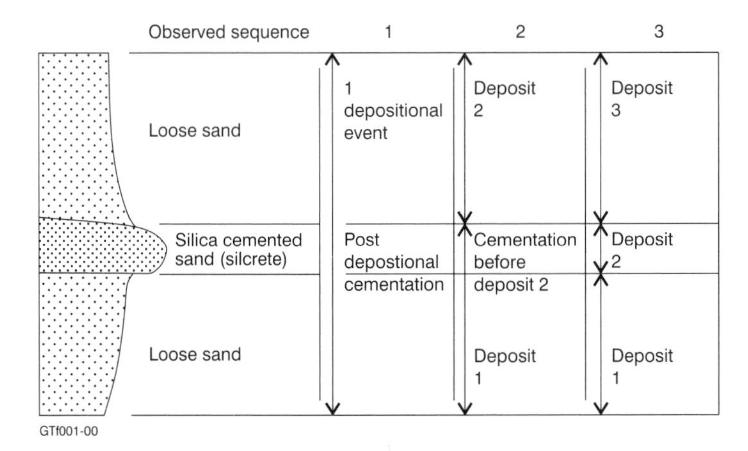

Figure 1.1 *There are many different interpretations that can be used to explain the observed sequence. Without more data it is difficult, but it is clear that the silcrete (silicified) layer is part of the regolith despite its being cemented. Three scenarios are presented to interpret the observations*

This definition is somewhat restrictive and does not address important areas in considering what is, and what is not, regolith.

One of the problems with Jackson's (1997) definition arises from the requirement that the regolith should be unconsolidated. Many examples of material such as cemented rock debris sandwiched between unconsolidated materials exist. One such example is where silica cements one layer of sand in a sequence of sands. Does this mean that the sands above and below the cemented layer are called regolith while the silicified layer is not? Not in our opinion (Figure 1.1).

Equally, are fresh basaltic lavas part of the regolith or are they bedrock? We have no clear answer, but suggest that its interpretation depends on the context of the basalts. Figure 1.2 shows a hypothetical, but often encountered scenario. Our suggestion would be that where fresh basalt (or other lava) conform to present landscapes in some way, and are sandwiched

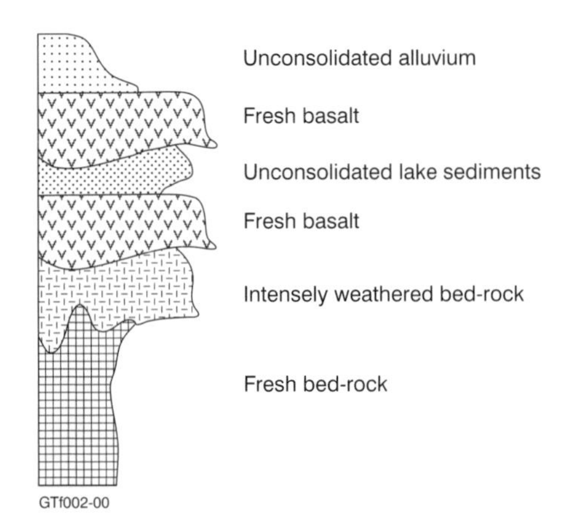

Figure 1.2 *A hypothetical section through a typical section from the Tertiary and Quaternary basaltic provinces of eastern Australia*

between unconsolidated or weathered rock materials, they should be considered as part of the regolith, despite Jackson's (1997) definition.

Although the age of materials does not form part of the definition of regolith, many would not consider it possible for Palaeozoic materials to be part of the regolith. This view probably emanates from the Northern Hemisphere where landscapes are predominantly of Quaternary origin. There are many examples of unconsolidated Palaeozoic material sitting at or near the present surface in Australia. The best known examples are Permian tills in northern and central Victoria. Macumber (1978) makes the point that the present drainage is cut into unconsolidated Permian tills in a landscape that has been established since the Neogene. The question here is – is the till regolith? Our answer is 'yes', because the tills are obviously related to the extant landscape and so are part of the regolith.

Pain *et al.* (1991) take the view that regolith is 'bedrock which has been altered by processes at or near the surface. This includes induration of the regolith . . .'. This is a pragmatic definition and is very similar to the working definition we use in this book for regolith: '*Regolith* is all the continental lithospheric materials above fresh bedrock and including fresh rocks where these are interbedded with or enclosed by unconsolidated or weathered rock materials. Regolith materials can be of any age.'

Regolith research

Study of the regolith includes its description, distribution, genesis and age. The techniques used in regolith research vary little from those used in conventional geological, pedological and geomorphological research. There are some problems peculiar to regolith work such as dating and mapping and these areas will be discussed in more detail later. Additionally, although researchers have been working for almost a century on regolith and regolith-related problems, in this area of Earth science, unlike most other branches, few genetic models exist for explaining the formation of the regolith. This lack of paradigm has held back advance in many areas of the science and we are only now getting to the point of developing robust genetic models against which to test ideas and to use as aids in understanding the regolith. Let us give an example. Igneous petrologists have models to explain the shape of plutons, so when mapping them there are some robust ideas to generate hypotheses as they map. On the other hand, regolith scientists have no robust and generally accepted model for the formation of 'lateritic profiles' so there is little help from a model when mapping. We have a long way to go. We think this book will take us some of the way and help to integrate and formulate ideas and eventually models.

Some regolith researchers and other geoscientists have asserted that regolith research is difficult and complex. There are several reasons for this including, most importantly, that most students of Earth science are never trained to work with surface and near-surface weathered or degraded rocks. How many professional Earth scientists can say they were trained to work with weathered and redistributed surface materials in their initial training? – very few. Most geologists look at weathered rocks and the first question they ask is 'what *was* this rock?' Few have any interest in what it is now or how it came to be that way. In this book we hope to make an understanding of regolith material available to all and to provide a framework for asking the right questions which will lead to knowledge.

Regolith studies in Australia

The study of the regolith has a long history in Australia, but not as an integrated subject in the Tertiary curriculum, yet regolithic materials have provided much of our wealth. Just as the regolith of the Yilgarn region in Western Australia is presently producing much gold, so it did in the past. The deep-leads of Kiandra were, in the 1850s, some of our richest gold workings. Opal is a world-renowned Australian gem produced entirely from deeply weathered regolith across much of New South Wales, Queensland and South Australia. Many other examples of regolith resource exploitation from the nineteenth century are available. Most were exploited because of their occurrence at or near the surface and the ease with which they could be worked. This early resource exploitation led to extensive knowledge on the part of prospectors, but added little to geology.

Regolith is interdisciplinary

The regolith is the material at the intersection of the lithosphere, hydrosphere, atmosphere and biosphere (Figure 1.3). To understand the features observed and the processes which form the regolith there is a need to consider the interactive effects of rocks, water, air and biota. In this sense the study is interdisciplinary. For example, the size and disposition of the mottles

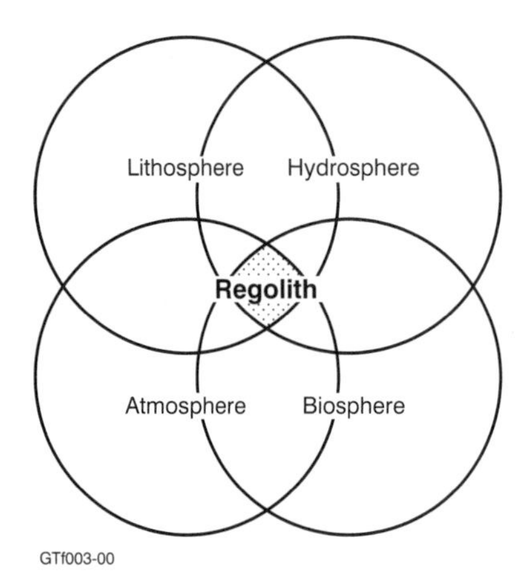

Figure 1.3 *Diagram illustrating the relative position of regolith in the global interaction between processes operating on Earth*

(patches of red to yellow in a white or grey background) in weathering profiles are dependent on:

- the clays produced by the weathering of the parent rocks;
- the rock structure;
- the state of oxidation of iron, which is in turn related to the amount of air in the water moving through the profile;
- the climate and porosity of the regolith; and, in many cases,
- patterns of biotic activity such as root channels or termite burrowings.

What is the role of biotic activity in rock weathering? Do algae and microbes play a role? Certainly, but the amount of evidence available is limited. There is no doubt that tree roots do play a role and we will discuss this. Water movement through weathering profiles plays one of the major roles in the actual process of weathering as well as controlling the degree to which rocks weather, and it is water which imports and exports components of the newly forming profile. It is thus obviously critical to understand water movement and composition if weathering processes are to be sensibly understood.

As well as the nature of the lithosphere materials being important, the shape of the land surface is critical in understanding the movement of regolithic materials, water and biotic residues (organic matter) across and through the regolith. Geomorphology is thus an important control on regolith formation and nature, and a deep understanding of it is a prerequisite to the interpretation of regolith evolution.

Understanding the processes at the surface and near-surface occurring in the soils is important in the transport of components, both particulate and in solution, downward through the entire regolith. This makes an appreciation for the nature and processes in soils an essential prerequisite to understanding regolith.

In this book we cannot cover all of these aspects in detail, but where such relationships do occur we will make reference to them. Ideally, research into regolith then should be interdisciplinary in order to extract maximum understanding.

Global regolith distribution

Strakhov (1967) produced a map and section (Figure 1.4) illustrating the distribution and depth of weathering (regolith) on a global scale. The major controlling variable on his models was climate, including temperature, precipitation, evaporation, organic production and relief. While it can be argued this is a very simple model, it is a first approximation to the distribution of regolith on a global scale (Figure 1.4). Strakhov's map also accounts for tectonic activity as a major influence on the thickness and consequently type of regolith preserved at various sites.

One major shortcoming of most studies of global regolith distribution is that they are climatically based, yet in many cases the regolith preserved is relict, and unrelated to contemporary climates, particularly in tectonically passive regions. Butt & Zeegers (1992b) do make this point and the map of Büdel (1982) encapsulates the concept of tectonic stability (degree of surface plantation) connected to age of regolith and the potential for relict features to be preserved (Figure 1.5). Büdel also inferred that where this occurs the regolith is potentially thicker than in his other morphoclimatic zones.

Strakhov's (1967) model of regolith evolution does not, however, incorporate any concept of relict features and his concept is relatively simple in the sense that where climates are wetter and warmer with high organic contributions regolith will be thicker than in areas of lesser rainfall, with cooler climates and reduced organics (Figure 1.6).

We are not aware of any factually based global maps of regolith. There are a number of world soil maps (e.g. US Department of Agriculture Soil Conservation Service, the 1 : 5 000 000 *FAO/Unesco Soils Map of the World* (Figure 1.6)), but these only reflect gross scale regolith surface variation and reveal little of the total regolith.

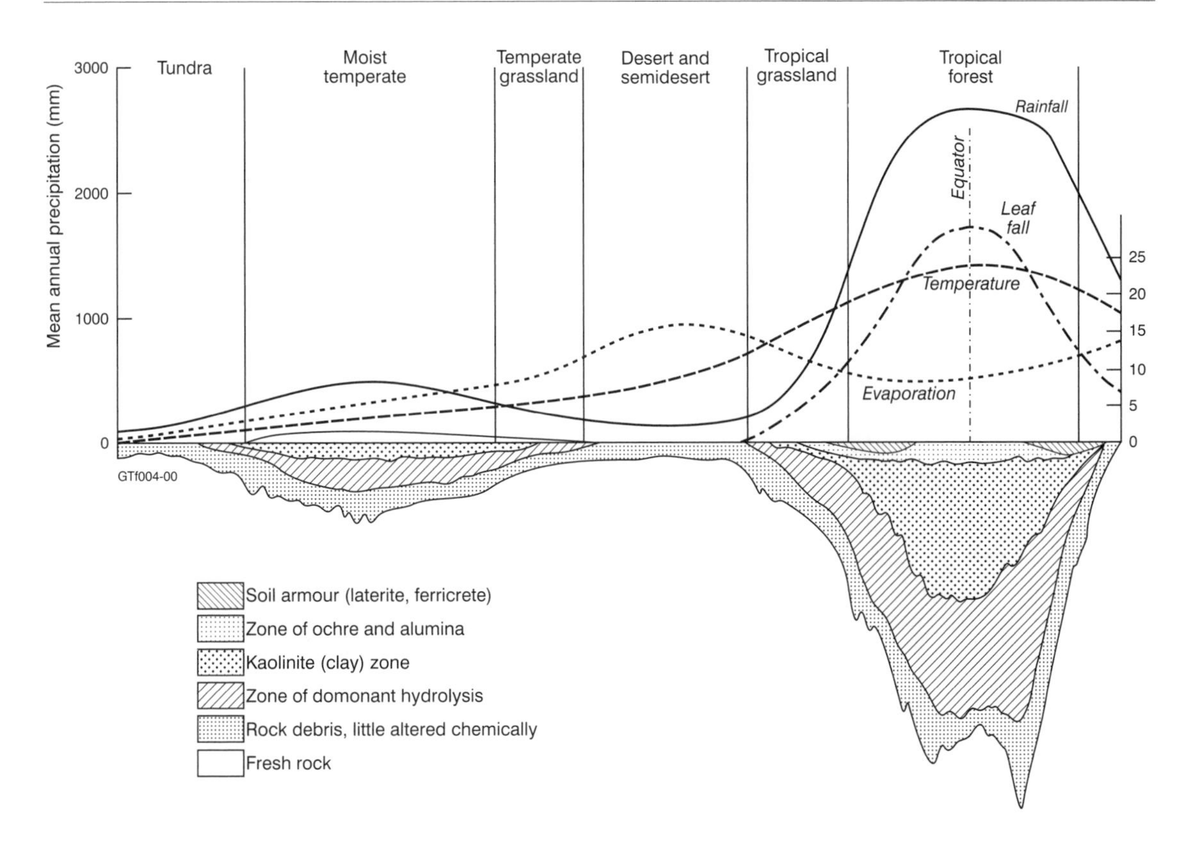

Figure 1.4 *Weathering mantles and climate. The effects of climatic change are ignored in this summary (after Strakhov 1967)*

There are many regolith maps of Australian regions, varying from local maps produced for a specific purpose (say at 1 : 10 000 scale) to the 1 : 10 000 000 scale map of the continent (Figure 1.7 in colour plates). This shows areas of extensive Cainozoic sedimentary cover (unconsolidated and consolidated), Cainozoic basalt fields, major areas of duricrust and deep weathering and Pre-Cainozoic surfaces.

The lithosphere

The lithosphere is chemically very simple with 98.5% of it consisting of seven elements (Table 1.2). Despite there being almost 3000 minerals known, these seven major elements combine in varying ways to make up 85% of the crust from only a few minerals or mineral groups (Table 1.3). However, underneath this mineralogical simplicity lie considerable complexities of solid solution and intergrowths. At the high temperatures of igneous and metamorphic crystallization, silicate structures are more flexible than at Earth surface temperatures. This leads to a wide compositional range in minerals of the same structure. In feldspars, K, Na and Ca can each be accommodated in one structural site. In amphiboles, the cations of Al, Mg, Ti and both ferrous and ferric iron are equally accepted. The minerals of the regolith are different, and simpler, in this regard, allowing little in the way of atom substitution in their structures.

This simplicity has two consequences for geological and more particularly regolith studies: (i) the Earth's crust and thus also its surface are relatively easy to understand at an overview level of observation, and (ii) that to understand much detail of the processes and materials at a more rigorous level we need to concentrate on examination of things which are rare, both chemically and mineralogically. For example, the Al content of goethite in regolith materials may be related to the conditions under which it forms, whereas the simple fact that goethite is present only indicates the precipitation of Fe under humid oxidizing conditions.

The composition of the lithosphere discussed above is for the whole crust, which is made up of about 95% igneous rocks, but regolith forms at the Earth's surface, and at the surface the majority ($\approx$70%) of the rocks

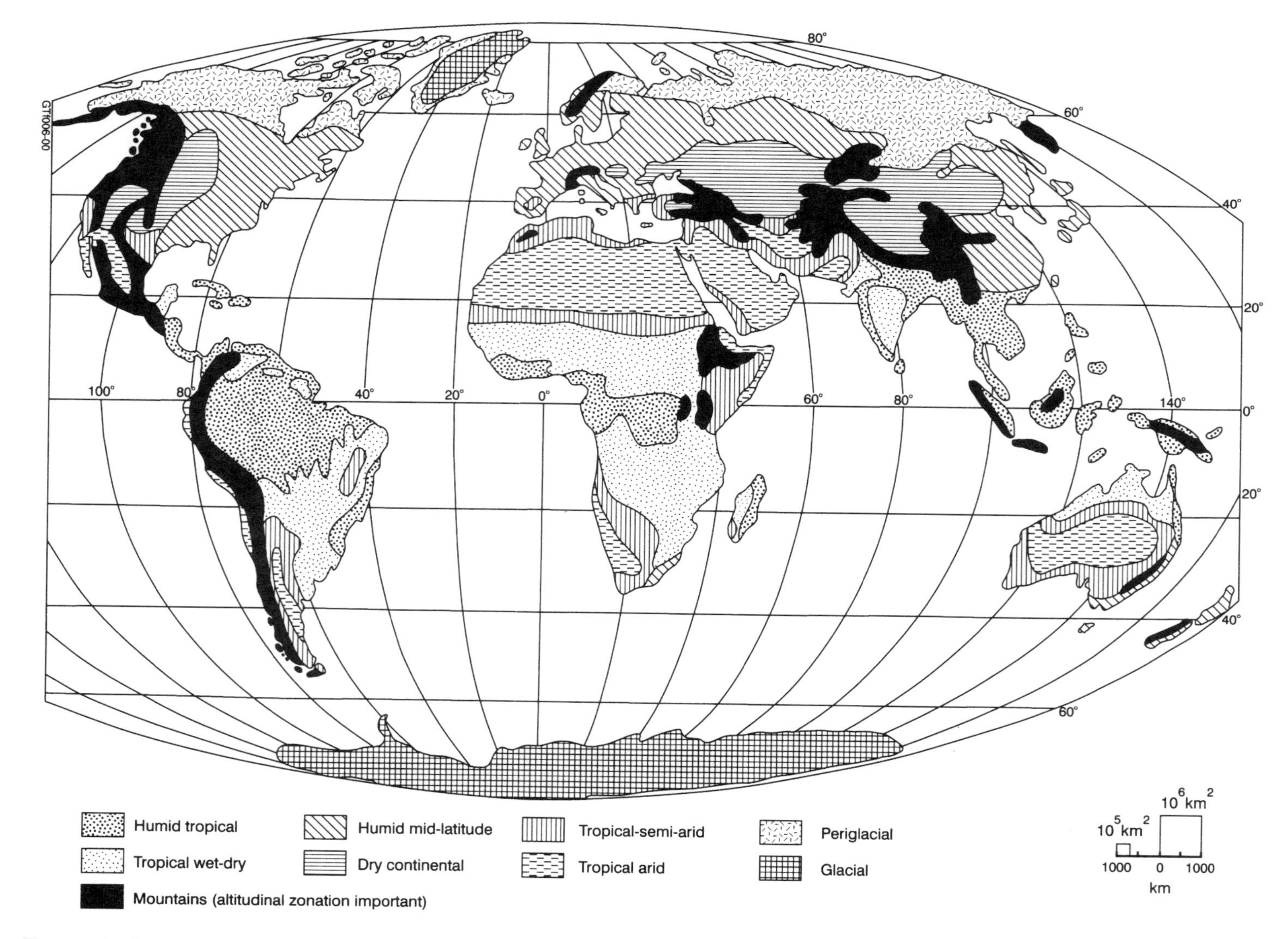

Figure 1.5 *Present morphoclimatic zones (after Tricart & Cailleux 1972)*

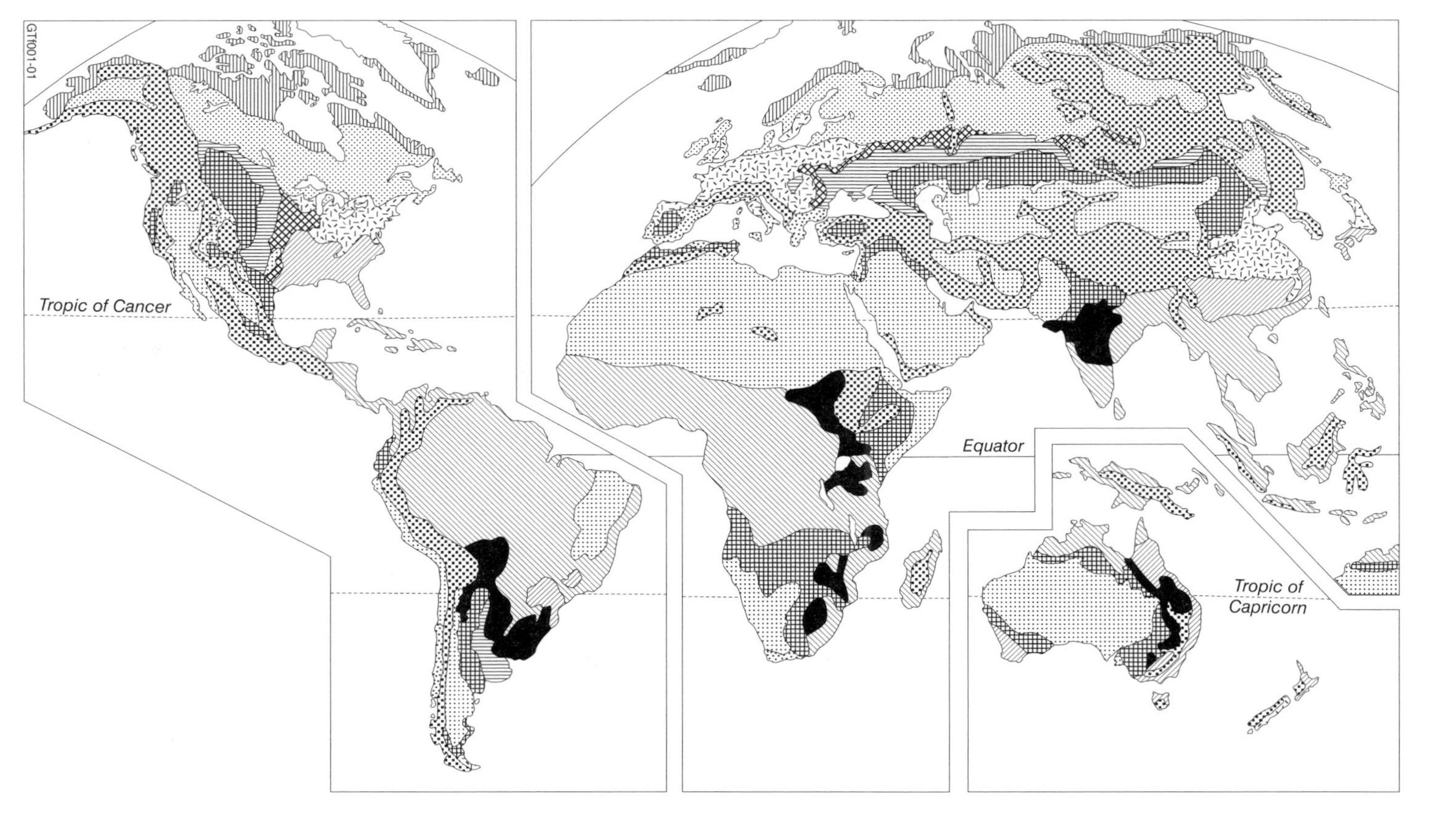

Figure 1.6 *Simplified world soil map (from FAO/Unesco* Soils Map of the World)

Table 1.2 *Lithospheric elemental abundances (%) (from Ollier 1984)*

O	64.6
Si	27.7
Al	8.1
Fe	5.0
Ca	3.6
Na	2.8
K	2.6
Mg	2.1
Subtotal	98.5
Remaining 95 elements	1.5

Table 1.3 *Lithospheric abundances of minerals and mineral groups (%) (from Ollier 1984)*

Feldspars	41.0
Quartz	12.0
Pyroxenes	11.0
Amphiboles	5.0
Micas	5.0
Clay minerals	4.6
Olivine	3.0
Calcite/dolomite	2.0
Magnetite	1.5

that crop out are sedimentary. Sediments have an even simpler composition than the whole crust, being essentially composed of materials that have been through at least one cycle of weathering and erosion. The average oxide composition of sedimentary rocks is shown in comparison to igneous (primary crustal) rocks in Table 1.4. These changes come about because rock weathering causes fractionation of the elements, essentially into water-soluble and insoluble species. The mineralogy of regolith is dominated by Al, ferric iron and Si, with Ca important in more arid regolith. Among minor elements, only Ti is particularly noteworthy.

With a restricted suite of available elements, and only one set of environmental conditions (25°C and 1 atm pressure), most regolith ends up as an aggregate of few major minerals: quartz, illite, kaolinite, gibbsite, goethite or haematite, anatase and calcite. Of these, quartz, illite, kaolinite, gibbsite and anatase have fairly constant compositions worldwide, goethite and haematite may include Al in the Fe site, and some regolith calcite contains Mg. Certainly there are many other minor minerals in the regolith: heavy metal carbonates around weathered sulfides, gypsum and halite in dry lakes, smectite clays in poorly drained parts, and unweathered resistant minerals from the parent rock such as magnetite, zircon and muscovite.

The data of Table 1.4 clearly show a reduction in alkalis, alkaline earths and ferrous iron with weathering, and an increase in ferric iron, CO_2 and water. These changes result from the processes of weathering that are precursors to the formation of sedimentary rocks. This trend is more clearly shown when average mineralogical shifts from igneous to sedimentary rocks are examined (Table 1.5). Here quartz, clay minerals and carbonates have increased at the expense of olivine, pyroxene and feldspar as a direct result of the sedimentary rocks having been through at least one weathering cycle.

Garrels & Mackenzie (1971) also summarize the effects of weathering as shown in Table 1.6, with results little different from those between igneous and sedimentary rocks, except carbonates are notable for their absence.

As can be seen from Tables 1.4 and 1.5, a sedimentary cycle decreases the variety of mineralogical

Table 1.4 *Oxide compositions of average igneous and sedimentary rocks (from Garrels & Mackenzie 1971)*

Composition	Average igneous rocks (%)[a]	Average shale (%)[b]	Average sedimentary rocks (%)[c]
SiO_2	63.5	61.9	78.0
Al_2O_3	15.9	16.9	7.2
Fe_2O_3	2.9	4.2	1.7
FeO	3.3	3.0	1.5
MgO	2.9	2.4	1.2
CaO	4.9	1.49	3.2
Na_2O	3.3	1.07	1.2
K_2O	3.3	3.7	1.3
CO_2	—	1.54	2.6
H_2O (110 °C)	—	3.9	2.2

From: [a] Brotzen (1966).
[b] Garrels & Mackenzie (1971).
[c] Pettijohn (1963).

Table 1.5 *Mineralogical composition of average igneous and sedimentary rocks*

Normative mineralogy (wt%)[a]	Average igneous	Average sedimentary
Olivine	13	
Pyroxene	16	
Wollastonite	3	
Plagioclase	46	
Feldspar	19	12
Haematite	3	4
Quartz	16	35
Calcite		7
Dolomite		4
Illite		27
Chlorite		7
Montmorillonite		3

From: [a] Garrels & Mackenzie (1971).

Table 1.7 *Hypothetical mineralogical composition of fresh and weathered granite illustrating the mineralogical changes between the weathered material and the original rock*

Fresh granite	Weathered granite	Very highly weathered granite
Quartz	Quartz	Quartz
Plagioclase	Orthoclase	Kaolinite
Orthoclase	Illite	Gibbsite
Biotite	Smectite	Haematite
Hornblende	Kaolinite	Goethite
Zircon	Haematite	Anatase
Magnetite	Goethite	Zircon
Sphene	Anatase	? Magnetite
	Zircon	? Pyrolusite
	Magnetite	
	Gypsum	
	Pyrolusite	
	Calcite	

Table 1.6 *Examples of % normative mineralogy. Calculations as applied to the alteration of granite, basalt and hornblende schist under tropical weathering conditions. The alteration condition is such that the major rock minerals are nearly completely gone, except for quartz (from Garrels & Mackenzie 1971)*

Normative mineralogy	Rock type and weathered residue								
	Granite				Basalt		Schist		
	Fresh	Residue I	Residue II	Residue III	Fresh	Residue	Fresh	Residue I	Residue II
Quartz	31	42	37	30			4	8	23
Orthoclase	17	9	6	1					
Plagioclase	28	2	2	1	60	5	38		
Olivine					14				
Augite					20	1			
Hornblende							49		
Muscovite	18	8	7	1					
Ilmenite	1	1	1	2	3	4	9	8	7
Kaolinite	3	37	44	63		72		32	44
Goethite	2					18		23	12
Gibbsite			2					27	2
Magnetite					3				6
Water				2				2	6

transformations that occur as sedimentary rocks weather compared to those involved in the weathering of igneous and metamorphic rocks. In all cases the weathering of rocks involves a transformation of many original minerals with shifts in the relative proportions of the new minerals depending on the degree of weathering (Table 1.7).

Accompanying these mineralogical transformations are fabric (arrangement of minerals and voids in the material) changes as well. Weathering granite, for example, may maintain its original fabric even though it is weathered, but as the degree of weathering increases, that fabric is lost and a new one developed. Similar fabric transformations occur as most rocks and other Earth materials weather. In addition to the profound mineralogical and fabric shifts that accompany weathering, large chemical changes also occur with the movement, of solutes mainly via water, both into and out of the active weathering environment.

Because water is the most active agent in these changes, and because water movement is controlled, among other things, by the slope of the land, geomorphological considerations are important in understanding the mineralogical and chemical changes that occur as rocks weather. Because water and regolith are also affected by gravity, landscape position is

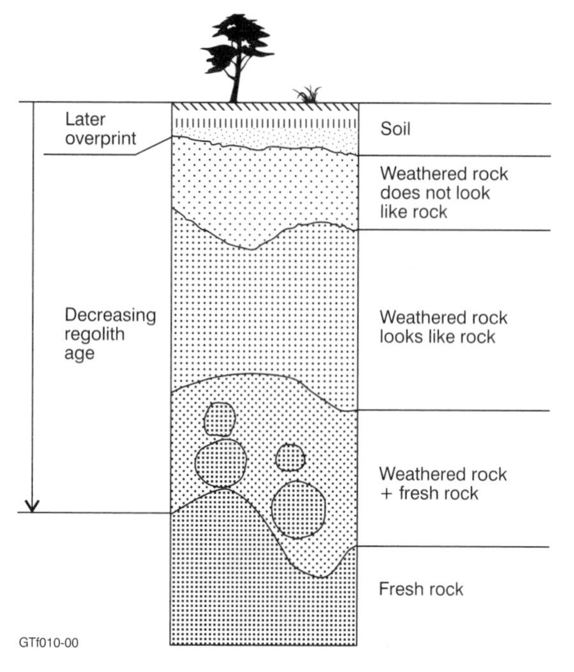

Figure 1.8 *The age of the regolith increasing from the surface to the weathering front between the fresh rock and weathered material. The uppermost part of the regolith, although much of the material may be old, is usually a soil that is the result of an overprint process on weathered rock materials*

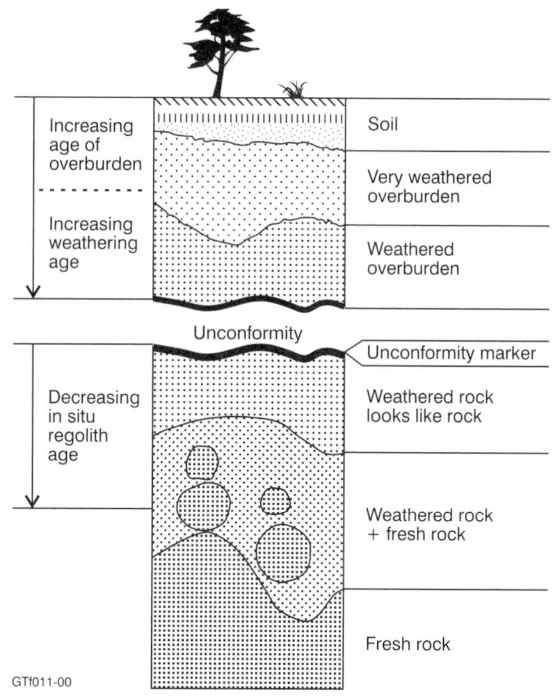

Figure 1.9 *Shows the same weathering (regolith) profile as in Figure 1.8, but with an unconformity in the sequence and the brief interpretations of this in temporal terms*

important in considering the cause and origin of fabric changes that occur, as is the biotic activity in the regolith and rocks.

All of these changes that occur as rock goes to regolith are treated in the following chapters.

Regolith and landscape

Regolithic materials are generally distributed in landscapes in particular ways and knowledge of this distribution leads to some understanding of processes operating in regolith formation, in others it merely confuses the situation. There are, however, a few simple observations we need to consider relating to regolith distribution and stratigraphy in a landscape context before moving on.

AGE OF REGOLITH

If one considers that weathering begins from the surface down, then there is an implied age sequence in the materials in a weathered rock profile (Figure 1.8). This concept, although generally true, is oversimplified. New materials are often added to the surface of weathering profiles by various processes, and many profiles can be shown to contain unconformities (Figure 1.9) of one type or another (see Chapter 11). Equally, the older near-surface materials are often removed by erosion leaving only partial remnants of profiles.

Apart from additions and unconformities, dating regolith is a difficult undertaking. It may be dated relatively by its relationships to landscape elements (Chapter 2) or by absolute methods such as isotopic dating, palaeomagnetic dating, or the fossil contents (Chapter 11).

REGOLITH IN A LANDSCAPE

Some general observations are pertinent at this stage but the subject is expanded later. Weathering of *in situ* Earth materials or from transported debris resulting, ultimately, from weathering, forms regolith. Thus the

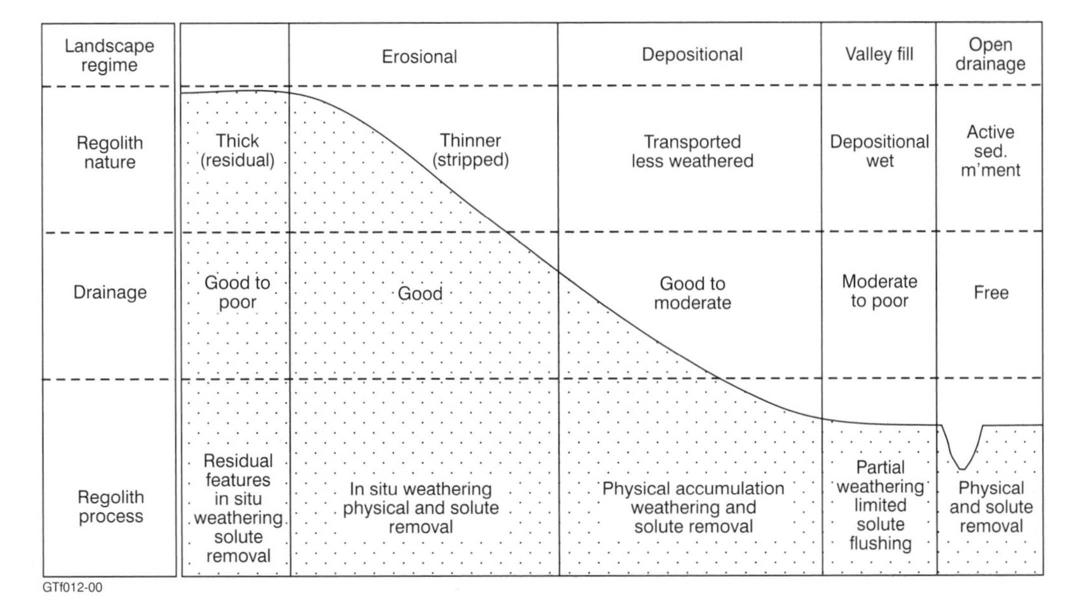

Figure 1.10 *Shows some simple relationships between landscape position, regolith type, drainage and regolith process*

nature of a regolith depends on two major landscape variables, topography and drainage (Figure 1.10). Each of these features will be treated in significantly more detail in later chapters; however, the fundamental relationships are needed at this stage. Topography basically controls erosion and thus regolith thickness as well as drainage, which in turn controls the degree of weathering and consequently regolith mineralogy. It also follows from Figure 1.10 that surface features higher in the landscape are, in general, older than those in lower situations.

Regolith, stability and its preservation

It is implied in Figure 1.10 that regolith is eroded, deposited and removed in solution. The regolith, whenever in contact with the processes which form it, is continually changing. Soils, which form the uppermost part of the regolith change hourly, for example as worms work through them, and they change on a longer timescale as rain erodes the surface. So, too the whole regolith changes as groundwater drains solutes from it, as infiltrating water carries colloid-sized solids deeper, as new rock minerals begin to dissolve at the weathering front and so on. In the landscape regolith tends to be more stable on elevated flat regions, erodes from upper slopes and is deposited (as sediment) on lower slopes and valley fills.

Preservation of regolith depends on either it not being eroded or being covered by materials too thick to allow continued regolith formation below them. We have just described an example of the former, and a good example of the latter is where a regolith is covered by lava flows or by thick sediment piles, so preserving the former regolith and saving it from further modification.

Factors involved in regolith formation

Many factors impinging on the development of regolith have been briefly discussed above. It is now time to summarize them in a systematic way and to delineate the portions of the regolith.

Factors important in regolith development are:

- tectonic activity;
- chemical, mineralogical and physical nature of the materials from which it is formed;
- topographic position;
- drainage conditions;
- landscape processes including erosion and deposition;
- biological activity;

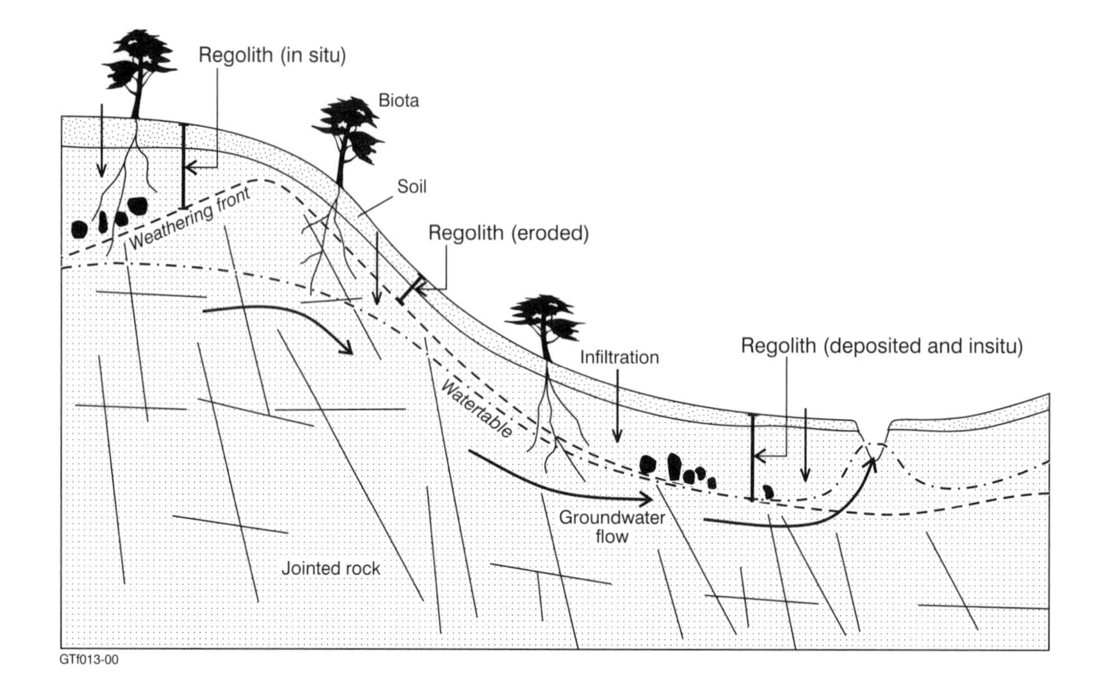

Figure 1.11 *Summary of the regolith features in a landscape and some of the factors important in its formation*

- climate, particularly precipitation and temperature; and,
- time.

Components of the regolith include all terrestrial materials above the weathering front in the lithosphere. These include:

- weathered rock and sediments;
- fragments or beds of fresh rock;
- biota; and,
- soils.

These features are illustrated in Figure 1.11.

2 Overview – landscapes and the evolution of the regolith

Introduction

There are many ways to consider landscapes; traditionally they are classified and conceptualized in relation to the factors that are presumed to have led to their formation or by the climatic zones in which they are observed or in which they are thought to have formed. It is these genetic factors which control the shape of the landscape and the nature of the regolith developed in the landscapes.

In Chapter 1 a brief overview of simple landscapes and the regolith was outlined. Virtually all landscapes, regardless of scale and no matter how flat they appear, can be considered as simply slopes. Hilltops are relatively stable, the upper portions of hillslopes are where erosion occurs and deposition occurs on the lower slopes and in the valley bottoms, but minor erosion is also common here as part of normal fluvial processes. A similar approach can be taken with regard to long sections downstream of river valleys.

Basically, high elevation elements of the landscape (hills, plateaux or mountains) exist where rocks are more resistant to weathering and/or erosion than those in the adjacent areas or where tectonic or volcanic forces have built them. It is essentially the interplay of erosion and tectonics that determines broad-scale landscape configuration.

Hills and valleys can be characterized in terms of four components (Figure 2.1).

Hilltops

Hilltops come in many shapes, but most have an area of relatively gently sloping or flat ground. Many hilltops, particularly in areas of arid or very old landscapes, have extensive areas of flat ground. The predominant processes on hilltops are infiltration of water with little and slow lateral runoff. This results in there being very little removal of surface material by water transport. However, wind can be a significant cause of both physical erosion and deposition on hilltops, as of course can ice in climates where this is a factor. Physical erosion of hilltops is thus minimal, but chemical erosion may be more significant.

Because infiltration of water falling on hilltops is often a significant process, chemical weathering and vertical movement of weathering products, both as particles and in solution, are important. This means that the regolith on hilltops can be thick, and on plateaux be over-thickened by deeper weathering. In many cases though, because hills form where the rocks are not very susceptible to weathering, outcrops of relatively fresh rock cap hilltops. Often, even though fresh rock crops out, it is there as a result of significant weathering and stripping of weathered material leaving the fresh rock exposed (Figure 2.2).

In the example of interbedded sandstone and shale the sandstones crop out because they are very resistant to both physical and chemical weathering compared to the shales. In the granite example, the exposed corestones do not weather as rapidly as the buried granite because their contact time with water, particularly groundwater, is significantly less, and they are thus relatively more resistant to weathering.

There is also abundant evidence from Australia and Africa that hardened regolith or duricrust that caps many hilltops or mesas (e.g. ferricrete, silcrete, see Chapter 13) protects the hilltops from erosion (see Figure 2.3). Many of these caps may not extend across the entire surface, but may owe their origin to the relatively flat hilltop and lateral migration of water and solutes. Water moves laterally through the regolith on the hilltop, dissolving minerals as it does so. When the

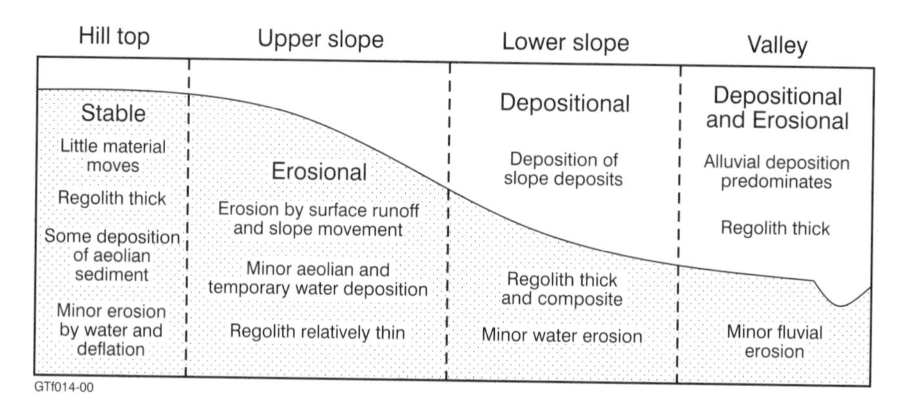

Figure 2.1 *A schematic sketch of generalized landscape elements, dominant processes and regolith nature*

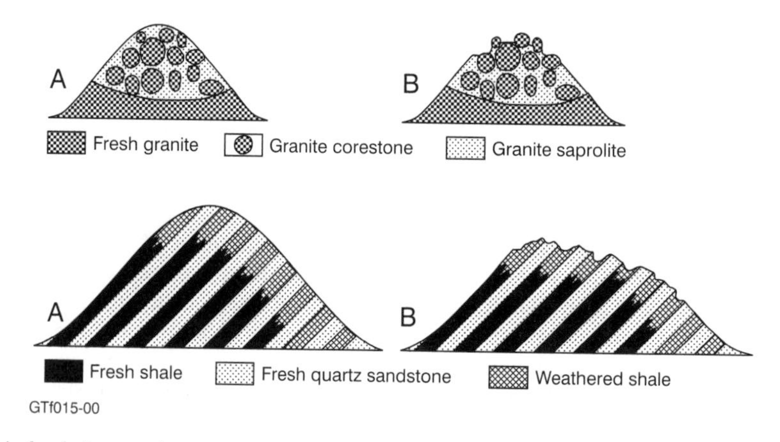

Figure 2.2 *This sketch diagram shows two examples of how fresh rocks can crop out at a hill crest, where weathered material may be present, but not readily visible. (a) Hill crest with weathering. (b) The same hill crest after some weathered material is eroded*

Figure 2.3 *Mesas and plateaux from the Hat Hill region of South Australia with a flat top protected from erosion by a duricrust. Although the duricrust 'protects' the mesa, it does erode around the margins as the duricrust is undercut and collapses*

water comes close to the margins of the hills the dissolved minerals precipitate to form a hard rim or skirt to the hilltop or mesa margin (Figure 2.4). This is well illustrated in many central Australian situations and in parts of the Eastern Highlands (Taylor & Ruxton 1987). Others (e.g. Milnes *et al.* 1985; Ollier & Galloway 1990; Pain & Ollier 1995) suggest the hills are present because the duricrust has protected the area from erosion that has proceeded in adjacent areas not previously covered by duricrust (Figure 2.4).

Upper slopes

The upper slopes of hills are significantly steeper than hilltops and are generally convex. As a result of the steeper slopes, erosion is much greater than at the hilltop. Upper slopes can vary in steepness from vertical (90°) to very low angles in subdued landscapes (Figure

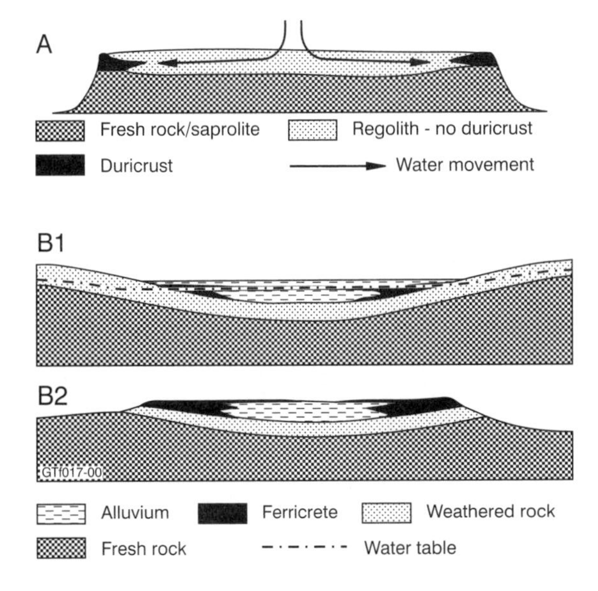

Figure 2.4 *Schematic models to illustrate the formation of (a) duricrust on mesa or hilltops and (b) the formation of a hill because of the presence of a duricrust. (a) Duricrust forms by lateral movement of solutes and their precipitation at the mesa margins by evaporation. (b1) The precipitation of duricrust at the foot of slopes where spring water evaporates. (b2) The duricrust protects the material underneath from erosion so that the topography is inverted to form 'skirts' of duricrust around the mesa*

2.5), nonetheless they are steeper than slopes above and below. The major processes operating in this part of the landscape are weathering and erosion. The shape and nature of the upper slope are controlled by the balance between weathering rates, erosion rates and the nature of the materials on which the slope is formed.

Chemical weathering on upper slopes generally occurs at slower rates than on lower slopes as water is resident in the profile a shorter time, but physical weathering may well be more intense due to the enhanced effect of gravity on steeper slopes. Erosion processes include regolith creep, flow, slide and fall as well as surface water erosion and removal of solutes and colloidal particles in groundwater. The rate of erosion depends on the nature of the regolith materials and how readily they are detachable from the slope, and this in turn depends on the degree of weathering and the strength of the regolith materials.

In this part of a hill slope the regolith is generally thinner than above or below it in the landscape. Because it is in this part of the landscape that most erosion occurs, the regolith here is resident on the landscape for shorter times and it is thus generally younger than on hilltops and often less well developed.

Lower slopes

Concave and concavo-convex slopes are predominant in the lower parts of the landscape. Slope angle decreases across this segment, being steeper near the hill and lower towards the valley axis. The major process is deposition, both from hill slope debris moved by creep, slide, fall, etc. under the influence primarily of gravity, and from sediment deposited from surface water flow. The area closer to the hill contains more mass movement deposits and the area nearer the valley more alluvium, often with no slope deposits at all, except in very small catchments.

The regolith in the lower slope region tends to be thick relative to the upper slope because it is thickened by deposition, but weathered bedrock may or may not be stripped, particularly in the axial part of the valley prior to accumulation of sediments. The deposits are weathered continuously, the degree of alteration being related to the rate at which they deposit and the amount and variability of groundwater flow through them. Unlike on the hilltop and upper slope, the regolith here is likely to be predominantly transported and may not reflect the character of the underlying rocks. Conversely, regolith on the top and high parts of the slopes will more closely reflect bedrock character, as it is close to being, if not actually, *in situ* there.

The shape of hills or slopes varies considerably. In Figure 2.6 the slope is a typical concavo-convex one, but others also occur, including concave, steep to vertical, and convex. Formation of these various shapes is said to relate to differences in formative process and/or climatic influences. Convex hills are largely thought to relate to landslip wasting of the steeper upper slopes under the influence of gravity. Steep to vertical slopes are the result of weathering resistant rocks occurring on the slope. Convex slopes are primarily related to the removal of weathered material from over crystalline or very weathering-resistant rocks (etchplanation of Büdel 1957, 1982) (Figure 2.6c).

While almost all landscapes are best described as a series of hills and valleys, or lows and highs, regardless of absolute relief, there are land surfaces with such little relief that they are best described as plains. Plains are also called a variety of other names including erosion surfaces, etchplains, peneplains, pediplains, all of which have genetic connotations associated with

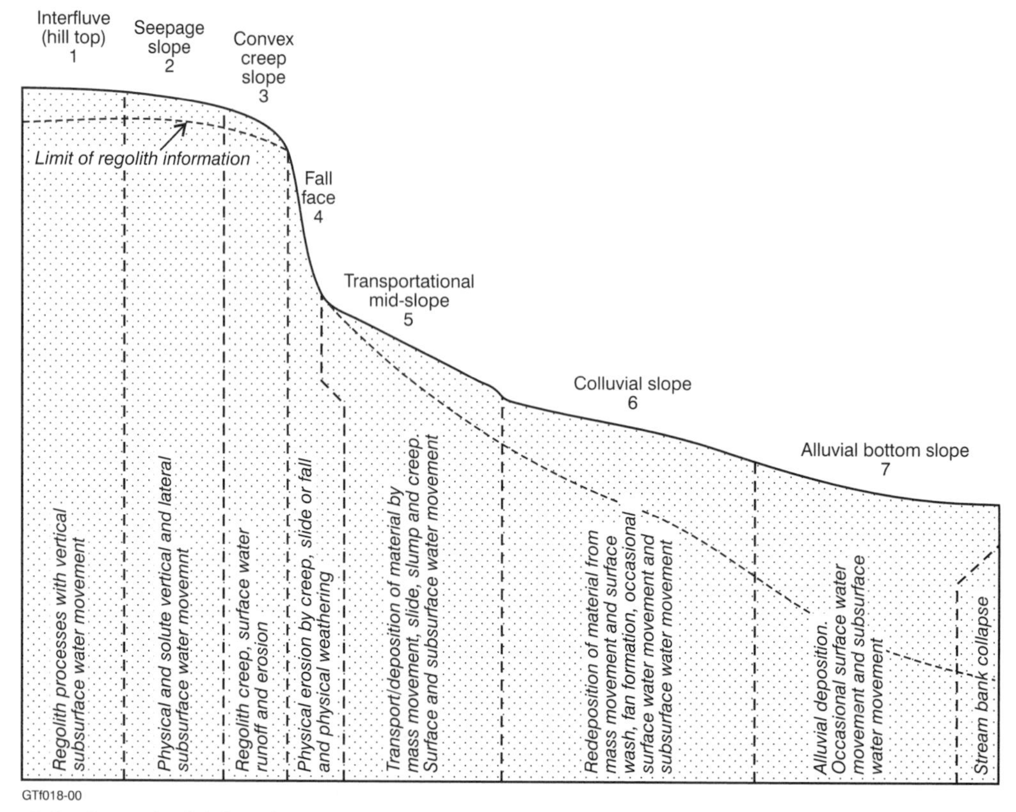

Figure 2.5 *A more detailed theoretical model of slope profiles including a comprehensive range of slope components and the processes operating in those segments. The various segments occur in nature in various possible sequences as illustrated by the data of Dalrymple* et al. *1968 (simplified from Selby 1993)*

them. Basically there are two genetic plain types, erosional and depositional.

Erosional plains have a variety of genetic explanations. Perhaps the earliest is that of Davis (1889). He suggested that landscapes evolve from flat surfaces through incision as shown in Figure 2.7a. The essentials of this landscape evolution model are that as erosion progresses, valleys widen and relief decreases (slope angles decline) until a more or less planar surface (a peneplain in his terminology) is achieved. Savigear (1953) and Pain (1986) suggested that slopes changed through time by a process of slope replacement where the slope becomes longer and angles decline (Figure 2.7c). Figure 2.7b represents the popular interpretation of Penck's (1924) model of parallel slope development where slope characteristics remain stable but valleys widen – basically a slope retreat model. These two models and the misinterpretation of Penck's are theoretical models based on little empirical work. King (1951, 1956, 1957), working in Southern Africa where hills bounded by very steep slopes or escarpments are common, developed his model of scarp retreat (Figure 2.7d). This process operates by rock fall from the steep slope (free face) and removal of the fall debris by water.

All these models, which have formed the basis of understanding of slope and erosional plain evolution for a long time, have very significant implications for the type of regolith that may be associated with the landscape, depending on which model is used to interpret landscape evolution. Whether the process of scarp retreat or slope decline can generate extensive plains such as are observed to cut across many types of basement rocks is another question. There is little doubt that in localized areas slope retreat produces low angle pediments surrounding higher plateaux (Figure 2.8), but can this lead to extensive plains such as are seen on the Yilgarn Craton in Western Australia or in regions of the Eastern Highlands of Australia near Goulburn (Figure 2.9). This question arises because in many cases there are no remnants of retreating scarps remaining in these landscapes.

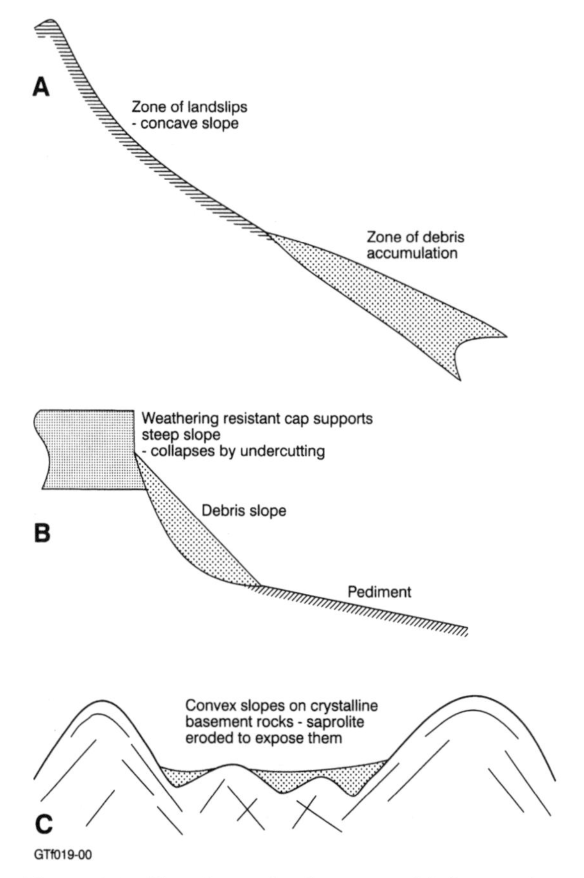

Figure 2.6 *Slope shapes related to process. (a) Concave slope; (b) free-face slope; (c) convex slopes*

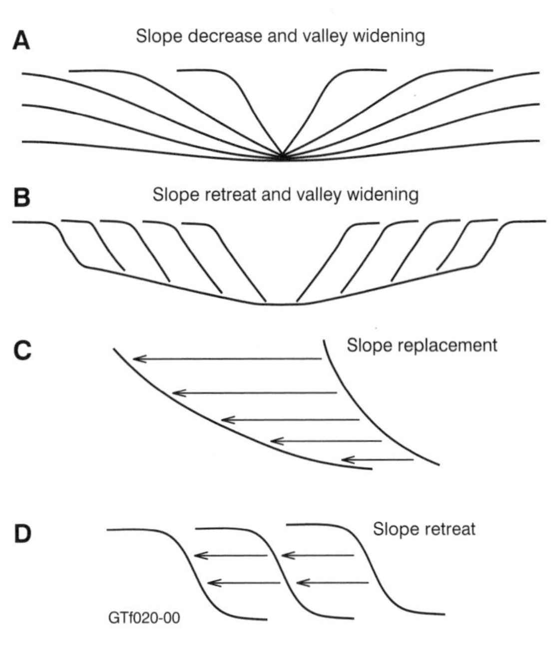

Figure 2.7 *Slope evolution models. (a) Davis's model explaining evolution from V-shaped youthful valleys to relatively flat peneplains indicative of senility of the landscape. (b) Penck's (1924) model of parallel slope retreat and the formation of a pediment. (c) Slope replacement (or slope retreat) accompanied by slope reduction. (d) King's model of the formation of erosion surfaces by slope retreat (cf. (b)) and valley widening*

How then do plains that cut across a variety of lithologies and geological structures form? Many authors have recognized the problems of the traditional models outlined above and have offered alternative explanations. Hills (1975) suggests that many of the plain lands on Palaeozoic (or older) rocks had formed by the beginning of the Late Palaeozoic or Early Mesozoic and that remnants of Permian glacial deposits and pavements which form part of today's landscape suggest that glaciation may have played a role. This is certainly possible for many of the areas glaciated during the Carboniferous to Permian, but there are large plains for which this explanation is not possible (e.g. the Victoria River region of the Northern Territory). One answer is etchplanation. This concept, first advanced by Wayland (1934) and subsequently developed and refined by many others, is that deeply weathered rock is partly or substantially removed by erosion leaving the fresh rock below the weathering front as convex hills sitting on a plain of saprolite and alluvium. A good summary of etchplain formation is provided by Thomas (1994). Uluru and Kata Tjuta are excellent examples of this form of weathering remnants (inselbergs) sitting above an etchplain (Figure 2.10). Another possibility for the surface cut across highly folded rocks in the lower-lying regions of the Northern Territory is that they are marine erosion surfaces cut during mid-Cretaceous high sea levels. The region does contain thin sequences of Cretaceous shallow marine sediments overlying the basement. A summary of the formation of many examples of erosional and exhumed land surfaces is provided in Widdowson (1997).

Pain (1986) has clearly shown that hill slopes and valleys form as a result of the operation of a combination of these various models. Figure 2.11 shows that parallel retreat, slope replacement and slope decline all contribute to the formation of slopes in the Razorback Range, New South Wales.

Since these models were developed, some research on hill slopes has developed a greater understanding based on processes operating on slopes in various geological situations. Selby (1993) summarizes the

Figure 2.8 *Photos of pediments from the Broken Hill region, New South Wales (photo GT)*

Figure 2.9 *Land surface near Hill End, eastern highlands, central New South Wales. This obvious land surface cuts across highly deformed Palaeozoic sedimentary rocks (photo GT)*

Figure 2.10 *Photo of Kata Tjuta, a typical example of an inselberg, from central Australia (photo GT)*

processes of slope development very extensively. One approach taken by him is to look at slopes and their development in terms of either being weathering-limited or transport-limited slopes.

Weathering-limited slopes are those not mantled by regolith, and whose development is controlled by the rate at which joints and other fissures open. In contrast, the development of slopes covered by regolith is controlled by the rate at which the regolith is moved off the slope, in other words their development is transport-limited. Weathering-limited slopes predominate where the rate of regolith production is less than the rate at which it is removed. In such cases slopes are dominated by *in situ* outcrop, often with very steep slope angles or near vertical free faces. The slopes developed on dolorite in the higher regions of Tasmania are good examples (Figure 2.12), with steep rock upper free faces and lower very steep depositional wedges consisting of the fallen debris.

Transport-limited slopes occur where regolith develops more quickly than it is removed. In this case the nature of the slope is controlled by the nature of the regolith and the processes that move it. Where weathering and transport rates are in equilibrium, stable transport-limited slopes are developed. Such slopes are similar to those shown in Figure 2.6a.

Depositional plains are far easier to understand. Some of the most common types of plains are those that occur across sites of marine regressions. In Australia the most extensive of these is the Eromanga Basin, occupying about 20% of the continent (Figure 2.13). Despite minor post-regression tectonism and incision, this plain stretches relatively undisturbed from the Gulf of Carpenteria to northern New South Wales and South Australia. Similar but less extensive plains exist in other Late Mesozoic and Tertiary basins in Australia and elsewhere. Similar marine regression plains occur in parts of North America, Africa and Europe in regions not affected by alpine tectonics during the Cainozoic (Bond 1978). In Australia, however, the lack of such mountain-building episodes

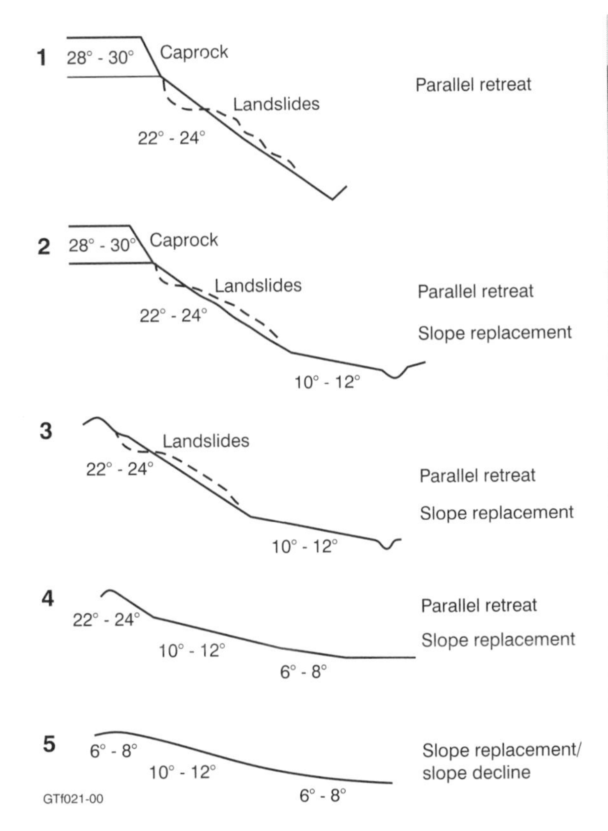

Figure 2.11 *Sequence of slope elevation at Picton, New South Wales (from Pain 1986). The stages fit observations made in the field. A complete model would include stages before 1 and after 5. Slopes are drawn with vertical exaggeration*

(a)

(b)

Figure 2.12 *Photo of (a) Debris slopes along the Grand Canyon of the Colorado, and (b) Dolerite hill slopes in north-east Tasmania showing steep debris slopes below tall free-faces (photos GT)*

has left much larger regions of marine regression surfaces than on other continents.

Alluvial plains, whether modern or ancient, are common throughout the world. Examples include the Mississippi, Amazon, Nile, Huang and other great river plains of the world. There is a major body of literature on these landforms and depositional environments well summarized by (Boggs 1987; Miall 1990, 1992, 1996; Walker 1984; Pye 1994; Marzo & Puigdefabregas 1993). The nature of these alluvial plains depends to a large degree on their gradient, because sediments that form the plain depend on that and their provenance.

The Riverine Plain of south-eastern Australia is a good example (Figure 2.14). This is an alluvial plain formed over marine and marginal marine sediments following a marine retreat during the latest Neogene (Brown 1989). The plain contains active streams and a plethora of abandoned stream systems, relics of former Quaternary climate shifts. It is also studded with lakes and their associated deposits. Butler *et al.* (1973) describe the plain as a series of very low angle coalescent alluvial fans deposited by ancient streams exiting the Eastern Highlands where the plains are dominated by clays cut across by sandy channel and levee deposits. The soils are of varying types reflecting differences in age, parent material and microtopographic variations. Common soils of the clay plains are red, brown and grey clay soils with carbonate in the upper part of the profile and gypsum in the lower part (Figure 2.14).

In the south-west the plain is still dominated by a moderately high succession of regressive beach ridges between which are deposited alluvial and lacustrine sediments. The beach ridges are sandy and contain significant concentrations of heavy minerals (Drexel 1988). They have been reworked and partially buried by aeolian, lacustrine and fluvial deposits.

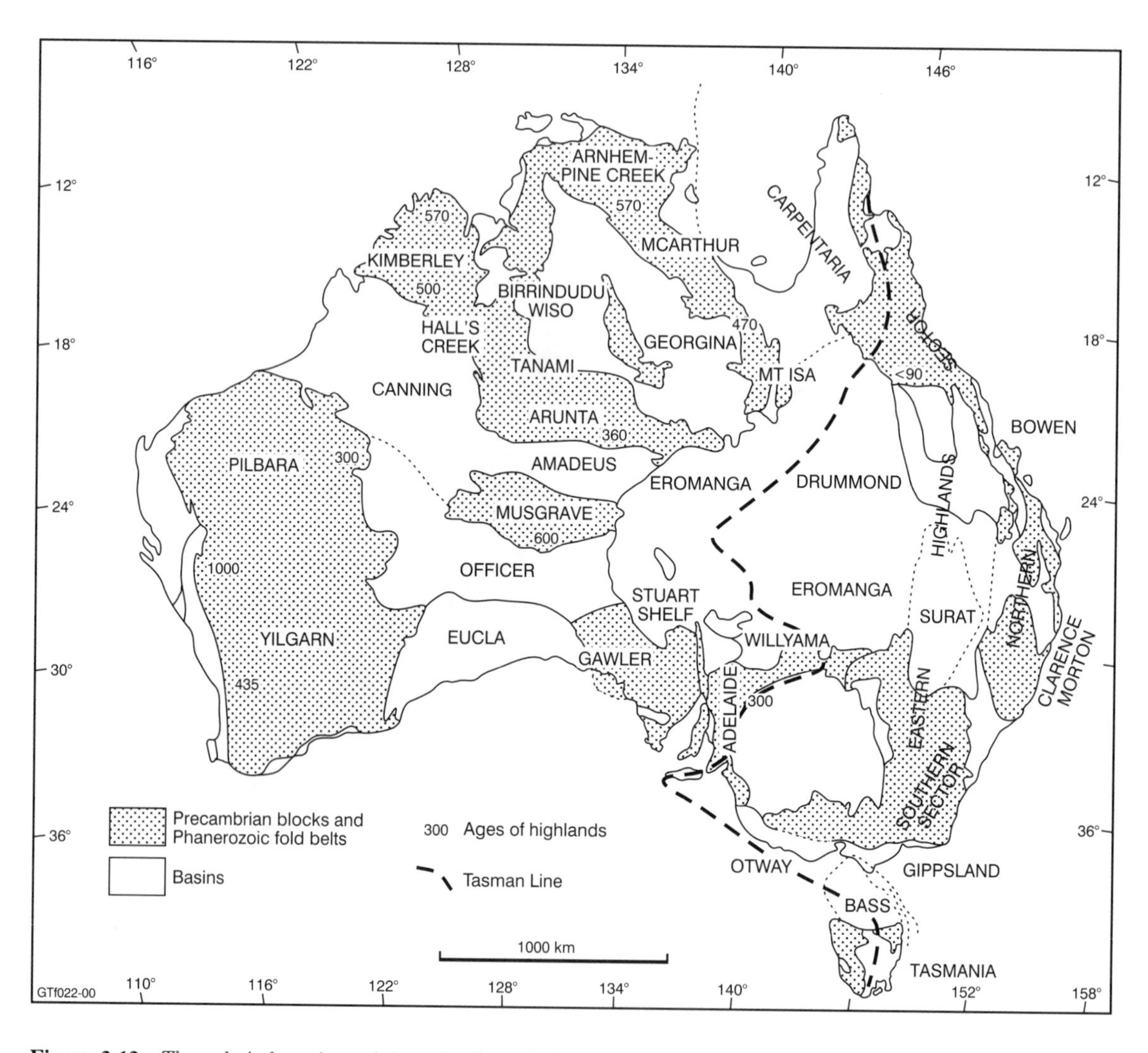

Figure 2.13 *The geological provinces of Australia (from Taylor & Butt 1998)*

Other Australian alluvial plain systems are those of the Cooper Creek and Diamantina River. These alluvial plains are deposited over gently deformed Cretaceous marine sediments of the Eromanga Basin (Figure 2.15). These streams occupy synclinal depressions and have accumulated a significant thickness of both sandy and more recent clay sediments. Their unique nature derives from large-scale braided flow patterns (Figure 2.16) which transport predominantly clays (Nanson *et al.* 1986). This occurs because the muds are sub-plastic and are thus transported as sand-sized fragments. The systems only operate as braided streams during very rare high flow events. Mostly they occupy narrow, deep, suspended load sinuous channels similar to those of the low-gradient Riverine Plain streams such as the Darling (Taylor & Woodyer 1978), Murrumbidgee, Lachlan and Murray rivers.

Lacustrine plains are often extensive, but more often of moderate size. They may be structurally controlled as is the case with Lake George in eastern Australia, Lake Eyre in central Australia or many of the rift lakes in Africa, South America and the Middle East. Deflation lakes such as many on the Riverine Plain of south-eastern Australia and in central Australia, occupy depressions in zones where salty groundwater cropped out during more arid phases of the Quaternary (Evans & Kellett 1989). Glacial lakes form part of extensive glacial plains in northern Europe, Asia and North America.

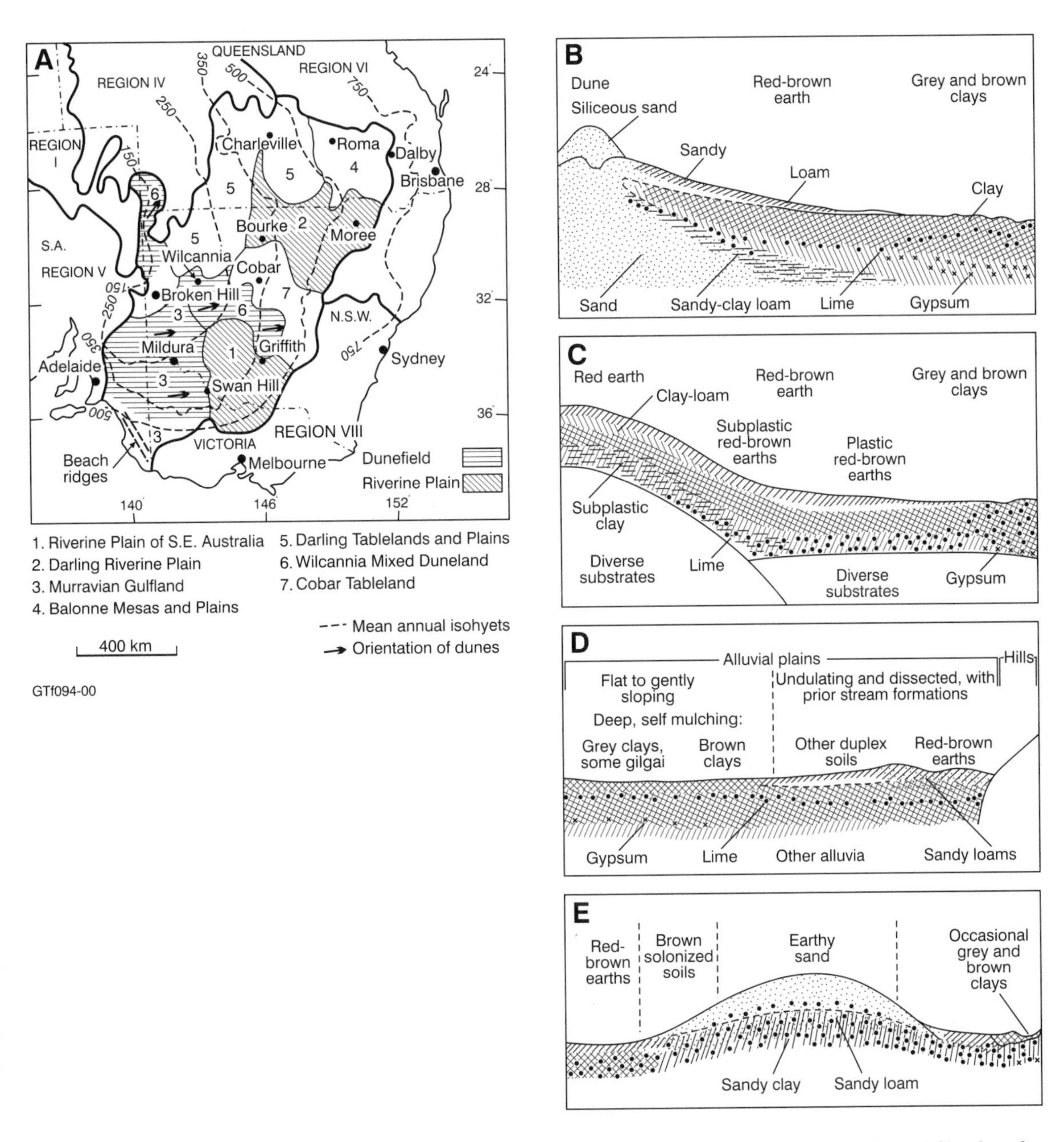

Figure 2.14 *Soil provinces (a) and typical soil–landscape associations (b–e) of the Murray Basin (from Butler and Churchward 1983)*

Lake George forms a good case study to illustrate many features of landscape and regolith evolution. Figure 2.17 illustrates the setting of this lake. Prior to the Miocene the country around Lake George was a rolling hill and valley landscape across which deeply weathered bedrocks formed the regolith. During the Late Miocene movement on the Lake George Fault dammed west-flowing streams creating the lake. For several millions of years the lake basin was progressively filled with alternating alluvial and lacustrine sediments that were deeply weathered, probably as a result of the continuation of wet and possibly warm conditions that also weathered the adjacent bedrock regions. More movement on the fault then converted

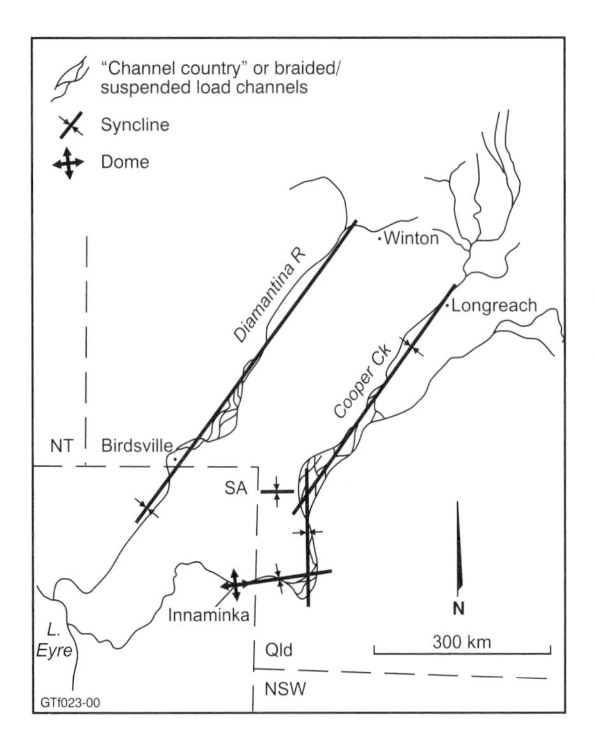

Figure 2.15 *The planform and structural control of the Cooper Creek and Diamantina River, central Australia*

Figure 2.16 *Photo of channel country, Cooper Creek near Karmona, western Queensland. Note the light-coloured braid islands (about 1 km long) and mid-grey braid channels that only flow at high stage. These are cut by the deeper sinuous channels of suspended load streams that flow at low stage. The alluvial plain here is about 25 km across (photo GT)*

the basin into an internal drainage depression and sediments accumulating were entirely lacustrine with only minor weathering events separating packets of lake sediment. This sequence of sediments separated by minor weathering or soil-forming events reflects the change to alternating wet warm and cool dry Quaternary climates. During this phase it is also likely that the deeply weathered bedrocks on the upthrown western margin of the lake were eroded except where they have been preserved beneath alluvial gravels and sands deposited by the formerly west-flowing drainage. These deep weathering profiles are well preserved in the downthrown basin below and adjacent to the lake plain.

Aeolian plains are widespread in arid climatic zones and in areas adjacent to them. Typical examples include many of the Australian sandy deserts and, of course, the Sahara. Most of these deserts are dominated by sand dunes with interdune clay pans. The sands are mainly transportational forms, that is, the shapes of the dunes represent the shape of mobile sand waves, and as they move little is left behind. The source of the sand and its transport distances are questionable. In the Namib of south-western Africa the dune sands originate from the rivers along the southern margin of the desert, and the sand is transported north until it again intercepts rivers which deliver the sand west to the sea. In Australia (Pell *et al.* 1997) have shown that the sands in the longitudinal dune system do not travel extensive distances from the source of the quartz grains making up the bulk of them. Sandy dune plains are all Quaternary although there is considerable evidence for multiple phases of activity of the dunes through this time (e.g. Wasson 1986).

Areas on the downwind side of aeolian sediment sources may develop extensive plains or have landscapes subdued by aeolian accession. The classical examples are the extensive loess plains of the USA, China, central Asia and Europe. These consist of predominantly silt-sized particles sourced from glacial outwash during the extensive glaciations of the Quaternary. Most are thought to be source-bordering, meaning they are close to and downwind of their sources. In Australia similar materials which blanket much of the Riverine Plain and Eastern Highlands, called 'parna', contain characteristic silt-sized grains in addition to significant clay-sized material. In these parts of Australia parna adds an unknown quantity of aeolian material to the regolith, but occasionally these sections of parna are revealed (Figure 2.18). Chartres & Walker (1988) have also shown that aeolian material contributes significantly to apparently *in situ* regolith in the Eastern Highlands. Gatehouse & Greene (1998) and Beattie (1970) have shown such deposits to be even more extensive on the western slopes of the highlands.

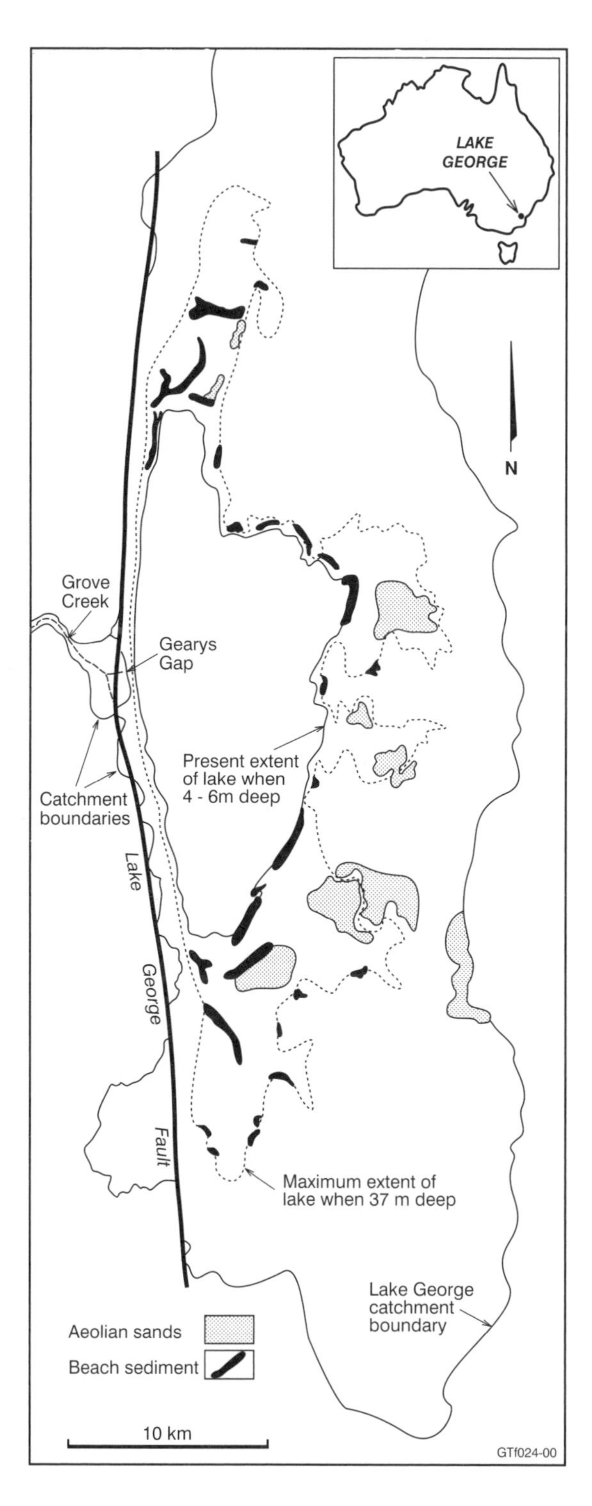

Figure 2.18 *Photo of parna at Junee, central New South Wales*

In the Australian deserts there are many tablelands, plateaux and mesas covered by stony pavements (Figure 2.19). These stones are called 'gibber' in Australia. They are composed of probably locally derived and sub-rounded silcrete and in many cases weathering-resistant rounded Precambrian or Palaeozoic quartzite and vein quartz. While the origin of the gibber may be somewhat enigmatic, it generally overlies a gibber-free 'desert loam' composed entirely of silt and clay. The clay minerals are mostly a proportion of smectite with kaolin and illite (Norrish & Pickering 1983) and are hence expansive. The desert loam overlies weathered bedrock. The gibber plains often have significant microrelief as a result of gilgai (Figure 2.20) formed by the shrinking and swelling of the desert loams. At a larger scale these stony desert surfaces are usually flat, and form clearly defined, if incised, plains. The desert loams are aeolian silt- and clay-sized accessions derived from windward lakes, dunes and dune swales that are washed into gibber on the surface, and as a result of their swelling nature lift the gibber as accession proceeds. As the thickness of loam varies from site to site it is interesting to speculate how important their presence is in creating the plain surfaces.

Figure 2.17 *Sketch of Lake George Basin illustrating the Lake George Fault (LGF) responsible for defeating west-flowing headwaters of Grove Creek (GC). The lake was dammed in the Miocene (Abell 1985). The early sediments in the basin are deeply weathered as are the bedrocks in the graben. Deeply weathered material has been removed from the upthrown block except at Geary's Gap (GG) where they are covered by fluvial sediments. The array of Quaternary features associated with the last high lake stand are also shown as beach ridges and aeolian sand sheets*

Figure 2.19 *Photo of gibber plains at Innaminka, northern South Australia (photo GT)*

Figure 2.20 *Photo of gilgai in gibber at Coober Pedy, central South Australia. Relief between the high ridges and the depressions is about 1.5 m. Spinifex is growing on the ridges*

In the Northern Hemisphere many plainlands are formed from glacial till and outwash plumes over glacially eroded plains. These are particularly common across the north of Russia and in the north of North America. All are Quaternary, are well summarized by Sugden (1978) and discussed further in Chapter 12.

Factors influencing the nature of landscapes

The shape of the hills and valleys and the regolith on them are primarily controlled by the lithology, rock strength, rock structure, local tectonic and volcanic features, the regional tectonic situation (plate tectonic setting) and climate and palaeoclimate. Which of these is important depends on the size of the area being considered and on the timescale over which the landscapes have developed.

These factors are to a greater or lesser degree dependent on each other and it is difficult to separate them, but for the purpose of rational discussion it is necessary to separate them. At the end of this section the relationships are discussed and illustrated. Because some of the factors logically occur under more than one major heading, some may be mentioned more than once. This is a direct result of the interrelationship between many of the genetic factors.

LITHOLOGY

Some rock types are easily erodible including limestone, shale and basic igneous rocks, while rocks such as quartz sandstone, acid igneous rocks and hornfels are relatively slow to erode. The reason for this is simply that rocks that weather readily are more easily eroded while those that do not are more resistant. It is, however, important to separate physical from chemical erosion, as some rocks which are physically strong (see rock strength below), for example marble, are chemically weak. Which is the dominant lithological control on landscapes depends on the weathering and climatic history of the rocks. The general idea of lithological control of landscapes is well illustrated in the Canberra region (Brown & Ollier 1975) where erosion-resistant acid igneous rock and sandstones remain as prominent hills, and shales and limestones form the valley systems.

Data on the erodibility of various rock types are few. Perhaps the best surrogates are the studies on rates of comminution of pebbles during transport. These data generally show that rocks containing an abundance of minerals with a low Mohs' hardness index (e.g. calcite) are more easily comminuted than those with a high index (e.g. quartz).

Most materials that are eroded in landscapes are regolithic. It is the weathered rocks and their products that are most readily moved and removed to sculpt the landscapes we observe. The weatherability of rocks is considered in more detail in Chapters 4, 9 and 10.

ROCK STRENGTH – DEGREE OF WEATHERING

This is the physical strength of the rock excluding fractures and joints. It is essentially controlled by the types of mineral present in the rock (these may include minerals produced by weathering) and their fabric (how the individual grains in the rock are related,

Table 2.1 *Fresh rock strength as measured by Schmidt hammer impact (simplified from Selby 1982)*

Strength	Character	Rock type examples
Very strong	Requires several hammer blows to break sample	Quartzite, dolerite, gabbro, basalt
Strong	Hand-held sample can be broken with a hammer	Marble, granite, gneiss
Moderately strong	Shallow indentations can be made by a firm hammer blow	Sandstone, shale, slate
Weak	Deep indentations can be made by a firm hammer blow	Coal, siltstone, schist

Table 2.2 *Examples of the physical strength of regolith materials*

Strength	Character	Regolith type examples
Very strong	Requires several sharp hammer blows to break sample	Silcrete, massive calcrete
Strong	Hand-held sample can be broken with a sharp hammer blow	Ferricrete, fine-grained saprock (siltstone, mica schist)
Moderately strong	Requires a hammer blow	Coarse-grained saprock (granite)
Weak	Can be broken by hand	Saprolite, clay B-horizons, soft calcrete
Very weak	Disperses or disintegrates in water	Sodic soils, highly saline saprolite

e.g. intergrown as in igneous and metamorphic rocks, or packed as in many clastic sedimentary rocks). Examples are given in Table 2.1. Physical rock strength is the main feature that controls erosion caused by abrading the surface of rocks exposed in the landscape.

As most materials eroded from a landscape originate from regolith material, it is also important to consider the physical erodibility of some regolith materials (Table 2.2). Engineering geologists generally measure the strength of soils and sediments (regolith for our purposes) by measuring such properties as unconfined compressive strength, triaxial compressive strength and resistance to shear.

As the majority of regolith materials contain clays, clay mineralogy is important in controlling the stability (or erodibility) of the regolith on slopes. Regolith rich in smectite has a shear strength about the same as the kaolin minerals at low water content ($\leq$20%), but as the water content rises smectitic soils lose strength more rapidly than kaolinitic ones, mainly due to their very different cation exchange capacities (Figure 2.21). In many weathering profiles smectite occurs low in the profile and kaolin increases in content up profile. This means failure that causes erosion mostly occurs in this smectitic zone unless one or more of the other factors operating is more significant.

Little work has been done on the erodibility of regolith materials, except as far as soils are concerned where soil scientists and engineering geologists have done extensive work relating to erosion, agricultural characteristics and stability during and after engineering constructions. The literature on this subject is massive, and since they form the upper part of the

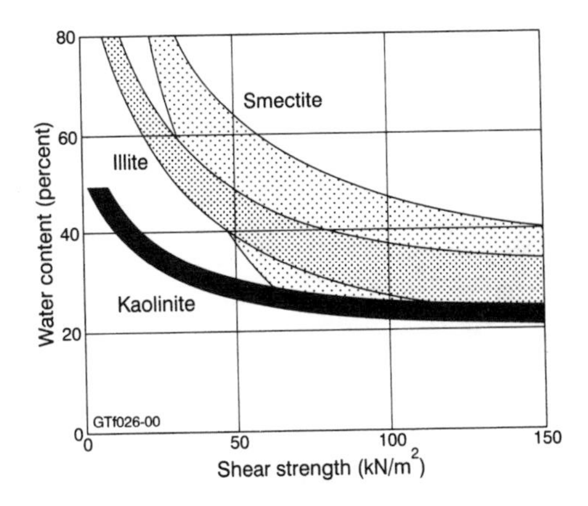

Figure 2.21 *The relationship between shear strength and water content for various clay minerals*

regolith their erodibility is critical to the stability of the whole regolith in landscapes. This is well illustrated by global maps of physical and chemical denudation rates derived from studies of oceanic sediment accumulation and chemistry of major continental drainage systems (Figure 2.22).

The erodibility of soils is a function of the soil properties and the environment in which they occur. Some of the soil factors important in controlling erodibility are:

- the nature of the profile. The nature of the A- and B-horizon will affect the infiltration capacity of the soil, whether perched water-tables are formed during rainfall events and the degree of runoff;

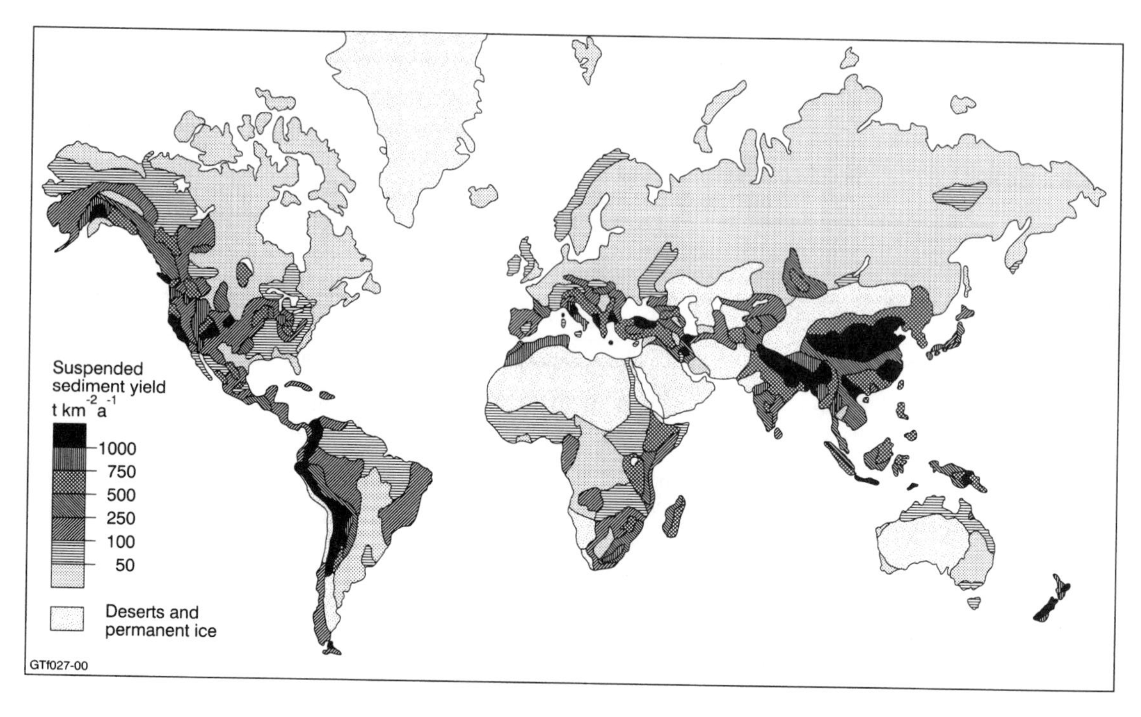

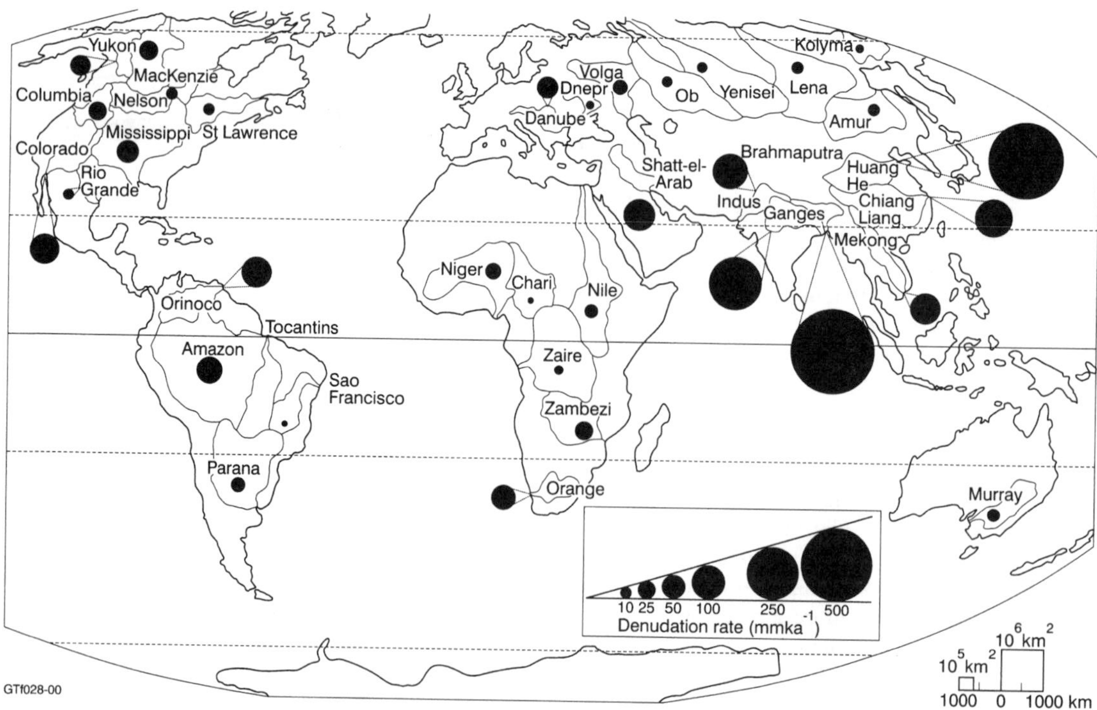

Figure 2.22 *(a) Global pattern of suspended sediment yield. The values relate to intermediate sized basins of 10^4–10^5 km^2 (Walling & Webb 1983 from Summerfield 1991). (b) Denudation rates for the world's 35 largest drainage basins based on solid and solute load data. Allowance has been made for the non-denudational component of solute load. Source rock density is assumed to be 2700 kg/m^3*

- soil structure (the degree of development of soil aggregates or peds and their arrangement) which determines the water infiltration capacity of soils;
- the nature and amount of exchangeable cations on the clays in the soil. Soils with, for example, Ca^{2+} saturation will be relatively stable, but if the clay minerals are Na^{+} rich then the soil will readily disperse on wetting and consequently erode;
- clay mineral types in the soil. The presence of smectitic clays will enhance erodibility compared to soils containing mainly kaolin group minerals due to the marked shrink–swell characteristics of the former; and,
- biotic disturbance and biotic 'knitting'.

The major environmental factors in controlling soil erodibility include:

- slope – the higher the slope angle the faster any runoff will be and thus the more erosive it will be (Figure 23a). The steeper the slopes the greater the effects of gravity (soil creep, mass flow, landslip);
- precipitation – in general terms the higher the precipitation the greater the physical erosion. This is only modified by the relationship between rainfall and vegetation cover, so the available data do not show this simple relationship (Figure 2.23b);
- tectonic relief is a major contributor to erosion rates. Although data to illustrate this relationship are not common, Figure 2.23c shows some data for Japan and New Zealand. This is somewhat different from slope steepness as it relates as much to height above erosional base levels as local relief and steepness; and,
- other factors such as runoff variability and aeolian erosion are also important.

ROCK STRUCTURE

Rock structure can be defined locally as fractures, joints and sedimentary structures as well as the attitude of bedding, and at a regional scale in terms of folds and faults. All these structures play an important role in landscape evolution, because they are a significant factor in affecting rock weathering, rock mass stability and in the regional shape and disposition of landscape elements, as fold and faults control the distribution of rock types. Because of their role in the control of weathering they also directly influence the nature and thickness of regolith development.

Joint and fracture control of both physical and chemical weathering is obvious. Water, one of the main agents of both types of weathering, is able to enter the rock along joints and cause disintegration that dislodges blocks of rock, and decomposition that removes solutes and produces new minerals. As the latter proceeds the rock around the joints is preferentially altered, but the centre of the joint blocks remains unweathered to form corestones, which when the weathered rock is eroded may be left as relics at the surface to form tors (Figure 2.24).

Joints, fractures and bedding structures in rock are a major factor in slope stability as they control the shear strength of the rock mass. Thus, generally, the more fractured or more closely bedded a rock is the less stable the slope will be and consequently the more it will be eroded, forming the lower parts of landscapes and lower slope angles. This effect is amplified as the fractures in the rock mass allow water to penetrate and chemically weather the rocks, causing a further increase in their erodibility. At a more regional scale, bedding and joints also have an influence on landscapes. The cross-sectional shape of hills may be controlled by bedding (Figure 2.25) as any geologist who has used aerial photographs will be aware (Figure 2.26). Flat-lying beds produce landscapes with horizontal stepped topography. As erosion progresses, relics of the flat-topped hills are left as mesas and buttes (Figures 2.25 and 2.26).

Inclined strata lead to a landscape where the erosion-resistant beds form prominent ridges along strike with intervening parallel valleys. The ridges or cuestas have a dip slope and a scarp or anti-dip slope. The dip slope is drained by sparsely spaced dip streams, while more closely spaced anti-dip streams drain the scarp slope. The stream flowing down the valley axis is a strike stream. The pattern formed by these streams is known as trellis drainage. Within such a landscape, regolith on the cuestas will be thin but thicken towards the valleys where it is composed of alluvium covering weathered rock profiles (Figures 2.25b and 26).

Jointing at a regional scale may also control the shape of landscapes. Joint lines may control drainage development, particularly in otherwise homogeneous rocks such as granites and sandstones. Typically the drainage forms rectangular or rhombic patterns (Figure 2.27). In regions where this type of drainage is formed, regolith is typically thin because once a thick regolith develops the joints originally in the rock are no longer able to control drainage development. Joints also play an important role in development of

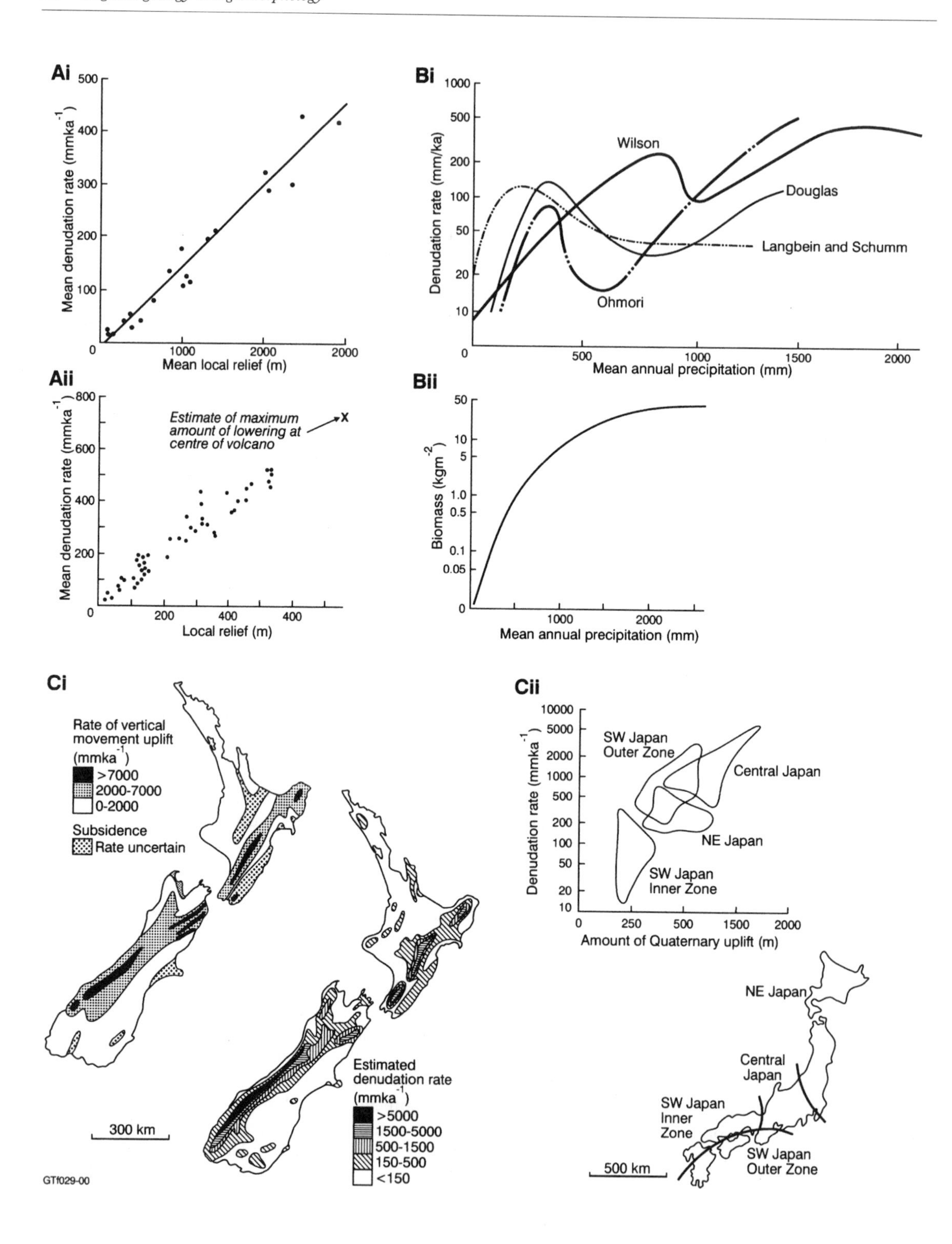
Ai
Mean denudation rate (mmka^{-1})
Mean local relief (m)
Aii
Estimate of maximum amount of lowering at centre of volcano
Mean denudation rate (mmka^{-1})
Local relief (m)
Bi
Wilson
Douglas
Langbein and Schumm
Ohmori
Denudation rate (mm/ka)
Mean annual precipitation (mm)
Bii
Biomass (kgm^{-2})
Mean annual precipitation (mm)
Ci
Rate of vertical movement uplift (mmka^{-1})
>7000
2000-7000
0-2000
Subsidence
Rate uncertain
300 km
Estimated denudation rate (mmka^{-1})
>5000
1500-5000
500-1500
150-500
<150
GTf029-00
Cii
SW Japan Outer Zone
Central Japan
NE Japan
SW Japan Inner Zone
Denudation rate (mmka^{-1})
Amount of Quaternary uplift (m)
NE Japan
Central Japan
SW Japan Inner Zone
SW Japan Outer Zone
500 km

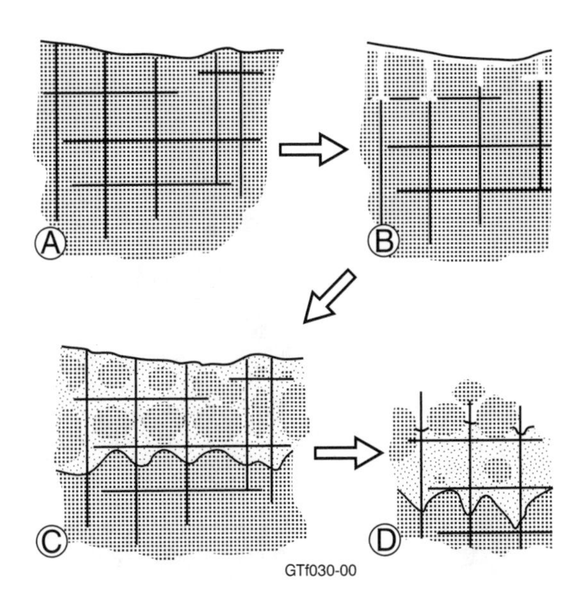

Figure 2.24 *A schematic sketch showing the evolution of corestones and tors in jointed rocks. (a) Jointed rock exposed at surface. (b) Weathering begins along joint planes. (c) Weathering proceeds along joints and into the blocks between them leaving fresh 'corestones'. (d) As weathering continues, some corestones below the surface are completely weathered but those left at the surface as 'tors' do not continue to weather at the same rate and remain intact at the surface*

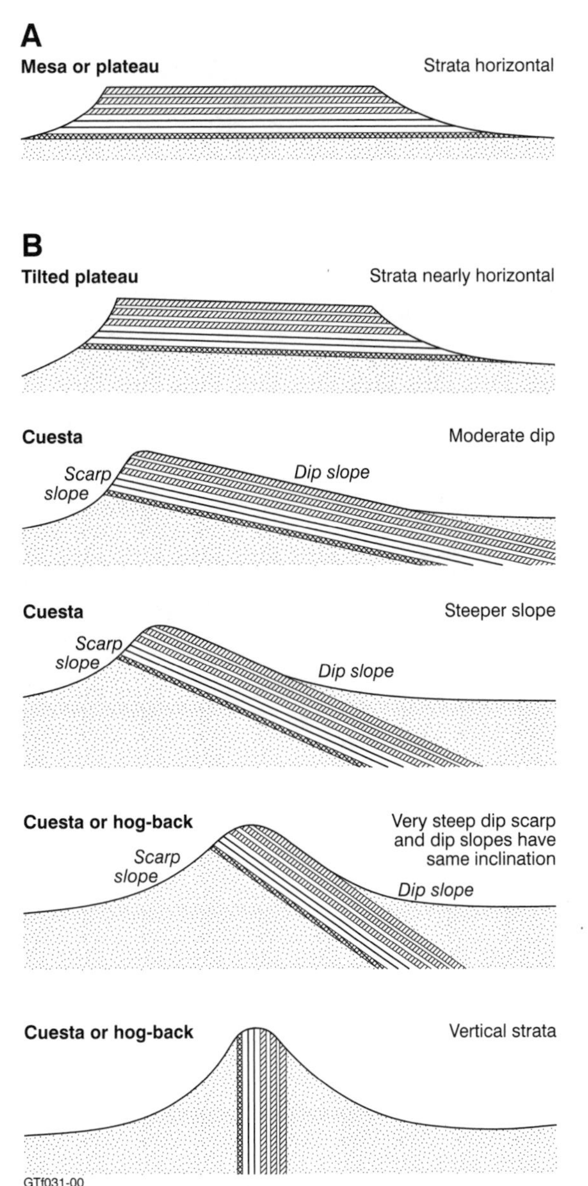

Figure 2.25 *A sketch showing the bedding control of landscapes*

cliffs and very steep slopes. The persistence and spacing of joints control the attitude and steepness of cliffs and slopes.

Jointing in bedrock can have a significant control on landscapes, even in cases where significant regolith occurs, as for example in the formation of tors and in the evolution of drainage. Bedrock jointing is controlled by a variety of factors including:

- proximity to the surface. Joints are more common within rocks close to the surface as confining pressure on the rock mass is less here (e.g. sheeting joints in granite bodies);
- stresses to which the rocks have been subjected. Rocks develop joints in response to brittle fracture during stressing. Such joints usually occur in conjugate sets (two joint sets at angles of 60°–90° to each other);
- texture of the rock. Coarse-grained rocks generally have more widely spaced joints than finer-grained ones; and,
- cooling history for igneous rocks. Rapidly cooled rocks, such as lavas, are often more closely jointed than are more slowly cooled intrusive rocks.

Figure 2.23 *(ai & ii) A mid-20° latitude relationship for denudation rate versus relief (from Ahnert 1970). (bi) Various estimates of the relationship between precipitation and denudation. (bii) The relationship between biomass (vegetation cover) and mean annual precipitation. (c) The relationships between rates of crustal uplift and estimated denudation rates for (ci) New Zealand (from Selby 1982) and (cii) uplift and present denudation in Japan*

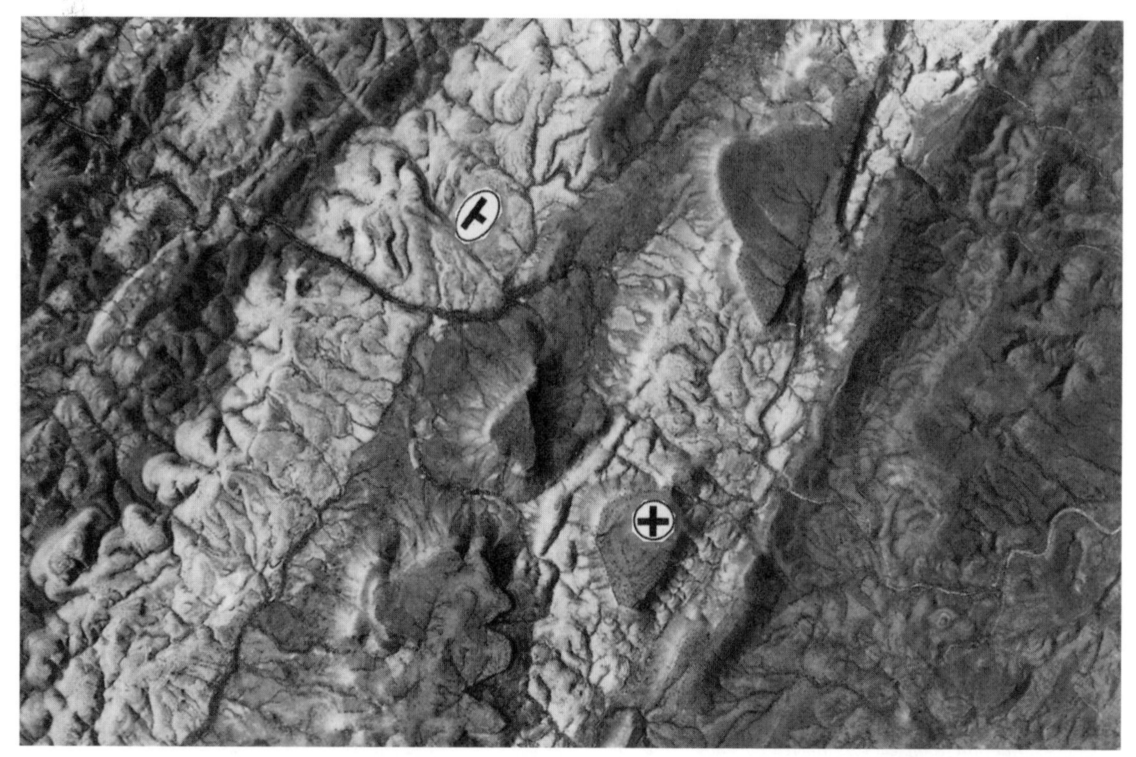

Figure 2.26 *An aerial photograph showing a synform with flat-lying beds (+) along the axis and cuestas dipping towards the axis (⊤). From the Victoria River region, north-west Australia*

Openly folded rocks produce landforms similar to those on dipping beds but exhibit closures. Regolith development is similar to dipping units in that it is concentrated in the valleys. The morphology of the hills is important, particularly in older landscapes where hill crests may be planed off, indicating the presence of older land surfaces. This variation is discussed below.

TECTONICS

Tectonic factors control the shape of land masses, the distribution of geological and landscape provinces in the land masses, and their elevation at a regional and continental scale. Figure 2.28 outlines some of these main features.

Shields are geological provinces not significantly deformed since the Precambrian and they are bounded by younger fold belts or are partly covered around their margins by younger sedimentary basins. Most are erosional regions and contain very few significant mountains. They may be composed of one or more cratons which form a tectonically stable regime composed of granitic, gneissic and greenstone rocks mixed with younger sedimentary and volcanic piles. Shields form the tectonically stable core against which younger fold belts accreted and sedimentary basins filled.

Marginal to the shields are fold belts, which consist of younger (Phanerozoic) folded and elevated sequences of sedimentary and volcanic rocks, which are intruded by granitic rocks, and which together form elongate mountain belts. They formed at convergent plate boundaries, some of which are still operative, for example along the western margin of the Americas. Older examples are not related to present plate motions (Figure 2.28) but rather to an earlier system of plate motions. These are areas of varying tectonic activity, the older belts tend to be relatively stable with little uplift like the eastern Australian Highlands or the Tasman Fold Belt, the Ural Fold Belt or the Appalachian Fold Belt. The younger are, however, much more active and undergo uplift, volcanism and seismic activity as in the Andean Fold Belt or the Himalayan Fold Belt. In a landscape sense these regions are of high elevation and are generally erosive terrains, whereas some of the older fold belts, while still probably responding to isostatic uplift (Lambeck & Stephenson 1986; Miller & Duddy 1989) are less so.

Figure 2.27 *Aerial photograph of rhombohedral joint controlled drainage in granite and Palaeozoic sediments at Mossman, north-eastern Australia*

Overlying the shields and fold belts are continental basins, which formed by down warping of the crust and its filling with thick sedimentary sequences, derived from the adjacent shields and fold belts. Consequently, they are formed mainly from mineralogically mature sediments. These areas make up the majority of continental land surfaces and tend to be of low relief and tectonically relatively stable.

Apart from the continental land masses the other significant land masses are a direct result of the present system of plate tectonic motions. These include mainly the island arcs and relict fragments of continents like New Zealand and the Indonesian Archipelago. Because these regions are along convergent plate margins they are tectonically very active and also experience considerable volcanic and seismic activity. They tend to have predominantly erosional landscapes shedding sediment into the seas. The majority of volcanic activity is andesitic and sedimentary sequences tend to be composed of labile sediments. A summary of the landscape types occurring at plate boundaries is given in Table 2.3.

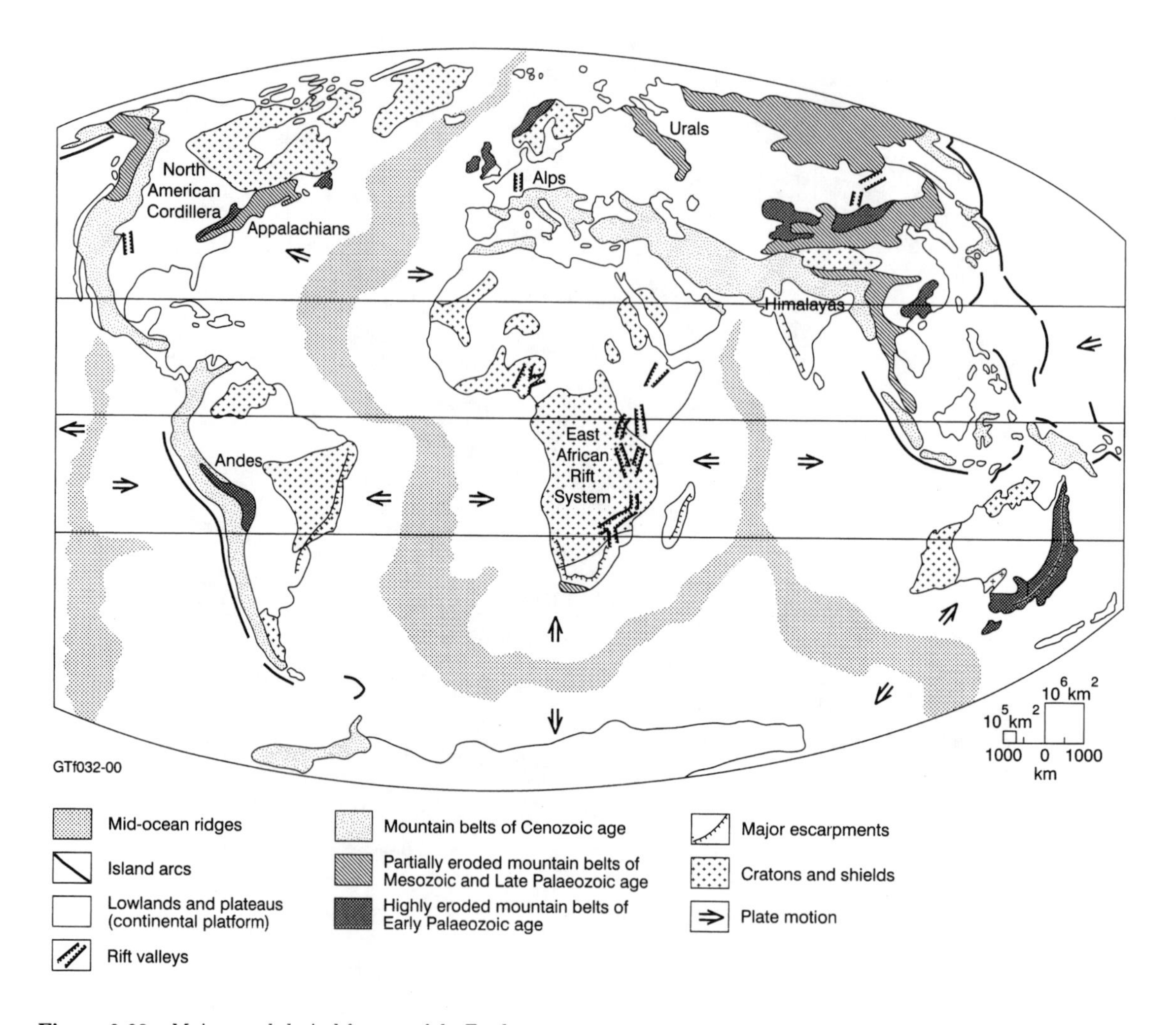

Figure 2.28 *Major morphological features of the Earth*

Table 2.3 *Classification and major landscapes associated with various plate boundaries (modified from Summerfield 1991)*

		Morphological and structural features		
Boundary type	Stress regime	Oceanic–oceanic crust	Oceanic–continental crust	Continent–continent crust
Divergent	Tensional	Mid-ocean ridge	–	Rift valley
Convergent	Compressional	Trench and island arc	Oceanic trench and continental margin mountain belt and volcanicity	–
		Complex island arc collision zone	Modified continental margin mountain belt	Mountain belt and limited volcanicity
Transform	Shear	Ridges and valleys normal to ridge axis	–	Fault zone, no volcanicity

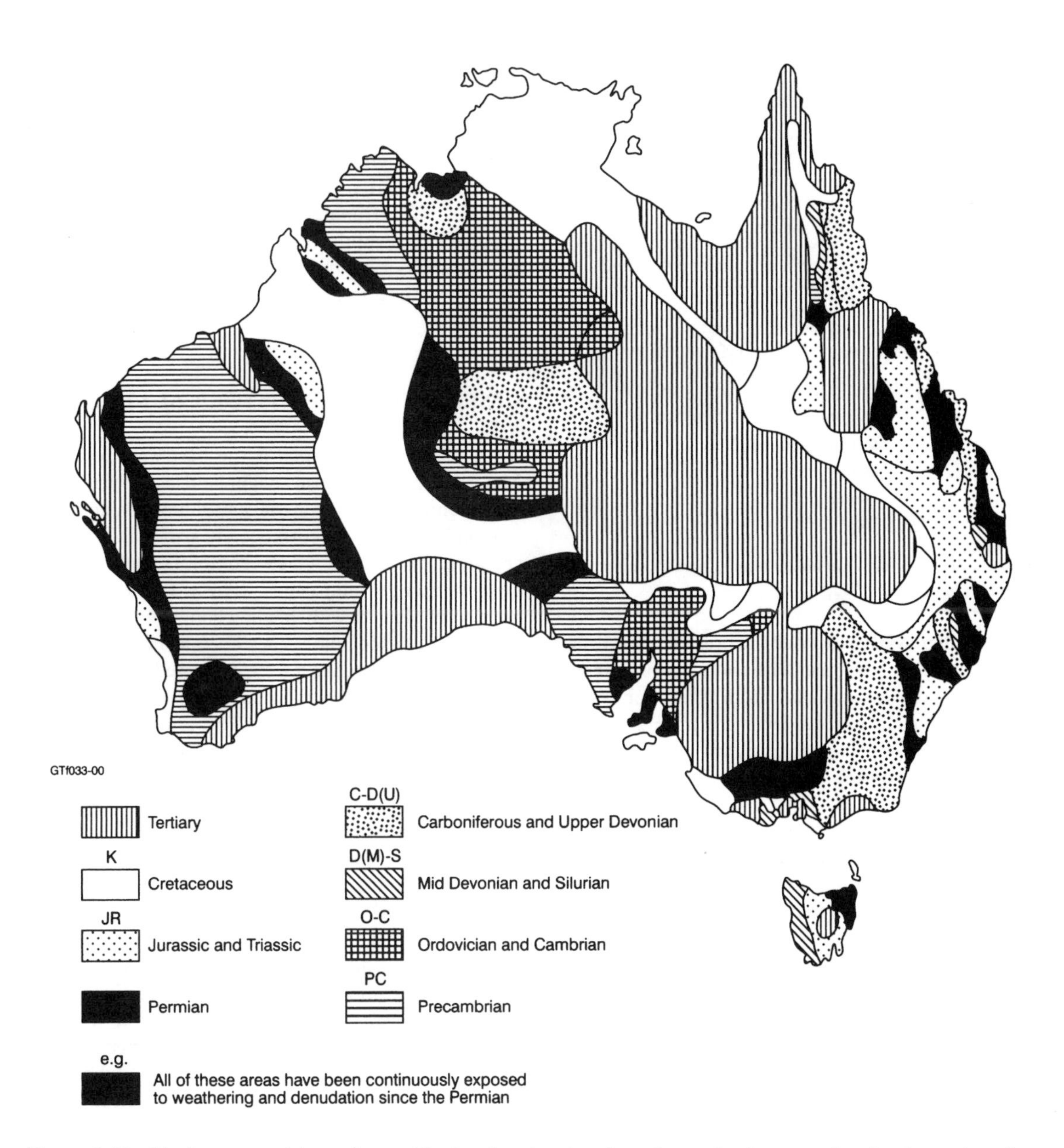

Figure 2.29 *Maximum potential age of exposed land surfaces based on the geology and palaeogeography of various regions of Australia (after Beckmann 1983)*

The ages of the landscapes and the regolith developed on them are also related to the age of the tectonic province on which they are developed. The landscapes cannot be older than the rocks on which they are developed. This is well illustrated by Beckmann (1983) (Figure 2.29) who drew a map of the maximum age of exposure for Australia based on geological and palaeogeographical maps of the continent. It is clear that the oldest terrains are potentially on the shields composed of predominantly Precambrian rocks. Other areas like the eastern seaboard are mainly composed of Palaeozoic fold belt rocks and Permian to Mesozoic sedimentary basin rocks, and much of the remainder of the continent is covered by Mesozoic and Tertiary sedimentary basins. In Beckmann's model an important feature is the extent of Permian glacial activity, as it

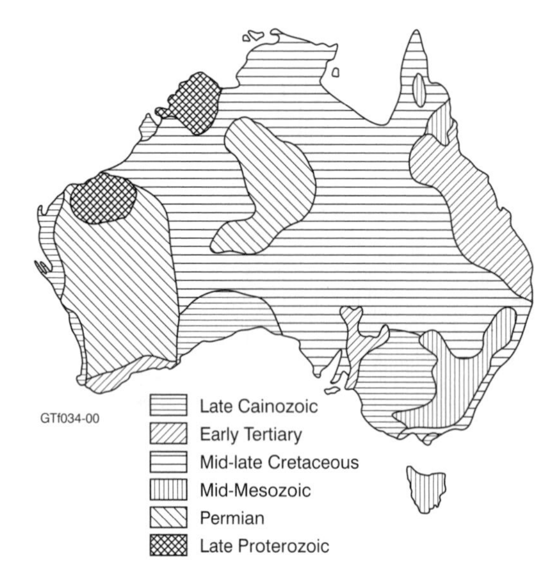

Figure 2.30 *The age of drainage systems in various parts of Australia (from Ollier 1991a)*

is this which may have exposed the present land surface south of his limit. Few other data about age or potential age of exposure are available from other continents, and the only estimates available are from geological maps or by using palaeogeographical reconstructions of land masses to estimate maximum ages of landscapes as Beckmann did.

Wilford (from Ollier 1991a) prepared a similar map (Figure 2.30) of the age of drainage systems for Australia. He based his model on the presumption that present drainage systems could only be as old as the last marine regression or retreat of the last glacial episode to cover the landscape. Wilford's map is very large scale and hence ignores minor changes and anomalies in drainage caused by climatic or minor tectonic events. It is incredible to see that many areas have drainage patterns that potentially date from the Proterozoic. These landscape ages are related, among other factors, to the age of the major tectonic provinces making up the continent.

Landscape and regolith ages

It is worth stating the obvious again; landscapes and regolith cannot be older than the rocks on which they occur. The age of a landscape and the regolith which sits below it may be different. Some of the controls on landscape development are discussed above and passing mention made of the age of features. The age of landscapes and regolith is determined by examining features involved in landscape formation. There are fundamentally three ways of looking at the age of landscape features and regolith materials, just as in conventional geological work: observation and historical records, relative methods and absolute methods. Unfortunately, most landscape events and processes of regolith formation are gradual and slow, so direct observation can only be used rarely, most times it is necessary to infer age using less direct methods.

OBSERVATION AND HISTORICAL RECORDS

To see something happen is the best way to date it. Major storms can cause major landscape changes in coastal, hill slope and alluvial environments, and plate motion leads to the occurrence of earthquakes and the eruption of volcanoes which can dramatically and very quickly change landscapes. These events are also usually associated with major changes or redistribution of regolithic materials. The occurrence of the Meckering earthquake in 1968 caused the development almost instantly of an escarpment about 1.1 m high across the country about 130 km NNE of Perth in Western Australia. Larger landscape modifications have occurred due to earthquakes (e.g. Iran in 1990, or T'ang in China in 1976), but the Meckering one is interesting in that it occurred within a shield area. Although not much damage was caused, it produced significant landscape modification in a 'tectonically stable platform'. Earthquakes can also lead to substantial modifications in the distribution of regolith materials by dislodgment of materials from hill slopes and their deposition on lower slopes. Many examples of volcanic eruption causing modification to landscapes and regolith are observed. It happens almost daily in Hawaii where gentle eruption of basalt is continually placing fresh basalt over weathered earlier flows. More spectacular are the explosive events such as occurred at Mt St Helens in the north-west USA in 1980 when the side of the mountain was blown out and distributed across some 600 km^2.

Other observational changes in landscapes are mainly related to major storms, causing erosion of coasts, shifting of river channels and collapse of hill slopes during periods of excess rain. Extreme examples are the shift of the Hwang Ho River from Bohai to its present position in 1851 or landslides in Hong Kong during typhoons (Figure 2.31). There are historic records of similar changes to both landscape

Figure 2.31 *Photo of landslides in Hong Kong which took down two apartment blocks with the loss of about 70 lives in 1973 (photo Hong Kong Geological Survey)*

and regolith, such as all the records from history of earthquake and volcanic events, the most studied and well known being the eruption of Vesuvius in AD 79. This changed the landscapes in the region of the volcano, and produced a new regolith burying the cities of Pompeii and Herculaneum and much of the pre-existing regolith. All this was recorded at the time by Pliny the Younger. We have historic records of floods causing change along the Nile, Euphrates and other rivers from ancient times, and of the sediments supplied by these rivers as it was these additions to the regolith which maintained the fertility of the soils on which the civilizations depended.

Beyond recorded history there are spoken records which document historical landscape events. Australian Aborigines have legends of mountains on fire from South Australia and northern Queensland. These presumably record volcanic eruptions as recently as 2000 years ago. Details of volcanic and earthquake activity are well known from pre-written histories of Andean cultures, Pacific islands and many other places through spoken history.

RELATIVE DATING

There are two main dating methods traditionally used in sequencing geological events.

Law of superposition

Nicholaus Steno first enunciated this principle in 1669 and it states that in an undeformed sequence of sedimentary rocks each bed is older than the one above it. This principle is based on the observation that in most sedimentary deposits, such as in lakes, the sea, rivers and caves, beds are laid down horizontally. The law still applies if sequences have been deformed, but it is necessary to determine which way the beds become progressively younger before it is applied. It is, however, only applied to layers formed by sedimentary or volcanic events where one layer is emplaced before the overlying one.

At most places there are a limited number of layers to be observed, and it is possible to correlate recognizable layers from place to place, and recognition of this fact formed the basis of regional interpretations of geological evolution of regions. It is possible to determine in a relative sense when deformation events occurred and when periods of non-deposition occurred by the presence of unconformities (Figure 2.32). There are situations where layering may be produced by processes other than sedimentation such as the development of soil layers or horizons (Figure 2.32c) or differentiation in weathering profiles as weathering proceeds deeper into the substrate with time (Figure 2.32d). These processes are discussed more fully in later chapters.

It is also worth recording a caution about the law of superposition here. Although at any one site a bed of sediment deposited on another is younger, the bed (defined as such by its lithology) may not be the same age at different sites within the depositional system. This is well illustrated in many depositional environments, but it is particularly well illustrated in deltas (Figure 2.33) although it equally applies in many fluvial, lacustrine and other terrestrial depositional systems.

Although when considering sedimentary sequences the law of superposition applies, there are other examples, besides soil and weathering profiles, where layering may not obey the law. One example is the occurrence of a layer of gibber (stones of pebble to granule size) overlying desert loam (Figure 2.34). Here the gibber is worked to the surface by the deposition of expansive clay minerals on the surface which wash through the pebbles, and as they expand and contract on wetting and drying they lift the pebbles to the surface. The pebbles may thus be older (were formed before) than the underlying loam, but achieved their present position after the loam was added.

Generally in depositional or volcanic terrains the law of superposition applies to regolith materials and buried landscapes; however, it may not apply to all extant landscapes. Much regolith shows layering (e.g.

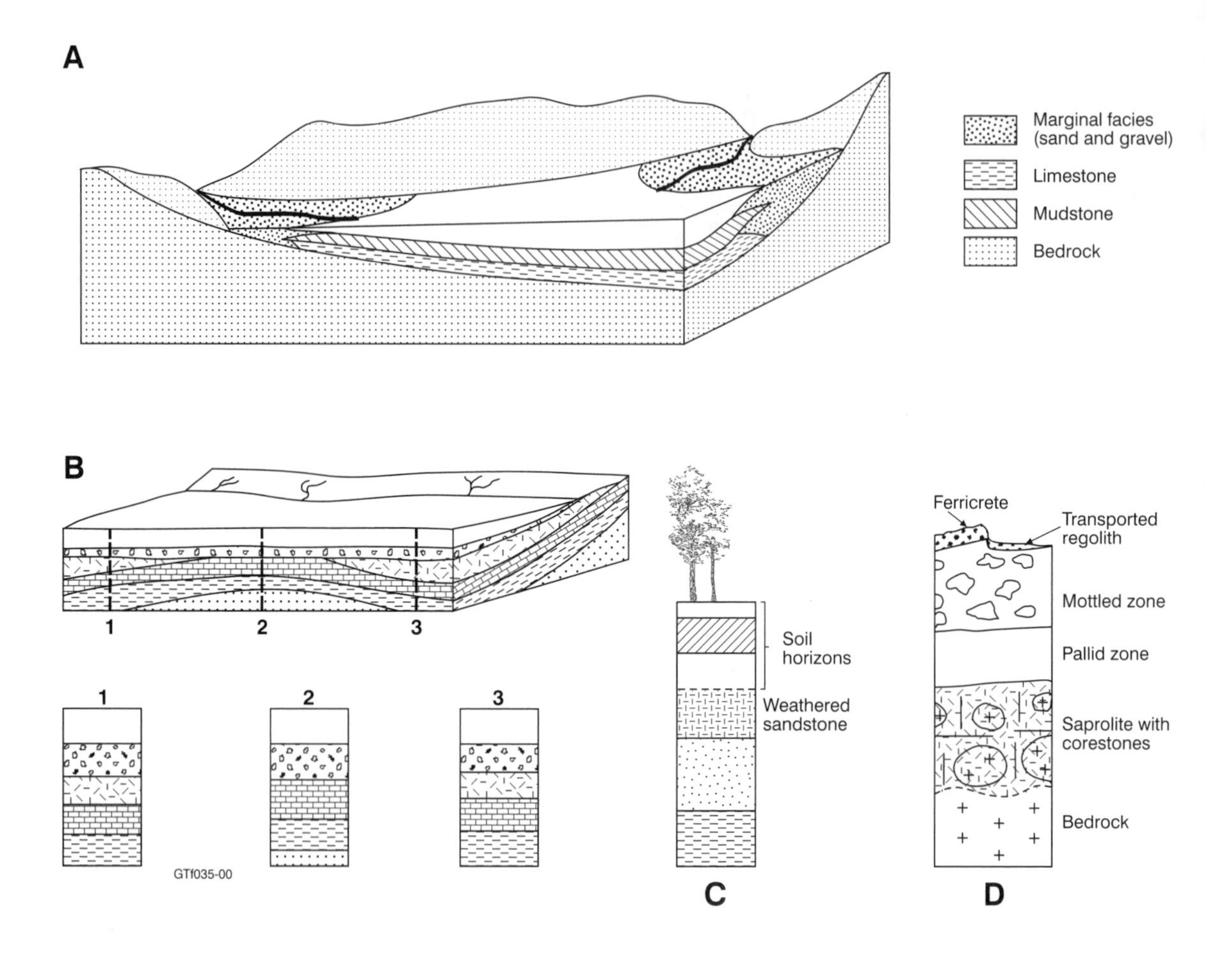

Figure 2.32 *Applications of the law of superposition. (a) Deposits in a lake. The limestone and younger mudstone show a typical superimposed relationship, but the sandstones deposited around the lake margins in the higher energy wave influenced area and close to the input points of detrital sediments are equivalent in age to both the limestone and mudstone. (b) Sections 1, 2 and 3 represent sequences observed at different places in a depositional basin. Correlation on lithological grounds allows a reconstruction similar to that shown. Note that the folding of the sandstone, limestone and siltstone must have occurred prior to the deposition of the conglomerate that is thus separated from the lower sequence by an unconformity. Unconformities represent periods of time during which deposition did not occur and during which other geological events such as folding or weathering and soil formation may have occurred, affecting the pre-existing rocks. (c) and (d) show examples of layering in regolith materials that do not obey the law of superposition. (c) The development of soil layers which are formed from the interaction of surface processes with the underlying materials. (d) Layered differentiation in a weathering profile again developed by the downward movement of a weathering front and the internal reorganization of materials by various processes*

in weathering profiles, see Chapter 11), but these are layers that originate from overprinting or post-depositional modification, and therefore the law of superposition is inapplicable.

Cross-cutting relationships

In traditional geological age interpretations where, for example, a granite body cuts across another rock body, this relationship shows that the granite post-dates the rocks it intrudes. Similar interpretations are made for dykes and other intrusive phenomena. Where any dyke cuts across another earlier one, the cross-cutting intrusive is younger.

A similar logic can be used in interpreting the relative age of landscapes. A land surface that cuts across folded rocks must post-date the folding of the rocks. Alluvial features which cut across others are

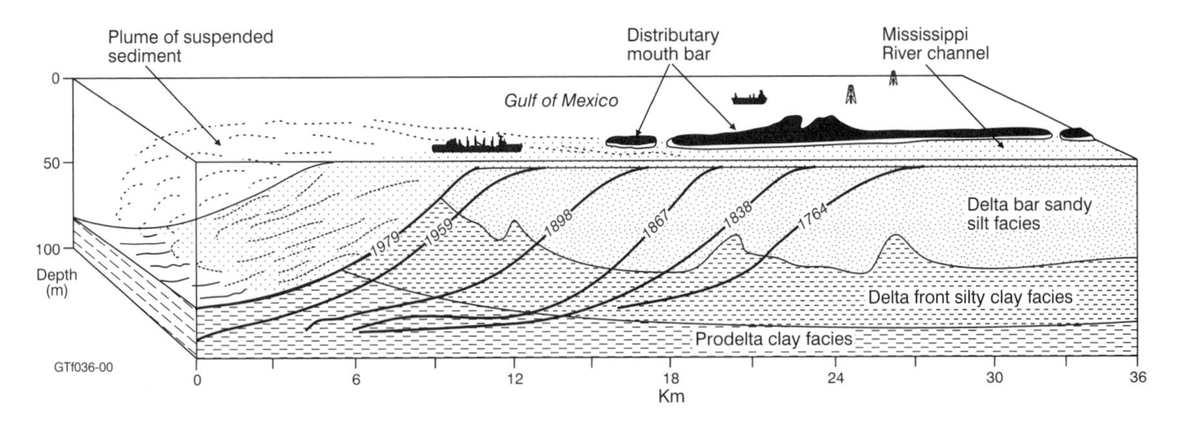

Figure 2.33 *A section through the Mississippi Delta showing 'time lines' as it grew out into the Gulf of Mexico. This clearly shows the non-parallelism of time and sedimentary facies (from Skinner & Porter 1987)*

Figure 2.34 *Gibber (pebbles) capping a 1.5 m thick desert loam sequence, Innamincka, South Australia (photo GT)*

younger (Figure 2.35) and similarly with other landscape features. Dunes cutting across a lake bed obviously post-date the lake as is observed in many central Australian salt pans.

As can be observed in Figure 2.35, the cross-cutting relationships of the present and former river systems are also reflected in the stratigraphic relationships presented by Bowler (1967). This is commonly the case, and although it is often very difficult to get the vertical dimension, landscape relationships are usually readily observable. It is also common that the nature of the regolith changes with the changes in landscape character as, for example, in the change from the sandy Katupna system on the very much more muddy deposits of the Goulburn system.

Tectonic deformation

Faults and neotectonic activity have significant effects on landscapes and regolith. Faults are fractures in the Earth's crust which vertically or laterally displace blocks of the crust. They must be younger than the rocks and the landscapes they displace or deform and hence provide a good means of relatively dating landscape sequences and the regolith on them.

Faults moving instantly or repeatedly can cause major change in landscapes and, as a result, in the regolith on them. Normal and reverse faults may cause the development of escarpments. The largest single movement observed was a 15 m uplift in 1899 at Yakutat Bay, Alaska. Thus to achieve displacements of the magnitude we can infer from geological records, faults must move repeatedly over extended periods. A common landscape modification produced by faulting is a series of parallel uplifted, downthrown and tilted blocks (Figure 2.36). Other commonly observed landscape modifications include dammed and diverted drainage. Damming may result in river beheading or the formation of lakes (Figure 2.36) and diversion causes rerouting of streams, the formation of relict channels and the temporary damming of the stream (Figure 2.36). Where streams are able to cut down at about the same rate as an upthrown fault block moves up, the drainage flowing on the former land surface is superimposed on a new incised surface. This is known as antecedent drainage.

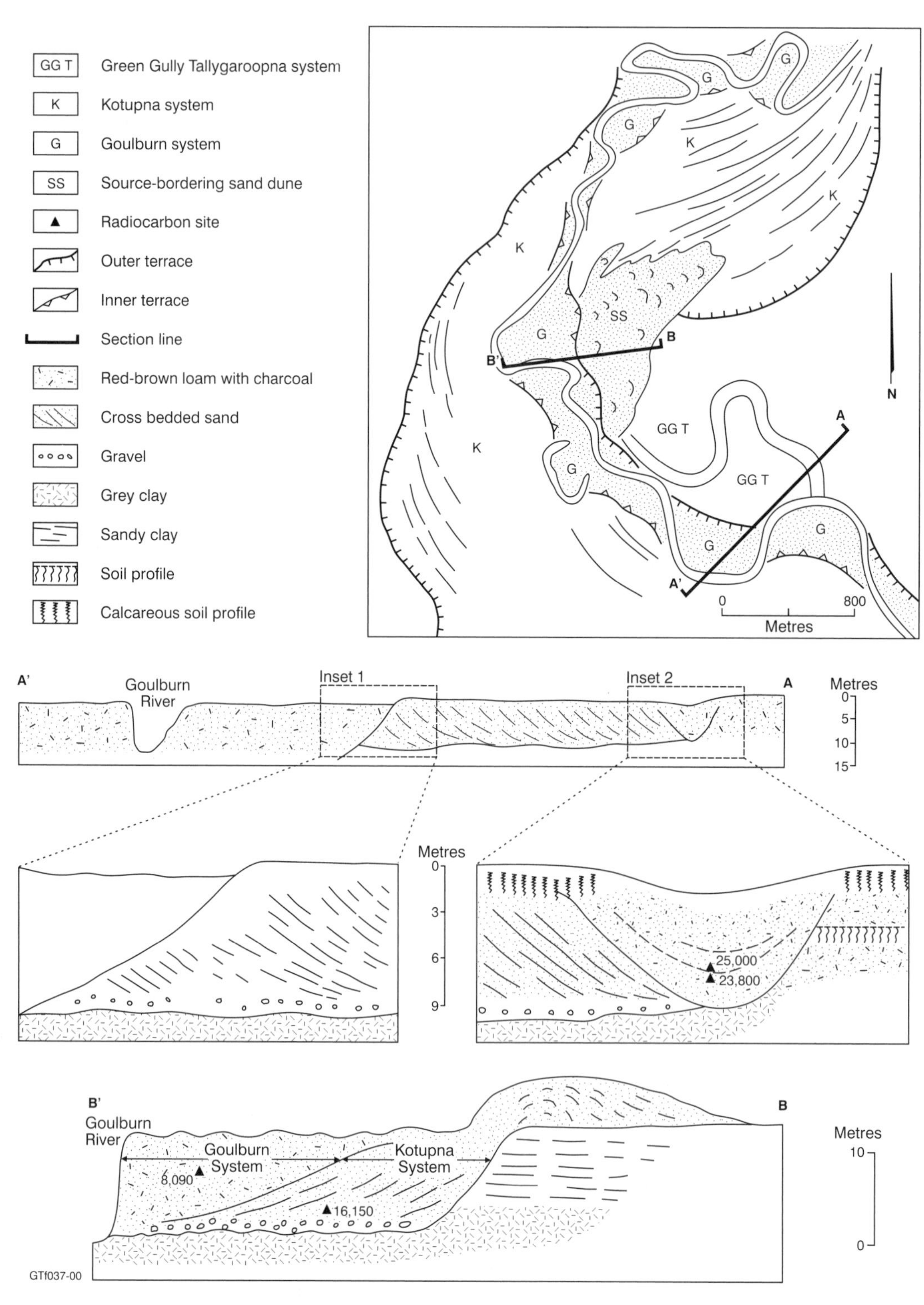

Figure 2.35 *The Goulburn River in northern Victoria. The plan shows the cross-cutting relationships of the modern Goulburn River and its earlier Katupna system. The sections show the internal stratigraphic relationships (from Bowler 1967)*

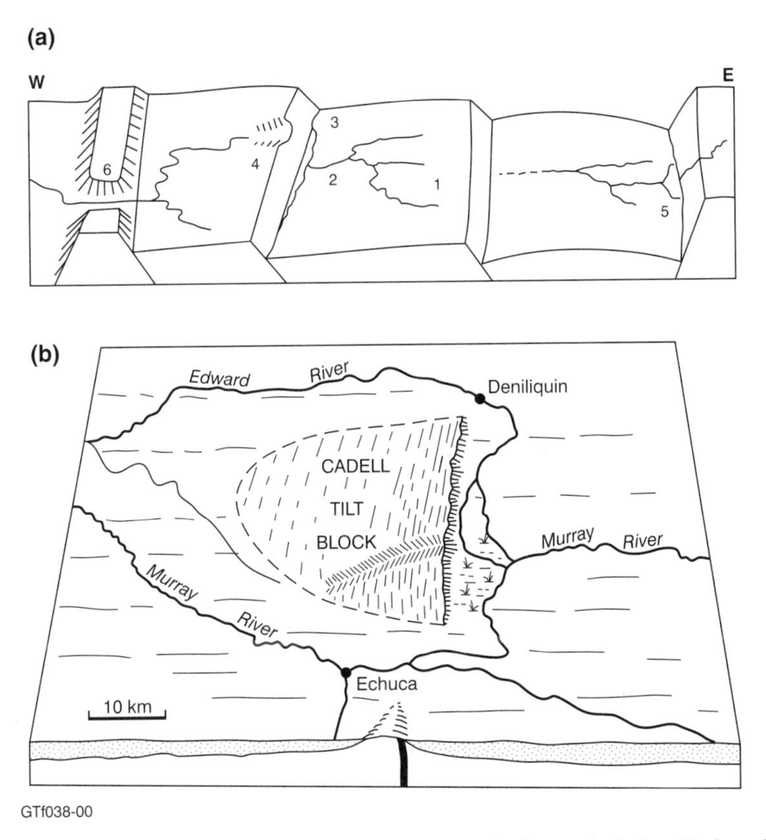

Figure 2.36 *(a) Sketch diagram showing the vertical fault movement on landscapes, including (1) damming, (2) beheading, (3) diversion and temporary damming, (4) antecedent drainage, (5) drainage reversal on a tilt block, (6) antecedent river usually flows through a gorge (from Ollier 1991a). (b) An example illustrating all these features from the Murray River near Echuca*

Examples of many of these landscape elements can be seen in the Canberra region of Australia. Canberra is sited in a graben (Figure 2.37) and is flanked both east and west by horsts. East of the Cullarin Horst sits Lake George, a lake formed by drainage defeated by movement on the Lake George Fault (see also Figure 2.17). Within the grabens, which date at least from in the Tertiary, deep weathering and thick sequences of unconsolidated and weathered Cainozoic sediments are preserved, but these are largely absent from the horst surfaces, only preserved in small and localized pockets which have escaped post-uplift erosion. East of the Shoalhaven Fault, Eocene sediments are preserved as are Early Tertiary basalts in the Shoalhaven Graben. Within the Lake George Graben, Miocene sediments are preserved and cover older, mid-Tertiary or earlier deep weathering (Abell 1985). The recent discovery of Miocene (22 Ma) weathering (Pillans, personal communication) on lake sediments in the Canberra Graben demonstrate its antiquity and that it was probably a landscape low by the Miocene, as was the Lake George Basin. The Shoalhaven Plain is much older and probably dates as a landscape similar to the present one from at least the earliest Tertiary.

In tectonically active areas, low-angle thrust faults have been able to push older rocks over regolith materials. Examples of this are recorded from many localities including the Himalayas where thrusting has included pods of unconsolidated sands in gneisses. Along the Tawonga Fault in Victoria, Beavis (1960) reports Palaeozoic rocks thrust over Pleistocene alluvial gravels and sands in a tunnel dug for the Kiewa Valley hydroelectric project.

Sag faults may cause ponding of drainage and often the formation of large lakes. Taylor & Walker (1986) report such an example from Cooma along the Murrumbidgee River where Miocene movement on the Murrumbidgee Fault dammed the river for about 20 Ma.

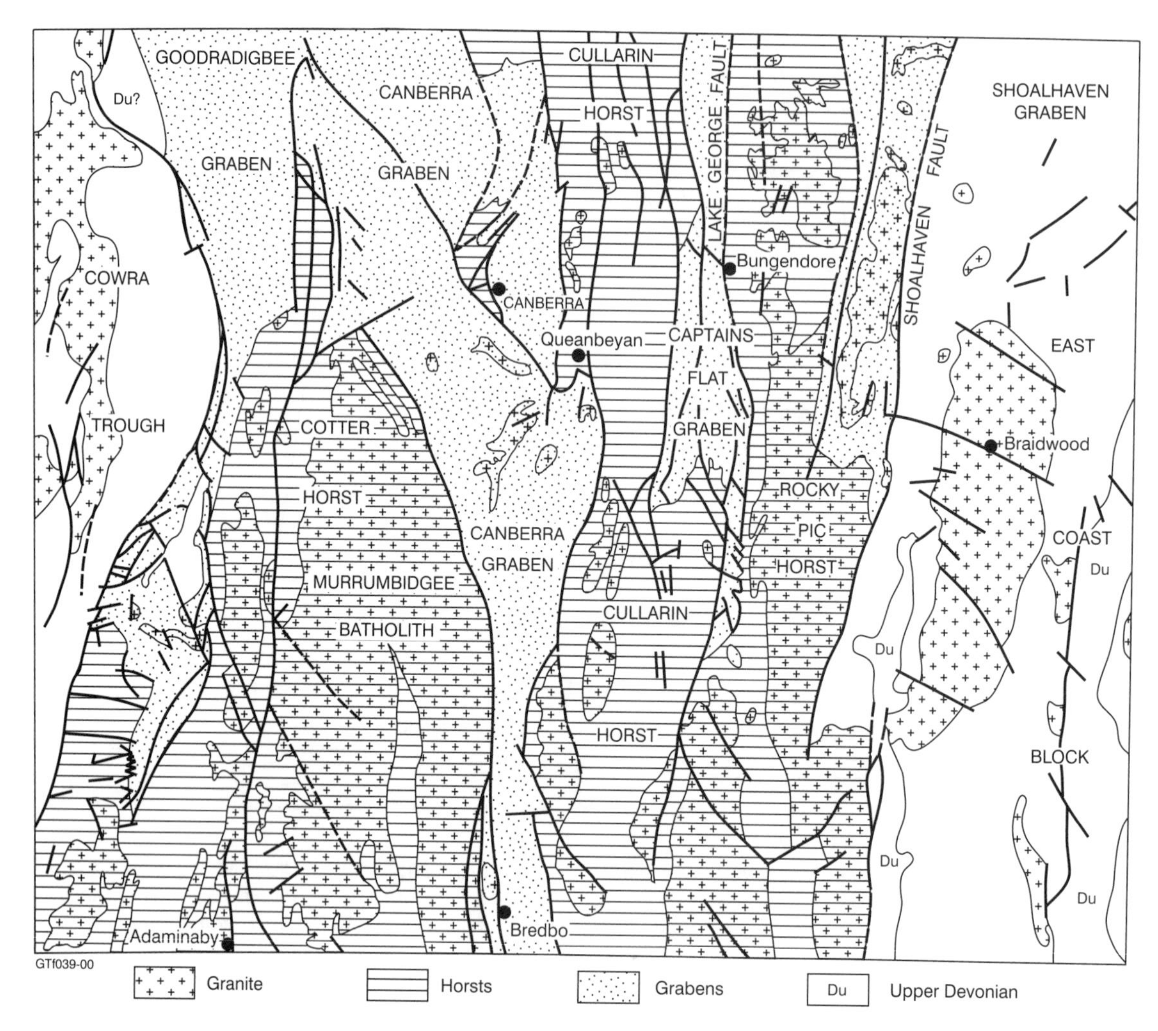

Figure 2.37 *A much simplified geological cross-section of the region around Canberra illustrating the fault control of landscapes and regolith distribution (from Strusz 1971)*

Strike–slip faults may cause a lateral displacement of drainage and ridge crests and displacement of regolith materials. Because strike–slip faults offset rock units across which they cut they also displace landscapes on a larger scale. For example, left-lateral movement on the Berridale Wrench Fault in southern New South Wales shifts subdued hilly landscapes developed on the deeply weathered Dalgety granodiorite some 11 km. Movement was probably mainly pre-Tertiary as the fault is overlain by Palaeocene to Oligocene basalts, but some vertical movement during and after the basalt eruption is suggested by displacement of the basalt sequence. Another fine example of fault-controlled landscapes is in the eastern part of the Snowy Mountains of south-eastern New South Wales as illustrated by White *et al.* (1977) (Figure 2.38).

In reality it is rare to observe the actual fault scarp in a landscape unless the fault has occurred very recently. Generally the upthrown block is weathered and eroded and the debris from this deposited at the foot of the scarp, thus concealing and masking the actual scarp. In cases of older fault scarps the scarp is eroded back into the upthrown block beyond the actual line of the fault, the line of the fault occurring some distance away from the escarpment.

Many other places illustrate the effects of faults on landscapes and regolith and their value in establishing age relations more dramatically than around Canberra, particularly in regions of current tectonic

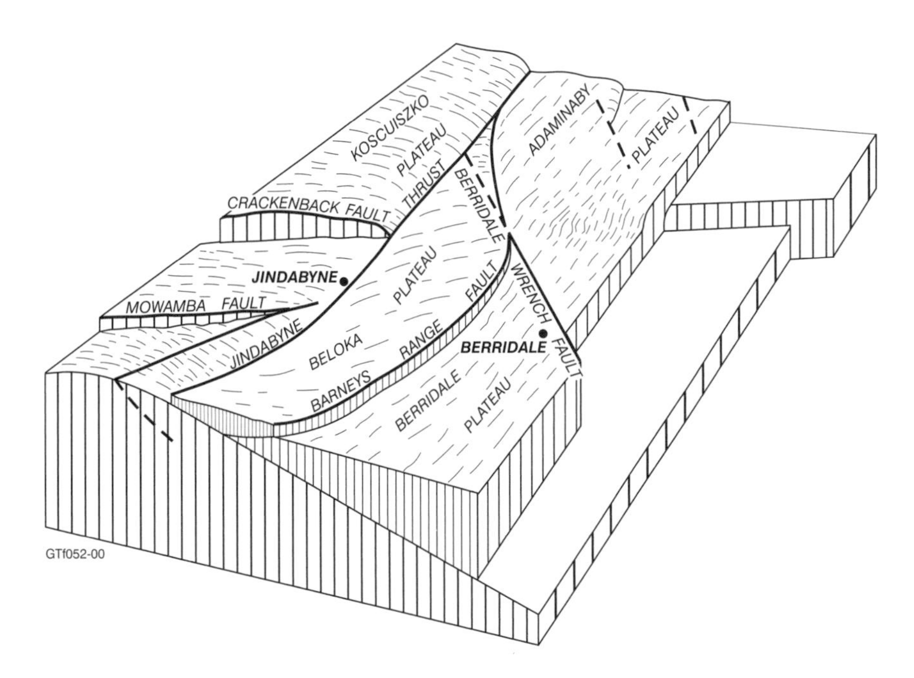

Figure 2.38 *Block diagram showing the relationships of faulting and landscapes in the Monaro–Kosciuszko region of southern New South Wales (from White* et al. *1976)*

activity such as active plate margins and rifts. The East African rift valleys epitomize the landscapes associated with active rifting in continental environments. These include the formation of terraced land surfaces marginal to the rifts, the building of volcanic edifices over a long period and the development of deep lakes in grabens.

In Iceland the Mid-Atlantic Rift is a shallow depression (50 m or so deep) with steep faulted margins from which blocks of basalt have fallen or rotated. The rift is marked by some shallow lakes and drainage diversion caused by the faulting. Volcanoes sit along the rift and their lavas also block or dam, and divert drainage. For such a major tectonic feature of the Earth's crust the rift valley itself is a relatively small feature despite its immense tectonic significance. Was this significance recognized by the early settlers of Iceland when they chose a site along the rift for their first parliament? In a land of spectacular landscapes it is remarkable that the Mid-Atlantic Rift was chosen for such an important activity in AD 930.

Neotectonism is defined (Jackson 1997) as post-Miocene deformation. For the purposes used here we define it as tectonic activity which is actively or can be interpreted to be actively involved in the development of extant landscapes. Many of the effects of faulting, both tectonic and neotectonic, have been discussed above, but now we need to discuss folding and how it affects landscapes and regolith development in them.

Folding which affects a landscape must post-date the development of the landscape and influence its continued development. A good example of this occurs in the Lake Eyre drainage basin, particularly along the Cooper Creek and Diamantina River (Figure 2.15). Here the rivers are confined to actively down-folding synclinal depressions with active basins and domes between (Senior 1975; Drexel & Preiss 1995). These folds are deforming the Cretaceous and Early Tertiary sediments and the oldest sediments recorded in the depressions are Late Oligocene (Drexel & Preiss 1995) so the tectonism dates from about the mid-Tertiary. The up-warping areas such as the Innaminka Dome are actively eroding and regolith is being stripped, the eroded material forming redistributed regolith on the lower slopes and in valleys.

It is also possible to date the tectonic events by examining the regolith and sedimentary sequences. Ollier & Pain (1995) describe the evolution of the Canobolis Divide across New South Wales dividing the northerly Eromanga Basin from the more southerly Murray Basin. Sediments in the Eromanga Basin are Early Cretaceous and older, but the oldest sediments in the Murray Basin are Early Tertiary, showing that the divide between them evolved during the Late Cretaceous.

Relative elevation and erosion features

Erosion is generally a gradual process that degrades landscapes and causes slopes to retreat through time. We have already discussed the lithological controls on this process, but there are important relative dating implications relating to erosion and the age of landscapes and the regolith on them. A valley cut into a land surface must be younger than the surface in to which it is cut, and the regolith formed on the valley walls and bottom must equally be younger than those on the land surface above the valley. The region around the Abercrombie River just west of the Great Divide in southern New South Wales (Figure 2.39) is a good example where the upper surface is a relatively flat surface cut across intensely folded Ordovician sedimentary rocks and is covered in places by Early Tertiary basalts. The surface must therefore pre-date the Eocene and post-date the Ordovician and is probably Triassic (Young 1981). Valleys had been cut into the surface to a depth of about 700 m by the Miocene when basalts erupted, again filling parts of those valleys (Young 1981).

Erosional terraces may result from a shift in local or regional base levels of erosion caused by tectonic movement, shifts in local base level by erosion or changes in sea or lake levels. Such is the case along much of the Shoalhaven River east of Canberra where episodic erosion (Schumm 1975) has resulted in a number of erosional bevels being preserved on valley slopes up to 100 m above the present stream. Thin deposits of alluvial gravels, from which placer gold has been won, cover these, and others are silicified to form silcrete (Ruxton & Taylor 1982). Their age is unknown but they probably relate to the draining of an Early Tertiary lake and the incision of the present valley during the period since (Nott *et al.* 1996; Craft 1931a–d).

Erosional terraces also occur along coasts and lakes. If landscapes are tectonically uplifted or sea/lake levels fall then erosional terraces may be preserved along these coasts. One of the classical examples of this occurs along the north coast of New Guinea at the Huon Peninsula (Chappell 1974). Here the terraces extend to 700 m above sea level (Figure 2.40) and each contains fringing reefs that have been dated using the uranium decay series on the corals and found to range from 70 to 320 ka. Similar sequences occur in Vanuatu on the island of Efate and on many other epirogenically uplifted islands of the south-west Pacific. Erosional shorelines are preserved on hillsides around some lakes, presumably formed during the last shrinking of the lake. A small but good example of this occurs around the western shores of Lake George (Figure 2.41) where a high water terrace is cut into the less consolidated alluvial fans at the margin of the lake.

Similar erosional terraces and landscape surfaces are generated in areas of flat-lying rocks where scarp retreat took place during the valley incision. Typical examples are seen in the Grand Canyon, the Karoo rocks of South Africa and in the Sydney Basin (Figure 2.42). Similar effects result where duricrusts above deeply weathered rocks create a prominent capped escarpment as valleys erode. In all cases the oldest surface is the uppermost and the youngest that occupied by the main drainage channel.

Depositional features

There are a large number of depositional features in the landscape where the site of deposition has moved systematically across the landscape, forming regolith of similar character and systematically younger age. Soils formed on such deposits have been termed 'chronosequences'.

One common depositional sequence, be it marginal marine, glacial or lacustrine, migrates laterally through time by the contraction of the depositional agent. These are referred to as regressive sequences. Generally only the last-formed sequence is preserved as any subsequent or previous transgressions erode materials deposited by earlier events.

As lakes shrink or the sea recedes coastal features are left relict as they form. Sequences of dune/beach ridges are common around shorelines of lakes and the sea. One of the most extensive such sequences occurs across the Murray Basin of south-eastern Australia (Figure 2.43). The oldest of these ridges is Miocene and occurs several hundred kilometres inland from the present coast (Mineral Industry Quarterly 1988). Similar but older (Eocene) ridges occur along the northern edge of the Nullarbor Plain in South Australia and Western Australia (Benbow 1990; Drexel & Preiss 1995).

Lake George provides another good example of a regressive sedimentary sequence of beach ridges deposited during the last 30 ka as the lake shrunk after the last time it was full to overflowing (37 m rather than the present 'lake-full' depth of 10 m). These ridges form a very good chronosequence with the oldest ridge (35–20 ka) being about 15 km from

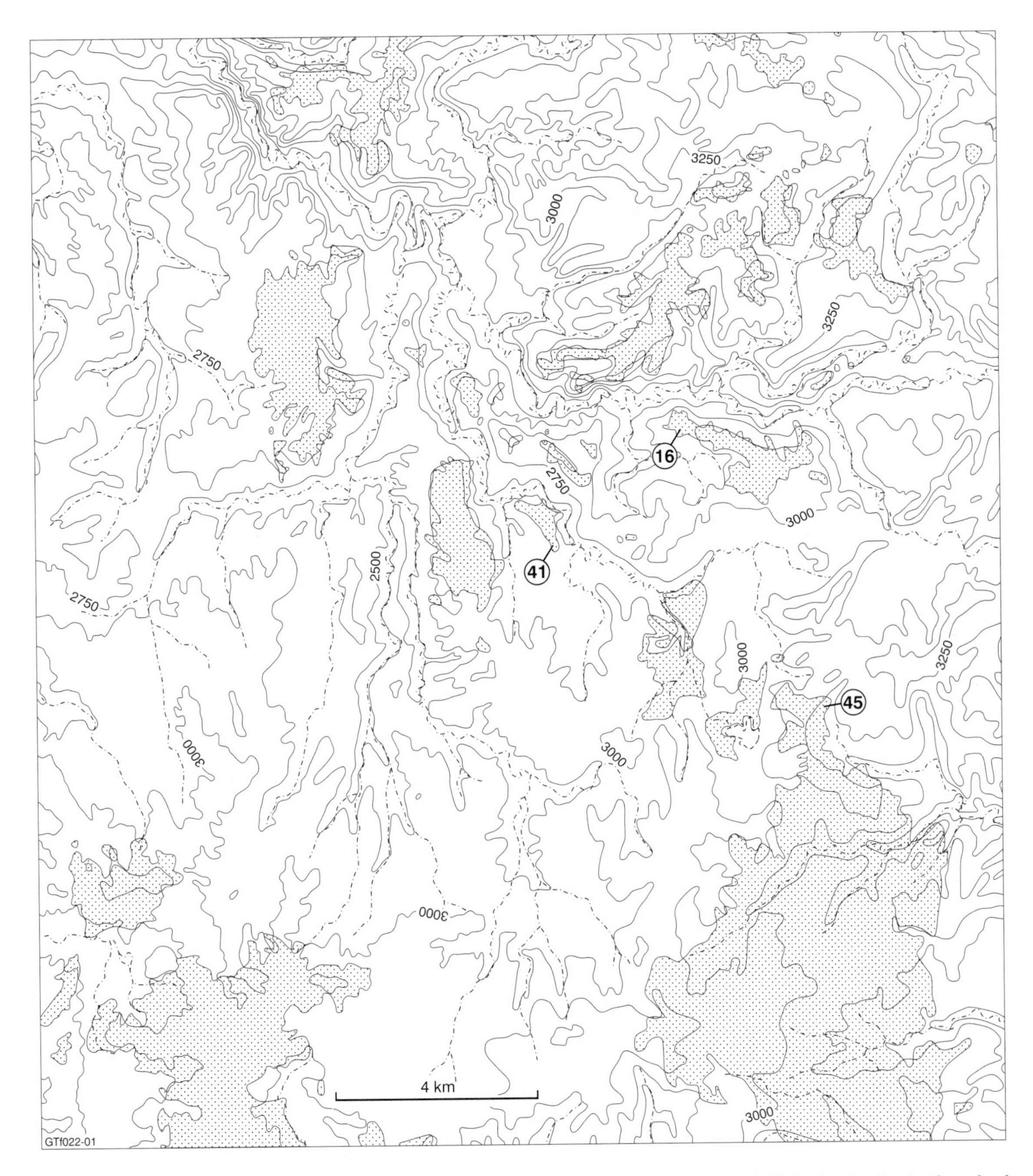

Figure 2.39 *Map of the Abercrombie River region west of the Great Divide in southern New South Wales showing the significant land surface at about 2800 feet cut by the Abercrombie Valley and the distribution of basalts and their ages*

Figure 2.40 *Coral reef-fronted terraces uplifted along the Huon Peninsula in Papua New Guinea (photo John Chappell)*

Figure 2.42 *A photograph of erosional terraces caused by scarp retreat along bedding in flat-lying sedimentary rocks in the Grand Canyon, Colorado (photo Glenys Eggleton)*

Figure 2.41 *Photograph of a 2.5 m high wave-cut terrace in alluvial fan gravels at the margin of Lake George, New South Wales. The terrace was formed during a temporary high stand after the last shrinking of the lake, and since the fan was deposited about 20 ka ago (photo GT)*

the present lake-full margin, with a number of younger ridges closer to the lake (Figure 2.15).

Glaciers as they retreat also commonly leave regressive sequences of till either as sheets or mounded in moraines consisting of unstratified till or mixtures of sediment with a wide range of grain size. The furthest moraine from the glacier's end is the oldest. Sediments of this type are not common in Australia, but do occur in Tasmania's north-west, as is clearly shown by the work of Kiernan (1990) (Figure 2.44).

Several types of depositional chronosequence are associated with rivers. As rivers migrate across their alluvium they leave traces of their former presence at various places. Meandering stream bends migrate laterally leaving depositional ridges or scroll bars in their wake. These ridges increase in age away from the channel. It is not unusual to see many examples of these on floodplains of major rivers, associated with the active stream and on the inside of abandoned channel bends (Figure 2.45). These features are preserved as surface patterns for as long as the stream takes to cover them by floodplain aggradation. In some localities they can date well back into the Quaternary (e.g. Goulburn River, Figure 2.35) or even the Tertiary (e.g. the Talliringa Palaeochannel in South Australia, Drexel & Preiss 1995). In both cases the channels forming the scroll bars have been abandoned, leaving the surface topography uncovered by subsequent sediments.

Another type of migratory feature is found associated with lakes and in some cases rivers. Sediments blown from the floor or margins of a lake or river deposit as dunes on the lee side of them. In the case of lakes the sediments deposited reflect the hydrological state of the lake. During lake-full stages sandy sediments from beaches on the lee shore are blown out and deposit as a lunette or dune. As the lake dries, muds and later evaporites may be blown out and accumulate (Butler & Churchward 1983; Cook 1997). In some lakes the formation of lunettes on the downwind shore of the lake basin cause the lake to migrate to windward as seen on the lakes occupying old Tertiary palaeochannels on the Yilgarn of Western Australia (Figure 2.46). Lunettes in both cases progressively get younger towards the lake, the oldest being furthest from the lake. In some lunettes (e.g. Lake Mungo, Bowler 1983) the lunettes only form a partial migratory sequence as each subsequent dune builds on previous ones, forming a stratigraphic sequence which reflects in reverse order the hydrological events in the lake.

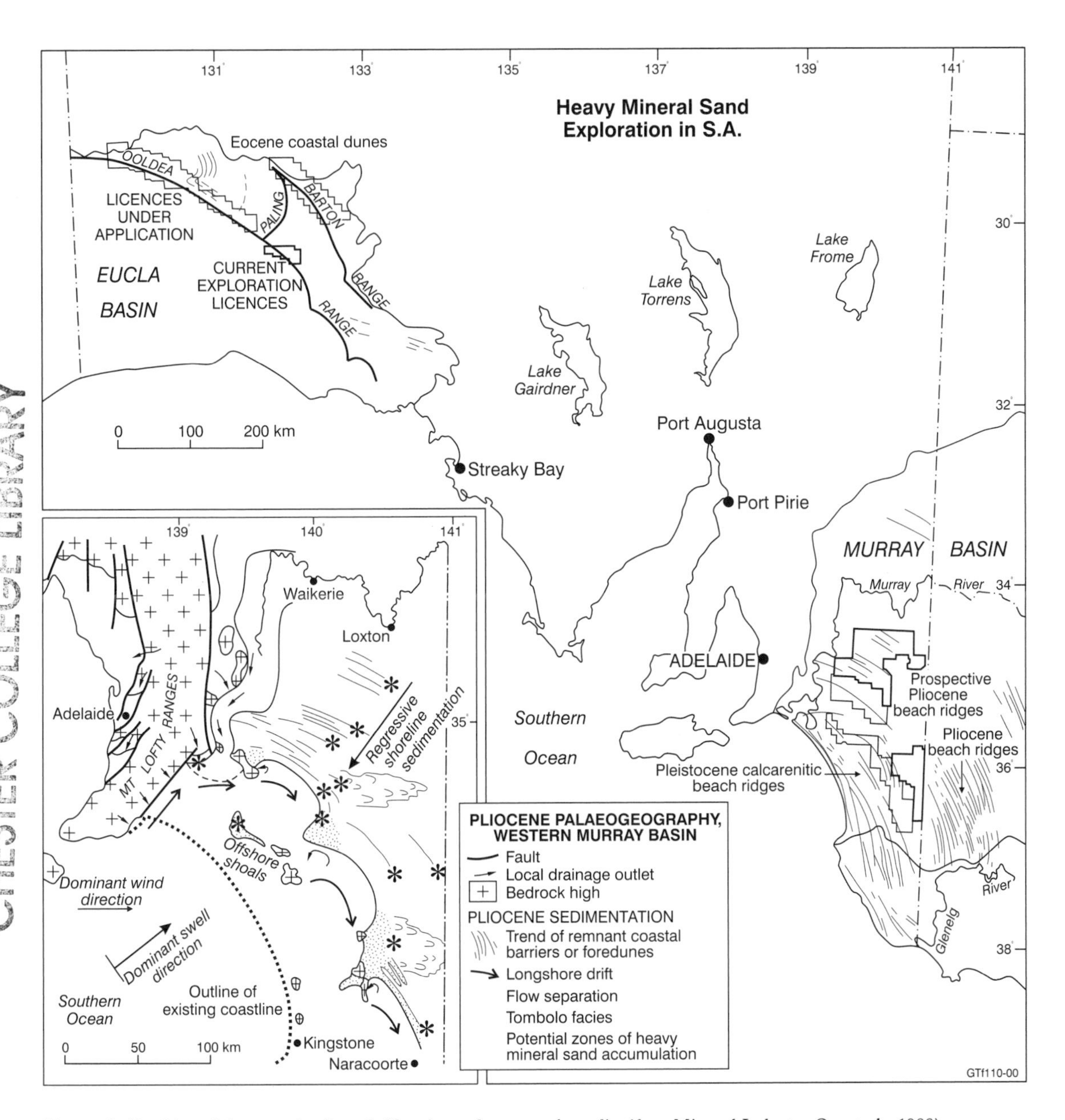

Figure 2.43 *Map of the regressive littoral ridges in south-eastern Australia (from* Mineral Industry Quarterly *1988)*

Depositional river terraces are another kind of age sequence which forms as the river cuts deeper into its existing deposits as a result of base level change or a regime change induced by climate change. They are common in Late Quaternary sequences and may date further back, but such cases are not well documented. Good Quaternary alluvial/soil terrace chronosequences are documented by Walker & Butler (1983) in Australia (Table 2.4) and numerous examples are described worldwide. Thomas & Thorp (1993) describe sequences similar in many respects to those of south-eastern Australia in Sierra Leone, even though Late Quaternary climatic changes, which drove the changes, were more subtle. The older terraces, however, have very much more mature soils than their equivalents in south-east Australia; lateritized soils in

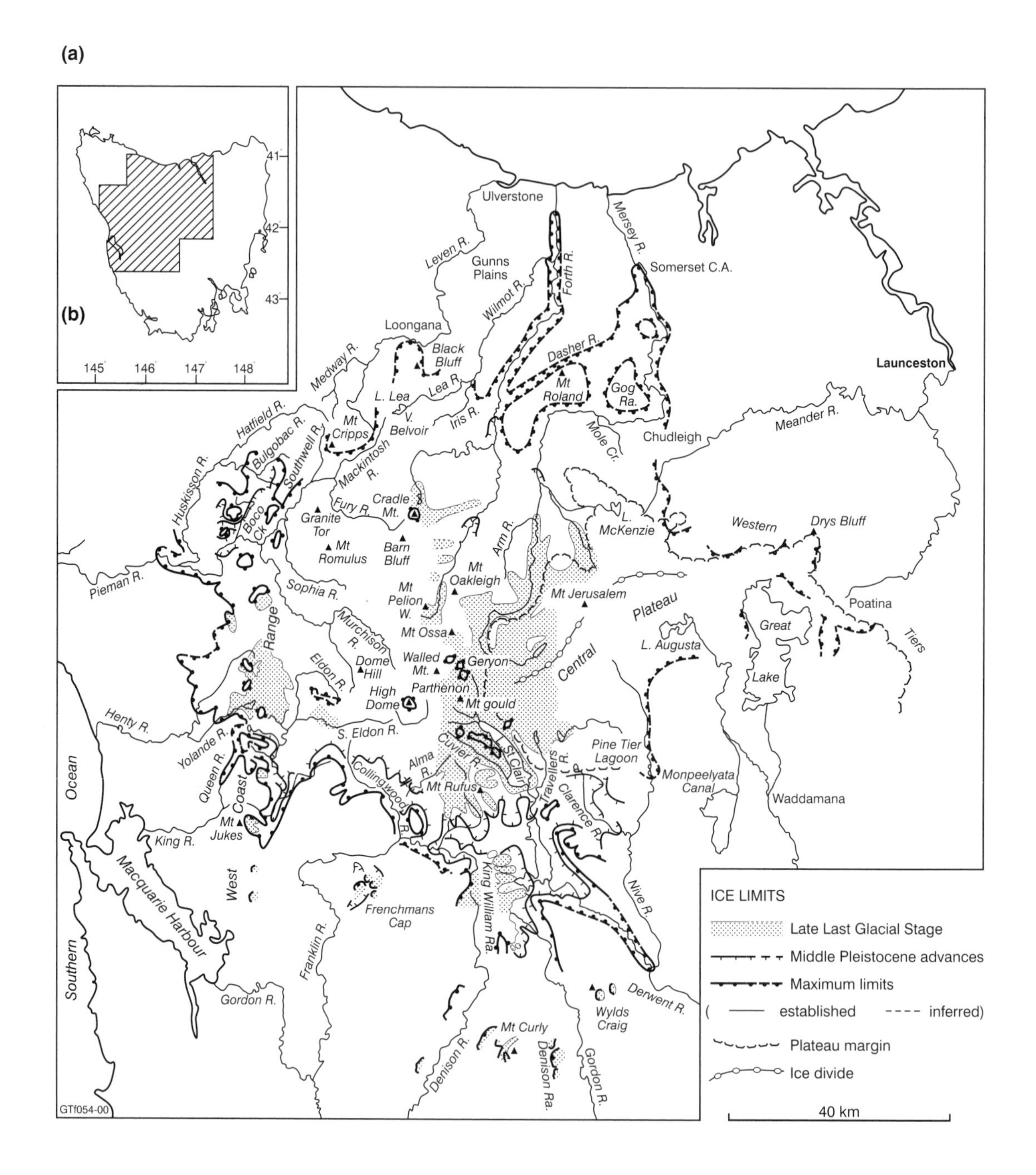

Figure 2.44 *The ice limits of the last maximum extent of the ice-sheet, tills of a Middle Pleistocene and late glacial stage deposits in north-western and central Tasmania (from Kiernan 1990)*

Figure 2.45 *Photograph of a floodplain with scroll bars and abandoned channels with scroll bars. The east–west river is the Murray forming the New South Wales–Victorian border and the river joining from the north is the Murrumbidgee. The southern part of the photo shows a prominent east–west linear dunefield*

Africa compared to red podzolic soils in Australia. This is not surprising given the 30° latitudinal difference in the sites.

Weathering and soil features

Weathering is generally considered to proceed from the surface downwards; however, much intense weathering is probably controlled by the groundwater regime. We discuss this in detail below, but it is possible to consider the degree of weathering as a relative dating tool. In general terms, surfaces exposed from longer times will be more intensely weathered, other things being equal. Ruxton (1968) has shown that the oldest lavas and ash deposits are more weathered than younger ones. Ollier (1991a) describes a similar situation on Western Samoa where the oldest volcanic rocks are deeply and completely weathered, while the youngest ones are still fresh. This approach, however, is difficult to apply in terrains in which many different rock types crop out as this confuses judgement as to which rock types are most weathered.

Soil development may be used in a similar way; the more mature soils are older than less mature ones. This principle is well illustrated in the classic study of Walker (1962) where the most mature red podzolic soils occur on the highest alluvial terrace and alluvial soils on the lowest and youngest surface. In a similar study near Brisbane, Thompson (1981) found that podzols developed on coastal sand dunes show increasing development with increasing age. The concept of soil maturity will be discussed later, but it is well illustrated by the deep red earth soils of north Queensland, where soils identical to those at the present surface are covered by basalts 2 Ma old (Coventry 1979). Soils associated occurring on various aged alluvial terraces are shown in Table 2.4.

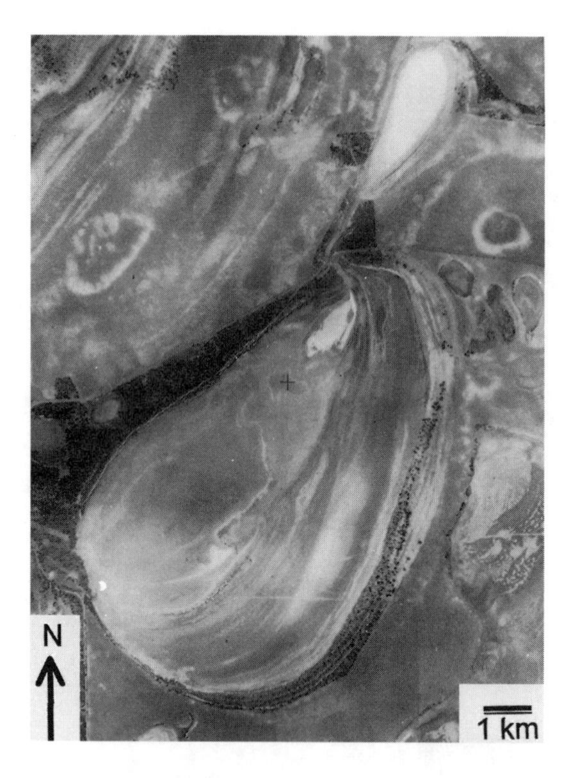

Figure 2.46 *Photograph of lakes and lunettes on the Yilgarn in Western Australia. The lunettes (dunes) form on the lee shore of the lake (east) and this accumulation causes the lake basin to move into the wind (west)*

The thickness of rinds and degree of weathering on pebbles and boulders in alluvial sediments are taken in many cases to indicate their age of deposition. This is based on the assumption that such rinds would not exist on them at the time of their deposition. Good examples of the use of weathering rinds to age sediments include McSaveney (1992), Kiernan (1990) and Colman & Pierce (1981).

Many researchers correlate materials using the characteristics of whole or particular zones or horizons of weathering profiles (e.g. Firman 1994). Since weathering and pedogenesis are processes of alteration of pre-existing materials or the development of an imprint on pre-existing materials, neither the normal lithostratigraphic nor chronostratigraphic principles can be applied to them. There is no reason to expect that weathering of a shale in one place will at the same time produce a weathering profile on a granite somewhere else similar to that on the shale. Equally, there is no reason to expect that the cementing of a granite saprolite by iron minerals to form a valley-side ferruginous duricrust will occur simultaneously in another place unless the landscape settings and groundwater regimes are similar at the same time. Quite possibly such cementation at one place may terminate long before conditions are suitable some distance down (or up) the same valley, let alone in another valley. It is therefore wrong to assume one can correlate over large or small distances using the weathered character of a material to infer anything about their relative time of formation. In fact there is no reason to use lithostratigraphy to infer synchronicity of deposition, let alone of weathering imprints. So the nature of a weathering profile does not aid in dating weathering.

Ages on individual surfaces

A theoretical concept that is important in determining relative ages of parts of the same land surface that has been formed by headward erosion of an escarpment is that the front edge of the surface will be covered by older regolith than the back. Such surfaces are said to be diachronous. As the scarp retreats it exposes progressively fresher rock from under the escarpment. Thus rocks at the back of the surface will be less weathered than those at the front, and are less likely to have mature weathering profiles or duricrusts developed on them.

The retreat of the 'great escarpments' in the Gondwanan continents after continental breakup are major, though complex, examples. These are very large-scale landscape features, encompassing a wide range of bedrock, with rather variable hilly terrain on the coastal lowlands. This has provided a challenge to the understanding of landscape evolution. Methods that have been applied to resolve the problem include apatite fission track and cosmic isotope dating (see below).

Similarly, the regolith formed on depositional surfaces may be diachronous. Areas on a floodplain remote from the river tend to receive less sediment on average than those areas closer to it. Thus, for example, on a floodplain soils are less well developed and the sediments are generally less weathered close to the river where regular increments of alluvial accretion occur, whereas farther from the river inundation and thus sedimentation are less frequent. This is very obvious on most large floodplains and Walker (1962) has demonstrated this well on an alluvial terrace sequence at Nowra in New South Wales.

Table 2.4 *Ages and characteristics of alluvial terraces in Australia (from Walker & Butler 1983)*

Location	Parent material	Dating (years BP)	Soils	Reference
New South Wales				
Coastal	Alluvium	0–120	Alluvial	Walker (1962)
		390	Minimal prairie	
		3 740	Minimal podzolic	
		29 000	Podzolic	
Tablelands	Alluvium	255	Alluvial	Coventry & Walker (1977)
		855	Minimal prairie	Walker & Gillespie (1978)
		2 500	Prairie	Walker & Green (1976)
		2 750	Shallow earth	
		27 040	Podzolic	
	Fan deposits	1 600–2 400	Minimal podzolic	Coventry & Walker (1977)
		26 800	Podzolic	
Australian Capital Territory				
Tablelands	Colluvium	17 170	Earth-podzolic	Walker & Gillespie (1978)
	Fan deposits	27 800	Podzolic	Costin & Polach (1973)
		31 000	Podzolic	
New South Wales and Victoria				
Riverine	Alluvium	<3 000	Alluvial	Butler (1958)
plain		4 000–13 000	Minimal prairie,	Pels (1971)
			Minimal red-brown earths	Lawrence (1976)
		18 000	Red-brown earths	
		20 000	Red-brown earths	
Victoria				
Coastal	Alluvium	<3 000	Alluvial	Ward (1977)
		42 000	Sandy red earth	
		210 000	Solodic	
South Australia				
	Alluvium	34 000	Red-brown earth	Williams (1969)
Northern Territory				
	Alluvium	Undated	Alluvial	Litchfield (1969)
		50 000–>60 000	Sandy red earth	
		100 000–130 000	Red earth	
Western Australia				
	Alluvium	4 000–6 000	Minimal prairie	McArthur & Bettenay (1960)
		10 000	Minimal prairie	
		17 000	Red podzolic	
		30 000	Yellow podzolic	

NUMERICAL DATING

Dating techniques applicable to the regolith and to determining landscape evolution are well summarized by Pillans (1998). He provides a diagram (Figure 2.47) showing the time spectrum covered by the various techniques. He also outlines the various methods useful in regolith research, and provides some key references to their use. Much of what appears here is taken from this summary.

Numerical dating methods provide numbers derived from an analytical technique applied to a mineral or other component of the material being dated. Their meaning is only applicable to the time of origin of the mineral being dated, and it is thus essential that to interpret the number derived we understand the paragenesis of that mineral. In the case of regolith materials particularly, this may not correspond to the age of the material enclosing that mineral. Hence we are in effect dating one component of the regolith that may pre- or post-date the formation of the regolith. In effect, all the dating is doing is confining the age of the regolith as a whole. In interpreting these ages it is also as well to be cognizant of the fact that

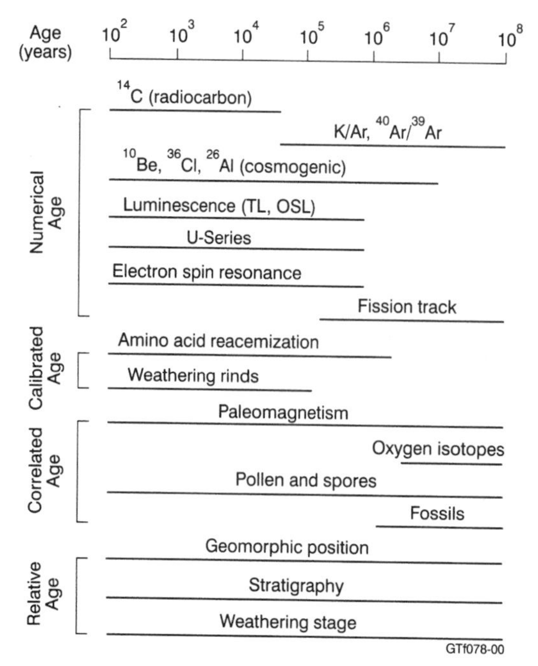

Figure 2.47 *Age ranges over which dating methods can be applied. Methods are grouped according to the type of age result produced (from Pillans 1998)*

many regolith materials are even now undergoing evolution in their composition, so dates need to be carefully interpreted.

Radiometric dating

Radioactive isotopes decay at a constant rate. Thus if we can measure the proportion of elemental isotopes in a material and if we know the rate of decay of them (their half-life) then it is possible to calculate their age. There are a large variety of isotopes that are amenable to this dating technique, but all are subject to problems, mainly relating to contamination, isotope loss or rates of production of the isotope. Detailed investigation of the nature of the sample used for dating is required (to determine whether it has been contaminated or has lost isotopes), its field relationships need to be well understood and contemporary processes at the collection site need to be known if the ages determined are to be used correctly.

Perhaps the best-known radiometric dating method is radiocarbon dating. It is based on the content of ^{14}C in a sample of material containing carbon; ^{14}C has a half-life of 5730 years so the technique is only applicable to materials formed over the last 40 ka or so. Carbon dating can be applied to any material containing C including organic materials, carbonate minerals, shells and bone.

Potassium–argon dating is based on the decay of ^{40}K to ^{40}Ar that has a very long half-life, making the technique applicable to regolith materials over 100 ka old. It is also possible to extend this technique to the $^{40}Ar/^{39}Ar$ method which gives more precise ages but also requires time and precision, hence it is expensive. Materials suitable to these techniques include any minerals containing K. It is readily applicable to fresh igneous rocks, particularly volcanic rocks, but when it comes to regolith materials contamination and Ar loss become problems, although minerals like alunite, K–Mn oxides and glauconite are usable. Weathered materials are susceptible to Ar loss and are consequently difficult to interpret.

Cosmic ray interactions with Earth materials produce ^{10}Be, ^{36}Cl and ^{26}Al. The longer the materials are exposed to cosmic radiation the higher the concentration of these isotopes becomes. They accumulate in the upper 1–2 m of the Earth's surface and can be readily measured. Thus they are useful for dating surfaces up to about 5 Ma, but there are a number of problems associated with the technique. Researchers are not sure of the rate at which these isotopes are produced, rock surfaces must be stable over the time period being dated (i.e. there must not have been erosion) and samples collected must be corrected for altitude and latitude, which also means generally they must be dated using another method.

Fleming *et al.* (1999) used ^{36}Cl and ^{10}Be to determine scarp retreat rates and summit lowering rates of the Drakensberg in Southern Africa. They demonstrate that escarpment retreat rates over the 10^4–10^6-year time span have been 50–90 and 6 m/Ma respectively. These rates of escarpment retreat are an order of magnitude less than previously thought. They argue on these rates of scarp retreat that since the breakup of Australia and Africa the scarp has only retreated some 10 km if the rates they measured are extrapolated back to 135 Ma, far less than its present distance inland. This being so, scarp initiation cannot be related to Gondwanan rifting. King (1945) argues that land surfaces formed prior to the breakup of Gondwana were preserved at the highest peaks of the Drakensberg. The isotopically determined erosion rates of Fleming *et al.* (1999) show this is not possible if their rates are extrapolated back in time. This raises the question of how valid are such extrapolations. Can rates determined over the last 10^6 years be validly extrapolated to 10^8 years? Problems of correcting for altitude are certainly a feature in the Fleming *et al.*

(1999) study, and it is also uncertain how much of the isotope accumulation has been lost through erosion. Nonetheless the results of their study provide useful results, calling into question the idea of longevity of land surfaces and the causes and rates of scarp retreat.

Researchers in Wyoming have used ^{36}Cl and ^{10}Be to date the surfaces of glacial and fluvial deposits (Phillips *et al.* 1997). They determined the oldest moraine to be >232 ka, but they quote quite large uncertainty ranges of 10–15%.

Uranium series dating is based on the parent/daughter isotopes in the U-series decay sequences, particularly $^{234}U/^{238}U$ and $^{230}Th/^{234}U$. It is useful for samples of speleothems, fossil corals and shoreline deposits. Problems with the method relate to the high mobility of U in the environment and it is unsuitable for calcrete because of Th contamination.

Palaeomagnetism

Palaeomagnetic dating techniques rely on three types of changes in the Earth's magnetic field:

- secular variation on timescales from 10 to 10 000 years;
- magnetic field reversals on a timescale of 10 ka–100+ Ma; and,
- apparent polar wander as continents move so magnetic declination changes.

Paleomagnetism is useful for timescales from 5 to 100+ Ma. The technique is applicable to regolith material containing Fe-oxides as well as most rocks.

Pillans *et al.* (1998) have used the apparent polar wander method to date weathering profiles at Northparkes Mine in central New South Wales. Here a deep weathering profile (30 m thick) is developed on the Ordovician quartz monzonite rocks containing the porphyry copper ore bodies. This has been incised to 25 m by a palaeochannel, which filled and was then also deeply weathered. The whole deeply weathered sequence is overlain by a thin (1–2 m) transported cover with red-brown earth soils developed in it. They determined that the older deep weathering has an apparent pole position at 51.2° S and 81.4° E. This lies west of the Australian Cainozoic apparent polar wander path (APAW), but lies on the Carboniferous APAW, demonstrating that these profiles formed during the Late Palaeozoic. The deeply weathered channel fills acquired their remnant magnetism during the Cainozoic, before 0.78 Ma. The transported overburden was not investigated magnetically in this study.

Fission track

The spontaneous fission of ^{238}U produces fragments that disrupt the crystal lattice creating damage tracks. The number of tracks is proportional to the U content, age and the thermal history of the mineral. The dates depend on the time of cooling below a mineral-dependent track-annealing temperature. Minerals used for this dating technique include zircon, sphene, apatite, volcanic glass and possibly gypsum.

The most commonly used mineral for this type of dating is apatite which has an annealing temperature between 120°C and about 100°C. As well as being useful for dating, apatite fission track formation enables the determination of when it reached temperatures below the annealing temperature in the Earth's crust. Thus if the geothermal gradients are known for the area the analysis gives us the date at which the mineral reached a particular depth within the crust. This has fundamental applications in the pursuit of unravelling landscape history. It is possible to determine how much material has been eroded from a particular landscape since the mineral passed through its annealing temperature, and thus how much crust has been eroded if that mineral is sampled at the surface. A corollary is that it also tells us the time period over which the erosion has occurred. This technique has been widely applied in Australia (e.g. O'Sullivan *et al.* 1998; Hill & Kohn 1999).

Hill & Kohn (1999) used apatite fission track thermochronology (AFTT) to constrain the amount and timing of movements on the Mundi Mundi Fault that forms the western margin escarpment of the Broken Hill Block in western New South Wales. The AFTT results show that the thermal histories of the blocks on either side of the fault are very different. Samples from the west (downthrown) of the fault record an initial cooling from palaeotemperatures >100–120°C during the Early Palaeozoic, whereas those from the east (upthrown) of the fault show significant cooling during the Late Palaeozoic. There is also evidence, although poorly constrained, of further movement on the fault during the Cainozoic. Local sedimentary and geomorphic evidence further constrains the movement on this fault. Cretaceous and Cainozoic sediments are widespread and thick west of it, but absent east of it. The existence of the fault (and escarpment) during the Palaeozoic suggests that not only were Cretaceous and Cainozoic sediments probably not deposited on the upthrown block but it contributed eroded material to the sediment pile to the west. Neogene sediments are >100 m thick to the west of the fault, suggesting the Cainozoic movement recorded by AFTT, though ill constrained, was probably during the Early Neogene.

The geomorphic evolution of the Eastern Highlands of Australia has been the subject of controversy for over 100 years. AFTT has been used in an attempt to constrain the age of uplift and the mode of evolution of the coastal plain (Moore *et al.* 1986; Gallagher *et al.* 1994; O'Sullivan *et al.* 1998). One of the models for the evolution of the coastal plains and escarpment dividing the lowlands from the highlands has been the downwarping of a formerly extensive palaeoplain (Ollier & Pain 1996, p. 213). O'Sullivan *et al.* (1998) tested this model using AFTT. Their conclusions clearly show that downwarping followed by headward erosion by coastal streams to form the escarpment could not occur, because the AFTT ages increase inland across the coastal plain. The reverse would be true if headward erosion had occurred to form the lowlands.

Stable isotopes (^{18}O and ^{13}C)

The O isotopic composition of kaolinite, gibbsite and other clay minerals is useful to broadly date weathered materials older than 5 Ma. This is because of the systematic variations in the isotopic composition of meteoric water from which these minerals have formed. The variations result from the latitudinal change as continents have drifted and global climate change. The isotopic composition of water is related to temperature so these changes affect the water isotopic ratios and consequently those of the minerals that form from them. This method can only date to an accuracy of 10^7 Ma.

Luminescence and optically stimulated luminescence dating (TL and OSL)

Natural radiation of α, β, γ and cosmic rays produces electron/hole pairs that can be trapped in crystal lattices. Crystals that contain such electron/hole pairs emit luminescence when heated (TL) or are exposed to visible light (OSL). The greater the luminescence the greater the radiation dose the sample has received. The age of the sample is determined by calibrating the luminescence with known radiation dose rates in combination with measurements of the samples' environmental dose rate. These techniques are useful for dating materials such as quartz, feldspars and gypsum up to about 1 Ma. It is essential that samples used in these techniques have not been exposed to visible light prior to sampling. One of the major problems with these methods is that samples may not always have been completely reset (all electron/hole pairs annihilated) by exposure to visible light prior to burial.

Electron spin resonance (ESR) dating

Another numerical method of dating quartz, gypsum, teeth, shells and coral is electron spin resonance (ESR) useful to 1+ Ma. Radiation produces unpaired electrons in crystal lattices and this radiation increases over time and with radiation dose. This method has the potential to date silcrete cements (Chapter 13).

Regolith and climate

Regolith is the product of the interaction between surface environments and geological processes. Climate refers to the atmospheric and hydrological features of an area, and it is known to have varied dramatically over time. The biosphere is to a large degree dependent on climate and the nature of the regolith, and it influences the nature of regolith formed. We know there is no necessary direct correlation between contemporary climates and regolith in any particular region, but we also know that present surface processes are related to climate among other controls. Herein lies a dilemma: to what extent can regolith products be used to infer the conditions of their formation? There is little doubt that glacial activity and the regolith it produces are restricted to cold climates. Conversely, while there is no doubt that deeply weathered rocks are common in contemporary wet and wet/dry tropical climates, this does not mean all deep saprolitic weathering was formed in tropical climates.

Weathering only requires water and time to occur. While temperature is important, it only controls the rate at which chemical reactions occur. Consequently in cold climates where water is available and physical erosion is very low, it is possible to deeply weather rocks, but slowly. Where temperatures are high, even though physical erosion rates may also be high, there is a greater chance for weathering rates to exceed erosion, so deeply weathered rocks can be preserved despite the higher erosion rate.

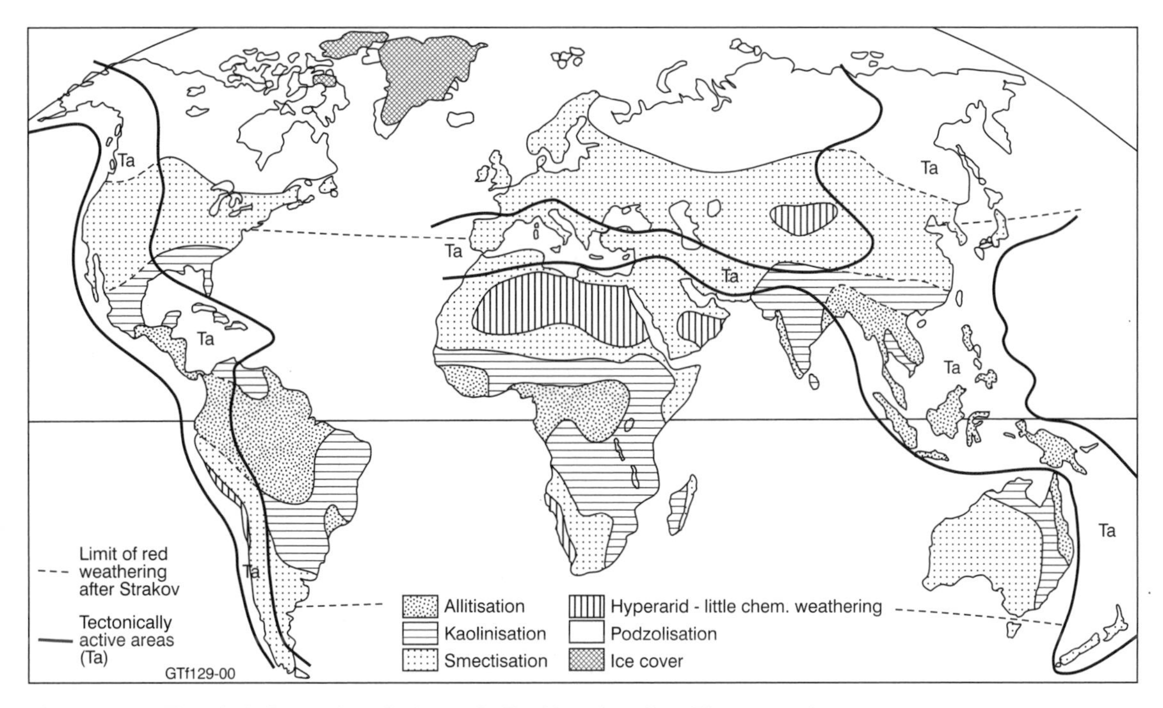

Figure 2.48 *The principal types of weathering on the Earth's surface (from Thomas 1994)*

Another dilemma relates to the occurrence of deeply weathered rocks in presently arid or cold climatic zones. Such examples may relate to past climatic regimes or they may have been formed under climates similar to those in which they now occur and be preserved because physical erosion has been minimal. Although the latter scenario is unlikely, evidence for climatic conditions at the time of their formation other than the weathering itself is required to confirm the conditions of their formation (e.g. Taylor *et al.* 1990a; Bird & Chivas 1988).

These arguments aside, there are some general relationships between weathering and regolith formation and climate, which are worth outlining here. One of the earliest global summaries of the effects of climate on weathering and formation of major soil clay minerals was by Strakhov (1967) (Figure 1.4). His distribution shows the degree of leaching (controlled by rainfall and temperature), evaporation and the degree of vegetation accumulation in the regolith. This is reflected in the depth of chemical weathering and the nature of the clay minerals formed. While it is a reasonable reflection of the global picture, Thomas (1994) makes the point that Strakhov overestimates the extent of tropical rainforest relative to savannah. Also Strakhov's model takes no account of relict landforms and regolith elements, so regolith formed in other than contemporary climates is not considered. Another map of the principal types of weathering was compiled by Thomas (1994) based on earlier work (Pedro 1968; Strakhov 1967) (Figure 2.48). This map is interesting in that it recognizes tectonically active areas where so-called climatically controlled regolith is modified. While this is accurate and significant there is no recognition of relict weathering and its significance on many continents, particularly outside the wet tropics.

Some attempt to recognize relict regolith or potentially relict regolith was made by Pedro (1968), based on the FAO *World Soil Map*, and Ollier & Pain (1996) (Figure 1.6) and Thomas (1994) similarly prepared a map (Figure 2.49). Many of the units shown on these maps show residual or relict soils or regolith that are formed under climates different from those in the area today.

Both physical and chemical weathering lead to regolith production, but each produces very different types. Climate and slope primarily control the balance between physical and chemical weathering. Generally in climates where water is not readily available chemical weathering is minimal and in areas of very steep terrain physical weathering dominates or is very significant. Rates of chemical weathering can be very

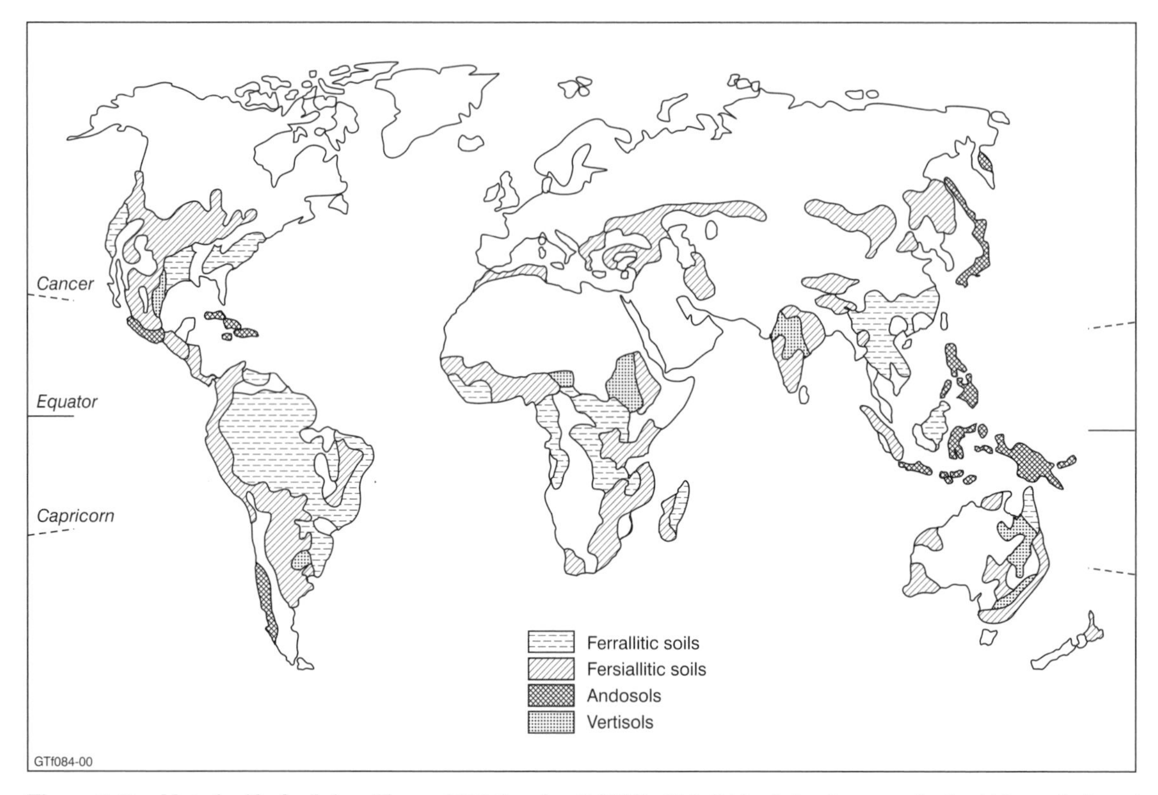

Figure 2.49 *Map of residual soils from Thomas 1994 (based on FAO* World Soil Map*) showing areas of soils which may be formed under palaeoclimates different from those in the area today*

high (up to 60 m/Ma, Thomas 1994) in wet tropical regions, but they can fall to almost nil in Arctic or very arid climates. Rates of physical weathering are harder to estimate, but in glacial areas most weathering is physical and erosion rates may be deduced from the duration of glacial episodes. Chemical weathering on most rock types produces a regolith of clay and/or quartz sand and Fe^{3+}-minerals, while physical weathering produces comminuted rock fragments, and generally there is a mixture of both in the regolith. Both are significant in different climates and although difficult to observe, except in karst terrains, chemical weathering is dominant in most climates and leads to the majority of erosion as well.

Regolith weathered in place versus transported regolith

When looking at a profile of weathered rock or regolith in the field there is a tendency to assume it has formed where we now see it; in other words it is *in situ*. On the other hand, we have no problem with accepting a pile of sediments as being transported regolith. As we described above, in most landscapes fragmental regolith material is moved by erosion and redeposited, and this process will occur many times before the particles finally find their way to their ultimate destiny in the sea, if they ever do! This means that in most parts of a landscape there is a fair chance that at least parts of the regolith will comprise both *in situ* and transported material. Chemically transported regolith components can also redeposit as they cycle through landscape in ground and surface water, dissolving at one place, precipitating at another and so on, until they too reach a sink outside the regolith regime. The differences between *in situ* and transported regolith are discussed in Chapter 11.

Scale

Another variable that is related to the factors controlling landscape evolution is the scale of landforms being considered. At a continental scale, plate tectonics is a fundamental control, whereas at a stream valley or hill slope scale lithological and climatic factors are probably more important. At a micro-scale the nature of materials making up the landscape is critical to understanding its evolution, and in Chapter 3 we begin to explore why. At this point scale of features will not be discussed further, but later a more detailed consideration will be given to scale, particularly in Chapter 14 on regolith mapping.

3 Earth materials

Mineral classification

Scientists construct classification schemes primarily to help in understanding the articles being classified. Early classifications of minerals were based on the morphology of their crystals, then as knowledge of chemistry grew, classification became based on their chemical composition, and use of this classification is still widespread. Chemical classification is useful because it separates minerals into groups which have somewhat common modes of origin (e.g. halides are mostly evaporites, many sulfides are localized together), and into groups which have similar reactions to weathering (halides mostly dissolve, sulfides oxidize).

The advent of X-ray diffraction and crystal structure analysis allowed a new classification based on the crystal structures of the minerals. This proved particularly useful for understanding the silicates, as it showed clear distinction between chemically similar minerals (such as amphiboles and pyroxenes) and gave a crystal chemical basis for the evolution of the mafic silicates during progressive crystallization of igneous rocks.

The structural classification of the silicates is based on the degree of polymerization of silica tetrahedra, and is essential for understanding physical properties and atomic bonding (Figure 3.1). Because the silica tetrahedra are unvarying except for Al↔Si substitution, most of the chemical variation in silicate minerals is ignored in the tetrahedral classification. Broadly speaking, the remaining elements occur in either octahedral coordination to the anions, or in rather open cavities, with coordination to the anions ranging from 8 to 12. The smaller cations (Mg and the transition metals) are mostly found in octahedral coordination, the alkalis and alkaline earths in the larger cavities (Figure 3.2).

Regolith mineralogy

In a magma, or during metamorphism, the minerals that crystallize do so from a melt or assemblage of minerals where the chemical activity of silica and of the other major rock-forming elements Na, K, Ca, Mg, Fe and Al, is high. Where silica is lowest and the temperature is highest, as in basaltic magmas, the first mineral to crystallize is olivine, a mineral in which the silica tetrahedra are separate and linked through the other cations. As the temperature falls and the magma silica content increases, progressive polymerization of the tetrahedra leads to the crystallization of pyroxenes (chain silicates). In more felsic, lower melting rocks, such as granites, the higher silica content yields double chain amphiboles and layer silicate micas.

By contrast, in the regolith, new minerals crystallize at earth surface temperatures from aqueous solutions, or, as in metamorphism, by solid state reactions. At these temperatures and in Si-saturated regolith waters, most Na-, K-, Ca- and Mg-silicates are not stable except when these species reach very high concentrations, as perhaps in evaporating lake deposits. Instead, highly polymerized silica minerals form, namely the clay minerals and silica itself.

In this chapter we describe the structures and ideal composition of regolith minerals. In Chapter 6, the solubility of the major regolith minerals is addressed, and in Chapter 7 their geochemistry is discussed. Their formation by weathering is much of the subject matter of Chapter 8.

The minerals making up the majority of the regolith are few in number. Excluding the residual minerals of parent rocks, 99% of regolith mineralogy is accounted for by the clay silicates, the iron and aluminium oxyhydroxides, anatase, carbonates and quartz. Both summary and detailed descriptions of minerals are commonplace and available in many texts.

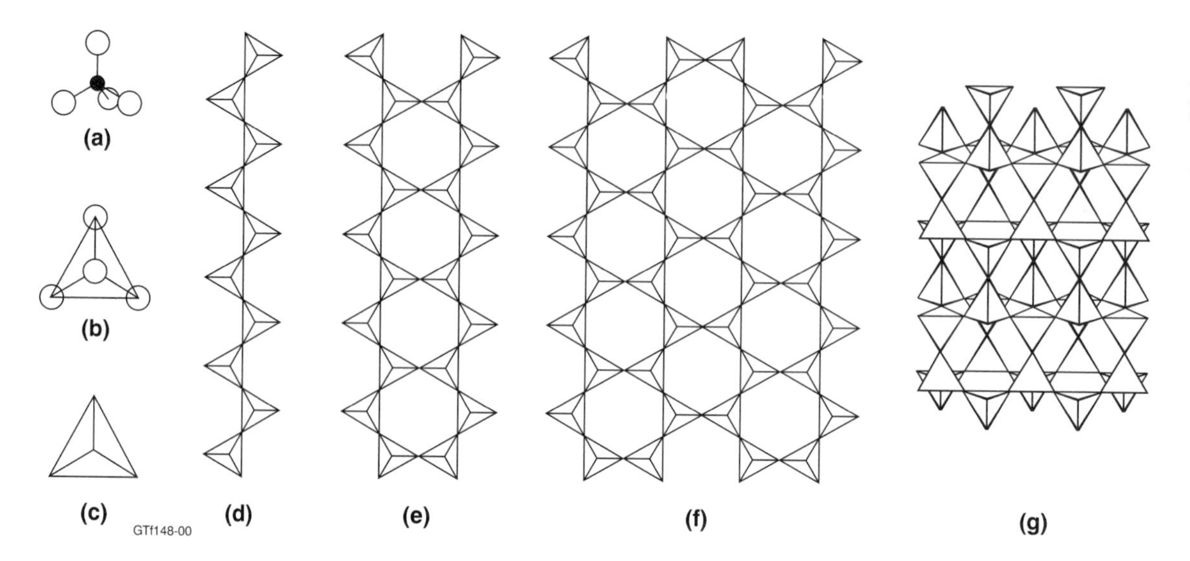

Figure 3.1 *Classification of silicates according to tetrahedral polymerization. (a) Silica tetrahedron viewed as four oxygens coordinated to a central silicon. (b) Tetrahedron viewed as a coordination tetrahedron with the oxygens at the apices. (c) Silica tetrahedron simplified. (d) Single chain polymer. (e) Double chain. (f) Tetrahedral sheet. (g) Tetrahedral framework*

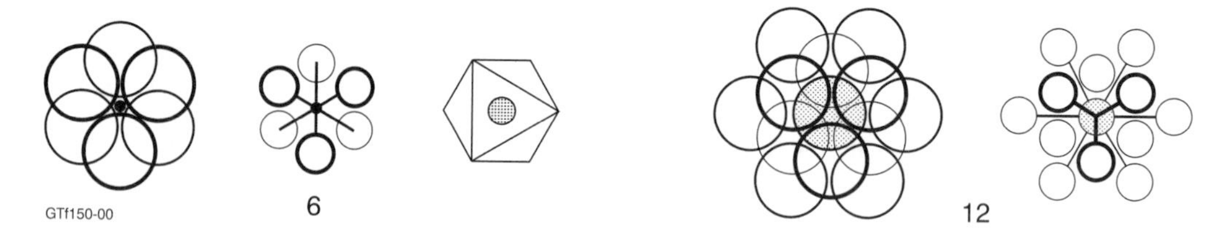

Figure 3.2 *Octahedral (6) and 12-fold coordinated sites in silicates. Outline circles are oxygen atoms and shaded circles are cations. Line figures shows a simplified octahedron with a cation at its centre*

SILICA MINERALS

The most abundant silica mineral in the regolith is quartz, almost all residual from weathered felsic rocks. Other forms of silica that are essentially regolith precipitates are chalcedony, moganite and opal. All three varieties are microcrystalline, and all have included water; some adsorbed on crystallite surfaces, some in larger cavities and some in structural sites. Chalcedony and moganite have structures based on multiple twinning of quartz. Opal has cristobalite and tridymite-like units in its structure. The colour of precious opal is the result of the regular packing of equal-sized spheres diffracting light. The much more common opaline silica or chert has a similar atomic structure, but with more irregular stacking of the cristobalite and tridymite units.

Silica precipitation in the regolith leads to hardpans, silica vein and crack fillings and to silcretes. Opaline silica is quite abundant in regolith over ultramafic rocks, as the low Al content of the parent rock does not provide for kaolinite as a host for silica. The origin of silcrete is the subject of argument, and discussed elsewhere in this book (Chapter 13). It appears to be a precipitate resulting from the saturation of the groundwater with silica, either by evaporation or by temperature change. The neutralization of alkaline silica-saturated groundwater can also lead to the precipitation of silica (Chapter 6).

THE CLAY SILICATES

Other than residual quartz, the bulk of the regolith is made up of clay-sized (<2 μm) layer silicates (also

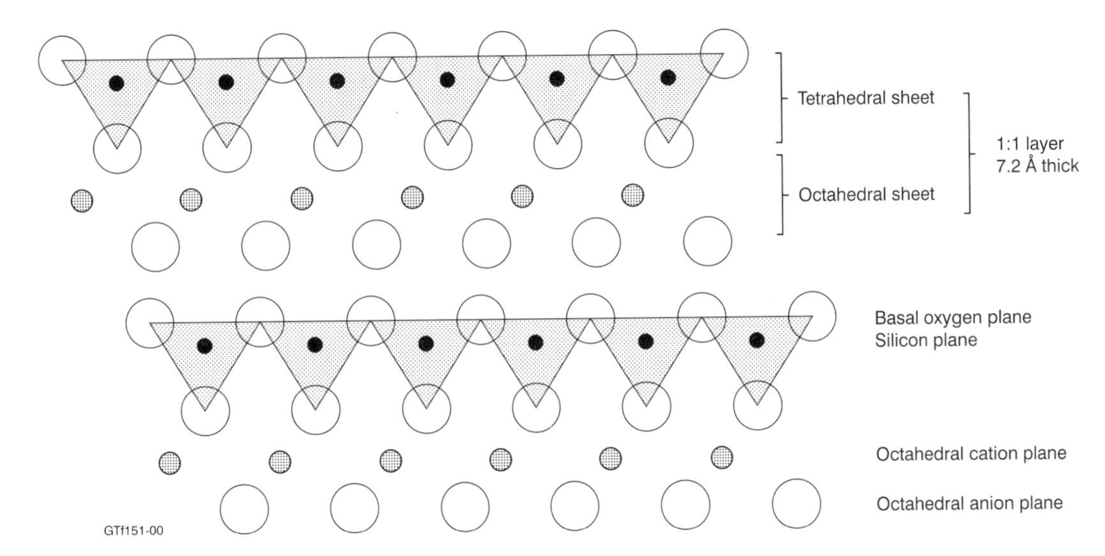

Figure 3.3 *Structure of kaolinite, a 1 : 1 dioctahedral layer silicate*

called phyllosilicates or sheet silicates). The basic crystal structure of the layer silicates was elucidated in the 1930s (Pauling 1930), and it is now recognized that all the minerals of the group have closely related structures. All have two structural units, an octahedral sheet and a tetrahedral sheet. The octahedral sheet comprises a plane of cations in octahedral coordination with planes of anions on either side. The tetrahedral sheet is formed of one plane of anions from the octahedral sheet, a plane of Si–Al cations, and a third anion plane of oxygens completing the tetrahedra (Figure 3.3). In all the layer silicates, the small Si-cations occur in tetrahedral coordination to oxygen, the tetrahedra being linked laterally at three of their corners to other tetrahedra in the form of a continuous hexagonal sheet (Figure 3.1f). The sheet-linking oxygens are referred to as the *basal oxygens*. These two planes of oxygen and cations are completed as polyhedra by oxygens of the adjacent octahedral sheet which provide the fourth, or apical oxygen of the tetrahedra (Figure 3.3).

A terminology for describing layer silicates has arisen from the work of the Clay Mineral Nomenclature Committee (Bailey *et al.* 1971). All components of the structure are planar:

- atoms are referred to as lying in *planes*;
- two planes of anions with a plane of cations coordinated between them to form linked polyhedra are referred to as *sheets*; and,
- sheets linked by common anion planes are referred to as *layers*.

Layer silicates are classified on two criteria. The first identifies the occupancy of the octahedral sheet. An isolated octahedral sheet, such as in the mineral brucite ($Mg_3(OH)_6$), has trigonal symmetry, and a unit cell containing three $Mg_3(OH)_6$ octahedra. By contrast, the mineral gibbsite ($Al_2(OH)_6$), while also having trigonal symmetry and three octahedra in its unit cell, has one of these vacant. Octahedral sheets having all three octahedra occupied are called *trioctahedral*; those with only two occupied are *dioctahedral* (Figure 3.4). The second classification criteria refers to the sequence of octahedral sheets and their flanking sheets of [SiAl] tetrahedra.

Known configurations for octahedral and tetrahedral sheet sequences are only three:

- one octahedral sheet with one flanking tetrahedral sheet (1 : 1 layer silicates) (Figure 3.3);
- one octahedral sheet with two flanking tetrahedral sheets (2 : 1 layer silicates) (Figure 3.5a); and,
- 2 : 1 layers with octahedral sheets between (2 : 2 layer silicates) (Figure 3.5b).

A single plane of oxygens has a thickness of about 2.6 Å. An overlying anion plane fits into hollows in the first, so that the effective thickness of each plane reduces to approximately 2.3 Å. 1 : 1 layer silicates have three anion planes, and so are about 7 Å thick; 2 : 1 layer silicates with four anion planes are about 9.4 Å thick (talc) or 10 Å if there is an alkali cation in the interlayer (micas); 2 : 2 layer silicates are

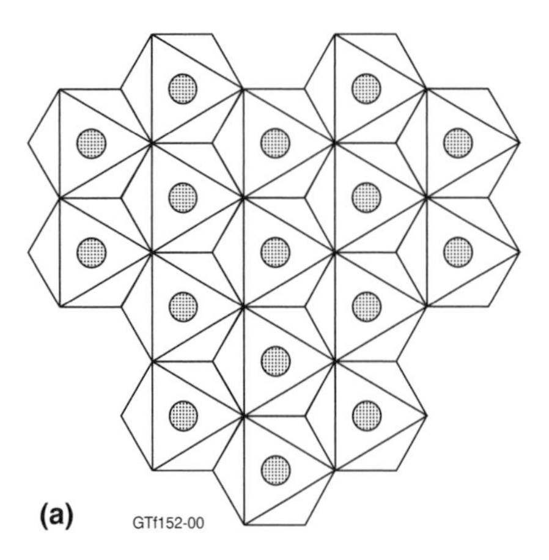

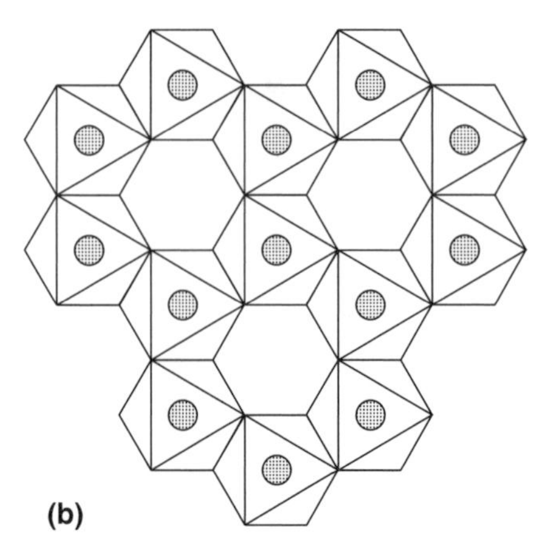

Figure 3.4 *(a) Trioctahedral sheet. (b) Dioctahedral sheet. Octahedra are represented as in Figure 3.2*

14.4–15.4 Å thick. These measurements, and other names applied from time to time, have led to multiple terminologies for clays:

1 : 1 layer silicate	7 Å layer silicate	kaolinite, 7 Å halloysite
2 : 1 layer silicate	10 Å layer silicate	pyrophyllite, talc, mica
2 : 2 layer silicate	14 Å layer silicate	chlorite, smectite (formerly montmorillonite group), vermiculite

Variations available to each layer type are:

- the nature of the octahedral cation, dominantly Al or Fe^{3+} in dioctahedral sheets, and Mg or Fe^{2+} in trioctahedral sheets; Mn^{2+}, Zn^{2+}, Mn^{3+}, Cr^{3+}, Ti^{4+} are common minor components in the octahedral sheet; and,
- substitution in the tetrahedral site of Si by Al, leading to a positive charge deficiency compensated by either a large low-charged cation between 2 : 1 layers (micas), or by positive charge generated in the octahedral sheet by substitution of a trivalent cation for a divalent, e.g. $Al^{3+} \leftrightarrow Mg^{2+}$.

The regolith is generally oxidizing, and ferrous iron is rare. Only vermiculite, an early weathering product of biotite or pyroxene, may carry Fe^{2+}. The clay silicates of the regolith can be classified simply according to whether Al, Fe^{3+} or Mg is in the octahedral sheet, and by layer type (Table 3.1).

Exchange sites

The clay silicates, particularly smectites and vermiculite, are important as temporary repositories for cations in the regolith. They exhibit the property of cation exchange, a topic taken up in more detail in Chapter 7. Cation exchange arises in minerals whose structure or surface has weakly bound cations. A cation with a stronger binding force may displace a more weakly bound cation, such as a Ca^{2+} compared with a resident K^{+}, or the influx of a high concentration of a soluble cation, such as Na, may swamp and so displace the resident cation. All mineral surfaces exhibit this property, and clays, because of their small particle size and therefore larger surface area, exhibit high exchange capacity. Surface cation exchange is pH dependent; kaolinite, for example shows cation exchange at pH 7, but at pH 4 the surface is neutral and the cation exchange capacity is zero.

Some minerals have sites within their structure where a weakly bound cation is essential, though non-specific. Such exchangeable cations can be displaced by other cations. The zeolites have such sites within silica cages, and among the clay silicates, smectites have exchange sites in their interlayer region. The difference between the two aluminous smectites – montmorillonite and beidellite – lies in the origin of their exchange capacity. For montmorillonite, interlayer cations (given as $Ca_{0.3}$ in the formulae above, Table 3.1) balance a positive charge deficiency that arises in the octahedral sheet ($Al_{1.8}$ compared to Al_2 required for a neutral 2 : 1 layer). For beidellite and nontronite, the interlayer cations balance a charge deficiency arising in the tetrahedral sheet as a result of $Al \leftrightarrow Si$ substitution ($Ca_{0.2}$ in the examples above,

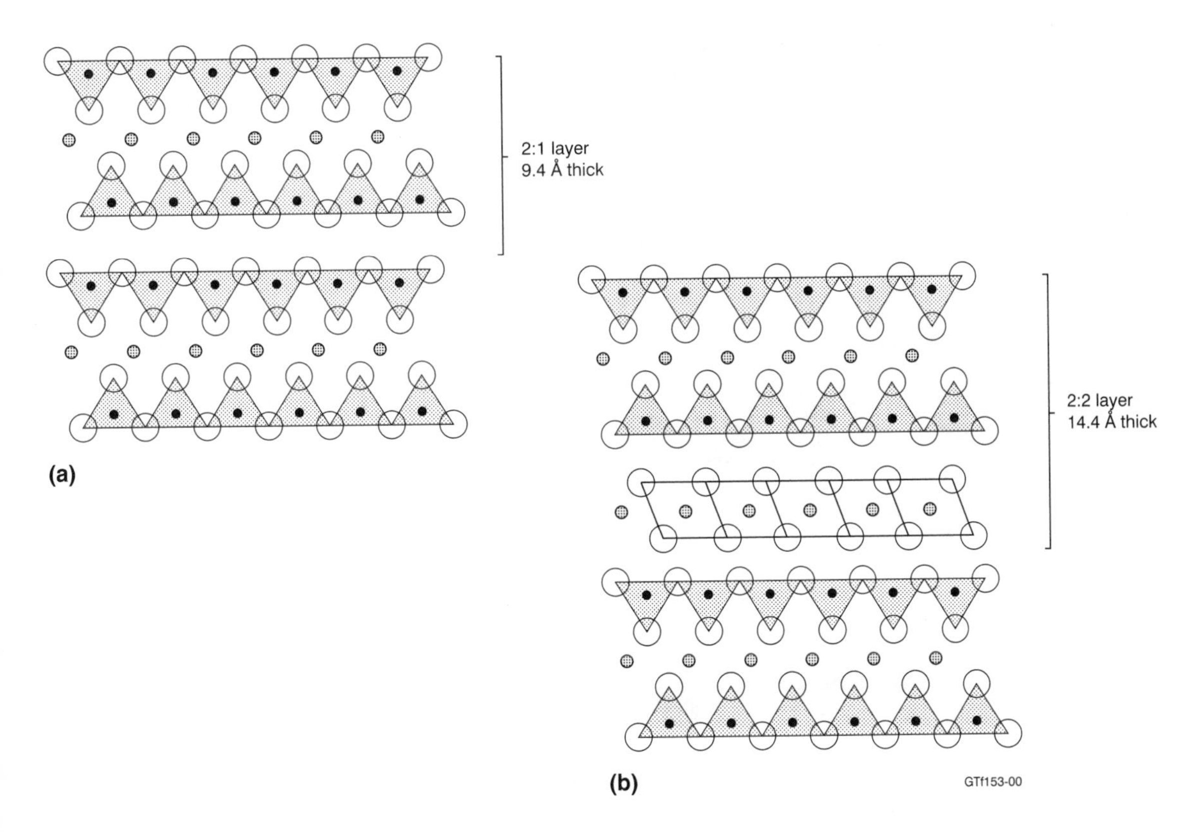

Figure 3.5 *(a) 2 : 1 layer silicate. (b) 2 : 2 layer silicate. Symbols are as in Figure 3.3*

Table 3.1 *Classification of regolith layer silicates*

Layer type	Al	Fe^{3+}	Mg
1 : 1	Kaolinite $Al_2Si_2O_5(OH)_4$ Halloysite $Al_2Si_2O_5(OH)_4.2H_2O$	Hisingerite $Fe_2Si_2O_5(OH)_4$	Serpentine $Mg_3Si_2O_5(OH)_4$
2 : 1	Illite $K_{0.9}Al_2[Si_{3.1}Al_{0.9}]O_{10}(OH)_2$	–	Talc $Mg_3Si_4O_{10}(OH)_2$
2 : 2	Montmorillonite $Ca_{0.3}Al_{1.8}[Si_4]O_{10}(OH)_2.2H_2O$ Beidellite $Ca_{0.2}Al_2[Si_{3.6}Al_{0.4}]O_{10}(OH)_2.2H_2O$	Nontronite $Ca_{0.2}Fe_2[Si_{3.6}Al_{0.4}]O_{10}(OH)_2.2H_2O$	Saponite $Ca_{0.2}Mg_3[Si_{3.6}Al_{0.4}]O_{10}(OH)_2.2H_2O$ Vermiculite $Mg_{0.3}(Mg_{2.4}Al_{0.2}Fe_{0.4})$ $[Si_{2.8}Al_{1.2}]O_{10}(OH)_2.nH_2O$

compensating for $[Si_{3.6}Al_{0.4}]$ when $[Si_4]$ would yield a neutral 2 : 1 layer).

Cation exchange capacity (CEC) is reported in terms of the molar quantity of exchangeable monovalent cation per kilogram of clay (though the older literature refers to 100 g of clay). Generally only a small weight of cations are exchangeable, and milliequivalent weight (one-thousandth of an equivalent weight, meq) is used. The equivalent weight of an ion (the exchange ion in this case) is the weight that could exchange with one gram of hydrogen. For Ca^{2+}, this is 0.5 mol or 20 g (atomic weight of Ca = 40). So an exchange capacity of 40 meq/kg means that 1 kg of clay can hold $40 \times 20/1000 = 0.8$ g of Ca on its exchange sites.

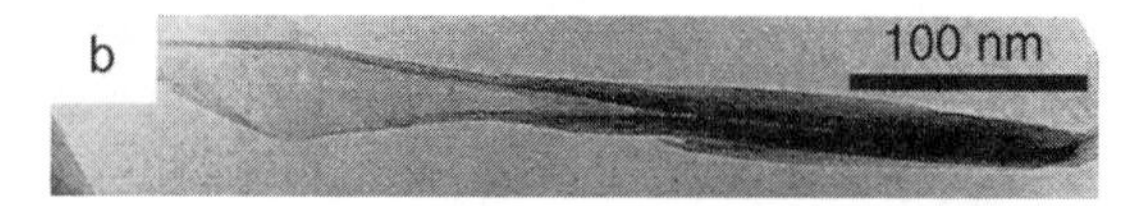

Figure 3.6 *Halloysite (a) Scanning electron micrograph of tubular halloysite with small feldspar fragments. (b) Transmission electron micrograph of a single, partly unrolled tube (photo R.A. Eggleton)*

Kaolinite and halloysite

Kaolinite is the most abundant regolith mineral, and it appears repeatedly throughout this book, including as the filler on every page of paper! It is an alumino-silicate, formed during weathering of all the alumino-silicate igneous minerals such as feldspars, muscovite, feldspathoids and zeolites. We will meet kaolinite and the chemically similar halloysite in discussion of chemical weathering, regolith chemistry, rock and mineral weathering, profile development and as an industrial mineral. Kaolin, which is a rock name for a material composed largely of kaolinite, is the main material for ceramic and china clay.

Kaolinite is platey, commonly forming hexagonal crystals 0.1–2 μm across, about one-tenth as thick. Halloysite has the same composition as kaolinite, but with water between the layers. Halloysite is a curled variety of kaolin, generally occurring either in tubes or spheres, although platey halloysite with curled edges has been described (Figure 3.6).

Kaolinite has a very low CEC, of the order of 20–40 meq/kg. The composition of kaolinite is simple and constant ($Al_2Si_2O_5(OH)_4$). There is no structural exchange site; cation exchange in kaolinite relies on surface and edge exchange sites (Figure 3.7).

Illite

Illite is the term used for clay-sized muscovite, probably very thin muscovite. Muscovite has K between 2 : 1 layers. The K is held there by electrostatic attraction to charge unsatisfied oxygens of the silicon–oxygen network (Figure 3.8); an oxygen that is bonded to one Si and one Al. Muscovite has one K and Al for every three Si (i.e. $KAl_2[AlSi_3]\ O_{10}(OH)_2$). Illite has less Al replacing Si in the tetrahedral sheet, and correspondingly less K. The K in illite is still sufficient to hold the 2 : 1 layers firmly together, so illite is a 10 Å layer silicate. Illite has a low CEC, similar to that of kaolinite.

Smectite (bentonite is a rock composed largely of smectite)

If there is even less Al in the tetrahedral sheet than in illite, and so less K, the 2 : 1 aluminosilicate layers are so weakly held together that the K becomes readily exchangeable (Figure 3.9), and water can also get in between the layers. This is the nature of the smectites, a group name for layer silicates formerly known as montmorillonite. That name is now only applied to one particular composition smectite. Though very much less abundant in most regolith than kaolinite, smectites are a most important mineral component of the regolith because of their physical and chemical properties. Smectites readily take water between their structural layers in the form of molecules bound to exchangeable Ca, K or Mg (or less commonly Na) (see below). Smectite crystals are extremely small and physically weak, thus bentonites have almost no wet strength, and roads built over bentonites commonly fail. Soils in Adelaide contain sufficient quantities of smectite that they have low strength when wet, and expand and contract with wetting and drying, leading to the term 'Bay of Biscay' soils – a term more evocative in the early days of Adelaide when migrants arrived by ship and recalled with revulsion their experiences as they sailed across the notoriously rough Bay of Biscay. The 'shrink–swell' character of smectites gives soil much of its mobility. In apparent contradiction of its physical weakness, small amounts of smectite can produce 'hard-setting' in soils. This

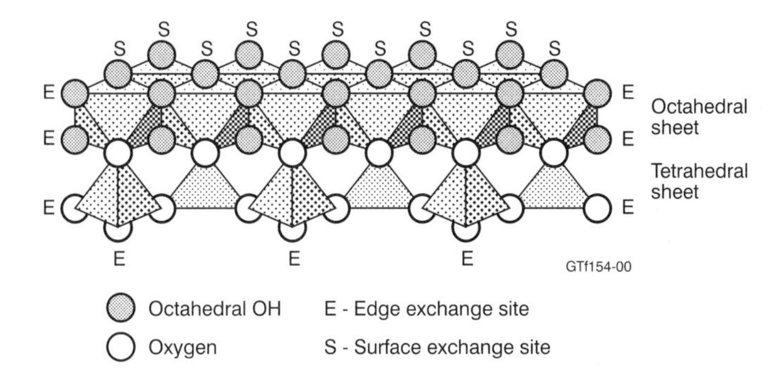

Figure 3.7 *Diagram of the structure of kaolinite showing exchange sites*

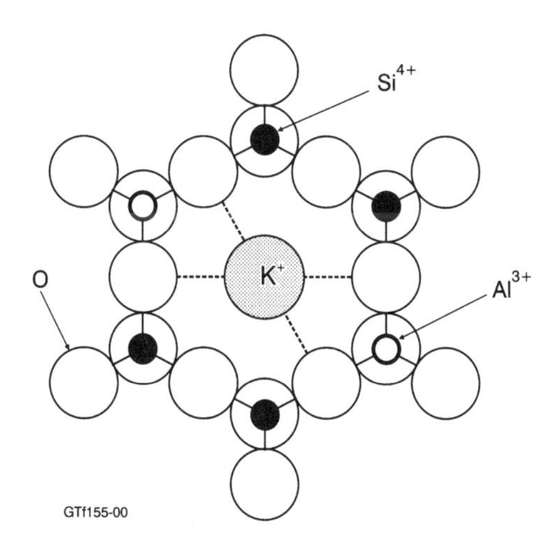

Figure 3.8 *Part of the tetrahedral sheet of mica showing the bonding of K to basal oxygens which are charge unsatisfied because of $Al^{3+} \leftrightarrow Si^{4+}$ substitution*

phenomenon appears to depend on smectite being intimately associated with other clays such as illite or kaolinite, in a form known as 'interstratified clays', described below.

Cations in solution become surrounded by water molecules because water is polar, and its negative region is attracted to the positive charge on the cation. The cation–water complex is still charged positively overall. The singly charged ions K^+ and Na^+ attract water less strongly than Ca^{2+}. The cation–water attraction is strong enough that when such a cation is adsorbed into the interlayer region of a smectite, the water is pulled in also. Na^+ smectites absorb one water layer at humidities above 5% and below about 50%, adding a second above 50% relative humidity. Ca^{2+} smectites keep one water layer unless completely dried, and add the second water layer at about 20% humidity. The thickness of a smectite packet is 9.6 Å without water, expanding to 12.5 Å with one water layer, and to 15.5 Å with two. Immersed totally in water, some smectites add a third layer, others expand indefinitely (the layers separate). Molecules that are even more strongly attracted to the interlayer cation can replace water in the interlayer. Such molecules include alcohol, glycerol, urea and many other organic molecules. Natural and synthetic organics such as pesticides and herbicides may enter the interlayer region of smectites.

Smectites typically form the finest particles in a soil. They may be no more than two or three layers thick (30–45 Å), and 0.1 μm across. They have very high CEC (800–1200 meq/kg), largely derived from exchange sites in the interlayer. Their small size also gives them a high edge exchange capacity. A small amount of smectite in a soil therefore has a considerable effect on the soil's properties.

Vermiculite

Vermiculite is structurally midway between biotite mica and chlorite. It has a trioctahedral 2 : 1 layer, and an interlayer of $Mg(OH)_2$ or $Al(OH)_3$. Both Mg^{2+} and Al^{3+} are able to hold two water molecules in the interlayer. Vermiculite has a higher layer charge than smectite and a high CEC (1000–1500 meq/kg). The layers are more strongly held together than are those of smectite, so the basal spacing is smaller (14.2 Å, compared with 15.5 Å for smectite). On heating, the water can be expelled, collapsing the structure to 12.5 Å and then to 10 Å. If a large vermiculite crystal is rapidly heated, the steam physically blows some of the layers apart and accordion-like 'worms' are formed, hence its name. The structural spacings,

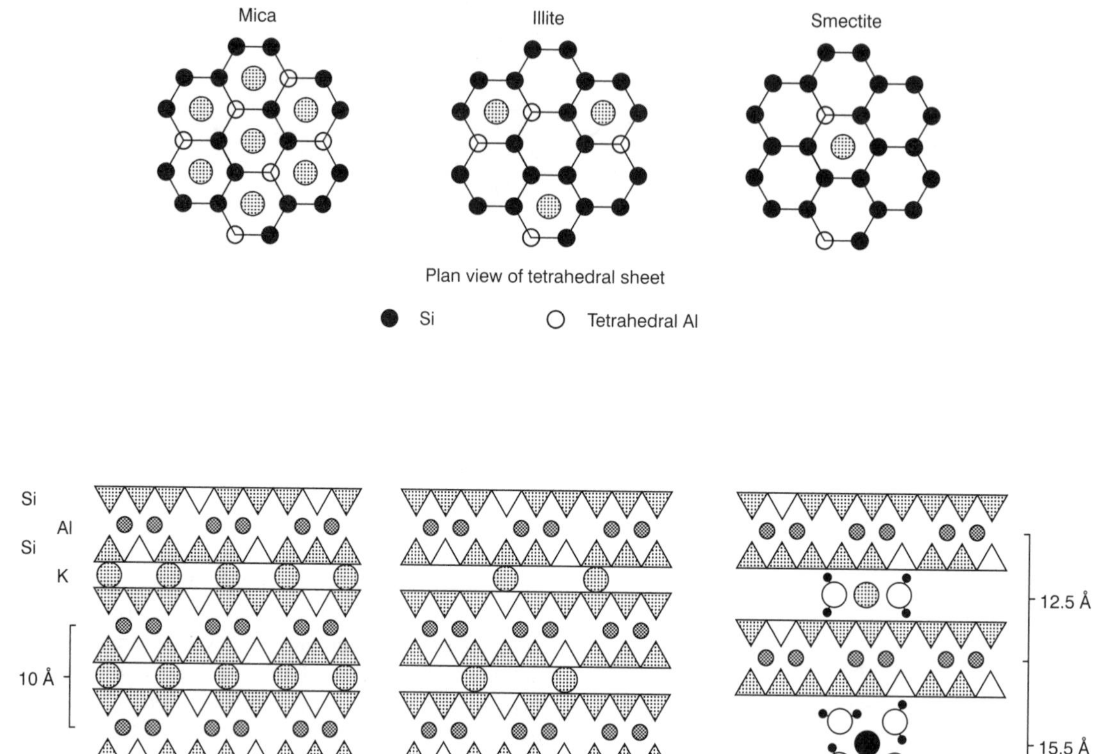

Figure 3.9 *Diagrammatic sketches of mica, illite and smectite, showing changes with decreasing tetrahedral Al*

nonetheless, collapse to 10 Å. These have use as an absorbent for potting soils, kitty litter and industrial clean up.

Interstratified clays

Also known (incorrectly) as 'mixed-layer' clays, interstratified clays generally make up only a small percentage of the clay fraction of regolith. Commonly, one component layer of the interstratified clay is a smectite, and this gives the clay a high exchange capacity and some 'shrink–swell' character.

In plan, all the clay silicates have the same structure: a hexagonal silica–oxygen sheet (or sheets) and a hexagonal octahedral sheet. Therefore, they have little difficulty stacking different layer types on top of one another. Illite and smectite layers may alternate, building up random sequences (ISISSIIISISSIIS) or regular sequences (ISISISIS or IISIISIISIIS). Kaolinite and smectite may alternate in soil clays, biotite weathers to vermiculite through an intermediate random interstratification

(BBBVBBBBVBVBVBBV)

gradually becoming semi-regular

(BVBVBVBBVBVBBVBVBVBVVB).

Sepiolite–palygorskite

Two fibrous Mg-chain-silicates that are found in the regolith are sepiolite and palygorskite. They both have a high exchange capacity of around 300 meq/kg and a high surface area (~900 m^2/g). They adsorb metal cations very effectively, and also absorb more than twice their own weight of water. Sepiolite and palygorskite form in saline evaporitic environments, both terrestrial and marine (see also Chapter 7, under the entry for Mg).

Fe-OXIDES AND HYDROXIDES

Iron, which is ferrous in most rocks, is introduced to the regolith by rock weathering. On exposure to oxygen and water the primary mineral structure breaks down. Initially Fe may be released as very soluble Fe^{2+} or it may be oxidized while still in the primary mineral to much more insoluble Fe^{3+}. See also the section on oxidation in Chapter 6 and the sections on regolith mineral hosts and Fe in Chapter 7.

Ferrihydrite

The approximate composition of ferrihydrite is $5Fe_2O_3.9H_2O$. Ferrihydrite is the brown rusty scum visible at springs, where water seeps from cracks in rocks, or as an 'oil slick' on some swamp water (especially swamps containing abandoned cars, farm machinery or iron roofing – the slick is not always seeping oil, but ferrihydrite). Ferrihydrite crystals range from about 20 Å in diameter to 75 Å (Figure 3.10). The degree of organization of these particles is low, and the X-ray pattern is very simple and weak and the lines are broad. Much ferrihydrite in soils is missed because it does not yield a marked diffraction pattern.

The surface area of ferrihydrite crystals ranges from 200 to 800 m^2/g (1 μm thick mylar (cling-wrap) has a surface area of about 2 m^2/g). Ferrihydrite is a strong adsorber of phosphate, silica, organic molecules and heavy metals. In the laboratory, ferrihydrite transforms to a more stable oxide–hydroxide (goethite, usually) over a period of a few years. In the soil it probably passes in and out of solution with the seasons. Ferrihydrite is of the order of 100 times more soluble in normal groundwater than the other Fe-oxides or oxyhydroxides (see Figure 3.11). Under reducing conditions, such as in the mud below the water of a swamp where organic matter has used up any available oxygen, Fe^{2+} (ferrous) is very soluble in water. As this water runs out of a swamp, say into a creek, or simply at the swamp water surface, the Fe^{2+} is oxidized to Fe^{3+} and precipitates as ferrihydrite. If

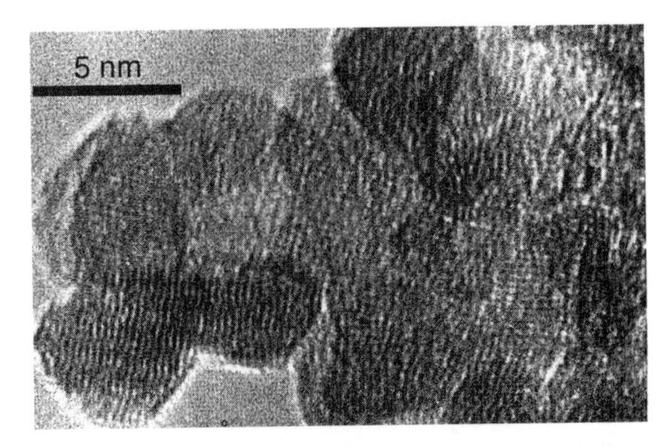

Figure 3.10 *Transmission electron micrograph of a cluster of ferrihydrite crystals*

the conditions become reducing again, the ferrihydrite will dissolve. If oxidizing conditions remain, it will slowly change to goethite.

Most ferrihydrite is associated with bacteria (*Gallionella* and *Lepthotrix*) which gain their energy from the oxidation reaction $Fe^{2+} \rightarrow Fe^{3+} + e^-$.

Ferrihydrite also precipitates from ferric iron solution as pH increases. Ferric iron is soluble at pH 2 (very acid), becoming less so with increasing pH. At pH 4 the solubility is negligible (about 1 in 10 million). Very acid waters (mine waters, some lakes such as Lake Tyrell in western Victoria) can hold appreciable ferric iron in solution, which precipitates as ferrihydrite on dilution (because the pH increases) or on input of alkaline water. Soils derived from pyritic (FeS_2) rocks or coastal muds are commonly sufficiently acid to mobilize Fe, whence it may rise to the surface and precipitate through rain dilution or oxidation.

Cyclic dissolution and precipitation of Fe by reduction/oxidation alternation or pH change move Fe away from reducing areas towards oxidizing areas and are responsible for most of the brown/yellow colour banding of soils and weathered rocks. Precipitation at the top of the water-table may yield an iron hardpan.

Goethite, FeO(OH)

The structure of goethite can be described as double chains of Fe–O octahedra linked laterally. The oxygens are in approximate hexagonal close packing.

The most common of the soil iron minerals, goethite is the first conversion product from ferrihydrite. It is a yellow–brown mineral, forming as needle-shaped crystals about 1 μm long in synthetic preparations,

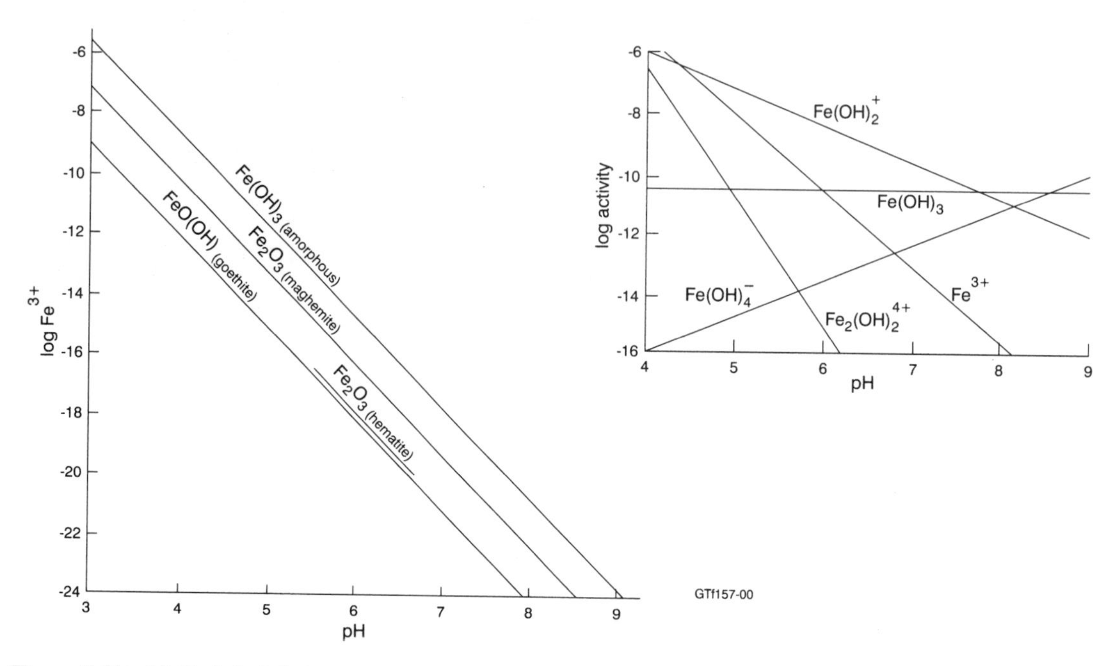

Figure 3.11 *(a) Ferric hydrolysis species in equilibrium with soil-Fe. (b) The activity of Fe^{3+} maintained by various ferric oxyhydroxides at varying pH (after Lindsay 1988)*

but typically more equant in soils. Together with ferrihydrite, goethite imparts most of the brown colour to soils.

The surface area of soil goethite ranges from 6 to 200 m^2/g, which gives goethite considerable adsorptive ability. Heavy metals such as Cu, Pb and Zn are adsorbed to the extent of about 1 $\mu M/m^2$ (20 $\mu mol/g$). Goethite is also an effective anion adsorber, notably of phosphate. At normal regolith pH, phosphate values of about 2–3 $\mu M/m^2$ have been measured both in laboratory and the field. Much of the superphosphate ($Ca(H_2PO_4)$) ploughed into fields becomes unavailable to plants in quite a short time because it is sequestered by goethite.

Lepidocrocite, FeO(OH)

In contrast to goethite, lepidocrocite is the FeO(OH) polymorph with the oxygens in approximate cubic close packing. Recognizable by its orange colour, lepidocrocite is a relatively uncommon mineral, forming in preference to goethite as a direct oxidation product of ferrous iron and in preference to ferrihydrite if oxidation is slow. It therefore indicates reductomorphic soils. It also seems to precipitate rather than goethite in the presence of Cl^-.

Haematite, Fe_2O_3

Haematite is very common in warm or arid regolith, and is red when fine-grained. Its intense colour may mask the presence of goethite. The surface area for soil haematite is about 100 m^2/g, much the same as goethite. The haematite–goethite ratio in soils increases with soil temperature (decreasing latitude) and decreases with soil moisture content. Locally, hilltops are richer in haematite, valleys in goethite; globally, the arid regions have haematite rather than goethite. Haematite has similar adsorption properties to goethite and can also be responsible for the fixation of phosphate.

Magnetite, Fe_3O_4

Magnetite is a member of the spinel group of minerals. It is not fully oxidized having one Fe^{2+} and two Fe^{3+}. It is not as common in regolith as other Fe^{3+} minerals.

Although some may be produced directly by bacteria, most is residual from magnetite in parent rocks.

Maghemite, Fe_2O_3

Maghemite is also a spinel, although its formula is the same as that of haematite. Written as a spinel, maghemite is Fe_8O_{12}, compared with magnetite's Fe_9O_{12}. Maghemite can form by the oxidation of magnetite, and some soil maghemite may result from the oxidation of 0.1 μm crystals of magnetite formed by bacteria. Most maghemite probably forms in soils by the dehydration of goethite or lepidocrocite during fires. Lepidocrocite can transform easily to maghemite, as both have cubic close-packed structures. Goethite has a structure based on hexagonal close packing, and normally dehydrates to (hexagonal) haematite. In the presence of organic matter it is thought that maghemite is the common dehydration product. Early in a bushfire, plant fragments in the soil burn, providing hot, reducing conditions capable of converting ferrihydrite or geothite to Fe_3O_4 or possibly FeO. As the fire passes and the carbon is consumed, the reduced oxides change to maghemite. Maghemite is strongly magnetic (it is used in magnetic tape), and a hand magnet is the quickest means of identification (assuming no magnetite is present).

Maghemite is greatest in abundance at the surface; hence it gives a strong magnetic signal to ground or airborne magnetic surveys. Geologists should be cautious about drilling a hilltop for base metals on the evidence of a magnetic anomaly (hilltops attract lightning and thus fires). Maghemite is dense (SG = 5) and stable in the weathering environment. Maghemite crystals, and maghemite-bearing ferruginous nodules and grains accumulate in stream channels where they are particularly obvious in magnetic surveys. Palaeochannels can be readily detected by their magnetic signature (Figure 3.12).

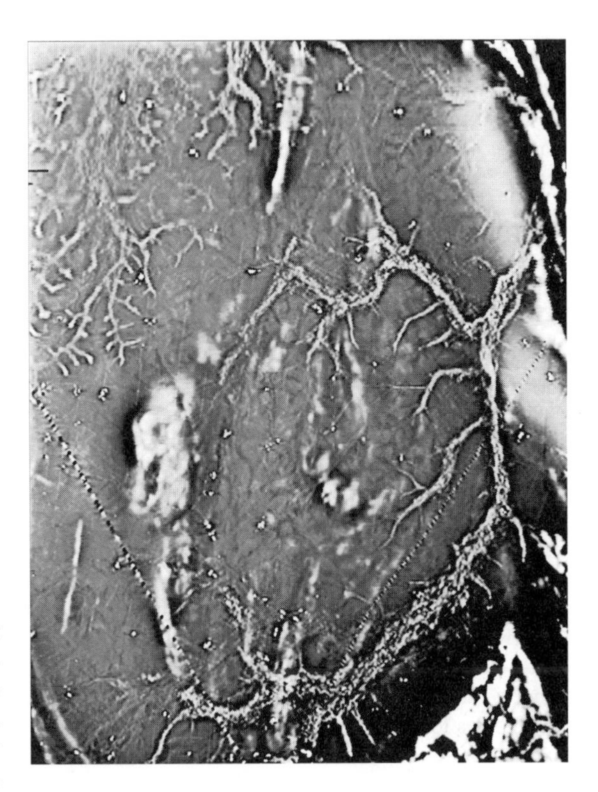

Figure 3.12 *Greyscale image of the first vertical derivative of magnetic data reduced to the pole. The data are from a portion of detailed survey flown by AGSO near West Wyalong in New South Wales, with 50 m line spacing and 60 m terrain clearance. The image covers an area of 17 by 22 km. The obvious dendritic patterns are characteristic of palaeochannels and volcanic flows. In this case they relate to maghemite-filled palaeochannels (Mackey 2000). Several cultural anomalies relating to buildings, fences and power lines are also evident in the image (image processed by John Wilford)*

ANATASE, TiO_2

The polymorph of TiO_2 most abundant in the regolith is anatase. It commonly is found as very small (0.1 μm) crystals, and is a major constituent of the fine-grained alteration assemblage known as leucoxene. Anatase has a cream-coloured appearance when it is concentrated, but mostly it is dispersed uniformly through silicate weathering products. It appears to concentrate in silcrete (Chapter 13).

Mn-OXIDES AND HYDROXIDES

In the weathering environment Mn becomes oxidized to the tetravalent state. The mineralogy of Mn-oxides and hydroxides is complex; the more common regolith species are the layer structures having cations other than Mn between MnO_6 octahedral sheets:

- vernadite (δ-MnO_2), incorporating Ba or K;
- lithiophorite ($(Al, Li)MnO_2(OH)_2$);
- birnessite ($(Na, K)_4Mn_{14}O_{27}.9H_2O$); and,
- the cryptomelane–coronadite–hollandite group ($(K, Pb, Ba)_{2-1}Mn_8O_{16}$), which have large cations in tunnels within columns of MnO_6 octahedra.

Other Mn-oxides and oxyhydroxides include pyrolusite and nsutite (MnO_2), romancheite (containing Ba), todorokite (containing Ca, Na and K), chalcophanite ($ZnMn_3O_7.3H_2O$) and asbolane (Ostwald 1992; Parc *et al.* 1989).

PHOSPHATES, SULFATES, CARBONATES

Crandallite group

The significance of the crandallite group of phosphates in the regolith was established by Norrish (1975) and Norrish & Rosser (1983). The group includes crandallite ($CaAl_3(PO_4)_2(OH)_5.H_2O$), gorceixite ($BaAl_3(PO_4)_2(OH)_5.H_2O$) and florencite ($CeAl_3(PO_4)_2(OH)_5.H_2O$), and this isomorphous series can host large divalent cations such as Ca, Ba, Sr, trivalent ions such as Al, Fe, Sc, Y and the rare earth elements (REE) and tetrahedrally coordinated groups such as PO_4, AsO_4, SO_4. Of these, the phosphates form a highly insoluble family of minerals that are quite stable in the weathering environment.

Sulfates

Gypsum is a common evaporite mineral in arid environments, occurring both in lake deposits and in the regolith over sulfides. The composition of gypsum varies little from $CaSO_4.2H_2O$.

Jarosite–natrojarosite ($(K\text{–}Na)Fe_3(SO_4)_2(OH)_6$) precipitate from a reaction between sulfuric acid formed by pyrite oxidation and surrounding silicates. These minerals are common in regolith where pyrite is weathering, and are particularly so in acid-sulfate soils and mine dumps. Brown (1971) showed that jarosite is only stable in the presence of goethite at pH below 3. Such extreme pH levels are reached during sulfide weathering and in saline lakes such as Lake Tyrrell, western Victoria, Australia, where sulfidic muds oxidize. The persistence of jarosite into environments of higher pH is attributed by Brown to the slowness of its conversion to goethite (see also the section on S in Chapter 7).

Alunite ($KAl_3(OH)_6(SO_4)_2$) is found in hydrothermal advanced argillic alteration, and associated with acid lake and groundwaters, where it crystallizes by reaction between clays and sulfuric acid from pyrite weathering (see also the section on Al in Chapter 7).

Carbonates

The major carbonate mineral of the regolith is calcite ($CaCO_3$). The other rhombohedral carbonates, dolomite ($CaMg(CO_3)_2$) and magnesite ($MgCO_3$), are less common, but may be abundant in high Mg terrains. Carbonate minerals form at alkaline pH (equilibrium pH is 8.4, Chapter 6). Metal carbonates such as malachite, azurite, cerussite and smithsonite are well known from the supergene region of weathered ore bodies, and the list can be expanded to include almost all of the mono- and divalent metals. Under semi-arid climates, carbonate precipitation is widespread in the regolith, commonly forming accumulations known as calcretes, which may actually be calcitic, dolomitic or magnesitic (see Chapter 13).

POORLY CRYSTALLINE MINERALS

Amorphous minerals were originally so called because they lacked a crystal shape. The term became extended to minerals that were not detectable by methods such as X-ray diffraction based on crystallographic structure (i.e. made up of units that repeated many times in three directions). Techniques such as scanning and transmission electron microscopy (TEM) have shown that many 'amorphous' minerals do have well-defined morphology. They may be composed of very small, or rather imperfect crystals, such as ferrihydrite, or they may have curved morphology, such as allophane. Both kinds of minerals yield X-ray diffraction patterns with broad, indistinct maxima and are better termed 'poorly diffracting'. Their presence has been long known to soil scientists (Gieseking 1975), and routinely estimated by chemical extractions. Examination of regolith minerals by TEM and differential X-ray diffraction has allowed mineralogical characterization of these materials, and their importance in regolith mineralogy and geochemistry is gradually being recognized (Tilley & Eggleton 1994, 1996).

Allophane is a hydrated aluminosilicate, formed as spheres about 50 Å across. It does not have a single composition, but ranges from $Al_2O_3.SiO_2$ to $Al_2O_3.2SiO_2$. It is difficult to recognize because it gives very poor X-ray reflections (broad bands centred at about 15, 3.4 and 2.5 Å). It is most common in soils derived from volcanic ash, and so is particularly abundant in Japan and New Zealand.

Imogolite is a thread-like mineral of composition about $Al_2O_3.SiO_2.2.5H_2O$, abundant in volcanic-derived soils. The threads are bundles of 20 Å diameter tubes. At pH 7, both these poorly crystalline clays have CECs of the order of 200–300 meq/kg.

HISINGERITE

Generally thought to be a rare amorphous alteration product of iron sulfides, carbonates and silicates, hisingerite has been recently shown to be a ferric form of spherical halloysite (Eggleton & Tilley 1998). Many specimens of hisingerite have come from mines at depths below the level normally regarded as within the regolith, though the mineral itself is the product of oxidation and hydration. Hisingerite has a formula close to $Fe_2Si_2O_5(OH)_4$, and other than the substitution of Mn, Mg and a small amount of Al for Fe, nothing is known about its chemistry. Studies by TEM of weathering amphibole (Wang 1988) and chlorite (Aspandiar 1992) suggest that hisingerite may be a common first weathering product. Its fabric of concentric 1 : 1 layers forming spheres about 140 Å in diameter gives it a high surface area and a high adsorption potential.

Al-Fe OXYHYDROXIDES

Pisolitic bauxites and laterites commonly yield very weak X-ray diffraction patterns. Tilley & Eggleton (1996) and Singh & Gilkes (1995) have shown that these near-surface regolith materials may contain a high percentage of ultra-fine-grained minerals occurring as crystals with diameter less than 10 nm, including χ- and ϵ-alumina, maghemite, akdaleite ($5Al_2O_3.H_2O$) and very fine goethite. These minerals have extremely high surface areas (~500 m^2/g), thus they may provide important sinks for adsorbed trace metals. No work has been done on their geochemistry.

Alternate classification of regolith silicates

The tetrahedral polymerization classification of silicates is useful because it is based on that part of the structure which undergoes the least variation within each group, and distinguishes the major structural and chemical variation between groups: the polymerization of silica tetrahedra and the ratio of Si to O. But when it comes to their behaviour during weathering, the silicate linkage classification obscures both similarities between minerals in different silicate groups, and differences between minerals in the same group. For example, the Mg–pyroxene, enstatite (single-chain silicate, $MgSiO_3$) can weather to Mg–amphibole (double chain), which in turn weathers to talc (layer silicate). The classical description of these changes is to emphasize the Si : O increase from 1 : 3 to 4 : 11 to 2 : 5, and the progressive polymerization of the tetrahedra. But in emphasizing change, similarities are concealed and the actual process masked. The reaction expressed more completely is:

Figure 3.13 *TOT unit of a silicate*

$$8Mg[SiO_3] + H_2O \rightarrow Mg_7[Si_8O_{22}](OH)_2 + MgO$$
enstatite — amphibole

$$Mg_7[Si_8O_{22}](OH)_2 + H_2O \rightarrow 2Mg_3[Si_4O_{10}](OH)_2 + MgO$$
amphibole — talc

Here there is no addition of silica, nor a change in silica to anion ratio. The chemical exchange is only two hydrogens for one magnesium at each step, the Si : (O,OH) ratio is constant at 1 : 3.

In an extension of the concept of polyhedral polymerization for silicate classification, Eggleton (1975), Veblen & Burnham (1978) and Thompson (1978) developed the recognition that for pyroxenes, amphiboles and biotite, the major structural unit was not simply the linked tetrahedra, but the linked triplet of a metal octahedron flanked by two silica tetrahedra. This triplet has become known as the TOT, and the three silicate groups are sometimes collectively referred to as the biopyriboles (Figure 3.13). When viewed along the axis of the linked TOT sequences, considerable similarity is evident in the structures of pyroxenes, amphiboles and layer silicates (Figure 3.14). It is clear from the diagrams that the three structures can join laterally without a structural discontinuity, and the truth of this is evident in high-resolution electron micrographs of weathered pyroxenes and amphiboles (Figure 3.15).

Viewed from a perspective of the TOT structure, the reaction from enstatite to talc is seen as one of

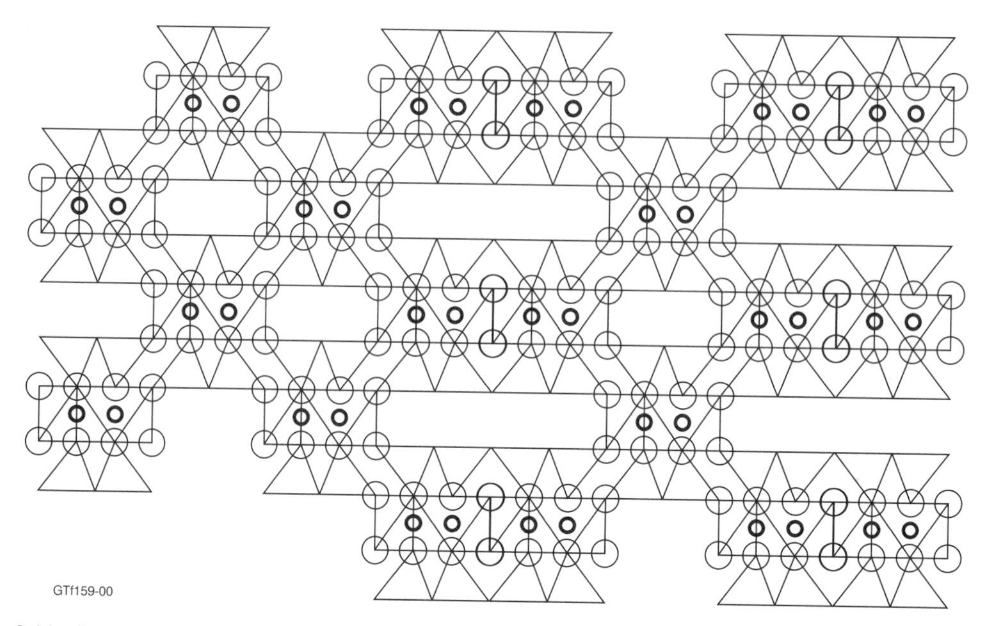

Figure 3.14 *Diagram showing the structural relation between pyroxene (left) and amphibole (right) linked as TOT units*

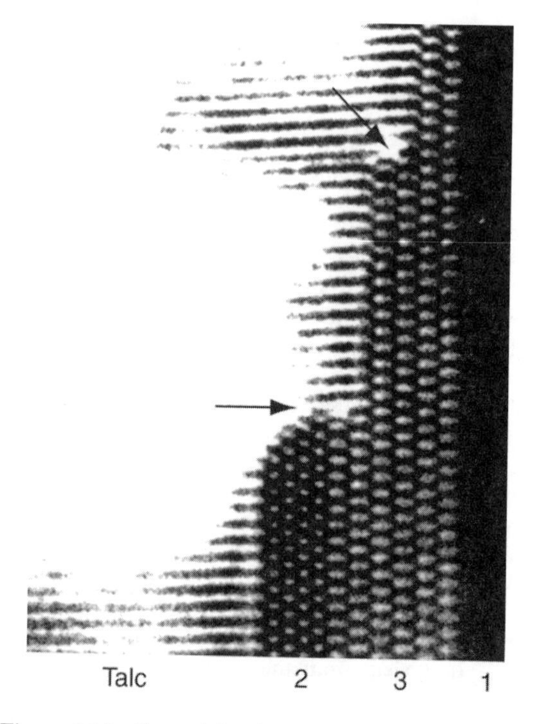

Figure 3.15 *Transmission electron micrograph showing structural coherence between enstatite, complex biopyriboles resulting from initial weathering, and talc. The numbers at the foot of each biopyribole are the numbers of aligned TOT units in that structural column*

rearrangement of parallel slabs of TOTs, with the progressive replacement by H of the Mg in the relatively open 8-coordinated site between the TOTs. There is in fact no change in the anion content at all; the 24 oxygens in the eight formula units of enstatite are still present in the two talc units.

Many minerals have a structure based on closest packing of their anions; included in this group are the Fe and Al-oxyhydroxides and many silicates, for example olivine. In the early stages of weathering, alteration can proceed apparently without change in the oxygen packing arrangements (Figure 3.16).

Zoltai & Stout (1984) pointed out that a number of silicates were more sensibly classified on the basis of their overall polyhedral linkages, rather than on their silica tetrahedral polymerization. Thus they regard cordierite and beryl as framework silicates, though they are traditionally regarded as cyclo-silicates, having six-rings of silica tetrahedra. When the Al-tetrahedra in cordierite and the Be-tetrahedra in beryl are considered, the linkages are seen to be those of a framework. Metal–oxygen octahedra may join at corners or edges, but rarely at faces; tetrahedra join only at corners in the mineral silicates. Broadly speaking, following Zoltai & Stout, structures dominated by corner-shared polyhedra may be described as frameworks, whereas those having significant polyhedral edge sharing may relate to close packing or to chain or sheet structures. Typical corner-shared

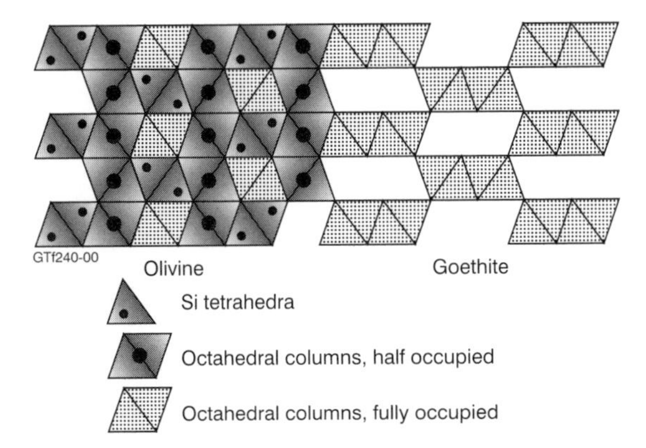

Figure 3.16 *Olivine and goethite hexagonal close packed parallels*

polyhedral structures are those of the feldspars and garnet. Edge sharing of octahedra occurs in andalusite and sillimanite and these structures, like the pyroxenes, are characterized by octahedral chains. Olivine, topaz and kyanite structures can be related to close packing of the anions.

Where parent and offspring mineral have a common structural character, it may help in understanding the processes of weathering to emphasize these similarities, rather than the chemical differences between the minerals (Table 3.2). This alternate classification of silicates helps to see the similarities in crystal structure inherited through a weathering reaction. For example, at first sight it may be surprising to find vermiculite, normally the weathering product of biotite, in the saprolite of an ultramafic rock containing, say, unweathered gabbro, olivine, pyroxene and plagioclase. The explanation lies in the relatively easy transition from pyroxene to vermiculite by rearrangement of TOT chains into sheets.

Table 3.2 *Alternate classification of silicates*

Primary	Regolith
TOT silicates	
Pyroxene	Vermiculite
Amphibole	Smectite
Mica	Illite
Chlorite	
Close-packed minerals	
Olivine	Goethite
Kyanite	Haematite
Spinels	Gibbsite
Topaz	Boehmite
Ilmenite	Lepidocrocite
	Maghemite
	Diaspore
Octahedral edge-shared	
Andalusite	Kaolinite
Sillimanite	Halloysite
Titanite	
Frameworks	
Quartz	Quartz
Garnet	Opal CT
Feldspar	
Zeolites	
Beryl	
Cordierite	

4 Weathering: agents and controls

Introduction

Weathering is a process of conversion of minerals making up rocks to an equilibrium state. Rocks that through the gradual process of erosion find themselves at, or near the surface are subjected to different physical, chemical and biological environments from those they experienced deeper in the crust, and will by various processes come to equilibrium with Earth's surface conditions. Here we wish to briefly introduce some of the characteristics of the Earth's surface that control that shift to equilibrium.

Physical environment

As rocks come closer to the surface by erosion of overlying materials one main factor that changes is the lithostatic stress on them (Figure 4.1). Lithostatic (or geostatic) stress at a point in a body of rock caused by the weight of the overlying column of rock is given by σ_z (N/m^2):

$$\sigma_z = \gamma z$$

where γ is the rock unit weight (N/m^3) and z is the depth (m).

Tectonic stresses are also released as rocks approach the surface where they are no longer confined as they are deeper in the crust. This, too, allows expansion and fracturing. Similarly, thermal stresses are released as rocks come closer to the surface. As the temperature decreases towards the surface so the rocks contract. This also contributes to the fracturing of rocks as they get closer to the surface.

The release of lithostatic, tectonic and thermal stresses allows the rocks to expand, joints and fractures open up and this permits water, air and biota to penetrate the rocks, changing their local environmental conditions compared with those they experienced deeper in the crust. Close to the surface other physical factors, discussed in Chapter 9, become important. These include such factors as frost wedging, salt wedging, exfoliation, biological wedging and similar mechanisms that open rocks further to entry of solutions, air and biota.

The lithological environment

The type of rock being weathered is obviously one important control on weathering. There are many types of rocks with an enormous range of mineralogy and thus chemistries. Each will influence the local weathering environment as it weathers by controlling the solutes released, and thus the chemistry of the solutions in the rock that will continue the process (or not). For example, the feldspars in a granite weather to produce a solution containing K^+, Na^+, Ca^{2+}, consuming hydrogen ions in the process, thus the weathering solution becomes more alkaline. Any sulfide minerals in the rock weather by oxidation, converting the sulfur to sulfuric acid with a marked increase in acidity of the water. Such reactions are discussed in more detail in later chapters, but it is already obvious that the type of rock weathering affects the weathering environment.

Landscape position

As we already know, water flows downhill. Thus weathering environments in lower landscape positions are likely to be wetter longer than higher ones. As

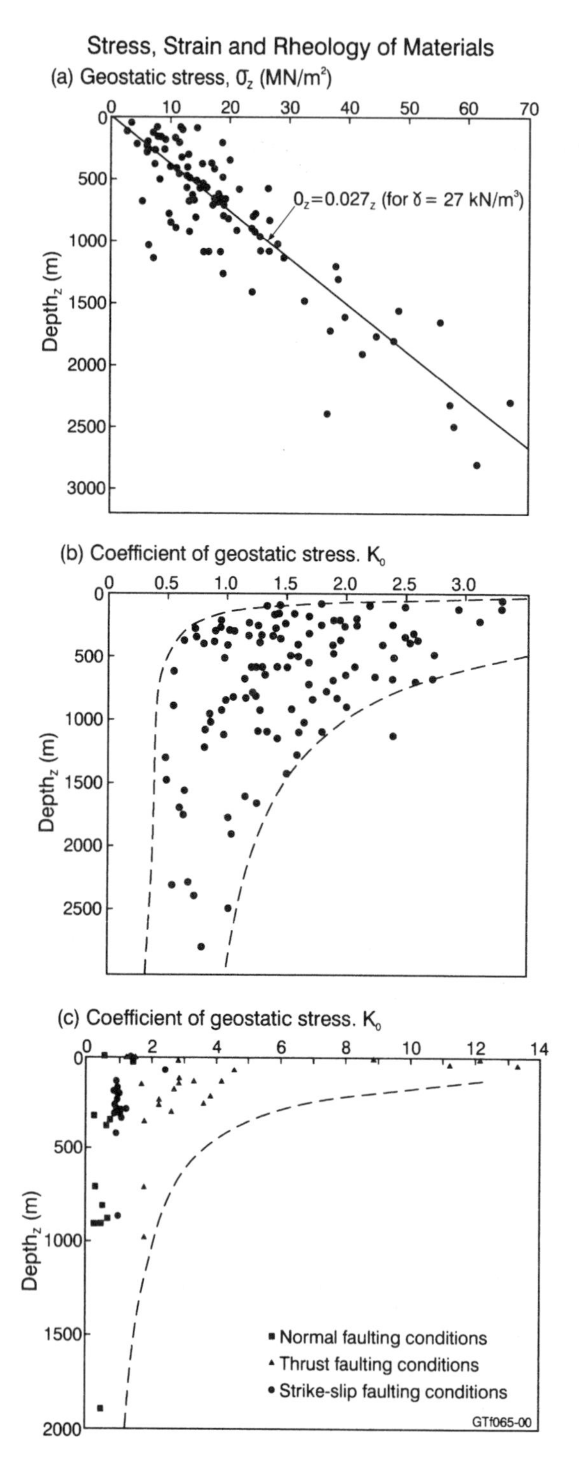

Figure 4.1 *Relationships between lithostatic stress and depth in the crust (from Selby 1993)*

water is one of the main weathering agents, weathering is probably going to be more intense where more water is available to do it. Thus rocks on hills weather more slowly than those in valleys, all else being equal, so that in most landscapes the depth of weathering is greater in valley bottoms than on the adjacent hill slopes.

An example of this is well illustrated at Lake George in southern New South Wales where an ancient stream bed (Late Miocene or earlier) cutting Ordovician turbidites is relict on a fault escarpment. Road construction through the palaeochannel revealed deeper weathering of the turbidites below the palaeovalley base than on the adjacent slopes (Figure 4.2).

Having said that, it is often the case that all other things are not equal. Drainage in valleys or other topographic lows is often more sluggish or groundwater flow is almost static. Under such circumstances chemical weathering is retarded or proceeds very slowly as the water comes into chemical equilibrium with the materials with which it is reacting. Thus weathering here can be impeded and different from that relative to adjacent slopes where drainage is better and solutes are readily removed. Similarly, in regions of poor drainage mineral weathering may not proceed beyond the early alteration products of primary minerals. For example, amphiboles may only alter to smectite and goethite because of the abundance of cation-rich water. Equally, more highly weathered products may undergo transformation to less weathered mineral species, for example kaolinite may be resilicified to form smectite and haematite rehydrated to form goethite or reduced to form soluble Fe^{2+} phases.

Thus, although landscape position is important, its importance depends on drainage. Soil catenas illustrate this well. Typically in regions of moderate rainfall soils at hilltops are redder than those on the lower slopes, which are often yellow-brown coloured, grading down to perhaps greenish-grey soils in valley bottoms. These changes reflect the residence time of water in the regolith. Similarly soils in valley bottoms under these conditions are often more sodic, again an artefact of poorer drainage in the valleys than on the slopes.

The local relief is important, as this controls the amount of erosion that occurs in most landscapes. Thus the higher the relief the higher the erosion rates are going to be, given other factors are the same. Consequently, in high relief areas erosion of the hill slopes will be rapid, comparatively fresh rock will be continually exposed to weathering agents and as a result physical weathering will predominate over chemical weathering. Conversely, in very low relief

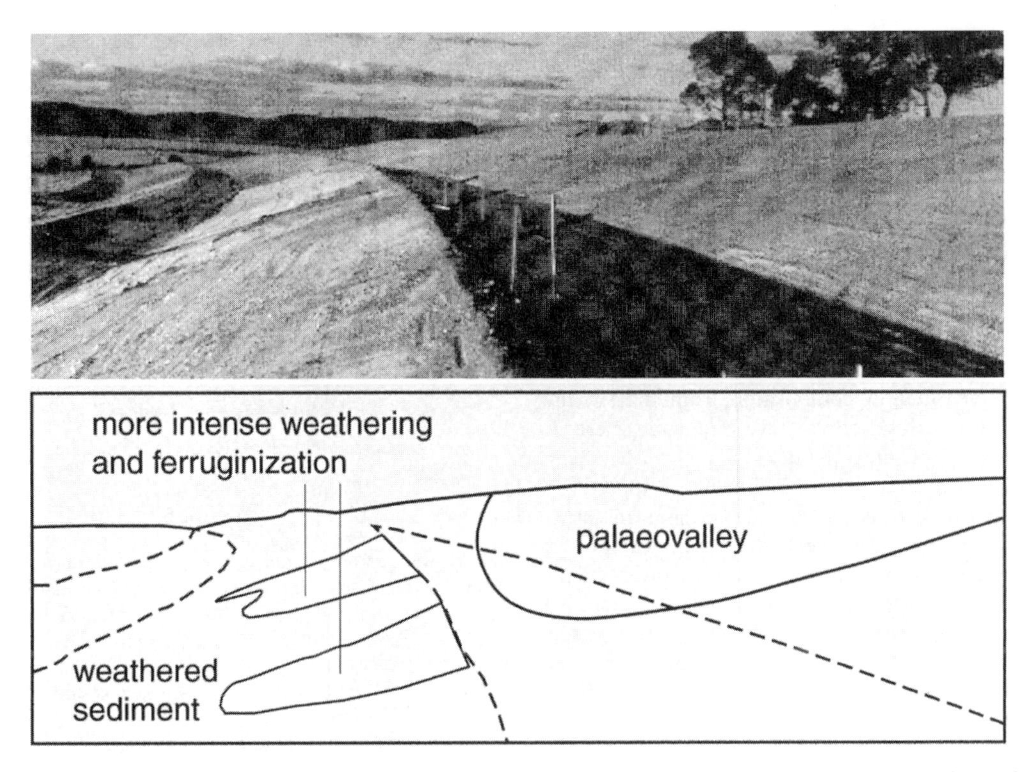

Figure 4.2 *Palaeovalley cut into weathered Ordovician turbidites at Lake George, New South Wales. Note how the weathering is deeper under the palaeovalley axis than on the palaeovalley slopes (photo GT)*

landscapes physical transport is low, water retention rates in the regolith are high and chemical weathering is likely to predominate. Somewhere in between these two extreme landscape situations maximal weathering will occur where water is retained in the regolith long enough to chemically alter the rocks, water flow is sufficient to remove solutes, and where erosion is sufficiently high to continually bring fresh rocks into the zone of weathering.

Other aspects of landscape position with respect to weathering are related to groundwater movement transporting solutes from higher parts of the landscape to lower. This is the main process of solute removal from reaction sites and it is this that enables weathering reactions to continue. Groundwater and its relationship to regolith evolution are considered in some depth in Chapter 5. Also related to landscape position is the depth in a weathering profile that materials occur. The deeper materials in a weathered terrain are generally more labile than those near the surface. Thus solutes are more available from deeper parts of the weathering profiles than near the surface.

This last point is important when considering chemical erosion in terrains with different weathering histories. Highly weathered rocks will yield very much less solutes than will less weathered or fresh rocks. Additionally, as the degree of weathering of the source rocks increases, so the weathering solutions will tend to maintain any initial acidity longer.

Climate

Climate can be considered in two ways: the scale at which we are looking at it; and the components of climate. By scale we mean macro- (or Earth surface) and micro- (or the climate on a scale of metres to micrometres, either relating to surface or internal regolith) climate.

In considering macroclimate the important environmental factors for weathering include precipitation, temperature, evaporation rates, wind, aspect and atmospheric composition. These factors are also important at the microscale, but we will point to differences between macro- and microclimates later.

Precipitation includes all water falling on the surface of the Earth. The amount and composition of water falling vary considerably from place to place. The amount varies with distance from the sea, sea temperature adjacent to the continent, orographic effects and vegetation cover, among other more localized variables. Only water as a liquid is effective as a weathering agent, except for the erosive power of wind-blown ice. As most chemical weathering and some physical weathering are related to the amount of water available to attack the rocks and to transport the soluble products, obviously then, at its simplest, the more water that is available the more weathering can occur.

The microclimate within the regolith is different from that at the surface. Water in this regime is generally of different composition from that at the surface, even though it may have originated there. It generally contains more solutes as the water has been in contact with rocks and regolith for comparatively longer times, thus it has dissolved more material. Additionally, organic activity in the regolith (microbes, roots, invertebrates) changes water chemistry. Generally the presence of organic materials causes a decrease in pH, increased solubility of silicate minerals and mobilization of metal cations. At least in the upper parts of the regolith water is O_2-rich so any Fe in solution will readily precipitate as Fe^{3+} oxides and oxyhydroxides, making the regolith materials a red–yellow colour. Within the zone of saturation below the water-table waters may be reducing, giving mottled blue-grey colours to the regolith, and areas of significant organic accumulation such as swamps and lake margins are generally chemically reducing.

TEMPERATURE

Temperature is said to be important in regolith evolution as reflected in the morphoclimatic school of geomorphology and landscape evolution (e.g. Büdel 1982; Butt & Zeegers 1992b). Temperature is certainly an important environmental variable in weathering, primarily because increased temperature increases the rate of chemical weathering (see Chapter 8). Change in temperature (within the range of surface water) does not, however, affect the processes of weathering. If a rock is so placed that it weathers to a bauxite in the tropics, the same rock in the same regolith situation but in, say, Iceland, may also weather to a bauxite – it just takes longer. Evidence that deep weathering is not dependent on high temperature comes from Taylor *et al.* (1992) who demonstrate the formation of bauxites in cool–cold seasonal climates and Bird & Chivas (1998) who show significant kaolinization at low temperatures.

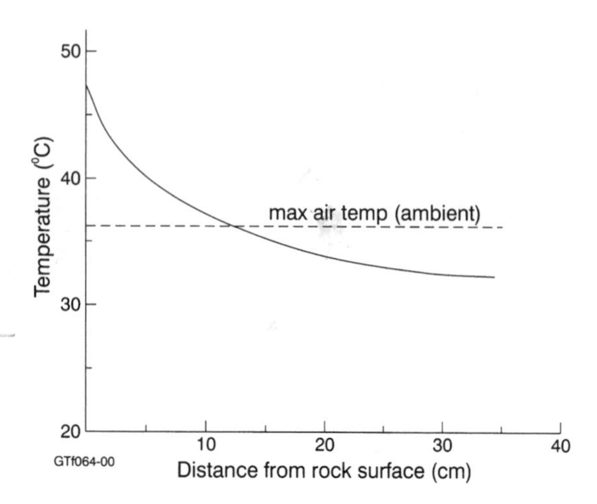

Figure 4.3 *Air temperature plotted against distance into a rock surface. The surface of the rock may well rise well above ambient air temperature, but by a depth of 10–20 cm rock temperatures are equal to or less than ambient. These different temperatures at differing distances into the rock set up stresses parallel to the rock's exposed surface that cause the rock to physically break*

Rocks are poor thermal conductors, so rocks exposed to the sun will heat rapidly at their surface, but that heat will not be transferred readily into the rock (Figure 4.3). Diurnal variations in temperature cause expansion and contraction that may induce fracturing between the warmer outer shell and the cooler inner rock, leading to exfoliation and resulting in effective physical breakdown of the rock. Quartz, for example, expands 0.15% on being heated from 20 to 50°C. The surface of a granite boulder 2 m in diameter undergoing linear expansion of 0.15% would lift by 1.5 mm on being heated over the same 30°C range. It is thus quite common for thermal effects to cause physical breakdown of larger rock and mineral materials sitting in sunlight simply as a result of persistent diurnal temperature changes.

Diurnal temperature variation is only important near the surface. At depths of about 0.5 m daily effects are smoothed out and the temperature approximates mean seasonal temperature. By 10 m depth, seasonal temperature variations are lost and temperatures reflect mean annual ambient ground temperatures. Fluctuations in mean annual temperature change on a much longer timescale (about 10^3 years) relating to the scale of climate change. Pollack & Chapman (1993) has shown such microclimate temperatures to be preserved at depths of up to several hundred metres. At this depth the local geothermal gradient

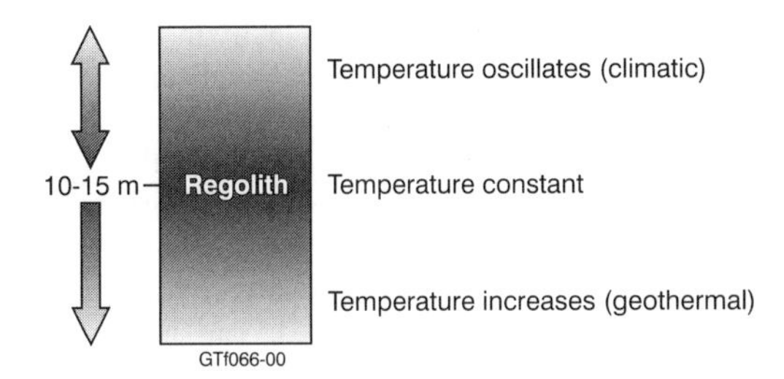

Figure 4.4 *Illustration of the microclimate temperatures in a deep regolith profile*

also has an influence. Figure 4.4 summarizes our view of microclimate temperatures in deep regolith. For many regolith processes (those occurring between 10 m and 100 m depth) only long-term (thousands to millions of years) temperature change, such as is controlled by global climate change or continental drift, will be significant.

Evaporation ratios are the ratio between evaporation (including evapotranspiration) and precipitation. The higher this ratio the less the amount of water there is available for weathering. In many climates (particularly monsoonal climates) this ratio changes annually from high to low, enabling chemical weathering to proceed during the wet season and the products (solutes) to be precipitated during the dry. In climates where the ratio is generally high, but with irregular storms, chemical weathering and mobilization of soluble components of the regolith proceed only briefly before the solutes are precipitated during the normal highly evaporative regimes. This often leads to the accumulation of minerals like calcite, gypsum and salt in regolith.

As well as affecting the rate of chemical processes, temperature has a profound effect on biological activity and so affects the nature and quantity of organic reactants as well as the O_2 and CO_2 levels of the regolith. Temperature also determines the stability of a number of hydrous regolith minerals, largely through its effect on relative humidity. For example, gypsum dehydrates to bassanite:

$$2CaSO_4.2H_2O \rightarrow (CaSO_4)_2.H_2O + 3H_2O$$

goethite converts to haematite:

$$2FeO(OH) \rightarrow Fe_2O_3 + H_2O$$

gibbsite converts to boehmite:

$$2Al(OH)_3 \rightarrow 2AlO(OH) + 2H_2O$$

and halloysite irreversibly loses its interlayer water. These changes may result from water availability rather than temperature alone, and are noticeable in catenas, where goethite may be found in the lower, wetter parts of the catena, with haematite in the higher parts.

Bush fires and lightning strikes also cause heating and dehydration, notably the conversion of Fe-oxyhydroxides to maghemite (γFe_2O_3) and of gibbsite to boehmite or even corundum. The production of maghemite may also result from an early stage of burning when C at and near the soil surface provides a reducing environment in which microcrystalline magnetite forms, followed by an oxidizing stage when the magnetite is converted to maghemite. Fire also can heat quartz past the temperature at which it inverts to β-quartz (573°C), leading to a sudden 3.7% volume expansion and cracking.

WIND

Wind is an important climatic variable in controlling the nature of regolith. Wind increases evaporation rates causing precipitation of solutes. Additionally, winds are the major transporting agent for aerosols carried across the continent from the oceans including such ions as Na^+, SO_4^{2-}, Cl^-, Ca^{2+} and other minor components. For example, Keywood *et al.* (1997) show the distribution of marine Cl^- across the western half of Australia (Figure 4.5).

Wind also transports considerable amounts of dust over large distances. Dust from Australia's interior and west are moved east, at least as far as New Zealand. This dust is added to the regolith across the continent and to the regolith in New Zealand, and with it many readily chemically mobilized elements that may be a factor in weathering. For example, salts blown from playas contribute significant Ca^{2+}, SO_4^{2-} and CO_3^{2-}. Similarly, clay minerals such as palygorskite from salt lakes, when they are removed from their lacustrine environments, will readily weather adding their cations to the depositional site.

Salt weathering

Moisture and salts combine to be most effective factors in rock weathering. Processes are complex, with physical weathering being very evident where the expansion of crystallizing salts causes flaking and exfoliation of a rock surface. The most prominent

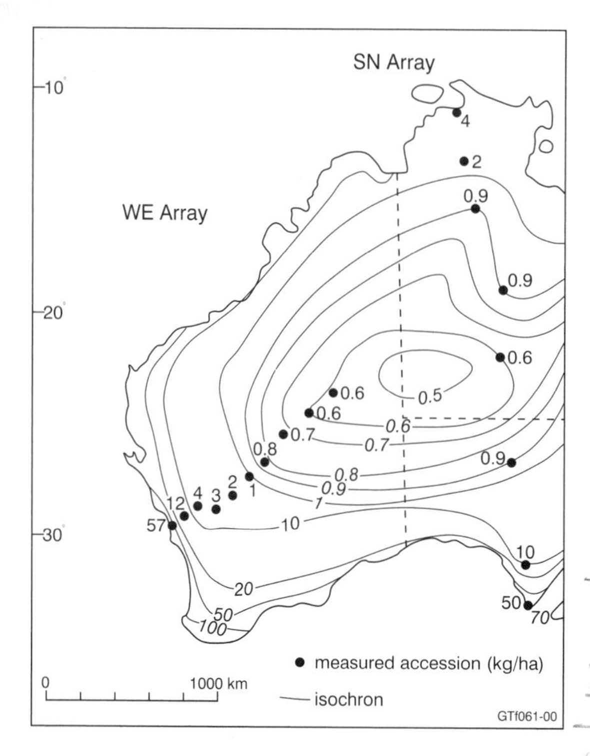

Figure 4.5 *Map of Cl⁻ accession to western and central Australia (from Keywood* et al. *1997)*

examples of salt weathering are tafoni and honeycomb weathered rocks at the seashore, where sea water is constantly wetting the rock. The steps that occur in salt weathering are:

- solution of the salt, commonly halite;
- transport through pore and microfissure networks by capillarity; and,
- crystallization in a confined space, causing volume increase and pressure (subflorescence heave).

Rodriguez-Navarro *et al.* (1999) have described the process whereby salt crystallization within a rock leads to the phenomenon of honeycomb weathering. Critical to the development of damage is that the saline solutions reach supersaturation behind the rock surface. If salts crystallize at the surface, an efflorescence appears which is relatively undamaging. If the crystallization occurs behind the surface, the pressure exerted by the growing crystals can break a surface flake off the rock. The conditions needed for this appear to be enhanced by rapid evaporation, and more damage occurs on surfaces exposed to prevailing winds than on those more sheltered. Once a small depression is initiated, the wind velocity in the hollow is greater than at the flat surface, a consequence of aerodynamics, and so evaporation from the hollow is increased, causing further crystallization behind that hollow. Eventually the hollows enlarge, leaving the intervening walls relatively undamaged. In addition, absorption of water by anhydrous salts (hydration) or thermal expansion of confined salt can both cause expansion and fracturing.

Salts can be in the original rock (e.g. sandstone), derived from chemical breakdown of minerals, precipitated from the atmosphere or borne on the wind. Non-destructive diffusion is typified by liesegang rings – low solubility ferric ions which precipitate in limonite bands wherever the pH of the pore fluid increases, lowering the solubility of ferric ions.

Frost weathering

The conversion of water to ice in a perfectly closed system yields a volume increase of 9%. Though such an ideal closed system is rarely achieved because of continuity of pores and the presence of air pockets, rock splitting is common in climates where freeze–thaw cycles are frequent. Small pore spaces lead to greater frost damage than larger ones, since there is less room to accommodate the expansion, but water in extremely thin capillaries may not freeze even at low temperatures.

Aspect

Slope aspect is not a very important control on weathering at low latitudes, but at high latitudes it is. In the Southern Hemisphere north-facing slopes at high latitudes tend to be drier than south-facing slopes because they are exposed to sunlight for much longer. The situation is reversed in the Northern Hemisphere.

Atmospheric composition

Atmospheric composition has a significant effect on weathering. At a global scale the level of CO_2 in the atmosphere is thought to have fallen sharply about 350 Ma ago because of the rise of land plants (Berner 1995). Generally the higher the CO_2 content of the atmosphere the higher the weathering rates, because of increasingly acid precipitation. The issue is not this simple however, because as partial pressure of CO_2

increases, so the floral biomass increases, and its degradation also increases water acidity at the Earth's surface thus increasing weathering. Since the Devonian when deeply rooted vascular plants evolved and exerted a major control on atmospheric CO_2, weathering of Ca and Mg silicates has increased fourfold (Moulton & Berner 1999).

Apart for CO_2, many other gases have been released into the atmosphere changing its composition, at least locally. Sulfur and nitrogen gases in particular have a marked effect on weathering as a result of the acid rain that their presence in significant concentrations causes. This acid rain leads to destruction of buildings, acidification of soils, increased local weathering rates and death of vegetation among other impacts. We see similar effects in the vicinity of volcanoes and hydrothermal vents where locally atmospheric compositions are changed.

Water

Water is the pre-eminent requirement for chemical weathering. As well as the obvious sources of water through surface rain and runoff, water can enter directly from the atmosphere. Cooling a mass of air may cause water to precipitate (raining inside the stone). At 60% relative humidity, air cooling from 30°C reaches 100% humidity at about 22°C, any further cooling deposits water (dew). Similarly, a hot dry (desert) air mass of 40% humidity at 40°C will form dew at about 25°C, i.e. during diurnal changes.

Groundwater is a second obvious and major source, and a particularly potent one as it carries soluble salts and may have a pH appreciably far from neutral. Entry of groundwater to a rock is largely via capillaries, whose extent depends on the rock type, and may vary from almost none in sound basalts, granites and limestones, to 10% of the rock volume in porous sandstones and partly weathered rock. Capillary water rises higher in smaller radius capillaries than in bigger:

$$\text{height} = 1.5/\text{radius}$$

At typical crack and pore diameters of 0.01 mm, water may rise about 1.5 m. Equally, capillaries can carry water laterally into a stone from its wet outer surface. Capillary movement is enhanced by air pressure increase, which may push water in the outer part of a stone further in, and by a temperature gradient; the water moving to the cooler parts, which generally means further in.

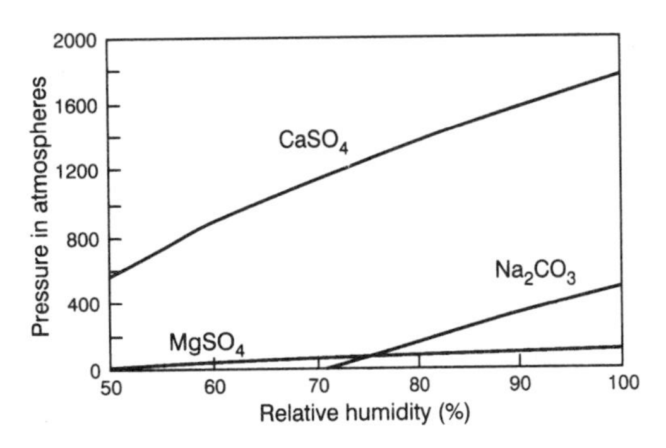

Figure 4.6 *Hydration pressure for: $2CaSO_4.H_2O$ to $CaSO_4.2H_2O$ (bassanite to gypsum); $MgSO_4.6H_2O$ to $MgSO_4.7H_2O$ (hexahydrite to epsomite); $Na_2CO_3.7H_2O$ to $Na_2CO_3.10H_2O$ (heptahydrite to natron) (data for graphs from Winkler & Wilhelm 1970)*

Hydration

Hydration is the addition of water without exchange or solution (reaction with water is hydrolysis, discussed in Chapter 6).

Hydration of salts increases volume, and therefore pressure if in a confined space such as a crack or pore. The production and then hydration of sulfates are very damaging to rock fabric or to structures using rock aggregate such as roads and dams. While most of the damage follows the chemical reaction that produces the sulfate (for example the weathering of pyrite to release sulfuric acid followed by reaction between Mg-rich chlorite and the acid to produce epsomite), further hydration of the original sulfate can create extra expansion. Such pressures can be important in arid climates, where the salts may precipitate to fill pores in a drier period, then if the humidity increases the salts may hydrate creating enough pressure to split off a surface flake (Figure 4.6).

As explained in Chapter 3, the exchangeable cation in smectite interlayers has a strong affinity for water, and considerable pressure can result from the expansion of the structure from 9.5 to 15.5 Å when dry smectite becomes wet (Table 4.1). These pressures can be compared to the maximum tensile strength of various rocks, as follows:

Basalt	28 MPa
Granite	40 MPa
Limestone	45 MPa
Sandstone	35 MPa
Quartzite	160 MPa

Table 4.1 *Swelling pressure of smectites (from Yatsu 1988)*

No. of water layers	Layer-spacing (Å)	Na-smectite pressure (MPa)	Ca-smectite pressure (MPa)
0	9.6		
		>300	>200
1	12.6		
		200	160
2	15.6		
		14	3
3	19		

However, when a clay body within a rock expands, it does so slowly, and is generally not entirely confined, as it is in experiments to measure swelling pressure. Further, the tensile strength of a rock is measured relatively quickly, whereas the expansion of clay may occur over some hours, giving a real rock some time to adjust and so resist fracture.

Biota

Water and atmospheric gases are clearly the primary agents of weathering, but their effect is magnified many times by the activities of the Earth's plants, animals and micro-organisms. Terrestrial biota affect weathering in a variety of ways. According to Barker *et al.* (1997), roots and fungal hyphae physically disrupt minerals, increasing the accessible surface area. By stabilizing the soil, plants increase water retention, allowing longer time for weathering. Their decay adds CO_2 to the regolith, increasing its acidity, while at the same time consuming O_2. In life they release a range of organic acids, notably acetic, citric, formic, fulvic, humic, malonic, oxalic and salicylic. Also, they introduce chelating compounds – organic molecules with specific ability to bind metal cations in such a way that they are isolated from the chemical environment of the regolith water. This changes chemical equilibrium, allowing ready dissolution of cations that under strict inorganic conditions would be almost insoluble (see Chapter 6).

The impact of biota on the regolith is incalculable. From the major disturbance caused by tree roots to the burrowing of ants, organisms constantly churn up the regolith. Chemically, they greatly increase the rate of mineral weathering through the introduction of acids and chelates. They selectively leach elements that they either need as part of their metabolism, or that they utilize by extracting energy from oxidation or reduction of primary minerals. Their catalytic action in the oxidation of pyrite, for example, produces huge quantities of sulfuric acid from abandoned base metal mines, where flooded shafts and tunnels provide abnormal quantities of water in contact with sulfides. Bacteria, fungi, lichens and plant roots all make contact with mineral surfaces, and use organic agents to dissolve the minerals. The actions of the regolith's biota increase the rate of weathering compared to simple inorganic processes by at least tenfold, and in some environments by up to 1000 times.

Conclusion

While physical weathering breaks up rocks and minerals, it is chemical weathering that alters the primary minerals and so exerts the most profound changes on the regolith. The main agent of chemical weathering is water, so in the next chapter we consider water in the context of the landscape.

5
Landscape and regolith hydrology

Introduction

Water is one of the main agents of physical and chemical weathering, of transport of solutes and fine-grained solids, and of surface erosion of the regolith. Some understanding of water movement on and below landscapes is thus essential to understanding regolith and landscape evolution. Water enters the regolith via infiltration from the surface (recharge), moving through it until it reaches the deep water-table and becomes part of the groundwater system. At this point it moves laterally until it is eventually discharged at the surface as springs, seeps into river or lake systems, or enters the sea. Water generally moves slowly through the regolith residing in the zone of infiltration and in the groundwater for periods ranging up to a million years, depending on the size of the system and the rock/regolith factors controlling its movement. Water on the ground surface may run off, evaporate or infiltrate. The water that runs off may transport particles and erode the surface. The evaporated water is returned to the atmosphere leaving behind any solutes it contained. The water that infiltrates may follow a complex course through the regolith (and rock), eventually being added to groundwater. This complex course is partly defined in Figure 5.1.

Water in the regolith

Water is held in various states in the regolith. Some is water bound in crystalline minerals (e.g. gypsum, goethite), but this does not play a role in regolith and landscape hydrology. Some water is adsorbed on to clay and oxide mineral surfaces, and again this plays no role in the hydrological development of the regolith, as it does not move freely. More water is adhered strongly by surface tension to particle surfaces as a thin layer (several molecules thick). This water can evaporate in the zone of aeration or the soil water zone leaving precipitates of solutes it held behind, but it too does not move freely in the regolith. However, if wetting occurs before this water evaporates, the solutes it contains are released into the greater regolith water system.

Water is also held within pores in the regolith by surface tension or capillary forces where it may form bridges between adjacent particles. This situation occurs above the water-table. When the amount of water a material is holding in this way is at a maximum it is called 'dead storage' by hydrologists or 'field capacity' by soil scientists. Water held by surface tension will not flow freely as the tensional forces exceed those of gravity. This water can however react with mineral and organic components of the regolith, and as it is not moving, solutes can achieve high concentrations. Surface tension water is also held more tightly in fine-grained regolith than it is in materials of larger sizes. Because water in smaller pores is in contact with a higher surface area of mineral than it would be in larger pores, it equilibrates more rapidly with the surrounding minerals. The effect is balanced by the lower water activity (a_w) in smaller pores (Nahon 1991). In very small pores, hydrated minerals may dehydrate, as the activity of water is low. Tardy & Nahon (1985) showed that in pores of 30–50 Å goethite dehydrated to haematite as a function of the a_w and Al_2O_3 concentration.

At water contents higher than dead storage (or field capacity) water is free to flow, initially as unsaturated flow and as the pores fill and become totally filled by water, as saturated flow. The total amount of water held in the regolith or rocks is dependent on porosity and the degree to which the water can move through them depends on its hydraulic conductivity.

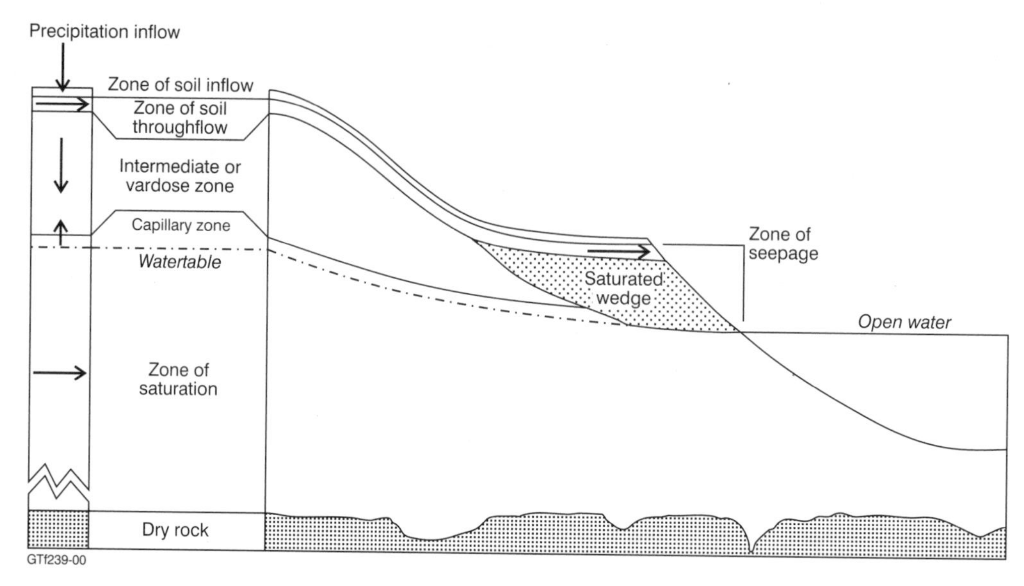

Figure 5.1 *Schematic sketch of the zones of subsurface water in a landscape. Arrows indicate the general directions of water movement*

Table 5.1 *Typical porosities and saturated hydraulic conductivities for common regolith and rock materials*

Regolith	Porosity (%)[a]	Saturated hydraulic conductivity (cm/s)[a]	Rock	Porosity (%)[a]	Saturated hydraulic conductivity (cm/s)[a]
Gravel	25	>1.0	Granite	0.5–1.5	
Sand	35	3.5×10^{-6}–0.02[b]	Basalt	0.1–1.0	
Coarse sand		3.5×10^{-6}–0.25[b]	Basalt	3–35[b]	0–1.6×10^{-5} [b]
Fine sand		1×10^{-5}–7×10^{-3}	Schist	4–49[b]	0–4.2×10^{-4} [b]
Silty sand		$\approx1\times10^{-4}$	Slate	0.1–0.5	0–1×10^{-7} [b]
Sandy clay		$\approx5\times10^{-5}$	Marble	0.5–2.0	
Silt	34–61[b]	1×10^{-7}–2.6×10^{-4} [b]	Limestone	7–56[b]	0–9.5×10^{-3} [b]
Silty clay		$\approx1\times10^{-6}$	Sandstone	2–25	
Clay	34–57[b]	$<1\times10^{-7}$ [b]	Fine-grained sandstone	2–40[b]	0–8.5×10^{-4} [b]
Aeolian sand	40–51[b]	3.9×10^{-4}–2.6×10^{-2} [b]	Medium-grained sandstone	12–41[b]	8×10^{-7}–3.9×10^{-3} [b]
Till		$<1\times10^{-7}$–4.6×10^{-2} [b]	Shale	10–30	
Granite (weathered)	34–57[b]	1×10^{-4}–2×10^{-3} [b]	Shale	1.4–9.7[b]	$<10^{-6}$ [b]
Gabbro (weathered)	41–45[b]	1.8×10^{-5}–1.4×10^{-4} [b]			

[a] Excluding macroporosity that can substantially increase porosity and hydraulic conductivity.
[b] Data from Morris & Johnson (1967).
Other data from Thomas (1994).

Porosity is a measure of total void space in the material and is a measure of the total fluid that a material may hold. As well as the intergranular spaces porosity may include macroporosity as a result of cracks, fractures and joints, and voids produced by biotic activity in regolith. In general, fine-grained regolith materials have a higher porosity than coarse-grained materials. Porosities of typical regolith materials and rocks are given in Table 5.1.

The amount of fluid a material can allow to pass through it is measured by saturated hydraulic conductivity. Flow may also occur in materials with water contents between dead storage (field capacity) and saturation and this type of flow is called unsaturated flow and its rate is measured by unsaturated hydraulic conductivity. The extent of hydraulic conductivity is important in the context of regolith processes. Flowing water meets varying mineralogies during its passage, establishing varying equilibria (or disequilibria) from place to place – here dissolving one mineral, later perhaps precipitating another. Stagnant water eventually reaches equilibrium with the minerals and weathering reactions cease. For weathering to continue solutes must be removed.

Zones of subsurface water

The zone of soil inflow is where that water which infiltrates begins to move into the ground. This part of the soil is where many plant roots concentrate and organic debris accumulates. As water percolates into the ground it may dissolve organic components which are typically concentrated in the upper parts of soils. It is also possible that elements ingested by plants and shed in their leaves, fruit and other bits, can be taken up in water percolating into the soils and transported downwards in the regolith. Additionally, solid particles may be entrained and moved downwards through the regolith, depositing again (illuviation) deeper in the profile. Fine-grained particles such as clays and Fe-oxides are most easily removed (leached or eluviated).

There are many impediments to infiltration in the topsoil. Commonly crusting, hard-setting surfaces and hydrophobism can delay infiltration. Once water does begin to infiltrate, it moves downwards under the influence of gravity. Many soils have B-horizons that have a much lower hydraulic conductivity than the topsoil. This means as water moves into the topsoil it is not able to continue its downward path, but accumulates in the topsoil. Eventually the topsoil becomes saturated and any additional water will not infiltrate but run off and may cause surface erosion. The water in the topsoil will move laterally downslope as soil throughflow. Throughflow will continue to the base of the slope or until B-horizon hydraulic conductivities increase downslope. It is these barriers which tend to retain much of the water entering the ground in the soil water zone. Water in this zone is almost always aerated (O_2-rich); consequently any Fe or Mn in solution precipitates rapidly. Exceptions are in areas of highly organic-rich soils (peat bogs, organosols) where O_2 is consumed by the decaying organics. Ultimately water will penetrate the soil layer of the regolith and enter the intermediate or vadose zone.

The vadose zone is an intermediate zone in the downward movement of water in the regolith/rock under gravitational force. This zone may be wet, often very wet, rarely saturated, or it may be dry and as a result, chemical alteration is considerable and weathering active. The water transports solutes derived from this and the soil zone downwards, except when its drying causes solutes in transport to be precipitated. Water in this zone is also generally well aerated. A result is the development of Fe-oxides as bands and mottles in this zone as well as precipitates of other materials carried in solution (e.g. gypsum, calcite, kaolin, opal). Eluviation of clays and Fe-oxides again occurs in this zone to be redeposited at permeability barriers lower in the profile. It is, however, basically a zone of depletion with minor accumulation of precipitates and illuviated fine-grained material.

The capillary fringe is a zone in which water from the water-table rises into the vadose zone by surface tension forces. The thickness of this zone depends almost entirely on the grain size of the materials. If they are fine-grained then the fringe will be thicker than if they are coarse-grained. When the vadose zone is dry evaporation occurs from the capillary fringe and it forms a gradual change in water content between the vadose and saturated zones. Evaporation also leads to precipitation of any solutes in the groundwater, and the lateral movement of the groundwater means that localized or point sources of elements are extended over larger areas.

The zone of saturation (phreatic zone) is where all the porosity is permanently filled with water. The surface partitioning the zone of saturation from the capillary fringe is called the water-table. The upper part of this water moves slowly down the hydraulic gradient, which for the most part follows the surface topography but with a lower relief. As shown on Figure 5.2 it may also move upwards under a hydraulic head. The upper part of this saturated zone, because it is moving and renewed, is generally aerobic and it is in this part of the zone that weathering is most effective. The soluble products of weathering are readily removed by its flow allowing weathering to proceed readily. Because the upper part is oxidizing, particularly near the water-table, Fe-oxyhydroxides mark the position of the water-table. The deeper water in the zone of saturation may sit stagnant and as the water reacts with the rocks/regolith and is gradually depleted in O_2 so the lower part of the saturated zone becomes anaerobic. Hydrolysis and hydration reactions still occur, but unless the water body is vast it will eventually come to equilibrium with the materials it is in and reactions slow to a stop. This also explains why most deep groundwater is highly saline. Mann & Ollier (1985), however, suggest that reactions occur due to ionic diffusion, and if this is rapid enough weathering

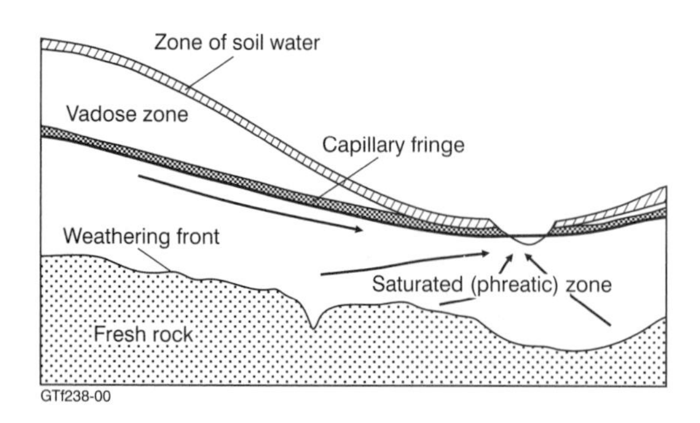

Figure 5.2 *Shows the weathering front in granite where it is likely that this also corresponds to the lower limit of groundwater penetration. Arrows are indicative of the direction of water movement*

Figure 5.3 *Photo of active ferricrete accumulations (dark) in a creek bed on the southern highlands of New South Wales where ferrihydrite leaching from the banks gradually dehydrates to goethite cementing bed and bank sediments. Rainfall is approximately 500 mm/yr (photo GT)*

may continue. They suggest this is how pockets of deeply weathered granite surrounded by fresh impermeable granite form.

At some depth there is a level below which water does not move because porosity is too low and the pores are not connected. Joints, fault planes and similar avenues allow deeper water penetration, so the boundary between regolith and fresh rock is always irregular. In some cases of very low porosity rocks (e.g. unjointed granites) the boundary between regolith and fresh rock (the weathering front) may correspond to this boundary between saturated and dry material (Figure 5.2).

The saturated wedge shown on Figure 5.1 results from the combined effects of the soil throughflow and the capillary fringe at the point of their discharge. During dry times when water-tables fall, the water, rich in elements leached from upslope, evaporates depositing the solutes. The most commonly observed cementation in such localities is Fe-oxides. Figure 5.3 illustrates such accumulations in a creek bank on the southern highlands of New South Wales where ferrihydrite leaching from the banks gradually dehydrates to goethite, cementing bed and bank sediments to form a valley ferricrete.

Movement of water-tables

Water-tables move up and down on various timescales. They move in response to climatic cycles and in response to longer-term climatic change. Floods during wet seasons cause rises on an annual or short-term basis; droughts and dry seasons cause them to fall. Because the water in the saturated zone flows down hydraulic gradients these changes in level are more marked in areas of highs in the water-table surface, and these usually occur under hills, while in the lows fluctuations are smaller in amplitude. Springs (where the water-table crops out) are generally in topographically low positions and they continue to flow long after water-tables become depressed under hills and drainage divides.

The zone through which water-tables fluctuate, whether in the short or long term, is a zone of intense weathering accompanied by removal and addition of minerals. In regions of seasonal shifts the effects are marked, as exemplified at Weipa on Cape York or on the Darling Escarpment west of Perth (see Chapter 15).

Longer-term changes are brought about by climate change over 10^4–10^7-year time spans. Such changes are within the time frame of weathering profile development (see Chapter 11) so the effects of such changes are also important in understanding the nature and origin of those profiles.

Other shifts in the water-table can be produced by downcutting of valley systems over long or short time periods. As valleys lower, so the water-table lowers as the local base level for the groundwater is lowered. In some instances this occurs over time spans as long as 10^6 years but it may also happen quickly. An obvious example is the rapid incision of stream beds following the European habitation of Australia. The clear felling of timber for grazing has resulted in widespread erosion of valley bottoms and channel incision by up to 10 m over a matter of 50 years or less. This has

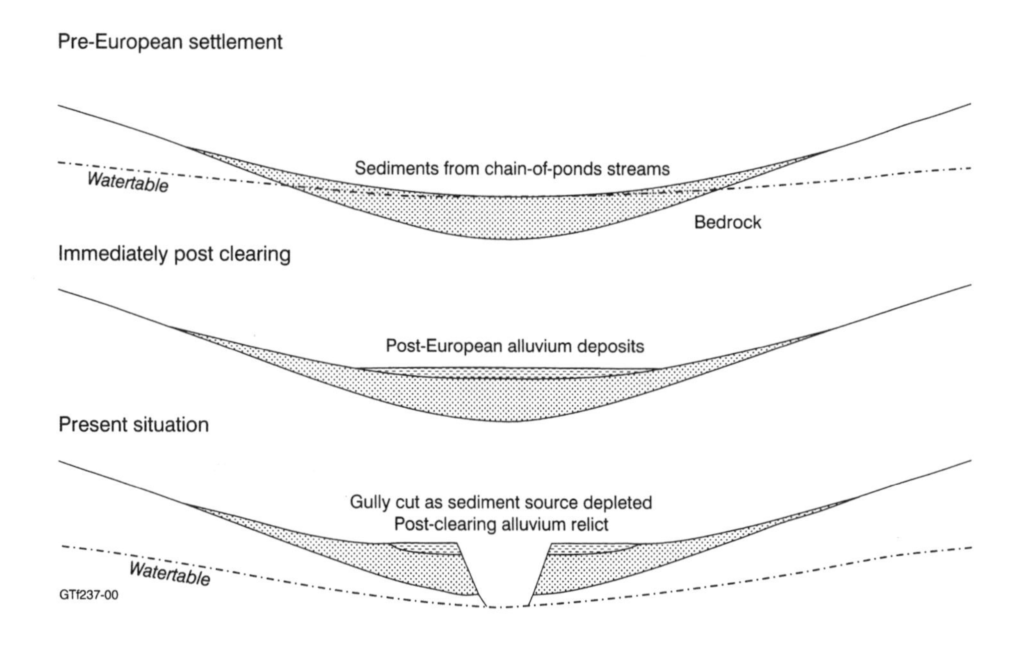

Figure 5.4 *Incision due to post-European settlement in Australia and the effects on groundwater*

resulted in rapid decline in water-table levels followed by their slower rise as infiltration increased due to evapotranspiration falls. A result of this process was the transfer of deeper saline groundwater to the surface and widespread surface salinization (Figure 5.4).

Groundwater levels may also be shifted in coastal regions by changes in sea levels. Transgressions will increase groundwater levels and regressions cause them to fall. Faulting, volcanic activity and deposition of sediments will also cause shifts in water-tables in the region affected.

Ancient water-table positions may be preserved in weathering profiles, most commonly marked by Fe-oxide cementation or Fe-oxide mottling. Another interesting phenomenon is that as water-tables fall, lateral movement is slowed as hydraulic gradients decrease. This results in decreased dispersion of elements derived from point sources (Figure 5.5).

Evidence for a falling water-table may well be preserved in the form of mottles on cemented horizons, but as water-tables rise they will drown and modify any previously formed palaeo-water-table indicators. Provided these remain below the new water-table long enough it is likely they will be destroyed by progressive reduction of the Fe-oxides in the anaerobic water in the saturated zone. Thus the most common case is that successive water-table indicators are only preserved when the last significant variation in water levels was a falling one.

Falling water-tables may also lead to 'overthickening' of duricrusts. For example, if a calcrete is forming at or near the water-table, as the latter falls so the precipitation occurs progressively deeper in the regolith resulting in an unusually thick duricrust. An illustration of this is the preservation of past water-table concentrations of Fe-oxides and Au^0 above or near primary gold ores across the Yilgarn Craton of Western Australia. Mann (1998) describes the widespread occurrence of supergene gold deposits in residual, erosional and depositional regolith. What cannot be done yet is to demonstrate the increasing age of the successively lower Fe-oxide/Au^0 concentrations, but theoretical considerations and the easy solution and precipitation of gold in these environments suggest such sequences represent falling water-table deposits (Figure 5.6).

Perched water-tables

The zone of soil throughflow is the result of water being temporarily held above the water-table by a

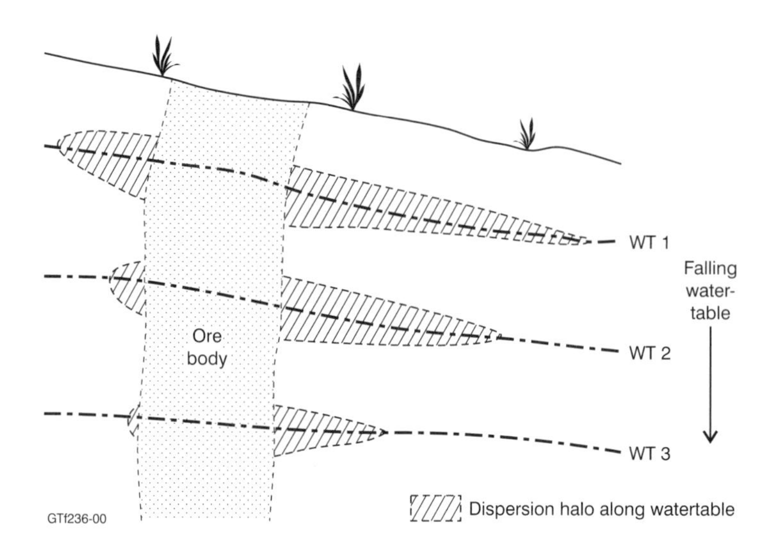

Figure 5.5 *Decreased dispersion of elements as water-tables fall through time*

permeability barrier (zone of reduced hydraulic conductivity). Similar features are common throughout the vadose zone. Water moving downwards through the vadose zone will be retarded or held up by decreases in hydraulic conductivity. Such impediments include clay bands in sediment sequences, soil B-horizons, already cemented zones of the regolith, increases in the amount of plasma in the weathering profile (see Chapter 11) and compaction.

Figure 5.7 illustrates the effect of perched water-tables in dune sands at Lake George. Sands and clays deflated from the lake have reorganized by the illuviation of clays in bands, which have in turn acted as permeability barriers and lead to the precipitation of Fe-oxides in ephemeral perched water-tables. Similar deposits may appear as mottles or lenses of precipitates (e.g. Fe-oxides, calcite, silica) in the vadose zone. Mann & Ollier (1985) describe such features from Armidale, New South Wales. Figure 5.8 shows wavy bands of Fe-oxides cementing jointed Eocene sediments. These they explain as precipitation of oxides at wetting fronts (i.e. they represent the limits of successive phases of wetting during infiltration events). An alternate explanation would be that they represent varying positions of a falling water-table. The decrease in wave amplitude reflects the increased hydraulic conductivity along joint planes so that when water-tables were higher the cones of depression along joints are greater than they are as it lowered.

Behaviour of groundwater in three dimensions

Figure 5.9 shows the pattern of groundwater flow below the landscape, illustrating the concept that the landscape shape has a considerable influence on the groundwater movement patterns. Under long straight slopes groundwater is contributed to the stream uniformly along the valley. At valley heads the groundwater flow is concentrated as shown by the convergence of flow lines (Figure 5.9). This tends also to concentrate solution, weathering and erosion by mass movement (due to higher pore water pressures) of the regolith in these sites. At spurs, on the other hand, flow lines diverge, meaning that groundwater will deliver less solutes to these sites and that erosion by mass movement will be less.

A consequence of this is that in hilly terrain most sediment moving in stream channels will be derived from the channel head, and least, if any, from spurs. It also means that weathering rates in valley heads are likely to be higher than elsewhere in the catchment due to the higher groundwater flows there. These simple observations are important in the context of understanding landscape evolution, sediment generation and chemical dispersion in transported materials.

Hydrochemical anomalies in groundwater may be traced to source if the patterns of water movement are known. This is well illustrated by the work of Andrew *et al.* (1997) who have shown that sulfur isotope ratios

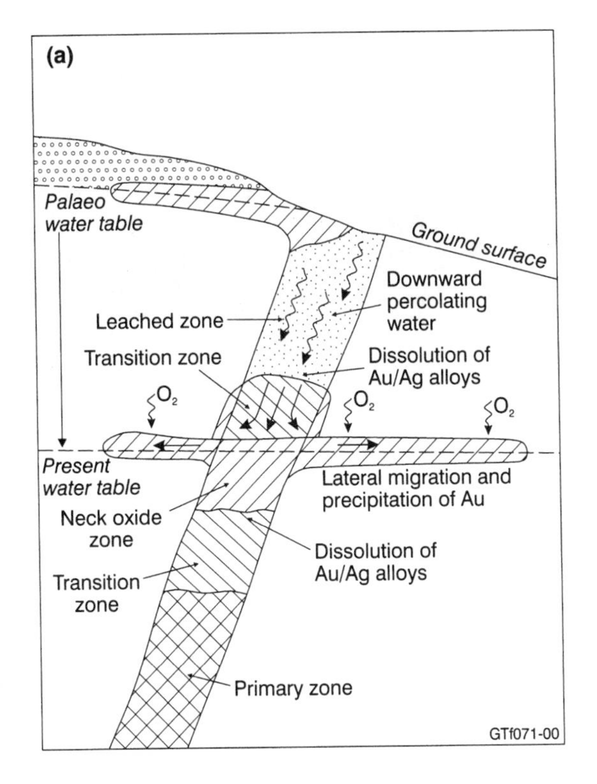

from mineralized rocks at Menninnie Dam in South Australia increase markedly as sulfur from weathering sulfides mixes with the regional values related to gypsum down groundwater gradients away from mineralization.

The same principle may be applied to mapping potential areas of salinization. If the directions of groundwater movement are known and the sources of the salts known then it should be a relatively easy matter to predict areas at risk of salinization, particularly if water-tables are rising. Currey (1977) showed clearly that most salts in the Wannon River groundwater catchment were coming from tributaries draining deeply weathered and ferricreted catchments, not from other tributaries and not from cyclic salts (derived from the ocean). He estimated that only about 8% of the salt present in the groundwater system comes from the latter resources, the remainder from within the catchment where it had been held in storage in the regolith for a long time. With point sources of salts like this it is predictable where salinization in the lower Wannon groundwater discharge zone might be.

Airborne and ground electromagnetic mapping (AEM) provides a valuable tool for mapping salt-affected country on a regional scale. Saline water is a better electrical conductor than is fresh water; AEM

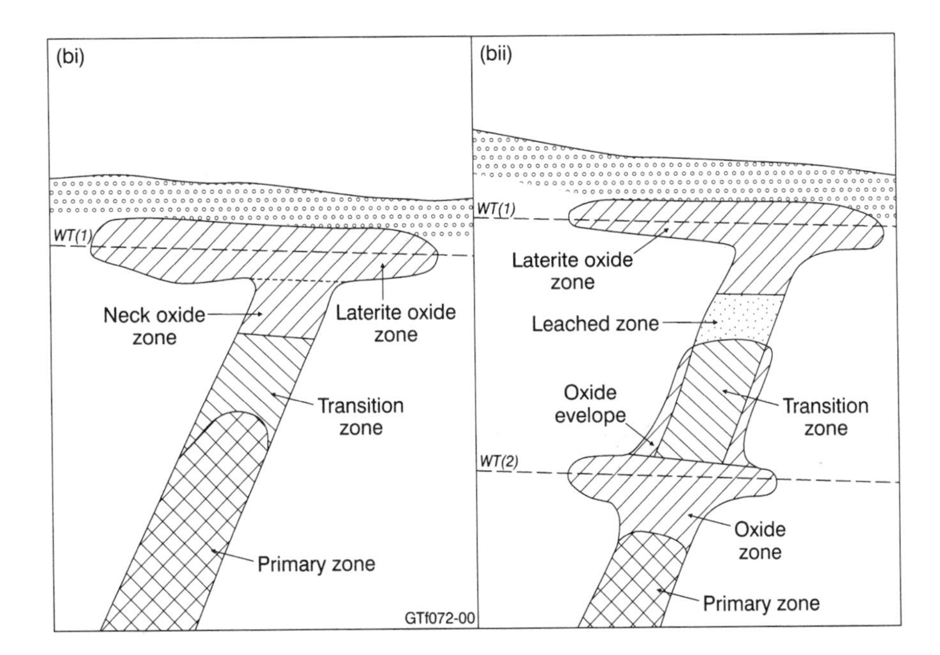

Figure 5.6 *(a) Generalized model for an oxidized gold deposit where oxidation can occur above and below the watertable. (b) Shows styles of Au deposits with (i) one high water-table, (ii) a relict Au deposit produced from a falling water-table (from Mann 1998)*

Figure 5.7 *Fe-oxide bands in stranded beach ridges of the Murray Basin near Kerang, Victoria (photo GT)*

Figure 5.8 *Photo of Fe-oxide bands which cut across bedding in Eocene sediments near the University of New England, Armidale, New South Wales. Note the coincidence of the low points in the bands and the joints and the decrease in amplitude of the waves down section (photo GT)*

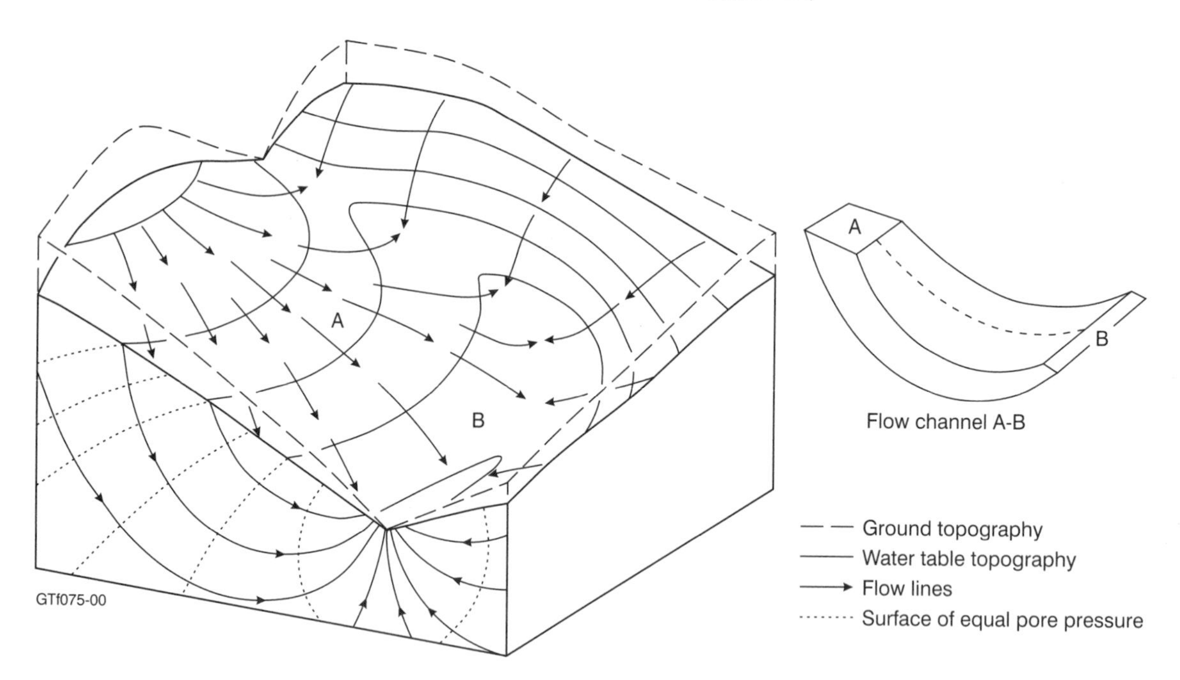

Figure 5.9 *Flow pattern of groundwater under a valley and ridge system (from Dennen & Moore 1986)*

responses are stronger where conductivity is higher, and have the potential to map saline water to a depth of about 80–90 m. It is also possible during still air conditions to use airborne radiometric surveys to map salinization from the radon plumes emitted from salinized ground. This method was discovered by the observation of ephemeral anomalies on radiometric images that disappeared as wind dispersed them.

Landscapes related to groundwater

It is not our intent here to review the extensive literature on limestone karst, perhaps the most obvious landscape feature produced by groundwater activity. While karst is an important aspect of regolith studies, it is well understood. Rather we wish to discuss some of the less obvious aspects of karst geology as they relate to regolith. Because of the authors' predilections it is, however, necessary to mention the importance of limestone regolith in relation to oenology. Many of the world's greatest vine-growing soils (*terra rossa*) are developed in thin regolith over limestone. *Terra rossa* the world over, provided climates are suitable, produce some of the best wine. Just part of the magic relationship between regolith, plants and human endeavour.

Several aspects of limestone karst are worth mentioning in respect of regolith. Firstly, in Guizho Province, People's Republic of China, clay minerals accumulated in karstic depressions, formed in the Early Permian Maokuo Group limestones, have been the sink for gold leached from adjacent volcanic rocks. Karstic landforms have a relief of about 300 m with depressions up to 1.5×10^6 m^3 filled with clays and angular gravels. The red clays (kaolin, illite, anatase, quartz ± chlorite) filling the depressions are very gold rich with values as high as 60 g/t being recorded from mud bricks in a farmer's pigsty (Mao Jian Quan, personal communication 1996). At least some of the gold occurs as Au^0. These deposits are currently being worked at an average grade of ≈5 g/t. Although reports on the locality are few, it would seem the Au, mobilized from Late Permian Danchang Formation (volcanogenic sedimentary and volcanic rocks), is mobile in the alkaline groundwater associated with the karst and on entering the more acid clay-filled basins the Au deposits (see Figure 7.16).

Another common groundwater karstic feature of landscapes is karstic depressions. These can form on a variety of rock types in a broad range of climates. McFarlane *et al.* (1995) have described siliceous karst forming a moderate sized, ephemeral lake-filled depression in the area of Darwin, Australia. These lakes form as a result of solution of Cretaceous sandstone causing surface depressions sealed by smectitic clay minerals. These karstic features are also very widespread in the Darwin area (Figure 5.10). They are currently in a tropical monsoonal climate, but the relationship between present climates and the depressions is open to question.

Similar lake-filled depressions are common across the basalts of the Monaro, southern New South Wales. They occur mostly on, but are not confined to, major drainage divides. These ephemeral lake basins are partly karstic in origin and maintained as depressions by deflation (Pillans 1987). The region is currently in a semi-arid cool climate, but solution channels within the basalt are commonly lined with gibbsite (Moore 1996a). This demonstrates the extensive solution that has taken place within the region, probably assisting to maintain the lake depressions. They, like the Darwin depressions, are lined by smectite.

Figure 5.10 *Photo of karstic solution channels filled by overburden in shallow marine Cretaceous sandstones and mudstones along the sea cliffs near Broome, Western Australia (photo GT)*

Figure 5.11 *Photo of the relics from karstic solution of quartzose sandstones in the Bungle Bungle Ranges, eastern Kimberley, Western Australia (photo Dean Hoatson)*

One of the requirements necessary to form karstic depressions in non-carbonate rocks and regolith would seem to be minimal erosion as a result of tectonic stability. This is borne out by the extensive development of karst in quartz sandstones across northern Australia (e.g. Bungle Bungle Ranges in Western Australia (Figure 5.11), and the Ruined Castle in the Arnhem Plateau of the Northern Territory).

A different landscape manifestation of groundwater activity are mound springs. They are an artefact of leakage to the surface of artesian aquifers and consist of mounds of sand and clay transported to the surface

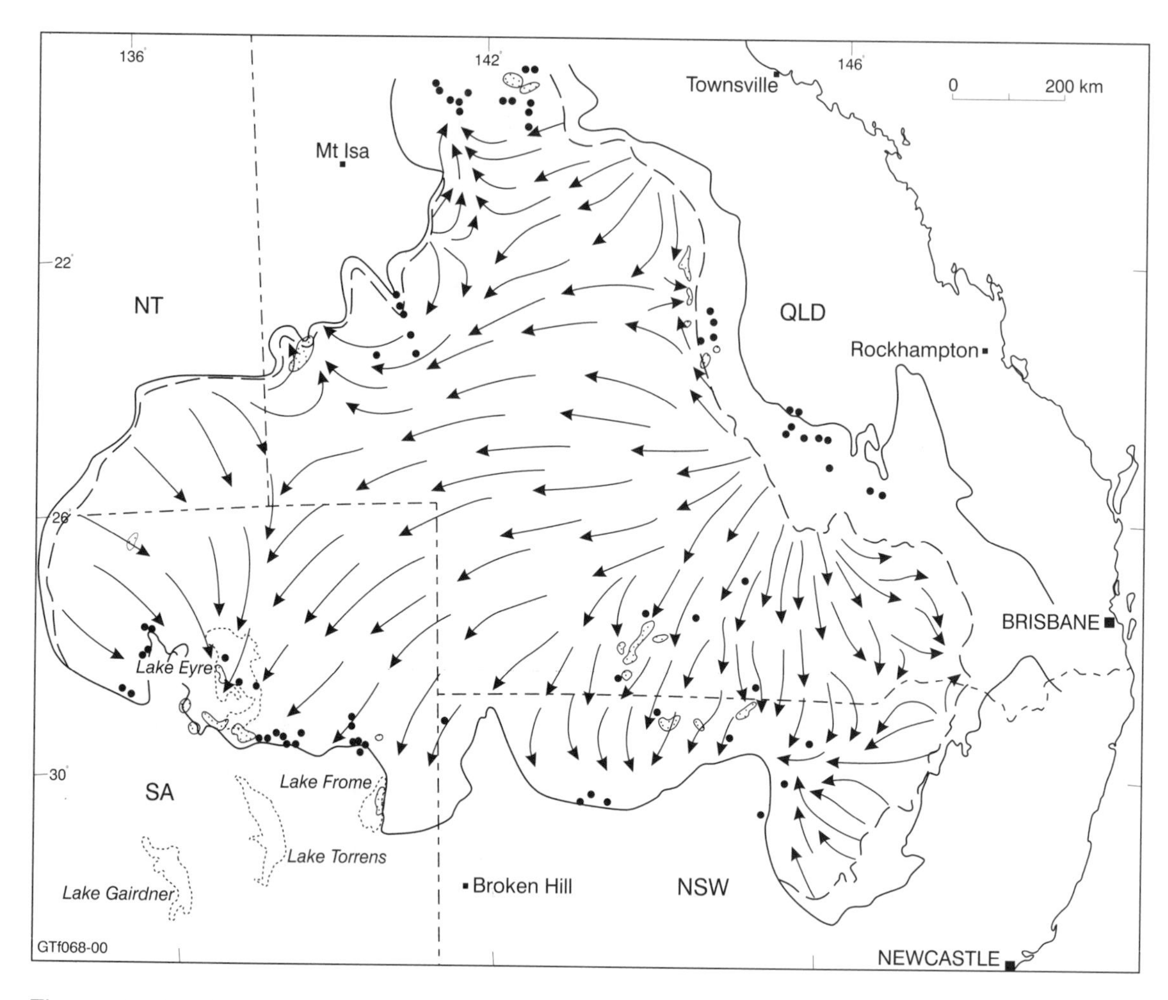

Figure 5.12 *The distribution of mound springs and flow paths in the Great Artesian Basin of Australia (from Habermehl 1980)*

from the walls of the water conduit. Together with the solid debris, salts are deposited, their composition depending on the composition of the water. Many mound springs have small lakes associated with them from which salts may also be deposited.

Mound springs are common throughout the Great Artesian Basin of north-eastern Australia, particularly where artesian water flow, often from 3000 m deep, is obstructed or encounters major fractures in the basin rocks. There are about 600 springs throughout the basin. They are particularly common around the basin margins where the artesian aquifers are shallower and obstruction to flow greater. Figure 5.12 shows their distribution in Australia, and it is interesting to note their frequency in the marginal parts of the basin. Figure 5.13 shows an example of one from near Eulo in south-western Queensland.

Figure 5.13 *Photo of active mound springs from Eulo in south-western Queensland (photo GT)*

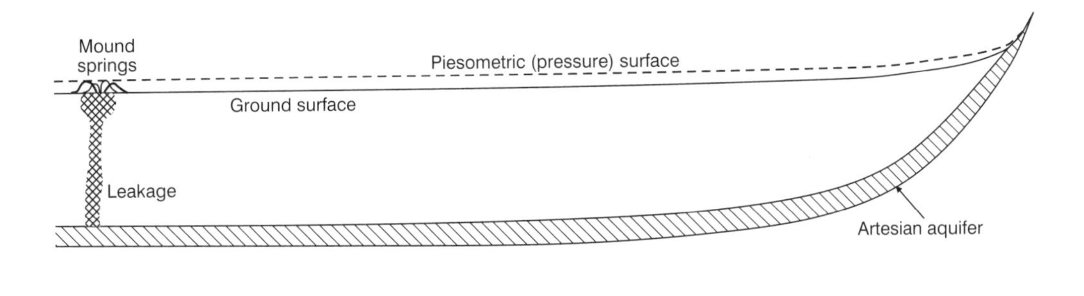

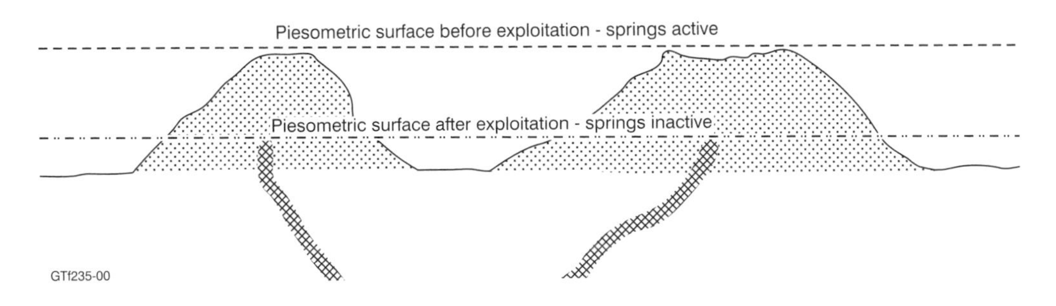

Figure 5.14 *The relation of the original piezometric surface of an artesian aquifer to the post-exploitation surface*

The mound springs locally have a significant effect on the regolith in that they contribute sediment (mainly colloids) and solutes to the surface gathered from a deep section of underlying rock and solutes from a very wide area through which the artesian water flowed. They are made of clay minerals, calcite and organic deposits. Wind-blown sand is also commonly incorporated with these materials which may significantly mask the local regolith characteristics. Additionally they enable the original piezometric surface (pressure surface) of the aquifer to be reconstructed by measuring the heights of the mound, even after the aquifer has been exploited (Figure 5.14). When first discovered at Cunnamulla in Queensland, in 1918, flows from the artesian aquifers were about 2×10^6 kl/day. Since increased exploitation of the aquifers this has fallen by 25% and many of the mound springs have consequently become inactive.

Ancient mound springs (said by Jessup & Norris 1971 to be Pleistocene) are also recorded in the basin. One example is Hamilton Hill in South Australia where a flat-topped hill 240 m across is composed of spring limestones with dense 'reed casts' sitting on a pedestal of weathered Cretaceous sediments of the basin. Clearly there has been erosion of the plains around the springs since their formation, and they have been formed after the Cretaceous sediments were weathered, but this could date them as anything from Palaeogene to Quaternary. Their age is as yet undetermined, although they are probably datable using palynology.

The Murray Basin

The Murray Basin has been the subject of intense study for many years because it supports some of Australia's main agricultural and pastoral land. Over the last 50 years or so soils in the region have been becoming salinized and they are still doing so at an accelerating rate. This presents a major problem for agriculture and for land managers, not just in the basin itself, but in all the catchment areas that feed surface and groundwater to it. This part of southern Australia in the westerly wind belt receives most of its rain from the Southern Ocean and with it considerable cyclic salt. This salt has been accumulating in the basin progressively as it has filled and it is still doing so.

The Murray Basin in south-eastern Australia is used here as an example of a relatively large groundwater system that has influenced landscape history and is still affecting the use of the land occupying the basin. It has

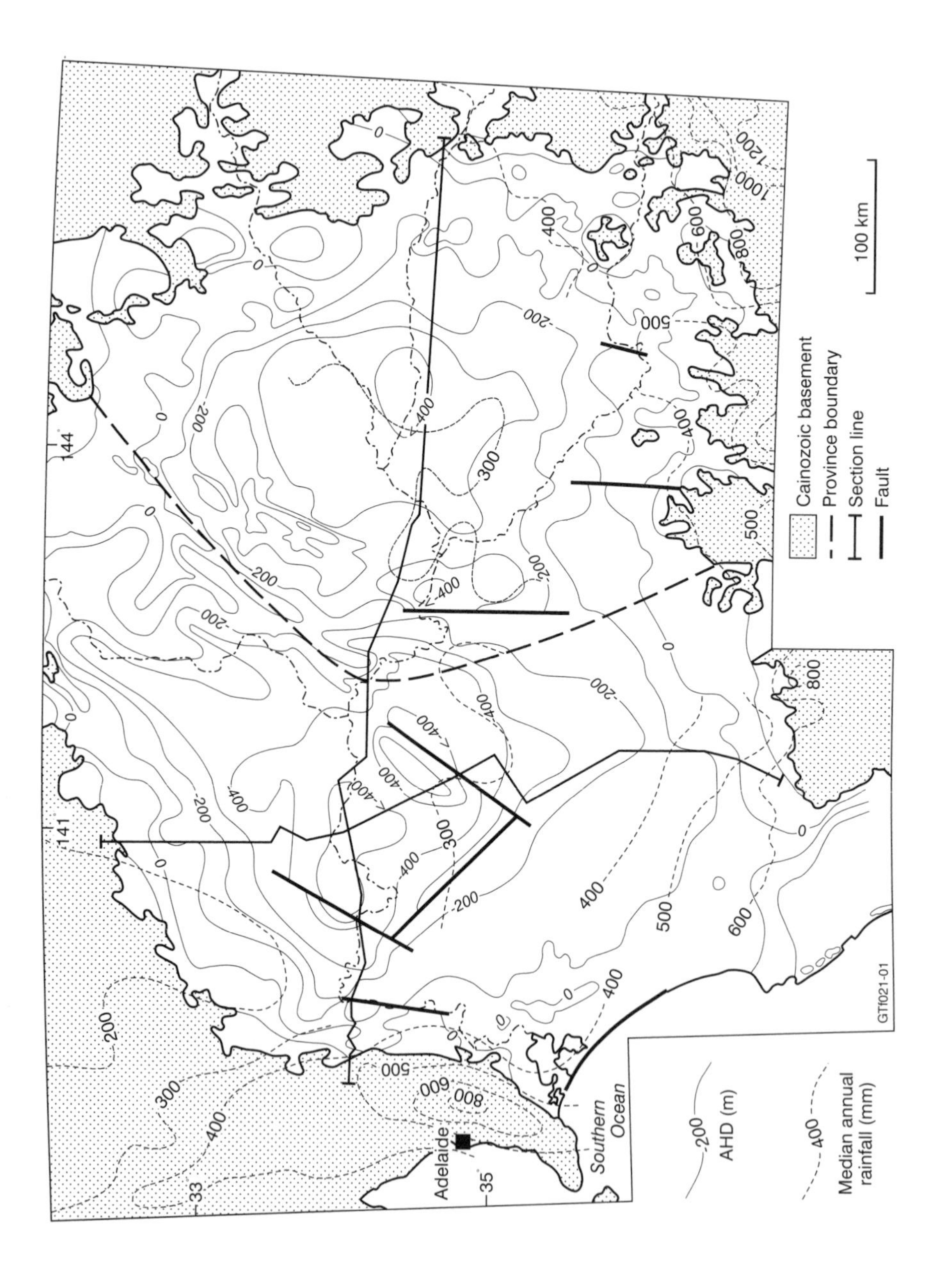

Figure 5.15 *Map of the Murray Basin showing the extent and thickness of the Cainozoic fill, the location of the Padthaway Ridge and Mallee-limestone and Riverine provinces (modified from Evans & Kellett 1989). Contemporary isohyets are also shown*

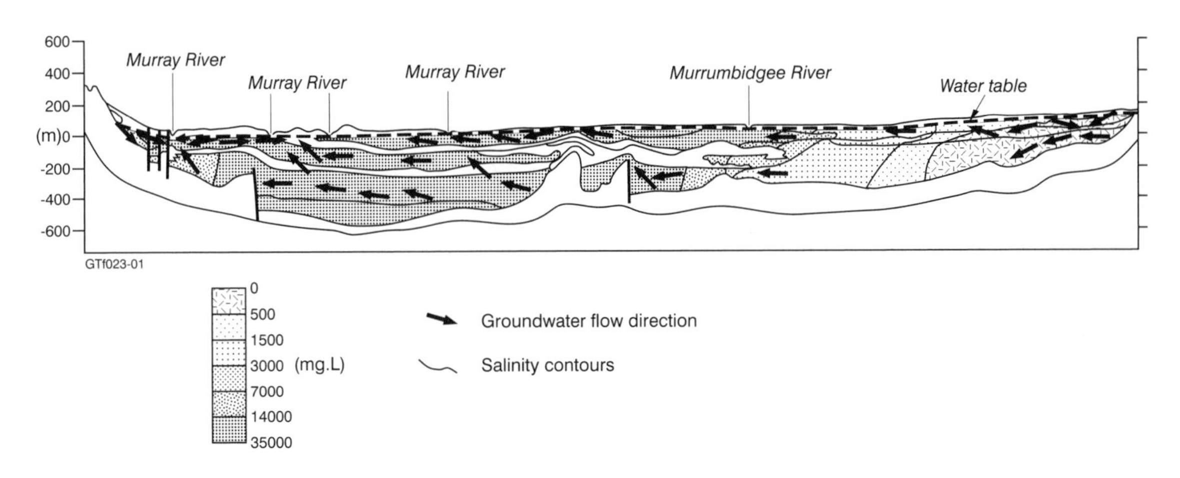

Figure 5.16 *An east–west section through the centre of the Murray Basin showing groundwater flow patterns and salinity. Note the outcrop of groundwater around the Murray River (modified from Evans & Kellett 1989)*

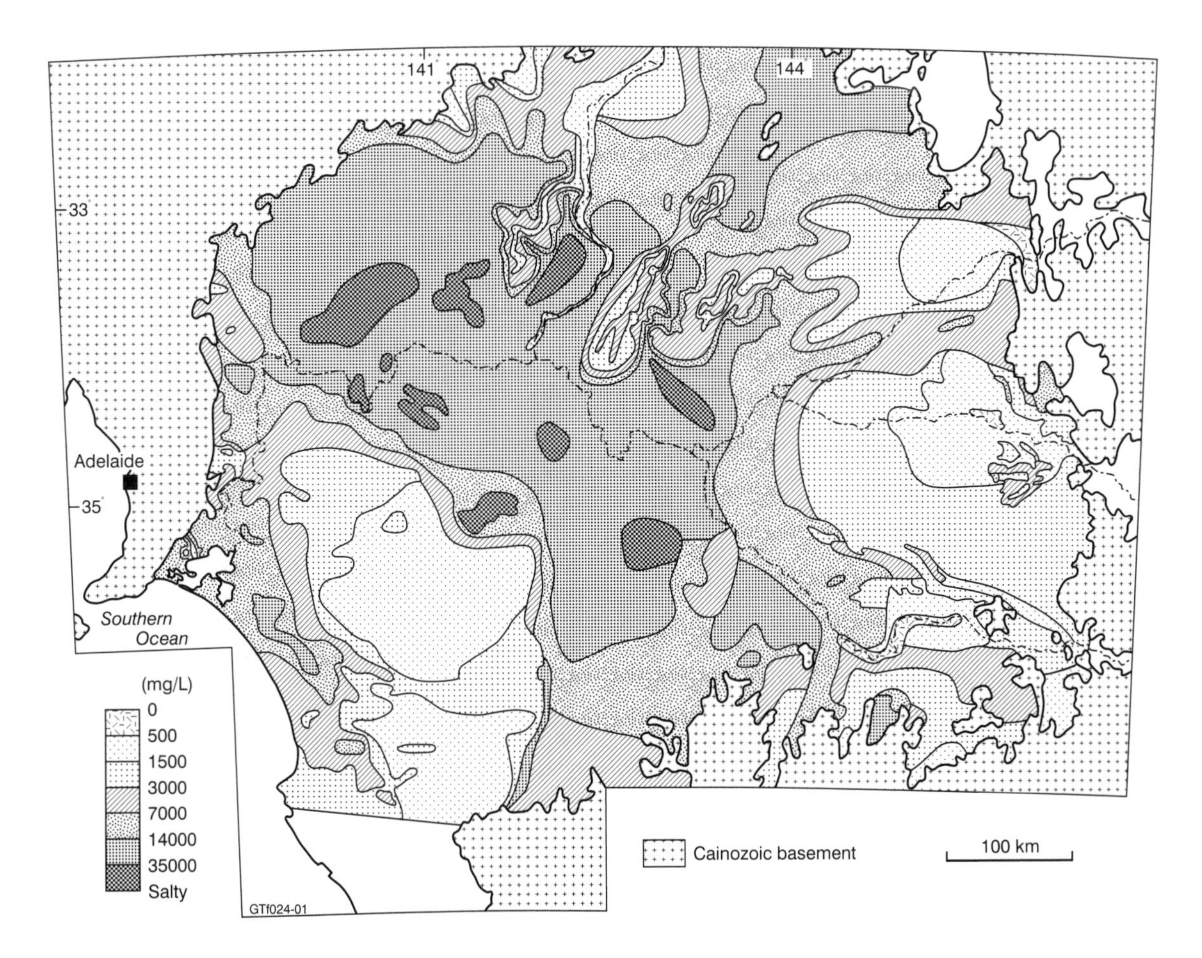

Figure 5.17 *The salinity distribution of shallow groundwater in the Murray Basin (modified from Evans & Kellett 1989)*

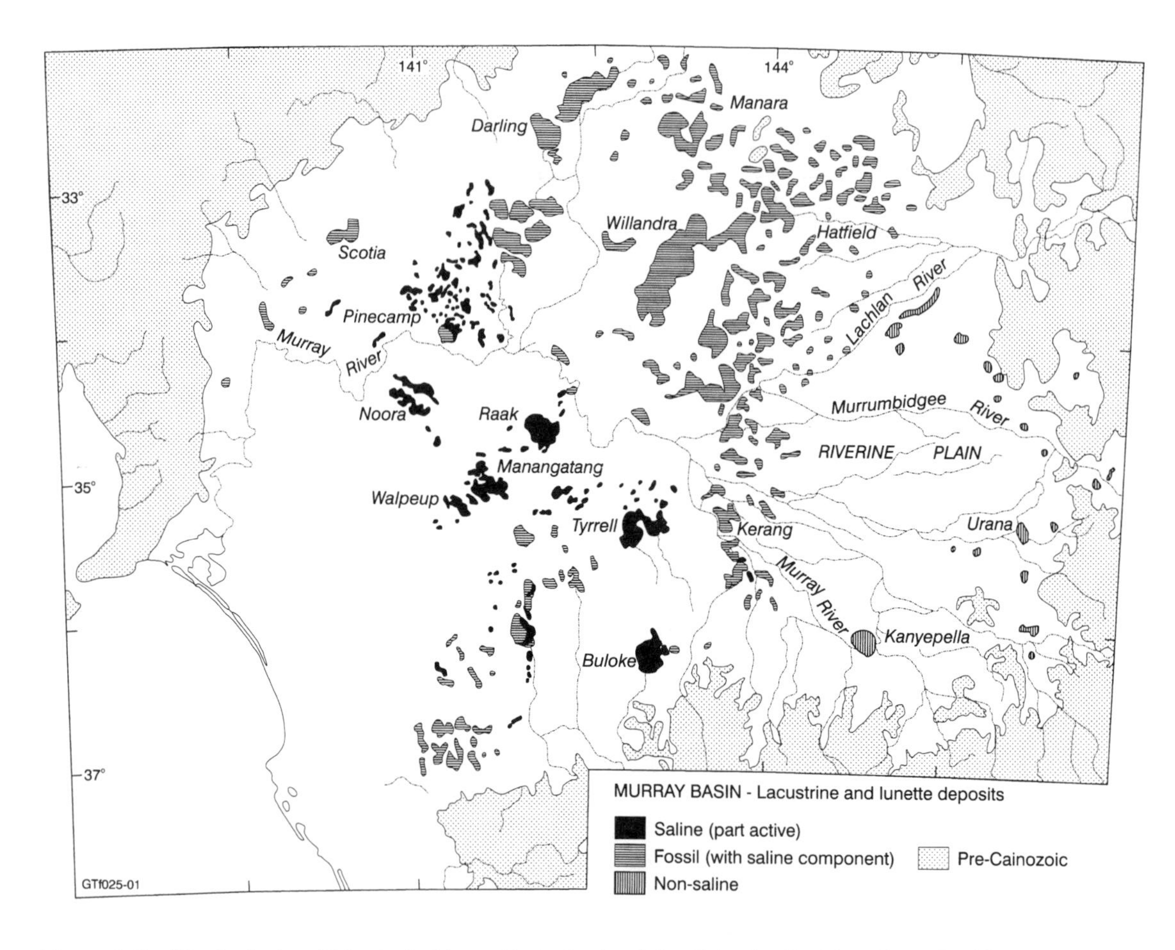

Figure 5.18 *The distribution of lakes in the Murray Basin showing their salinity (modified from Evans & Kellett 1989)*

an area of approximately 700×550 km and it is filled with between 200 and 600 m of Cainozoic sediments (Figure 5.15), and it is still accumulating them. At its mouth to the Southern Ocean is a horst, the Padthaway Ridge (Figure 5.15) that blocks major groundwater discharge from entering the ocean. The fill is effectively divided into two, a riverine province in the east and a limestone province in the west. The former consists of alluvial sands and gravels grading to more clayey sediments westerly before the sequence is dominated by limestones encased in confining clay-rich sediments. In the western part, limestones are overlain by extensive clay-rich confining beds.

The presence of the clayey confining beds between the riverine and mallee–limestone provinces causes shallow groundwater, driven by a hydraulic head from the intake beds of the shallow aquifers, to reach the surface (Figure 15.16). Figure 5.17 shows the distribution of saline shallow groundwater with the highest salt levels occurring in a north–south strip across the centre of the basin corresponding to the presence of the clayey confining beds for the lower aquifers.

The eastern two-thirds of the Murray Basin has a large number of lakes dotted around it, and these increase in salinity in a westerly direction reflecting the increase in salinity of the shallow groundwater in that direction. The lakes near the intake beds around the basin's margin are fresh, those in the remainder of the eastern half of the basin are fossil lakes with a saline component and those across the zone of groundwater outcrop (Figure 5.16) are highly saline and partly active (Figure 5.18). The distribution and activity of these lakes are a direct consequence of the nature of the groundwater and its outcrop pattern.

During glacial maximum climates through the Quaternary the extent eastwards of saline lakes increased as a consequence of decreased fresh inflow to the aquifer systems of the basin. The extent of surface salinity increased significantly into the zone of presently partially saline lake systems, in a similar fashion to the increasing area of salinity today, the difference being that during glacial climates intake was naturally reduced because of drier climates. Vegetation, particularly large trees that used much of the recharge water, were also reduced by significantly lower temperatures (by 10° mean annual). Today the causes are increased intake to shallow aquifers as a result of widespread irrigation and deforestation, particularly of the catchment and intake areas of the basin. Formerly inactive lakes are becoming active and saline, soils are being salinized and valuable agricultural land is being taken out of production.

An understanding of the relationship between the distribution of lakes, the shallow groundwater flow systems, their saline nature and the Quaternary history of the region could have prevented the major environmental disaster this part of Australia is currently faced with.

6
The chemistry of weathering

Introduction

This chapter provides an outline of cation solubility and mineral stability in regolith water; broadly speaking, the subject of aqueous geochemistry. To understand the concepts properly, much deeper reading will be needed than is provided here. A selection of recommended texts is given at the end of this chapter.

It is commonly asked: what is the solubility of, say, Al in water? Or: what is the saturation concentration of Al in water? Or: how will, say, silica react in water containing Al in solution? Such questions can be partly answered by applying equilibrium thermodynamics (see the last sentence in the paragraph above!). In this chapter we will present some of the results of such theory and compare them with observations. To do this, we will need to introduce a few concepts.

The *concentration* of a species (ion, or hydrated ion) is the amount actually present and measurable, usually expressed in moles per litre. A one-tenth molar solution of NaCl contains 5.85 g NaCl/litre.

The *activity* of an ion in solution is effectively the available amount of that ion (moles per litre), which is always less than the concentration. Activity, which is close to being equal to concentration in very dilute solutions, is written in square brackets, for example $[H^+]$. In solutions where an element is part of the structure of a soluble non-dissociating molecule, the concentration of the element may be high, but its activity low, because the ion is not available for reactions with other minerals.

Most elements in regolith water are present in very small concentrations, and for general discussion it is acceptable to treat activity and concentration as equivalent. Though concentration levels in regolith water vary widely, much of this water reaches streams and rivers, from whose analyses we can gauge an approximate composition for regolith water (Table 6.1).

Solubility and hydrolysis

Salts such as NaCl (strong electrolytes), dissolve in water and exist in solution as ions (Na^+, Cl^-); the water causes the ionic solid to dissociate without itself being affected. By contrast, many minerals, for example quartz or wollastonite, react with water to form new species in solution:

$$SiO_2 + 2H_2O = H_4SiO_4 \text{ (or } Si(OH)_4)$$

$$CaSiO_3 + 3H_2O = Ca(OH)_2 + H_4SiO_4$$

This process is termed 'hydrolysis', and its essential character is that the water is itself changed. Many elements in solution have more than one hydrolysis species, whose abundance depends on the pH of the solution. The equations for two of silica's other hydrolysis products are:

$$SiO_2 + 2H_2O = H^+ + H_3SiO_4^-$$

$$SiO_2 + 2H_2O = 2H^+ + H_2SiO_4^{2-}$$

and these differ from H_4SiO_4 in being negatively charged. The total amount of an element in solution will be the sum of all its hydrolysis products.

HYDROLYSIS OF METALS

The divalent metal (M) oxides and hydroxides hydrolyse in water, yielding a variety of hydrolysis products, such as M^{2+}, $M(OH)^+$, $M(OH)_2$, $M(OH)_3^-$. For most, the dominant species in solution below pH 9 is M^{2+}. The reaction:

$$M(OH)_2 = M^{2+} + 2(OH)^-$$

involves hydroxyls, and is therefore pH dependent, the concentration of M^{2+} decreasing with increasing

Table 6.1 *Median concentration of selected elements in river water (from Bowen 1979)*

Li	2	μg/l	K	Variable		Zn	15	μg/l
Be	0.3	μg/l	Ca	Variable		Ga	0.1	μg/l
B	15	μg/l	Sc	0.01	μg/l	As	1	μg/l
F	100	μg/l	Ti	5	μg/l	Se	0.2	μg/l
Na	Variable		V	0.5	μg/l	Mo	0.5	μg/l
Mg	4	mg/l	Cr	1	μg/l	Zr	0.8	μg/l
Al	300	μg/l	Mn	8	μg/l	Cd	0.1	μg/l
Si	6	mg/l	Fe	500	μg/l	Au	0.002	μg/l
P	20	μg/l	Co	0.2	μg/l	Pb	3	μg/l
S	Variable		Ni	0.5	μg/l	Th	0.03	μg/l

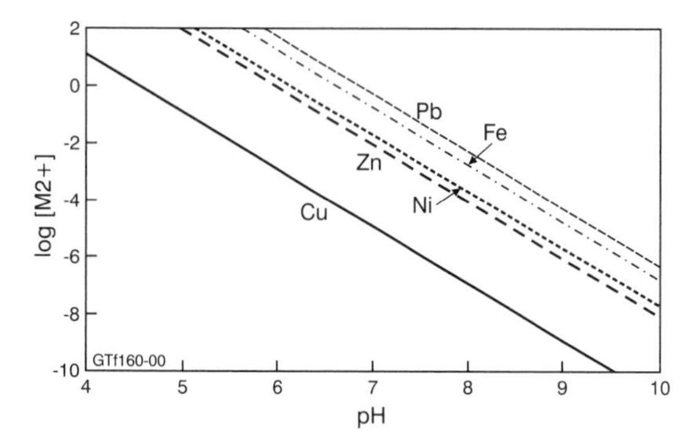

Figure 6.1 *Concentration of M^{2+} at varying pH in equilibrium with corresponding hydroxide (data from Yatsu 1988)*

pH (see Figure 6.1). For example, wollastonite ($CaSiO_3$) hydrolyses according to

$$CaSiO_3 + 3H_2O = Ca(OH)_2 + H_4SiO_4$$

$$CaSiO_3 + 3H_2O = Ca(OH)_2 + H^+ + H_3SiO_4^-$$

as well as other hydrolysis species. Calcium hydroxide itself will dissociate in water:

$$Ca(OH)_2 = Ca^{2+} + 2OH^-$$

Of these, at normal regolith pH, the most abundant hydrolysis products are Ca^{2+} and H_4SiO_4. Because Ca^{2+} is an abundant product, the charge balancing $(OH)^-$ must also be abundant, giving an alkaline solution.

EQUILIBRIUM CONSTANT (K) AND SOLUBILITY PRODUCT (K_{sp})

When a reaction is in equilibrium, the activities of all the chemicals involved become constant, for if any were changing, the reaction would not be at equilibrium. Since all the activities are constant, any mathematical expression involving them only must be a constant. The equilibrium constant (K) for a reaction is defined as the product (multiplication) of the activities of the products (chemicals) divided by the product (multiplication) of the activities of the starting reactants. For example, for the reaction

$$CaSiO_3 + 3H_2O = Ca(OH)_2 + H_4SiO_4$$

$$K = \frac{[Ca(OH)_2][H_4SiO_4]}{[CaSiO_3][H_2O]^3}$$

In the case of solution reactions, there are three kinds of chemicals: solids, water and dissolved species. The activity of any solid and of water is unity (1), so these phases do not affect the equilibrium constant. Since only the soluble species are involved in the equilibrium constant, it is also called the solubility product. For example, for the reaction

$$Ca(OH)_2 = Ca^{2+} + 2(OH)^-$$

$$K = [Ca^{2+}][OH^-]^2/[Ca(OH)_2]$$

Since $[Ca(OH)_2] = 1$,

$$K = [Ca^{2+}][OH^-]^2$$

This only involves the soluble species, and it is the product (multiplication) of their activities in the solution, so a new constant is used, called the *solubility product* (K_{sp}). Some authors equate the solubility product with the equilibrium constant for the reaction, others take the negative log of the value, as will be done here:

$$K_{sp} = -\log_{10}[Ca^{2+}][OH^-]^2,$$
$$\text{which in this case} = -\log_{10}K$$

Hydrolysis equilibria are usually written in terms of the solid, water, the cation and H^+. For example, consider the reaction

$$Ca(OH)_2 + 2H^+ = Ca^{2+} + 2H_2O$$

the equilibrium constant (K) for this reaction can be calculated from knowledge of the thermodynamic properties (Gibbs free energies in this case) of all the reactants, and it is $10^{22.8}$ (Stumm & Morgan 1995). Though we say 'K can be calculated', the data for the calculation were originally obtained by experiment, so in the long run the equilibrium constant is an experimentally based figure:

$$K(=10^{22.8}) = [Ca^{2+}][H_2O]/[Ca(OH)_2][H^+]^2$$

The activities of the solid phase and of water are both unity, hence

$$10^{22.8} = [Ca^{2+}]/[H^+]^2$$

$$\log 10^{22.8} = \log[Ca^{2+}] - 2\log[H^+]$$

$$22.8 = \log[Ca^{2+}] - 2\log[H^+]$$

One of the best known equilibrium constants is that for the dissociation of water:

$$H_2O = H^+ + OH^-$$

$$K = [H^+][OH^-] = 10^{-14}$$

$$\log[H] + \log[OH] = -14$$

Returning to the calculation for $Ca(OH)_2$ hydrolysis, since

$$2\log[H] = -28 - 2\log[OH]$$

then

$$22.8 = \log[Ca^{2+}] + 28 + 2\log[OH]$$

$$-5.2 = \log[Ca^{2+}] + 2\log[OH]$$

$$10^{-5.2} = [Ca^{2+}][OH^-]^2$$

i.e. $-\log([Ca^{2+}][OH^-]^2)$ $(=K_{sp})$ in this case is 5.2. The solubility product for $Ca(OH)_2$ is 5.2.

This relation implies that if the pH of a solution changes (and therefore $[OH^-]$, since $[H^+][OH^-] = 10^{-14}$), then so must the activity of the metal ion in solution. The lines of Figure 6.1 are determined from solubility products. The solubility product can be used to indicate relative solubility of different elements, provided the dominant species in solution is the ion used for the solubility product determination (Stumm and Morgan 1970).

If, for example, acid groundwater containing zinc in solution should emerge from a weathering sulfide orebody and then meet water of higher pH, zinc hydroxide may precipitate, because in that higher pH water it is less soluble.

Provided the reaction is this simple (i.e. nothing else is dissolved in the water), there is an equilibrium concentration of M^{2+} in a solution in contact with the solid hydroxide, and a corresponding equilibrium pH, known as the pH of hydrolysis. The pH of hydrolysis is computed from the three equations:

$$[H^+][OH^-] = 10^{-14} \quad \text{water equilibrium}$$

$$K = [M^{2+}][OH^-]^2 \quad \text{equilibrium constant}$$

$$[OH^-] = 2[M^{2+}] + [H^+] \quad \text{charge balance}$$

Combining these:

$$[OH^-] = 2[M^{2+}] + 10^{-14}/[OH^-]$$

$$\therefore [OH^-] = 2K/[OH^-]^2 + 10^{-14}/[OH^-]$$

$$\therefore [OH^-]^3 = 2K + 10^{-14}[OH^-]$$

If $[OH^-]$ is small (say 10^{-7}), $10^{-14}[OH^-]$ $(=10^{-21})$ is very small compared with 10^{-K} and can be ignored.

$$\therefore [OH^-] = \sqrt[3]{2K}$$

For Zn, $K_{sp} = 16.1$.

$$\therefore 2K = 10^{-15.8}$$

$$[OH^-] = 10^{-5.3}$$

$$\therefore pH = 8.7$$

The equilibrium concentration of the cation in solution can be readily calculated if the pH is known. For example at pH 7, $[OH^-] = 10^{-7}$, hence if K_{sp} for Zn, say, is 16.1, we have (in the presence of $Zn(OH)_2$):

$$16.1 = -\log[Zn^{2+}] - 2\log[OH^-]$$

$$\log[Zn^{2+}] = -16.1 - 2[-7]$$

$$\log[Zn^{2+}] = -2.1$$

$$[Zn^{2+}] = 10^{-2.1} \text{ molar} \quad \text{or} \quad 65.4/126 = 0.52 \text{ g/l}$$

It is important to note that this gives only the concentration of Zn^{2+} in solution, not the total Zn. If other hydrolysis species are present, such as $Zn(OH)^+$ or $Zn(OH)_2$, the total Zn in solution will exceed 0.52 g/l at pH 7 (Figure 6.2). In the case of Al, the dominant species in solution at low pH is Al^{3+}, but at pH 7 $Al(OH)_3$ is more abundant while in alkaline solution $Al(OH)_4^-$ is dominant (see Figure 6.5a). At pH 8, the total Al in solution is 10^{-7} molar, whereas the concentration of Al^{3+} derived from the solubility product is 10^{-16} molar.

Although the derivation of these solubility relations is all in accord with solubility theory, the application to natural systems has to take into account reality. On the one hand, few regolith waters have no other anions that might form species with the divalent metals; $(CO_3)^{2-}$, and $(SO_4)^{2-}$ are particularly common, and Cl^- and F^- may be involved. On the other hand, few of the divalent metal hydroxides are common regolith minerals, although the results are similar if the oxide is in equilibrium with the solution. In the case of Cu, the equilibrium concentration of Cu^{2+} with CuO (tenorite) is $10^{-7.1}$ mol/l, compared with $10^{-6.5}$ mol/l for equilibrium with $Cu(OH)_2$. Thus tenorite is the stable solid in this case, because it establishes equilibrium at a lower Cu^{2+} concentration.

If, as is common, the regolith water is in equilibrium with atmospheric CO_2, the solubility of these divalent cations changes again, establishing equilibrium with carbonate species. Being a gas, the chemical concentration of CO_2 is reported as its partial pressure – its share of the total gas pressure, written pCO_2. Stumm & Morgan (1970) show that at pCO_2 above about 10^{-6}, siderite rather than $Fe(OH)_2$ controls the stability of Fe^{2+}. In equilibrium with air, hydrozincite ($Zn_5(OH)_6(CO_3)_2$) controls the Zn^{2+} concentration in water, while above a pCO_2 of 10^{-3}, smithsonite ($ZnCO_3$) is the stable zinc carbonate.

Atmospheric water in equilibrium with calcite has a pH of 8.4 (see below). At this pH, in equilibrium with atmospheric CO_2, siderite, would dissolve:

$$FeCO_3 + 2H^+ = Fe^{2+} + CO_2 + H_2O$$

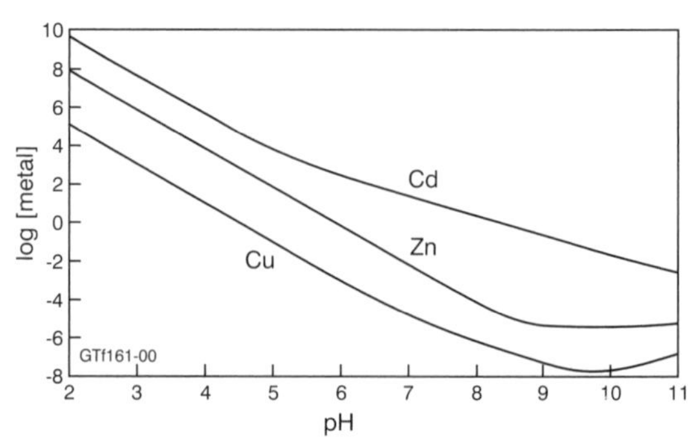

Figure 6.2 *Total concentration in solution for Cu, Zn and Cd (= sum of the hydrolysis products for $Cu(OH)_2$, $Zn(OH)_2$ and $Cd(OH)_2$) with change in solution pH*

Table 6.2 *Divalent metal hydroxides: solubility product, equilibrium pH of hydrolysis and concentration of M^{2+} in solution at that pH (computed from Gibbs free energy data in Yatsu 1988)*

	K_{sp}	pH	mg/l
Cd	14.3	9.3	1.24
Pb	14.3	9.3	2.13
Fe	14.7	9.2	0.43
Ni	15.7	8.9	0.21
Zn	16.1	8.7	0.18
Cu	18.9	7.8	0.02

until the concentration of Fe^{2+} in solution reached 0.12 mg/l, when equilibrium would be established between siderite, CO_2 and Fe^{2+}. This can be compared with the data in Table 6.2, which shows that at its hydrolysis pH of 9.2, $Fe(OH)_2$ supports 0.43 mg/l of Fe^{2+}. Thus siderite is the stable solid for Fe^{2+} in the regolith rather than $Fe(OH)_2$, because if the Fe^{2+} content was as high as 0.43 mg/l, more siderite would precipitate and remove the extra Fe^{2+}, trying to get Fe^{2+} in solution down to 0.12 mg/l. This would make more $Fe(OH)_2$ dissolve, which would make more siderite precipitate until all the $Fe(OH)_2$ had dissolved. The stable mineral is the one that coexists, at equilibrium, with the lowest concentration of its constituent ions in the solution.

SILICA IN SOLUTION

Silica (either quartz or amorphous silica) dissolves in water to form H_4SiO_4 (or $Si(OH)_4$). The reaction

$$SiO_2 + 2H_2O = H_4SiO_4$$

does not involve the gain or loss of hydrogen ions, so this reaction is pH independent. Amorphous silica has a solubility in water of $10^{-2.74}$ M (= 109 mg/l SiO_2); quartz 10^{-4} M; (about 6 mg/l). Note that silica may be reported as Si, SiO_2 or H_4SiO_4. In each case, the molar concentration will be the same, but values reported as mg/l will depend on the species:

Concentration (mol)	Si	SiO_2 (mg/l)	H_4SiO_4
10^{-4}	3	6	10
$10^{-2.74}$	51	109	175

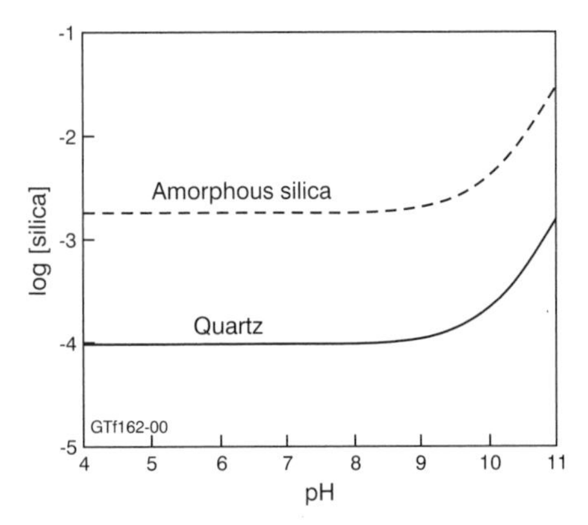

Figure 6.3 *Solubility of silica (data from Lindsay 1979)*

The second and third hydrolysis products, formed by the reactions

$$SiO_2 + 2H_2O = H^+ + H_3SiO_4^-$$

$$SiO_2 + 2H_2O = 2H^+ + H_2SiO_4^{2-}$$

involve hydrogen ions, and so are pH dependent. Increasing the pH (removing H^+) pushes the reactions to the right, increasing the concentration of the negatively charged silica species, thus the solubility of silica increases with pH. It is evident that the answer to the question: 'What is the solubility of silica?' depends both on pH and on the phase that is dissolving, i.e. it depends on the mineral species in equilibrium with the solution. When all the hydrolysis products are included, the solubility of quartz and amorphous silica is as shown in Figure 6.3.

At very high pH, the solubilities of $H_3SiO_4^-$ and $H_2SiO_4^{2-}$ surpass that of H_4SiO_4, and the total dissolved silica increases markedly. For normal regolith water, pH is between 4 and 8, and silica concentration in solution remains fixed by equilibrium with the solid silica species: at 10^{-4} M in the presence of quartz, or at a somewhat higher concentration if, as is commonly the case, poorly crystalline or amorphous silica is present.

Bottomley *et al.* (1990) present data on groundwater chemistry from boreholes into a gabbro-anorthosite pluton in Ontario, Canada. Figure 6.4 plots the log of the silica concentration versus pH for samples taken from shallow depths; it shows that the Si concentration is essentially independent of pH, with a mean of 6.6 mg/l (log[Si] = –3.6) (equivalent to an SiO_2 concentration of 14 mg/l). Neither the host rock nor its alteration products include quartz, consistent with the higher levels of silica in solution than equilibrium with quartz would allow.

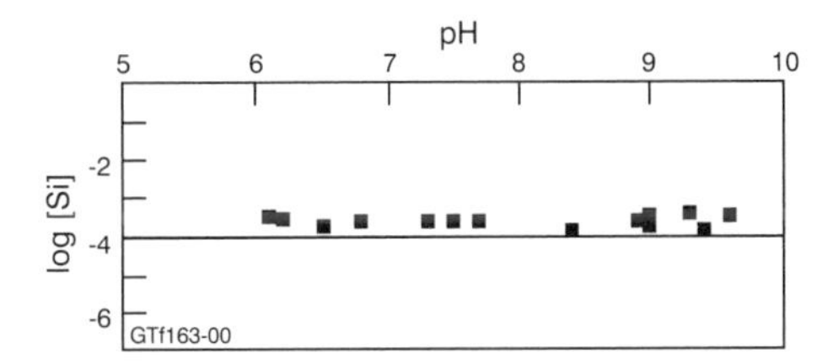

Figure 6.4 *Silicon in groundwater of the East Bull Lake pluton, Ontario, Canada (data from Bottomley* et al. *1990). The line at log[Si] = –4 is the concentration of silica in equilibrium with quartz*

ALUMINIUM IN SOLUTION IN EQUILIBRIUM WITH GIBBSITE

Gibbsite dissolves in water to yield Al^{3+} in solution, and this cation undergoes hydrolysis, the dominant species varying with pH (Figure 6.5a). At pH from about 3 to 5, the dominant species is Al^{3+} (surrounded by six waters). From pH 6–7, $Al(OH)^{2+}$ and $Al(OH)_3$ are the most abundant hydrolysis products, while above pH 7, $Al(OH)_4^-$ dominates. The sum of all the hydrolysis products yields the total Al activity.

The total Al in solution in the range of most weathering solutions (pH 6–8) is very small. Only in very acid and very alkaline waters does the Al content begin to become appreciable. Even at pH 4, where the solubility has gone right off scale in Figure 6.5b, the concentration of Al is only about 3 mg/l. But if the conditions become extreme, such as in acid mine drainage or in highly saline alkaline lakes, the aluminosilicate minerals dissolve quite readily and rock weathering is greatly accelerated. Similarly, where

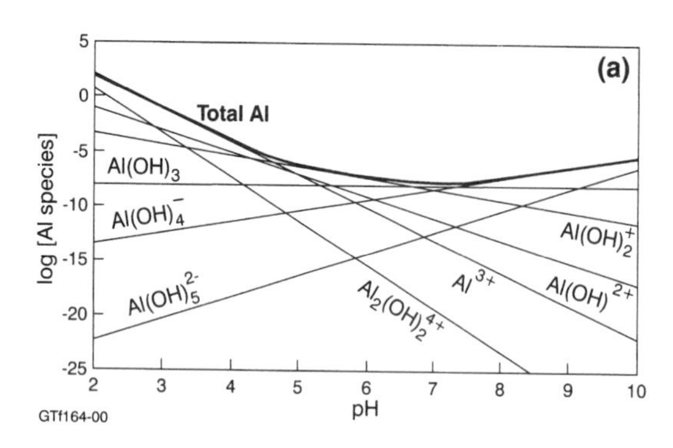

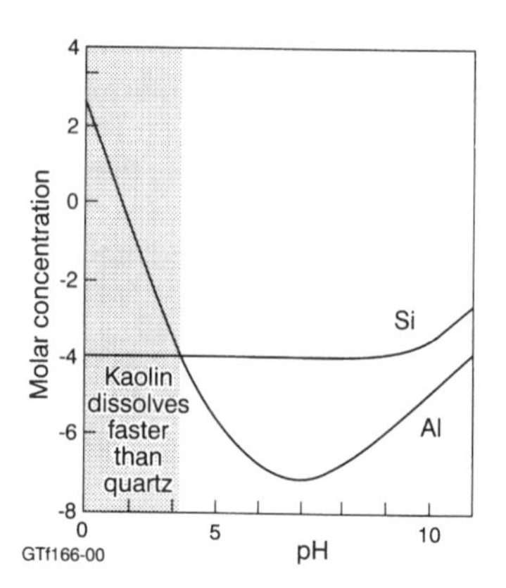

Figure 6.6 *Silicon and Al solubility*

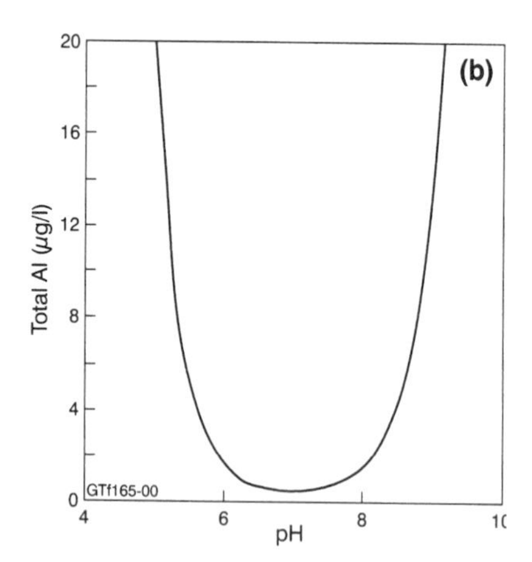

Figure 6.5 *(a) Aluminium hydrolysis species and total Al (thick line) expressed as log[activity]. (b) Total of all Al hydrolysis phases in equilibrium with gibbsite expressed as parts per billion (or micrograms per litre), assuming activity equals concentration*

organic acids are high, such as in the first rains of the monsoon season where a pH as low as 4.5 can result from washing out acid vapours accumulated over the dry season, there will be an initial rise in silicate weathering.

Throughout arid Australia, silica cemented sands and gravels known as silcretes occur in the regolith. Many of these appear to have replaced clay-rich sediments or granite saprolite, and so require a chemical environment where Al dissolves while quartz does not. Equilibrium diagrams (Figure 6.6) suggest that at pH<4, kaolinite would dissolve faster than quartz, allowing preferential leaching of clays and retention of quartz. Cementing the remnant quartz with silica would then require a different mechanism to precipitate silica, possibly simple drying out of silica-rich solutions.

IRON HYDROLYSIS

Ferric iron is soluble in acid solution. At pH 1, the most abundant species is the Fe^{3+} ion, which is surrounded by six water molecules, the oxygens of the water forming an octahedron about the iron ion. If such an acid solution is diluted by fresh water, the pH is increased, and pairs of $Fe(H_2O)_6^{3+}$ polymers combine, releasing two hydrogen ions:

$$2Fe(H_2O)_6^{3+} = Fe_2(OH)_2(H_2O)_8^{4+} + 2H^+ + 2H_2O$$

By pH 3, $Fe_2(OH)_2(H_2O)_8^{4+}$ has become the dominant species in solution. With further dilution, polymerization of the ferric iron hydrates eventually leads to formation of large polymers of $Fe(OH)_3$, the dominant species at pH 7. These polymers are typically 2.5–5 nm in diameter, and are the common ferric hydroxide precipitate, mineralogically known as ferrihydrite (Chapter 3).

The hydrolysis sequence of ferric iron with increasing pH, and the total of the Fe species is summarized in Figure 6.7. In very acid waters, such as drains from sulfide mine dumps or in acid sulfate soils, upwards of some hundreds of milligrammes per litre of Fe may be dissolved. As the pH rises the total Fe in solution falls, reaching a minimum of 0.004 mg/l in alkaline groundwater (pH 8). Some lime-rich soils may have so little Fe in solution that plants develop an Fe deficiency.

DISSOLUTION OF SILICATES

Another level of complexity is added when silicate dissolution involves the possibility of several solid

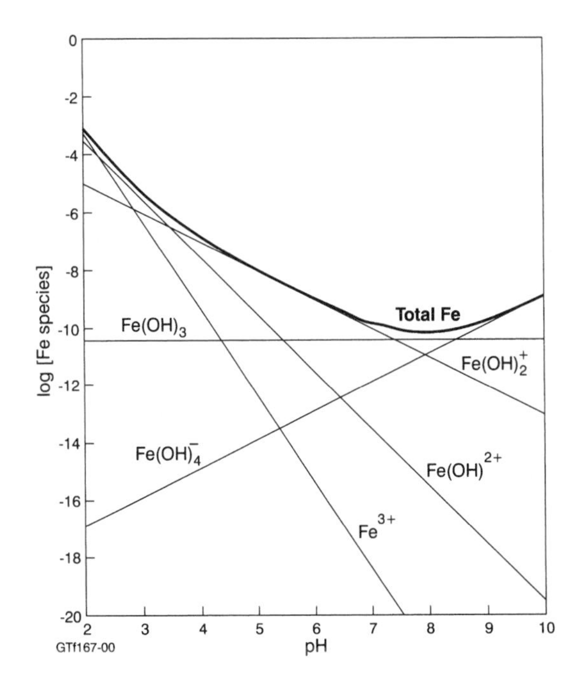

Figure 6.7 *Variation in the solubility of ferric iron hydrolysis species with pH, and the total Fe in solution*

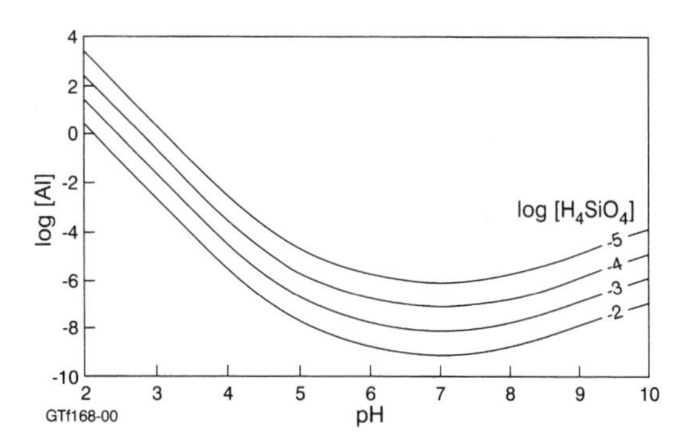

Figure 6.8 *Dissolution curves for kaolinite: logarithm of the total Al in solution versus pH for several different levels of log[H_4SiO_4]*

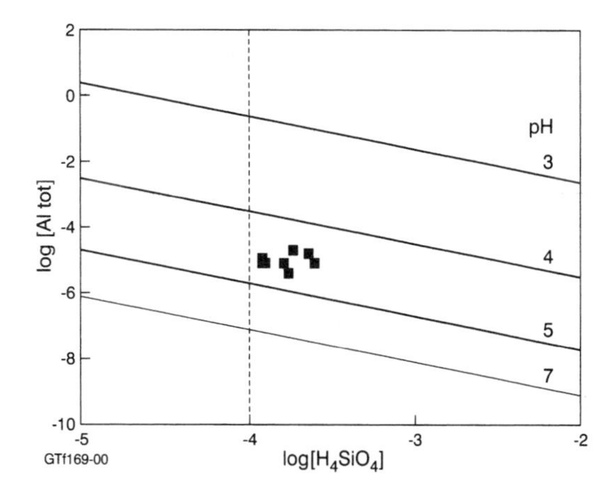

Figure 6.9 *Stability relations for kaolinite at varying [H_4SiO_4], for different pH values. The plotted points are Si/Al measurements for groundwater at Weipa, Queensland, Australia (data courtesy Comalco Ltd). Vertical line is the silica/quartz equilibrium value*

phases. For example, kaolinite, $Al_2Si_2O_5(OH)_4$, would release silica and alumina upon dissolution, therefore the possibility has to be considered that either or both of quartz and gibbsite might precipitate. This mineral also brings to three the number of solution variables that have to be included in the equilibrium diagram ([Al], [Si], pH), making two-dimensional representation difficult. Commonly, one variable is held constant, or contours are used to show the third variable (Figures 6.8 and 6.9).

Comalco Aluminium operates a major bauxite and kaolinite mine at Weipa, north Queensland in Australia. Groundwaters from the quartz–kaolin levels of the deeply weathered Cretaceous sediments have a pH of around 4.5, and yield [Al] and [H_4SiO_4] values consistent with equilibrium between these minerals and solution (see data points in Figure 6.9).

EQUILIBRIUM ASSEMBLAGES

The principles of aqueous geochemistry are applied to regolith mineral–water reactions on the assumption that equilibrium is maintained. This is probably a fiction, certainly so over the short term (days). Over the longer term (the timescale of the residence of groundwater in a weathering profile, which may span years), local equilibrium is a reasonable assumption, and the two data sets depicted in Figures 6.9 and 6.12 support this assumption. An equilibrium assemblage is a group of minerals and their bathing solution, all of which interreact and maintain a constant solution composition, with neither dissolution nor crystallization of the minerals. Such assemblages are commonly depicted graphically, with solution parameters providing the graph axes, and regions of mineral stability being defined by equation lines describing the way the assemblage reacts to change in conditions.

Continuing with the example of quartz and kaolinite, since at low pH (<6) the main Al species

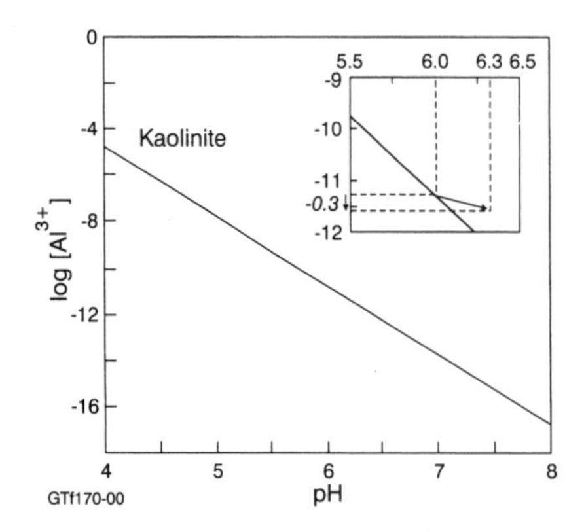

Figure 6.10 *Equilibrium concentration of Al^{3+} at varying pH in the presence of quartz and kaolinite ($\log[H_4SiO_4] = -4$) (data from Lindsay 1979). Inset shows the effect of pH change (see text)*

in solution is Al^{3+}, the principles of silicate dissolution can be explored by, for the moment, ignoring the other Al species. The relevant equation is

$$\underset{\text{kaolinite}}{Al_2Si_2O_5(OH)_4} + 6H^+ = 2Al^{3+} + 2H_4SiO_4 + H_2O$$

Since hydrogen ions are involved in the reaction, the solubility relations are pH dependent; a graph of the Al^{3+} concentration and pH in equilibrium with kaolinite and quartz is shown in Figure 6.10, where the sloping line shows the equilibrium concentrations of the solution and minerals. At solution compositions above the line, kaolinite will precipitate. In doing so, Al would be taken from the water (as well as silica, which would lead to some quartz dissolving to restore silica saturation), and the solution composition would move downwards. Similarly, if the solution composition is below the line, kaolinite will dissolve and release Al to the solution with the precipitation of quartz. If hydrogen ions are added to this assemblage at equilibrium (water becomes more acid), the equilibrium is upset, and the effect will be to create more Al^{3+} in solution and precipitate silicic acid as quartz at the expense of kaolinite. The result will be the same mineral assemblage (although in changed proportions), still at equilibrium, but at a lower pH and in a solution containing a higher activity of Al^{3+}. In the same way, if the solution is suddenly diluted by the introduction of an equal volume of fresh water, the pH ($= -\log_{10}[H^+]$) will increase by 0.3 (pH and activity are on a log scale; $\log_{10}(2) = 0.3$), and the Al^{3+} concentration will change by -0.3. The inset to Figure 6.10 shows the region of the graph around pH 6. These changes move the solution composition away from the equilibrium line (in the direction of the arrow). To restore equilibrium, the Al^{3+} activity must decrease and the H^+ activity increase (pH decrease), i.e. the reaction is driven to the left and some kaolinite is precipitated.

EFFECT OF THE SOLID PHASES ON SOLUBILITY

The H_4SiO_4 content in solution in the presence of quartz is 10^{-4}M, or 10 mg/l ($\log[H_4SiO_4] = -4$), and is independent of pH in the range pH 4–8. Figure 6.11 shows that at pH 6, if the activity of H_4SiO_4 is 10^{-4} M ($\log(H_4SiO_4) = -4$) set by the presence of quartz, then $\log[Al^{3+}] = -11.3$, i.e. the Al^{3+} concentration in solution is $10^{-11.3}$ M, or five parts in 10^{12}. At pH 5, $\log(Al^{3+}) = -8.3$, or 1000 times higher Al concentration than at pH 6.

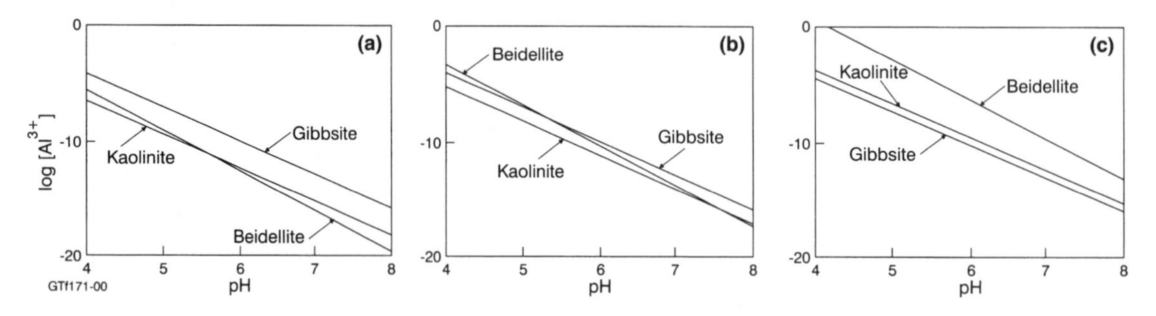

Figure 6.11 *Stability relations of gibbsite, kaolinite and beidellite: (a) In the presence of amorphous silica; (b) in the presence of quartz; (c) in the absence of a solid silica phase (data from Lindsay 1979)*

Figure 6.11a, b, c shows the Al^{3+} activity (approximately = concentration) in solution at varying pH in the presence of amorphous silica (Figure 11a), quartz (Figure 11b) and at very low silica activity (Figure 11c). Figure 6.11a reveals that in the presence of amorphous silica, at pH<5.5, kaolin is the stable mineral (maintains the lowest Al^{3+} activity). If the Al content of the groundwater were to increase, a reaction between silica and Al would occur to create more kaolinite and reduce the Al^{3+} activity. At pH = 5.5, silica, kaolinite and a beidellitic smectite coexist in equilibrium. At pH>5.5, beidellite is the stable phase. Changing pH from 5 to 6 would therefore cause kaolinite to alter to beidellite. In the presence of quartz (Figure 6.11b), kaolinite is the stable phase up to pH 7.5, at which value kaolinite and beidellite coexist at equilibrium.

If there is no free silica in the system (e.g. under conditions of high leaching), the silica activity $\log[H_4SiO_4]$ may be lowered until gibbsite becomes the stable phase (Figure 6.11c: $\log[H_4SiO_4] = -6$). Gibbsite and kaolin coexist at equilibrium at a silica activity of about $10^{-5.5}$ M.

Johnson *et al.* (1981) analysed stream waters from a catchment in New Hampshire, USA, draining till composed of fragments of the underlying high-grade metamorphic rocks and granites. The minerals in the soils of this catchment are the minerals of the till plus pedogenic kaolinite, vermiculite and amorphous alumina. Silica content of the stream water is approximately 6 mg/l, thus $\log[H_4SiO_4]$ is −4, indicating equilibrium with quartz. Johnson *et al.* conclude that the water composition could be explained by equilibrium with either natural gibbsite or kaolinite, or indeed several other phases they considered (e.g. synthetic gibbsite, halloysite). In Figure 6.12, the May and October data of Johnson *et al.* are plotted, with stability lines for natural gibbsite and kaolinite, using the equilibrium relationships they considered. On this figure their data are entirely consistent with their observed mineralogy of kaolinite plus an alumina phase.

These examples show that an expression such as 'solubility of aluminium' is only meaningful when the conditions are fully specified. In the presence of kaolinite and gibbsite at pH 7 in the absence of quartz, the Al^{3+} activity is 10^{-13} M, increasing with decreasing pH. In the presence of quartz and kaolinite at pH 7, $Al^{3+} = 10^{-14.3}$ M, or about 20 times less. In the presence of beidellite and amorphous silica, $Al^{3+} = 10^{-16}$ M, i.e. there would be 1000 times less Al in solution than if no silica were present.

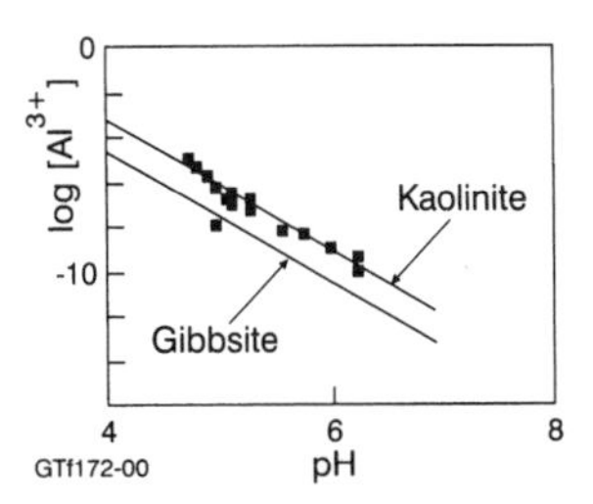

Figure 6.12 *Composition of surface waters draining felsic rocks in New Hampshire from Johnson* et al. *(1981), with the stability relations of natural gibbsite and kaolinite used by those authors (gibbsite line from May* et al. *1979, kaolinite line from Stumm & Morgan 1970)*

CONGRUENT AND INCONGRUENT DISSOLUTION

Some minerals, for example halite, dissolve entirely, or *congruently*. Others break down with the release of some components and the formation of a new, less soluble mineral. Such minerals are said to dissolve *incongruently*. Of the minerals considered in this chapter, quartz dissolves congruently:

$$SiO_2 + 2H_2O = H_4SiO_4$$

and kaolinite dissolves incongruently:

$$\underset{\text{kaolinite}}{Al_2Si_2O_5(OH)_4} + 5H_2O = \underset{\text{gibbsite}}{2Al(OH)_3} + 2H_4SiO_4$$

yielding silica in solution and leaving a residue of gibbsite. Most silicates dissolve incongruently: feldspars to smectite or kaolinite with the release of K, Na or Ca and silica, muscovite loses K and leaves kaolinite and so on. The dissolution reaction stops when the concentration of the soluble components reaches equilibrium with the minerals present.

CHANGES IN MINERALOGY WITH WEATHERING

Figure 6.13 shows a way such equilibrium diagrams can be used to predict the change in mineralogy as weathering proceeds. In this block diagram, the variation in concentration of Al^{3+}, H_4SiO_4 and H^+ (pH) are shown as a surface describing the Al^3 concentration. The clay minerals beidellite, kaolinite

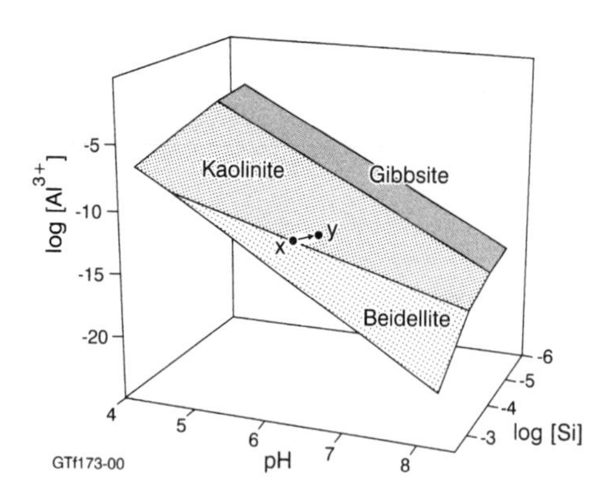

Figure 6.13 *Stability fields of beidellite, kaolinite and gibbsite*

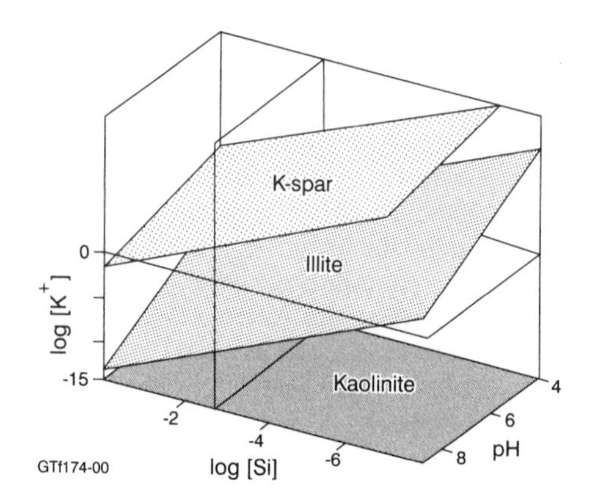

Figure 6.14 *Stability relations between microcline (K-spar), illite and kaolinite. The illite has 0.25 Mg per three octahedral sites; the activity of Mg^{2+} is fixed at 10^{-3} (data from Lindsay 1979). The rectangle at about log[Si] = 2.7 shows the silica concentration in equilibrium with amorphous silica*

and gibbsite are stable in differing solutions, and their stability fields are indicated on the figure. Suppose beidellite and kaolinite are coexisting. The solution composition must lie somewhere along the boundary line between their stability fields, say at point X. The addition of fresh water (i.e. rainfall) will dilute all the solutions, in particular moving the H_4SiO_4 concentration to Y, where kaolinite is the stable phase. Beidellite will dissolve and kaolinite will crystallize until equilibrium is again achieved. Eventually, repeated introduction of fresh water will dissolve all the beidellite, leaving only kaolinite. This process, essentially desilication, is a well-known phenomenon during weathering.

In terms of mineralogy, desilication changes the clay minerals from the 2 : 1 mineral smectite to the 1 : 1 mineral kaolinite and then to the alumina only mineral gibbsite. Weathered rocks dominated by smectite or illite have therefore been described as bisialitic (two Si sheets per Al octahedral sheet), kaolinitic rocks as monosialitic and gibbsitic rocks as allitic (Pedro 1966).

The scenario becomes more complex as other ions in the solution are included in the theory. Consider the weathering of K-feldspar, say microcline. It may dissolve in acid solution according to the reaction:

$$KAlSi_3O_8 + 4H^+ + 4H_2O = K^+ + Al^{3+} + 3H_4SiO_4$$

The alteration of microcline to, say, illite, involves all of the four soluble species H^+, K^+, Al^{3+} and H_4SiO_4, but a diagram involving all four becomes difficult to visualize. Many such weathering reactions, at least in the early stages, proceed at essentially constant Al, that is, the Al is largely locked in the solid phases, and there is so little in solution that it can be ignored.

Figure 6.14 shows the stability fields for microcline (K-spar), illite and kaolinite in terms of pH, silica activity and K^+ activity. While the upper part of this three-dimensional diagram is unrealistic (K concentrations of 10 000 molar and beyond!), the figure shows that at normal regolith pH around 6, and in the presence of amorphous silica (the vertical plane at $[H_4SiO_4] = 10^{-2.7}$), a concentration of K^+ in solution of 105 mol would be needed to maintain microcline as a stable mineral. Illite remains the stable weathering product until the K^+ concentration drops below about 10^{-3}, where kaolinite becomes the stable weathering product. Put another way, illite, kaolinite and amorphous silica maintain an equilibrium with each other in a solution at pH 6, having a silica concentration of $10^{-2.7}$ mol (125 mg/l SiO_2) and a K^+ concentration of 10^{-3}.

It is common to see stability relationships portrayed on a graph of $\log[H_4SiO_4]$ versus $\log([K^+]/[H^+])$, or $[Na^+]/[H^+]$ or $[Ca^{2+}]/[H^+]^2$. This approach is used because in the reactions such as feldspar → smectite → kaolinite [alkalis] and $[H^+]$ must balance, as these are the only ions (other than Si) released or consumed, and their ratio can be used as a single variable. Bottomley *et al.* (1990) use Drever's 1988 diagram to demonstrate that the shallower groundwater compositions in the East Bull Lake gabbroic pluton are in equilibrium with kaolinite and Ca-smectite, and they conclude that the 'waters are a product of incongruent dissolution of primary silicate minerals in which Si and Al are largely conserved by formation of a clay mineral, primarily kaolinite' (Figure 6.15).

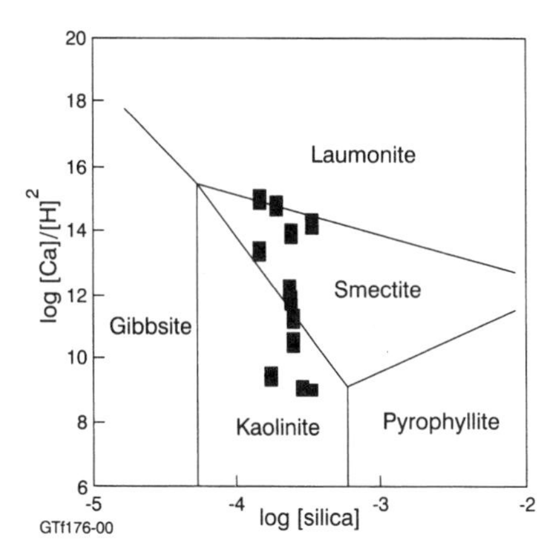

Figure 6.15 *Stability diagram for phases in the system CaO–Al_2O_3–SiO_2–H_2O (Drever 1988), with the shallow groundwater chemistry of Bottomley* et al. *(1990). The waters fall in the stability fields of kaolinite and smectite*

Tardy (1971) took a much broader view, analysing many water samples taken from crystalline massifs from Norway, France and North Africa. These data cover a wide climatic range, and Tardy compares the compositions of more dilute waters (taken from humid environments), with more concentrated waters (from dry climates). Waters from rocks weathering under humid climates plot in the stability field of kaolinite ('wet' field in Figure 6.16), whereas those from drier climates overlap the kaolin/smectite boundary.

Another way to view these observations is to note that the silica concentration in the groundwater from quartz-bearing granitic rocks is close to the thermodynamically established figure for silica solubility ($\log[H_4SiO_4] = -4$). The low silica content of such water favours the low silica clay mineral kaolinite (Si : Al = 1 : 1). For the basic rocks, silica in solution is higher, leading to the precipitation of the more silica-rich clay mineral, smectite (Si : Al = 2 : 1 approximately). This result may seem counter-intuitive; the more silica-rich rocks, granites, yielding silica-poor water compared to basic rocks. Such observations reinforce the fundamental concepts of element solubility in weathering rocks – that the concentration of each element depends on the equilibrium mineral assemblage, not simply on one or two parameters of the solution (Table 6.3).

Oxidation and reduction: *redox*

As the name implies, oxidation during weathering involves a reaction with the oxygen of the atmosphere. Whenever oxidation occurs, there is a balancing reduction event. In the simple reaction:

$$4FeO + O_2 = 2Fe_2O_3$$

the iron is oxidized from ferrous (Fe^{2+}) to ferric (Fe^{3+}). In the process, the molecular oxygen is reduced to O^{2-}. Including all the ionic charges in this equation shows:

$$4Fe^{2+}O^{2-} + O_2^0 = 2Fe^{3+}{}_2O_3^{2-}$$

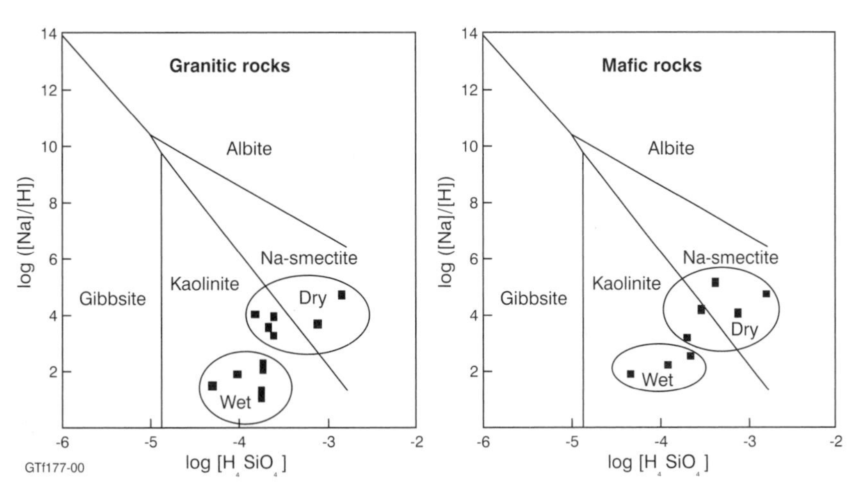

Figure 6.16 *Stability diagram for minerals in the system Na_2O–Al_2O_3–SiO_2–H_2O, with granitic and mafic rock groundwater chemistry, from Tardy (1971). Dilute solutions from humid climates (wet) are in equilibrium with kaolinite; concentrated waters (dry) lie near the kaolinite/smectite boundary*

Table 6.3 *Concentration of ions in solution with various weathering assemblages (mg/l)*

	SiO_2	Al	Fe	MnO	Mg	Ca	Na	K
Felsic igneous								
Kaolinite								
Tardy (1971)	8.4				0.6	1.7	2.0	0.7
Stauffer & Wittchen (1991)	9.7				0.7	2.0	1.1	0.4
Jones *et al.* (1974)	22.2	1.5	3.2					
Smectite								
Tardy (1971)	6.9				2.4	10.6	12.7	1.8
Andesite								
Smectite								
Miller & Drever (1977)	16.9				1.7	5.9	6.3	0.5
Mafic igneous								
Kaolinite								
Tardy (1971)	9.6				1.3	3.0	3.4	0.5
Bottomley *et al.* (1990)	14.0		0.1		3.2	17.0	17.6	3.1
Stauffer & Wittchen (1991)	9.4				2.2	3.4	1.1	0.5
Smectite								
Tardy (1971)	48.3				20.0	29.5	19.7	3.4
Shales and sandstones								
Kaolinite/quartz								
Weipa bauxite average	10	0.3	0.5		<1	<1	5	<1
Limestone								
Smectite								
White *et al.* (1963)	8–25	0–0.2	0–3.5	0–0.3	1–40	8–25	1–61	1–7
World mean river	13.1	0.7		4.1	15.0	6.3	2.3	

In becoming oxidized, the Fe ion has undergone electron loss. In becoming reduced, the oxygen molecule has undergone electron gain to become an oxygen anion. This then becomes another way to describe oxidation and reduction: oxidation is electron loss, reduction is electron gain.

As an example, consider the oxidation of ferrous ions in acid solution:

$4Fe^{2+} - 4e^{-} = 4Fe^{3+}$ this expresses oxidation as electron loss

$O_2 + 4H^{+} + 4e^{-} = 2H_2O$ this expresses reduction as electron gain

$4Fe^{2+} + O_2 + 4H^{+} = 4Fe^{3+} + H_2O$: this expresses the overall reaction

Different ions hold their outer electrons with differing strength, hence there are differences in the electric potential required to remove an electron, i.e. required to oxidize the ion. $Fe^0 \rightarrow Fe^{3+}$ has an oxidation potential of –0.04 V, $Cu^0 \rightarrow Cu^{2+}$ an oxidation potential of 0.34 V. If Cu^{2+} in solution meets Fe metal, the Fe, having the lower oxidation potential, has electrons taken from it by the Cu, oxidizing the Fe^0 to Fe^{3+}, and reducing the Cu^{2+} to Cu^0. So an old car body submerged in Cu-bearing mine water will gradually become converted to a Cu-plated car body and a mass of rust. A reaction between auriferous groundwater and Fe may occur in a weathering profile at a level where reduced (oxygen-poor) groundwater becomes oxidizing by interaction with the atmosphere (the redox front). Here, the reaction

$$Fe^{2+} + Au^{+} = Fe^{3+} + Au^0$$

may produce gold nuggets from soluble Au^{+}.

Iron is the only major element in rocks that undergoes oxidation during weathering. Meteorites contain *metallic* iron (Fe^0), igneous rocks contain Fe in the ferrous state, and oxidized surface minerals contain ferric iron. Important organic elements in the weathering profile that may be oxidized are C and N. These elements are concentrated in the soil and regolith by organisms, and then on their death, or by photosynthesis or respiration, become oxidized. Sulfur, occurring in rocks predominantly as pyrite, is oxidized to sulfate, and Mn^{2+} in many silicates is oxidized to a variety of Mn-oxides such as birnessite and pyrolusite (MnO_2).

In summary, the major, minor and trace elements in crustal rocks which undergo redox reactions in the weathering profile include

Fe^0	$\rightarrow$	Fe^{2+}
Fe^{2+}	$\rightarrow$	Fe^{3+}
C	$\rightarrow$	carbonate $(CO_3)^{2-}$
N	$\rightarrow$	nitrate $(NO_3)^-$
S	$\rightarrow$	sulfate $(SO_4)^{2-}$ ($S^0 \rightarrow S^{6+}$)
Mn^{2+}	$\rightarrow$	Mn^{3+}
As^-	$\rightarrow$	arsenate $(AsO_4)^{3-}$
U^{4+}	$\rightarrow$	U^{6+}
Cu^0	$\rightarrow$	Cu^+
Cu^+	$\rightarrow$	Cu^{2+}
Au^0	$\rightarrow$	Au^+

U^{6+} is soluble in oxidizing conditions, whereas U^{4+} is insoluble in reducing conditions and precipitates as UO_2, uraninite.

The tendency of an environment to be oxidizing or reducing is measured in terms of its electron activity (pe), or electron potential (E_h). The higher the E_h, the lower the electron activity:

$$pe = -\log_{10}(\text{electron concentration}),$$

just as

$$pH = -\log_{10}(H^+ \text{ concentration})$$

Different minerals and phases have different stability fields in terms of pe and pH; however, unlike equilibrium diagrams for solutions of various ions and hydration species, pe/pH diagrams are indicative only of the way the predominant species change with change in redox state. According to Drever (1988), pe is not a quantity that is able to be uniquely defined for a given sample. The best that can be said is that the redox condition is that maintained by an observed pair of reactants, such as $Fe^{2+}/Fe(OH)_3$. The redox condition will remain approximately constant as long as there is a large amount of an oxidizable or reducible species present. Thus oxygenated water has a high pe, which remains high as long as the reacting pair

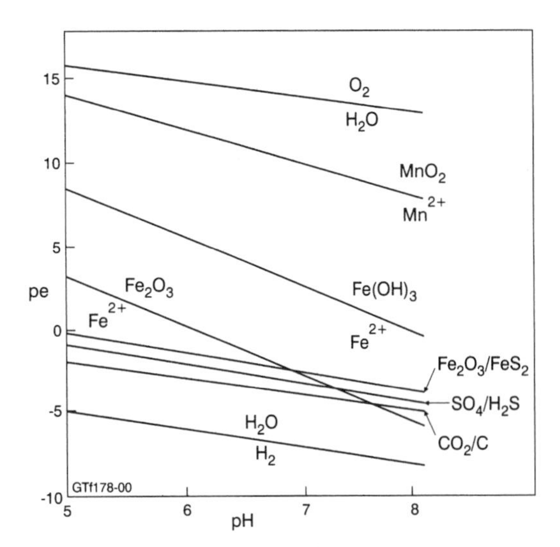

Figure 6.17 *pe versus pH relationships for several redox reactions (concept and data from Drever 1988)*

$$O_2 + 4H^+ + 4e^- = 2H_2O$$

are controlling the redox state. If such water is reacting with ferrous iron in unweathered silicates, a time will come when the oxygen has been consumed by reaction with Fe^{2+}, and the pe will fall to a value controlled by the reaction:

$$Fe(OH)_3 + 3H^+ + e^- = Fe^{2+} + 3H_2O$$

Reaction pairs of this type are termed redox buffers. All of the relatively abundant elements in rocks and groundwater that undergo redox reactions (Fe, Mn, S, C) may provide redox buffers. Figure 6.17 shows the pe/pH regions where particular redox reactions buffer pe. Each boundary shows pe/pH conditions at which the paired species are at equal activity. Thus just below the $Fe^{2+}/Fe(OH)_3$ boundary, Fe^{2+} is the predominant species, but $Fe(OH)_3$ continues to be present. The diagram also shows that the redox conditions at which a ferric mineral predominates over ferrous iron in solution depends on the equilibrium ferric mineral: haematite (Fe_2O_3) dominating over Fe^{2+} at lower pe than would ferrihydrite (represented here as $Fe(OH)_3$).

FERROLYSIS

The reaction

$$Fe^{2+} + 3H_2O = Fe(OH)_3 + 3H^+ + e^-$$

releases hydrogen ions; that is, oxidation of ferrous iron lowers pH. The released H-ion may attach to clay at exchange sites, forming a short-lived hydrogenated clay, but other cations in the groundwater soon replace the hydrogen ion, and its rerelease acidifies the water, leading to the dissolution of clay. Brinkman (1970) referred to this process as *ferrolysis* (van Breeman 1988).

In weathering profiles, the water-table may be the boundary between reducing (below) and oxidizing conditions (above). Ferrous iron diffusing from the reduced rock/saprock becomes oxidized at or about the water-table, with the release of H-ions causing the pH to fall. This immediately changes the solubility and hydrolysis species of all elements, leading to increased chemical weathering as well as fixation of some cations by adsorption to the precipitating iron hydroxide (see Chapter 7).

McArthur *et al.* (1991) describe discharge zones at the margins of playas in the arid Yilgarn region of Western Australia. The water is very acid (pH as low as 3) which the authors ascribe to ferrolysis as ferrous iron is oxidized at the interface between aerated and bacterially reduced groundwater. The low pH causes kaolin dissolution which leads to high Al in solution. The waters are also sulfate-rich, so the aluminium sulfate alunite precipitates in the playa sediments.

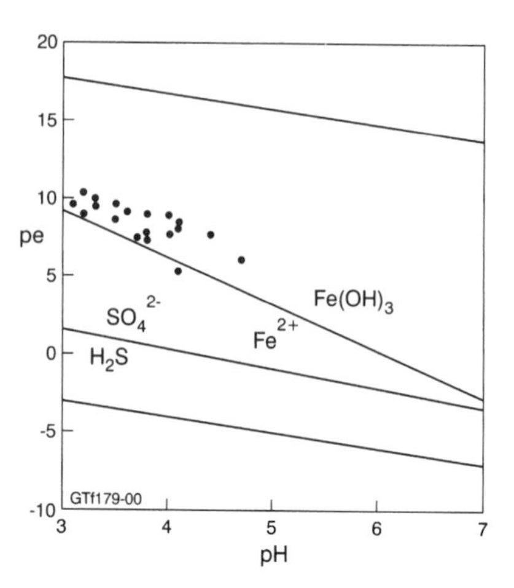

Figure 6.18 *pH/pe properties of Lake Tyrrell spring waters (data from Macumber 1991)*

The weathering solutions

Pure distilled water never reaches rocks. The solutions that effect mineral weathering contain a range of chemicals acquired from the atmosphere, soil and regolith.

Sulfur

The most abundant source of S to the regolith is from pyrite (FeS_2). Oxidation of pyrite is a strongly acidifying reaction:

$$2FeS_2 + 7O_2 + 2H_2O = 2Fe^{2+} + 4H^+ + 4SO_4^{2-}$$

Water draining sulfide-rich rocks or mine dumps may have a pH as low as 2 (e.g. Brukunga pyrite mine near Nairne, in South Australia). Groundwaters draining through Tertiary pyritic sediments into Lake Tyrrell in western Victoria become extremely acid as they reach the oxidizing environment near the surface. Figure 6.18 shows measured pH/pe data from Lake Tyrrell springs (Macumber 1991). Just below the sediment surface in the lake are black pyritic muds indicative of microbial activity and a local reducing environment, whereas at the surface and throughout sands flanking the lake, ferrihydrite and goethite are common precipitates.

CARBON DIOXIDE AND CARBONATE

Carbon dioxide makes up about 0.03% of the atmosphere, with a partial pressure of $P_{CO2} = 10^{-3.5}$. It dissolves in rainwater according to the equation

$$H_2O+CO_2 = H_2CO_3$$

Carbonic acid is a weak acid. It dissociates as

$$H_2CO_3 = H^+ + HCO_3^-$$

yielding a solution of pH 5.6.

Soil and regolith contain organic material which, as it undergoes oxidation, produces CO_2. The partial pressure of CO_2 in the soil is of the order of 10^{-2}, which if this were the only constituent, would yield a more acid groundwater. However, once the rain has entered the soil, reactions with silicates occur which put other cations into solution (and H-ions into clays) increasing the pH.

The formula for the solution of limestone (calcite) is

$$\begin{aligned} CaCO_3 + H_2O + CO_2 &= CaCO_3 + H^+ + HCO_3^- \\ &= Ca^{2+} + 2HCO_3^- \end{aligned}$$

Here, the H-ion from the dissociation of carbonic acid has formed another bicarbonate ion, leading to a rise in pH. Atmospheric water in equilibrium with calcite has a pH of 8.4, soil water at P_{CO2} of 0.02, a pH of 7.3 (Drever 1988; Stumm & Morgan 1970).

The solubility of calcite in water at differing levels of CO_2 partial pressure can be calculated (for detail again see Drever 1988 or Stumm & Morgan 1970). In soils with P_{CO2} of 10^{-2}, the solubility of calcite is about 140 mg/l, whereas rain (P_{CO2} of $10^{-3.5}$) can only dissolve about 40 mg/l. Solution weathering of limestone is thus more effective on those parts of boulders buried in the soil.

ORGANIC ACIDS

Although carbonic acid is commonly referred to as the main agent of chemical weathering, the solubility of the silicate minerals is not dramatically increased by the change in pH from 7 to 5.6 caused by atmospheric CO_2. Of much greater importance is the presence of organic acids produced by biotic processes. Both in life and in death, plant and animal material produces a range of organic acids, prominent among which are oxalic, fulvic and humic acids. Huang & Keller (1971) compared the dissolution of the common clay minerals in distilled water and various 0.01 molar organic acids. They found that, compared to distilled water, in tannic acid, clay minerals released about 10 times as much silica and 100 times as much alumina, typically reaching levels of 10^{-3} mol/l. Barker *et al.* (1997) show results indicating that at pH 6, oxalic acid increases the rate of feldspar dissolution by up to 100 times compared to inorganic solutions, and Furrer & Stumm (1986) showed a similar effect on alumina.

Antweiler & Drever (1983) determined the concentration of dissolved organic C and the major inorganic cations in groundwaters from the weathering of volcanic ash. At pH of the water ranging between 4.3 and 6.5, they found that the concentrations of Al and Fe correlated with the amount of dissolved organic C, and were orders of magnitude greater than equilibrium chemistry at that pH would suggest. Aluminium levels reached 6 mg/l which is 12 times as much Al as equilibrium with aluminium hydroxide at pH 4.3 would indicate, and Fe reached 5 mg/l, 1000 times more than the 25° equilibrium value. They conclude that organic matter plays a major role in the transport of Al and Fe in the soil.

CHELATION

Organic acids are weak acids, and the overall effect they have on regolith water pH is slight. Though plants and other organisms may provide high organic acid concentrations locally, their overall concentration is low (rarely more than a few micromoles per litre). Thus organic acids may be effective in dissolving minerals in contact with biota, but they cannot maintain the enhanced solubility needed to transport relatively insoluble ions like Al^{3+} or Fe^{3+} away from the site simply by virtue of their acidity.

Some organic molecules bind very strongly to cations, because they have a molecular structure capable of forming multiple bonds to the cation. They are termed 'chelating' agents or compounds, because they clasp the cation, and 'chelate' is derived from a Greek word for 'claw'. In water, the (OH) ion forms a single bond, and the attached cation is readily extracted by another charged entity, such as a clay mineral surface or root tip. A chelating agent capable of forming two or more ligands (bonding anions or molecules) to a cation can hold the ion more strongly than can water, isolating it from the surrounding media (for an extended discussion of this topic see e.g. Stone 1997). Oxalate is a chelating agent commonly released by lichens, fungi and plant roots, and it is a bidentate ligand (Figure 6.19) (forms two bonds with a cation). The presence of oxalate can therefore increase the amount of an ion in solution over its 'equilibrium' level in water of that pH; in effect, the water does not 'know' there is any Al^{3+}, say, in the solution when it is chelated by the oxalate. Ferric iron can be chelated by citric acid, and by fulvic or humic acids. Some organic chelating compounds have such an affinity for metal ions that they can extract metals from minerals in the regolith

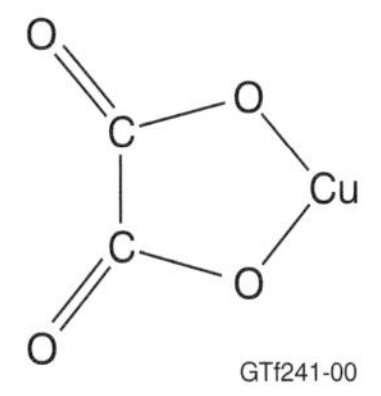

Figure 6.19 *Diagram of oxalate bonding to a cation with two ligands*

(Thornber 1992). Plant roots provide a reducing environment where ferric iron is changed to ferrous, causing the Fe-chelate to dissociate, whereupon the iron is taken into the plant, and the chelating molecule diffuses away to sequester more ferric iron from the oxidized parts of the regolith. Such a process provides a mechanism for Fe movement in an essentially oxidizing environment. Chelation is not essential to the process, but increases the amount of Fe so mobilized (Lindsay 1988).

It is clear from these examples that organic acids play an important role in the dissolution of minerals during weathering. They provide the acid environment needed to dissolve significant quantities of cations, and they accelerate the dissolution process. Extensive coverage of the role of organisms in mineral weathering is given by Yatsu (1988) and Banfield & Nealson (1997).

For further reading see Drever (1988) or Stumm & Morgan (1995)

7
Regolith geochemistry

The geochemical behaviour of the elements in the regolith results from the interplay between groundwater and minerals. In Chapter 6 we discussed this from the point of view of equilibrium chemistry, concentrating on the major minerals of the regolith: quartz, clays and the iron oxyhydroxides. Many elements cannot find accommodation to a significant extent in the structures of these major minerals, for unlike crystallization at high temperature, there is little structural flexibility at 25°C (see Chapter 1). Regolith mineral compositions tend to be simple, rarely accepting more than 1 mol % of 'foreign' ions in their structure. Goethite and haematite are exceptions, incorporating appreciable Al substituting for Fe, and the interlayer exchange site of smectite is very adaptable, but in general, appreciable solid solution does not occur in the minerals that form by regolith processes. Though we will see that a variety of elements may occupy the octahedral sites of smectite, they do so largely in separate species: Al in beidellite, Mg in saponite, Fe^{3+} in nontronite.

In the first part of this chapter we will look at the way the elements partition between solution and solid, with an emphasis on where they come to reside in regolith minerals. We also need to bear in mind the importance of pH and redox in understanding regolith geochemistry, and the way the primary host mineral for the element weathers. Mineral weathering is not discussed in detail until Chapter 8, yet the elements do not enter the realm of regolith geochemistry until they have been weathered from their parent. For example, Mg in clinopyroxene will be released into the regolith early in the weathering of a parent basalt, whereas Mg in pyrope garnet may never be chemically mobilized, and ultimately be eroded still in the mineral grains.

In the second part of the chapter we discuss the regolith geochemistry of 33 elements, selected on the basis of their abundance or their significance to mineral exploration or agriculture, or their value in helping us to understand regolith processes.

In both primary rocks and the regolith, there are three modes of occurrence for an element:

- as an essential element of a mineral, e.g. Na, Ca, Si and Al in plagioclase, or Zr in zircon; and/or
- as an accidental element in a mineral (camouflaged), e.g. Cu in goethite (Figure 7.1), Ba in plagioclase and Hf in zircon; and/or
- adsorbed on to the surface of another mineral, e.g. P on goethite.

Lelong *et al.* (1976) classify at four levels: elements in essential minerals, elements camouflaged in essential minerals, elements in accessory minerals and elements inside fissures in other minerals.

Once released by weathering from their primary source, elements are either lost in solution from the weathering profile, or retained in the minerals making up the regolith. Certainly those that are lost may be lost gradually, or in stages, thus retention includes the process of temporary retention, either in metastable regolith minerals or adsorbed non-permanently. For example, under intense weathering, smectites are transient minerals. They form early in weathering, but later are broken down to kaolin, gibbsite and Fe-oxides and hydroxides. While smectites exist, there is a retention site for many large cations in the smectite interlayer (Ca, K, etc.), but as the smectites in turn weather, their interlayer elements may be leached from the profile.

Mineral hosts for elements

SMECTITES

The 2 : 1 layer of smectites can include a variety of cations having comparable size with the major octahedrally coordinating cations Mg^{2+}, Al^{3+} and Fe^{3+}. Paquet *et al.* (1987) detail the involvement of

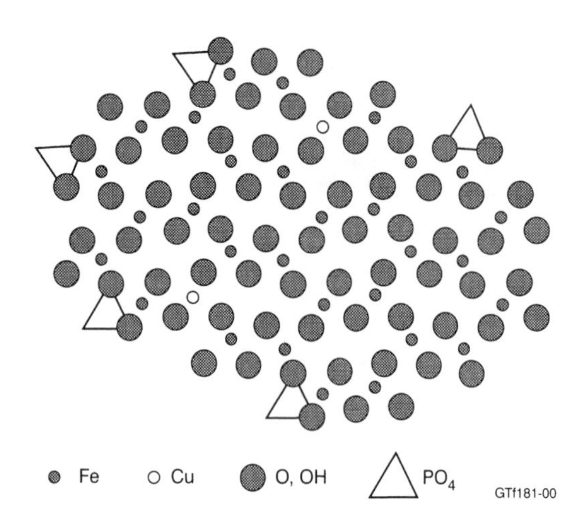

Figure 7.1 *Sites for elements in minerals, illustrated for the mineral goethite (FeO(OH)). Fe, O and OH are essential elements in the mineral, Cu is accidental (camouflaged) in the structure, whereas PO_4 is adsorbed on the surface*

smectites as early hosts for selected trace elements during the development of a weathering profile. Trace elements accommodated in the octahedral sheet include Li, V, Mn, Co, Ni, Cu and Zn. The tetrahedral sheet is not known to include any elements other than Si and Al. The smectite interlayer is an important site for larger ions; besides the major elements Na, Mg, Al, K and Ca, ions that may be exchanged in the interlayer include Pb and Cu, which can be adsorbed in relatively high amounts (Helios-Rybicka *et al.* 1995; Sikora & Budek 1994). It appears that Cd, Ni and Zn are adsorbed at non-exchange sites, presumably in the octahedral sheet; Ni-rich smectite is called stevensite, Zn-rich sauconite.

VERMICULITE

Vermiculite has a structure similar to that of the smectite group, and can be expected to include the same octahedral cations. The interlayer region normally hosts octahedral cations such as Mg, Fe or Al and may admit any similar-sized cation. According to Marker & De Oliveira (1990) (see also the section on REE), vermiculite scavenges Ba and REE.

KAOLINITE

Kaolinite has little structural flexibility; the tetrahedral sheet is pure Si, the octahedral sheet is dominated by Al and there is no interlayer region. Thus kaolin contains low amounts of trace elements in its structure, and having a low cation exchange capacity (40 meq/kg), low amounts of adsorbed elements also. Minor and trace elements reported in the octahedral sheet of kaolinite are Fe (up to 3% Fe_2O_3, Ma 1996) 0.9% Cr_2O_3 (Singh & Gilkes 1991) and 0.2% Cu (Mosser & Zeegers 1988).

Many trace elements have been found associated with kaolin, but it is rarely certain whether these are incorporated, adsorbed, or in other clay-sized minerals. Yeliseyeva & Omel'yanenko (1988) described masses of kaolin with high U, but showed that the U concentrated in regions of high anatase and iron hydroxides.

HALLOYSITE

Halloysite has the same chemistry as kaolinite but with additional water. Its cation exchange capacity is moderate (300 meq/kg), thus it may adsorb cations from solution more extensively than kaolinite. Halloysite can incorporate up to 12% Fe_2O_3 (Ma 1996), but little else is known about its trace and minor element composition.

SERPENTINE

The serpentine minerals are common products of the hydrothermal alteration of ultramafic rocks, but are less common as regolith minerals. Nepouite is Ni-rich serpentine, found in the weathering profiles of ultramafics.

ALLOPHANE

Allophane-rich soils have been shown in experiment to strongly adsorb phosphate, and extraction of the order of 100 mmol/kg phosphorus from allophanic soils has been reported (Parfitt 1990). Their small particle size (5 nm) suggests high adsorptive capacity, but because allophane is always intimately mixed with other clays, isolating its trace-element content has not been possible.

QUARTZ (CHERT, OPAL, ETC.)

Most quartz in the regolith is detrital thus, although it may contain trace elements, these will relate to source rocks rather than the weathering environment. Quartz is known to accommodate Al, P, Ti, Fe, Ge and Ga substituting for Si. The charge deficiency generated by an R^{3+} substitution for Si^{4+} may be balanced by H, Li, Na, K, Cu, Ag and possibly many other ions held in tunnels in the silica framework (Götze & Plötze 1997). Quartz crystals commonly contain inclusions, both mineral and fluid, and these account for most high levels of impurity elements. Levels range up to about 600 g/t for Li and Al, while TiO_2 and Fe_2O_3 levels reach 0.005% and 0.3% respectively (Frondel 1962). Quartz deposited in the regolith is commonly in the form of very finely crystalline silica, either as quartz in silcretes, or as opal and porcellanite.

IRON OXIDES AND OXYHYDROXIDES

Iron oxides and oxyhydroxides are the dominant products of the weathering of Fe-bearing minerals under most regolith conditions. The early precipitate, ferrihydrite, is a highly reactive mineral because of its large surface area (350 m^2/g), and is capable of adsorbing a wide range of trace elements. Ferrihydrite slowly converts to goethite or haematite, and many of its adsorbed ions remain trapped in the better crystalline mineral, thus Fe-oxides and hydroxides become important hosts for many elements in the regolith. Goethite can also precipitate directly from solution, and it may incorporate foreign metals in its structure or adsorb them as it precipitates. The importance of Fe in the weathering process is emphasized by Thornber & Wildman (1984), who studied the co-precipitation of several transition metals with Fe. They write:

> The solubilities of Cu, Ni, Zn, Co and Pb are all controlled to some degree by oxidising Fe(II) if it is present. The amount of Fe relative to base metal is important in determining the composition of the precipitate and the oxidising Fe also controls the pH of the environment where the precipitate is forming.

There are many natural examples of adsorption and co-precipitation of heavy metals on iron oxides and oxyhydroxides, summarized and exemplified in several chapters in Butt & Zeegers (1992a).

Both natural and synthetic goethite and haematite have been investigated for their ability to sequester elements other than Fe.

a) Within the mineral structure

Aluminium

Aluminium occurs in goethite substituting for Fe, up to 32 mol%, and up to about 15 mol% in haematite (Fitzpatrick & Schwertmann 1982). In the regolith, goethite formed in hydromorphic environments, such as mottles, concretions and ferricretes tend to have lower Al substitution (0–15 mol%), whereas freely drained regolith such as saprolites and bauxites have Al substitution ranging from 15 to 32 mol%. Fitzpatrick & Schwertmann explain the difference as resulting from lower pH and therefore higher Al activity in the more freely drained regolith. Substitution of Al is readily estimated from the X-ray diffraction pattern of goethite and haematite; substitution of Al reduces the unit cell dimension (Schulze 1984, Stanjek & Schwertmann 1992) as well as reducing the mean crystallite dimension.

Transition metals

Schwertmann *et al.* (1989) synthesized goethite with up to 10 mol% Cr, and Gerth *et al.* (1985) were able to incorporate similar amounts of other transition metals in the goethite structure (Ni : Fe = 0.12, Co : Fe = 0.12, Cu : Fe = 0.05, Zn : Fe = 0.11). In their study of the dissolution kinetics of naturally occurring goethites, Singh & Gilkes (1992) concluded that much of the contained Co, Cr, Cu, Mn, Ni and Zn was present within the crystals (Table 7.1). Haematite similarly can include transition metals in its structure, though to a lesser extent than goethite; for example, Wells & Gilkes (1999) incorporated Ni in haematite to the extent of 0.06 Ni : Fe.

Table 7.1 *Median (rounded) content of major (%) and trace (μg/g) elements in Fe-oxide concentrates from soils on three parent materials (from Singh & Gilkes 1992)*

	Felsic	Mafic	Alluvial
Fe(%)	31	38	29
Al(%)	7	7	5
Cd	7	8	6
Co	60	220	170
Cr	230	420	240
Cu	30	60	70
Mn	90	230	120
Ni	80	90	70
V	520	670	610
Zn	40	60	70

Table 7.2 *Adsorption of metals to goethite (Thornber 1992)*

M^{2+}	Cu	Pb	Zn	Co	Ni	Cd	Mn
pH	4.5	5	6.2	6.5	6.7	7	7.5

b) Adsorbed

Transition metals

Because of its importance in mineral exploration and agriculture, there is a very large literature on trace element adsorption on goethite. The solubilities of the transition metals are pH dependent, as are their adsorption characteristics on precipitated Fe-oxides and oxyhydroxides (Thornber & Wildman 1984). Many studies attest to the scavenging ability of Fe-oxides and oxyhydroxides, particularly studies of gossans.

Thornber (1992) provides a sequence for the pH at which adsorption of metals begins on goethite with increasing pH (Table 7.2). Rose & Bianchi-Mosquera (1993) compared the adsorption of Cu, Pb, Zn, Co, Ni and Ag on goethite and haematite, finding that at pH 7, goethite adsorbed all the available metal, whereas haematite, while adsorbing all the Pb, Zn, Co and Ni, held only 70% of the Cu and Ag.

In the Mt Gibson district of Western Australia, Anand *et al.* (1991) found a strong correlation between Cr, V, Pb, Sb, Mo, Ga and Zr in haematite/maghemite magnetic gravels, whereas Ni and Cu were more associated with clays and goethite.

Copper

Copper, for example, co-precipitates with goethite as tenorite or cuprite above pH 7, whereas at pH 4.5, where its solubility is 10 000 times greater, it adsorbs on precipitating Fe-oxides.

Gold

The association of Au with Fe-oxides and oxyhydroxides is well documented and its recognition has led to some major gold discoveries (e.g. Boddington, Mt Gibson, see Butt & Zeegers 1992a). Greffie *et al.* (1993) found that in experimental systems, Au was incorporated as isolated 5–14 Å masses in the amorphous Fe-matrices (ferrihydrite) with some clusters up to 60 Å, whereas with goethite Au occurred as particles up to 300 nm in size. da Costa (1993) summarizes exploration problems and opportunities of this association in tropical laterites.

Rare earth elements

In acid solutions, REE may be adsorbed on to precipitating ferric oxyhydroxides. Fee *et al.* (1992) observed that REE introduced into Lake Tyrrell in western Victoria, Australia, were scavenged from solution by ferric oxyhydroxides precipitated as the pH of the solution rose. Carvalho *et al.* (1991) found by contrast that lateritic material over gabbros had been depleted of rare earths by acid leaching, but were REE enriched in laterite formed over more basic carbonate.

Phosphorus

Because of its importance in agriculture, the association of P with Fe-oxides and oxyhydroxides is an entire field of research in itself. Phosphate is thought to sorb by ligand exchange with goethite surface (OH) groups (Parfitt 1978). In both natural and synthetic goethites and haematites, Torrent *et al.* (1992) report of the order of 2–3 $\mu mol/m^2$ P adsorbed from 1 mg/l phosphate solutions, however, P adsorption to haematite was slower than to goethite.

Uranium

Botryoidal haematite with up to 600 g/t U was reported by Wernicke & Lippolt (1993); von Gunter *et al.* (1999) found that over 80% of U released during weathering of rocks at the Ranger uranium mine (Northern Territory, Australia) became adsorbed to Fe-oxides and oxyhydroxides.

MANGANESE OXIDES AND OXYHYDROXIDES

These minerals provide structural sites for small ions replacing Mn^{4+} (ionic radius (IR) = 0.54 Å in octahedral coordination), for larger ions in the tunnel structures, or a structural environment for atom clusters (e.g. Co, Manceau *et al.* 1987). Nicholson (1992) divides the terrestrial supergene Mn deposits into 'dhubites' (the Mn-equivalent of Fe-gossans), weathering deposits (surface crusts, laterites), freshwater deposits (lacustrine nodules, groundwater veins and stream-deposited coatings), and bogs and soils. He reports that Mn-oxides of dhubites and weathering deposits incorporate Ag, Ba, Ce, Co, Cr, Cu, La, Mo, Ni, Pb, Sb, Sr, V, Y and Zn. Diagnostic enrichments for these environments are Ba in the weathering environment, and Pb–Zn in dhubites. Pracejus & Bolton (1992), on the basis of correlation between trace element chemistry and normative mineralogy in the supergene Groote Eylandt Mn deposit, found that pyrolusite incorporated REE, whereas cryptomelane and todorokite incorporated Mo, U, V, Zn and Nb, with cryptomelane also including Cu and Co. They emphasize the need to check these results by analyses of individual minerals.

PHOSPHATES

There are a number of relatively insoluble phosphate minerals that host elements in the regolith, among them the plumbogummite group which may host Pb, Ag, Ba, Sr and REE, and rhabdophane (REE $[PO_4].H_2O$), discussed below under yttrium and the REE. At Reaphook Hill in South Australia, Zn is hosted by parahopeite, scholzite and several other phosphates (Hill & Milnes 1974). There are literally hundreds of secondary phosphate minerals, and almost every element found in the crust may occur in the regolith in a phosphate.

SULFATES

Both Pb and Ag can substitute for the monovalent cation in jarosites, leading to the formation of plumbojarosite and argentojarosite respectively, in the oxidized zone of Pb- or Ag-bearing sulfides.

ANATASE

Anatase is a common mineral in the regolith, where it typically occurs as submicron crystals. The minor and trace element chemistry of regolith anatase is largely unknown because its extremely fine size prevents concentration or electron microprobe analysis. Coarsely crystalline anatase occurs in hydrothermal veins, and most knowledge of its chemical substitutions comes from studying such material. Elements known to substitute for Ti in natural anatase include Nb, Fe (Edenharter *et al.* 1980), Sn (Dana 1985), Mg, Zr and Cr and V (Anand & Gilkes 1984) and Ce^{4+} (Angelica & da Costa 1993).

CARBONATES

Besides Ca and Mg, many transition metals are found in carbonates, either as divalent cations substituting for alkaline earths in the rhombohedral carbonate structure, or as other carbonate minerals. In regolith carbonate deposits, carnotite is one common associate ($K_2(UO_2)_2(VO_4)_2.3H_2O$), which may reach ore-level concentrations of U (Carlisle 1983). Gold has been found associated with pedogenic calcrete (Lintern & Butt 1998), but it is not known if the metal occurs *in* the carbonate mineral or *associated* with it, thus its presence has as yet no chemical or geological explanation.

Process of adsorption

Adsorption refers to the attachment of ions to a mineral surface, or in sites within the mineral accessible to weathering solutions. Adsorbed ions are accidental to the structure of the host mineral, and in many instances adsorbed ions are exchangeable. Extensive coverage of the mechanisms and processes of surface adsorption on soil materials is provided by Sposito (1984, 1989), Davis & Kent (1990) and Schindler (1990).

The outer surface ions of the common regolith minerals are oxygens. In acid solutions, the mineral surface is protonated (the surface layer becomes hydroxyls) and positively charged. As the surrounding solution becomes more alkaline, the surface oxygens tend to lose their protons, leaving a negatively charged surface. The change from a positive to a negative surface depends on the mineral, occurring at different pH for different minerals. The pH at which the net surface charge is zero is called the point of zero charge (PZC). At pH below the PZC, the surface is positively charged and is an anion adsorber, at pH above the PZC the surface is a cation adsorber. Surface chemists define several PZCs, depending on the method of determination, including a point of zero net charge (PZNC) and a pristine point of zero charge (PPZC) (see Sposito (1989) and Davis & Kent (1990) for an extended coverage of this topic). Some values of PZC for the major regolith minerals are listed in Table 7.3.

The adsorptive character of inorganic compounds can be determined by direct experiment with natural or synthetic adsorbers under varying conditions of pH and concentration of adsorbing cation in solution. The amount of a cation that can be adsorbed depends not only on its concentration in solution, but also on the pH of the solution. Typically, divalent metal oxides in low concentration (10^{-4} M) are only slightly adsorbed on goethite below about pH 4, and are

Table 7.3 *PZC for common regolith minerals*

	S&F PZC	S&M 81 PZC	D&K PPZC	MT
Allophane				6–7
Birnessite			2.2	
Quartz	1	2	2.9	
Montmorillonite	2	2.5		2.5–6
MnO_2	4	2–4.5		
Kaolinite		4.6		
Anatase			5.8	
Gibbsite		5		9.5
Goethite	6.5	7.8	7.3	7.6–8.1
Haematite		6.7	8.5	6.5–8.6
Ferrihydrite		8.5		6.9
Corundum	8.5	9.1	9.1	

S&F: Salomons & Förstner (1984).
S&M: Stumm & Morgan (1981).
D&K: Davis & Kent (1990).
MT: Thornber M.R. (1992).

completely adsorbed by pH 6. The order found by McKenzie (1980) was that Cu started to adsorb at lower pH than Pb, followed by Zn, Co, Ni and Mn. See also Table 7.2.

Increased adsorption with increasing pH has important implications for element migration within the regolith. Under acid conditions, such as develop during sulfide weathering, most metal cations are soluble, occurring as the simple ion. Pyrite is the commonest sulfide, and although the strongly adsorbing Fe-oxides and hydroxides are the dominant regolith mineral produced by sulfide weathering, as long as the conditions are acidic, there will be relatively little metal-oxide adsorption. However, if the sulfide has associated carbonate gangue, the acidity of the regolith water will be quickly neutralized, and the Fe-oxides and oxyhydroxides will become important scavengers of metals in solution. Examples of such occurrences are given under the discussion of individual elements (see, for example, Cu).

Equally important is the reverse process. Should the groundwater pH decrease, heavy metals which had been isolated from the environment by their adsorption to regolith minerals such as goethite or Mn-oxides may be desorbed. Such a pH change commonly results from anthropogenic disturbance, such as draining swamps, which allows air to oxidize sulfidic muds with the consequent release of sulfuric acid.

Neutral species, such as silica, also adsorb on oxides. Herbillon & Stone (1985) explains the adsorption of silica on goethite as chemisorbtion of silica to the goethite:

$$
\begin{array}{ll|l}
 & OH_2 & + \\
Fe & & \\
 & OH & \\
Fe & & \\
 & OH &
\end{array}
+ H_4SiO_4 \leftrightarrow
\begin{array}{ll|l}
 & O\text{-}Si(OH)_3 & 0 \\
Fe & & \\
 & OH & \\
Fe & & \\
 & OH &
\end{array}
+ H_2O + H^+
$$

At the site of silica adsorption, the surface then behaves like silica gel rather than goethite. As silica gel has a pH_{pzc} of 2 and that of goethite is 8, the pH_{pzc} of the surface is lowered.

During the growth of minerals in the regolith, adsorbed ions may become incorporated into the structure of the mineral, in which case they would cease to be exchangeable. They become either ions in isomorphous substitution for the major cations (e.g. V for Fe^{3+} in goethite), or trapped in interstitial sites and defects. In real examples, it may not be possible to distinguish surface adsorbed ions from internally trapped ions simply by chemical analysis. Sequential dissolution experiments have been used to discriminate surface from interior trace elements.

Cations in solution are invariably surrounded by a coordination shell of water molecules. Adsorption sites for ion complexes on silicates and oxides can be grouped into three. If the surface of the mineral has a stronger attraction for the cation than have the water molecules, the cation adsorbs directly to the mineral surface, and is termed 'inner-sphere' adsorption (Figure 7.2a). Adsorbed cations in inner-sphere sites are strongly held and difficult to displace. Examples are the interlayer site in vermiculite (10 Å), where K^+ is bonded directly to the basal oxygens of two adjacent tetrahedral sheets, divalent metals adsorbed on oxide surfaces and anions such as $[HPO_4]^{2-}$ adsorbed to the surface of goethite. If Sur represents the mineral surface, the adsorption of a cupric ion can be written as

$$\text{Sur} - (OH)_2 + Cu^{2+} \rightarrow \text{Sur} - O_2Cu + 2H^+$$

Inner-sphere adsorbed ions may not be readily exchangeable, and for this reason they are of major importance in geochemical prospecting because they can remain adsorbed through extremes of other ion concentrations.

Where the mineral surface anions attract the hydrated cation less strongly, the water shell is retained, and the adsorption is termed 'outer sphere' (Figure 7.2b). Such adsorbed cations remain bound to the mineral surface as long as there is no competing

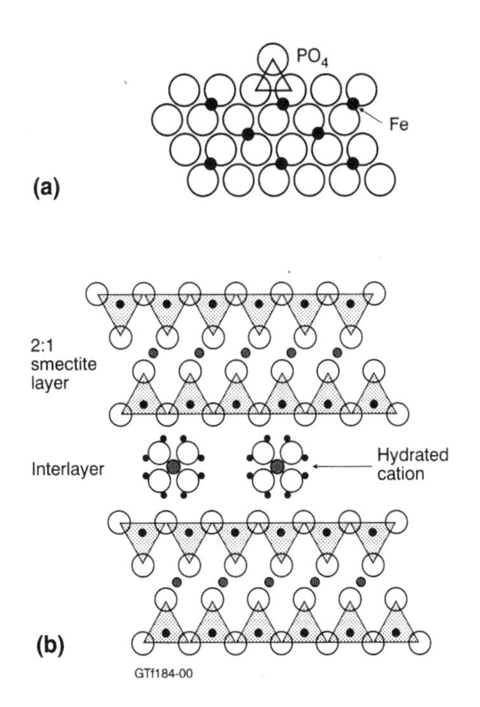

Figure 7.2 *Adsorption sites on mineral surfaces. (a) Inner sphere adsorption of [PO_4] on goethite (after Parfitt 1978); (b) outer sphere adsorption of hydrated cations in the smectite interlayer*

ion of higher concentration. Outer-sphere sites are typical ion exchange sites such as the hydrated Na^+, K^+, Ca^{2+} and Mg^{2+} ions in the smectite interlayer (Davis & Kent 1990).

Outer-sphere complexes bond to similar parts of the mineral surface as do inner-sphere complexes, but the cations, with a surrounding hydration shell of water molecules, are relatively weakly bonded through the electrostatic attraction of the hydration complex for the negative charge excess on the mineral.

Any charged mineral surface attracts a cloud of weakly associated hydrated ions. Where these are not specifically bound to the mineral surface, they are termed diffuse-ion swarms. The more readily exchangeable outer-sphere ions and diffuse-ion swarms are important in soil science and for bio-availability. Diffuse ions are not directly associated with any site in the mineral, but are retained at the surface by weak electrostatic attraction of the charged ion aqueous complex for the surface of the mineral.

Table 7.4 summarizes the mineral hosts for selected elements in crustal rocks and in the regolith and Table 7.5 summarizes the major regolith hosts for some trace elements.

Element distribution in the landscape

The associations between trace elements and regolith mineralogy leads to a predictable distribution of some trace elements in a weathered landscape. In deeply weathered terrains, Fe-cemented duricrusts commonly form erosion-resistant high points, with clay-rich regolith beneath the duricrusts. Further downslope, eroded fragments from the eminences may mantle weathered bedrock, while in the valleys, weathering-resistant detrital minerals accumulate. In arid climates, calcium carbonate-rich duricrusts (calcretes) are common, and under any climate localized manganocretes can occur. Each of these regimes has a characteristic mineralogy, and consequently, a characteristic trace element signature.

Tardy (1969) and Lelong *et al.* (1976) established the selective concentration of certain trace elements with different mineral weathering products. Roquin *et al.* (1990) present the results of factor analysis of laterite mineralogy and the concentrations of trace elements significant in mineral exploration. Butt & Smith (1992) provided a series of case histories concerning element dispersion in the regolith. Table 7.6 summarizes these authors' findings.

PATHFINDER ELEMENTS

A pathfinder element is one that is associated with an exploration target commodity, and which is more readily detected than the target. In this context 'ready detection' may mean the pathfinder is simply more abundant, or it may be cheaper to analyse for, or it may be more uniformly dispersed and so more certain to be found by a structured search plan. For example, Au is commonly hosted by sulfide-rich rocks which also contain appreciable arsenopyrite. On weathering, the As is dispersed more widely than Au, is much more abundant than Au in the first place, and is cheaper to analyse for than Au. All these properties make it a good pathfinder.

These factors alone are insufficient. The pathfinder element must also remain in the regolith in the general vicinity of the ore-body, so it can provide a broad exploration target, thus a pathfinder must have a particular kind of chemistry. For example, the oxidative weathering of sulfides, such as pyrite and arsenopyrite, creates acidity in the groundwater, as we saw in Chapter 6. At low pH, ferric iron and most metals are more soluble than they are at neutral pH (see Figure 6.7), thus the acid groundwater near the

Table 7.4a *Distribution of the elements according to their role in common rocks*

	In major mineral as			In accessory mineral as	
	Major element	Trace element	S	Major element	Trace element
Li		Micas		Spodumene	Tourmaline
Be		Plagioclase, amphibole, muscovite		Beryl	
B		Muscovite, plagioclase		Tourmaline	
C	Calcite				
F		Mica, amphibole		Apatite, fluorite, topaz	Sphene
Na	Plagioclase, zeolite				
Mg	Olivine, garnet, pyroxene, amphibole, biotite, chlorite				
Al	Feldspar, mica, zeolites, chlorite				
Si	All silicates				
P				Apatite, monazite	
S			X		
Cl		Biotite, amphibole		Scapolite, sodalite	Apatite
K	Mica, K-feldspar				
Ca	Plagioclase, amphibole, pyroxene				
Sc		Pyroxene, biotite amphibole			Cassiterite, rutile, ilmenite, titanite, zircon
Ti		Garnet, pyroxene, amphibole, biotite		Ilmenite, rutile, titanite, perovskite	
V		Pyroxene, biotite			Magnetite, titanite, ilmenite
Cr		Garnet, pyroxene, amphibole, biotite		Chromite	
Mn		Fe-silicates			Ilmenite, magnetite
Fe	Olivine, pyroxene, amphibole, biotite, chlorite		X		
Co		Olivine, chlorite, biotite			Spinel
Ni		Olivine, pyroxene	X		
Cu			X		
Zn		Garnet, olivine, pyroxene, amphibole, biotite	X		Magnetite
Ga		With Al			
Ge		With Si			
As			X		
Se			X		
Sr		Feldspar			
Y				Monazite, allanite	
Zr				Zircon, baddeleyite	
Mo			X		
Ag			X		
Cd			X		
Sn				Cassiterite	
Sb			X		
Cs		Feldspar			
Ba		Feldspar			
REE				Allanite, monazite	Apatite, titanite, epidote
Hf					Zircon
W				Cassiterite	
Au				Au	
Pb		Feldspar	X		
Bi			X		
Th					
U					

Table 7.4b *Distribution of the elements according to their role in the regolith*

	In major mineral as		In minor mineral as	
	Major element	Trace element	Major element	Trace element
Li			Lithiophorite	
B	Borax			
C	Calcite			
F		Mica	Fluorite	
Na	Lost, halite			
Mg	Smectite, palygorskite			
Al	Kaolin, smectite, illite, halloysite, gibbsite, amorphous		Alunite	
Si	All silicates			
P		Goethite	Apatite, monazite, crandallite group	
S	Gypsum		Alunite, jarosite	
Cl	Halite			
K	Illite, jarosite		Alunite	
Ca	Calcite, gypsum			
Sc		Fe oxides	?Anatase	
Ti			Anatase, ilmenite, rutile	
V		Smectite	Carnotite, montroseite	Mn-oxides
Cr		Smectite	Chromite	Mn-oxides
Mn	Pyrolusite, birnessite			
Fe	Goethite, haematite, ferrihydrite			
Co				Mn-oxides
Ni		Smectite, Fe-oxides	Pimelite, garnierite	Mn-oxides
Cu		Smectite, Fe-oxides	Malachite, azurite	Mn-oxides
Zn		Smectite, Fe-oxides	Smithsonite	Mn-oxides
Ga		With Al		
Ge		With Si		
As		Fe-oxides	Various arsenates	
Se		Fe-oxides		
Sr				Mn-oxides
Y			Phosphates	Mn-oxides
Zr			Zircon, baddeleyite	
Mo		Fe-oxides		Mn-oxides
Ag				Mn-oxides
Cd				
Sn			Cassiterite	
Sb				Mn-oxides
Cs		Clays		
Ba			Barite, gorceixite	Mn-oxides
REE			Rhabdophane, florencite, gorceixite, lanthanite	Mn-oxides
W				Cassiterite
Au			Au	
Pb			Cerussite, anglesite	Mn-oxides
Bi				
Th			Thorite?	
U			Carnotite, phosphates	Zircon

sulfides will carry a considerable quantity of dissolved elements characteristic of the ore-body. As that water moves away from the sulfides, it reacts with other minerals, mainly the silicates of the host rocks, and its pH rises. This in turn promotes the precipitation of ferric hydroxide. Earlier in this chapter we noted that transition elements, which may be quite soluble at mildly acid pH, are adsorbed strongly to precipitating Fe-oxyhydroxides. Thus as the pH rises, many elements are fixed in the regolith by adsorption, among them As, Cu, Pb, Sb, W, Mo, Bi and Zn (Thornber 1992). The S itself remains quite soluble as sulfate

Table 7.5 *Summary of major regolith mineral hosts for selected trace elements*

Host mineral	Trace elements																		
Fe-oxides and oxyhydroxides		P	Sc		Cr		Ni	Cu	Zn	As		Se	Mo						
Mn-oxides and oxyhydroxides	Li			V	Cr	Co	Ni	Cu	Zn		Sr		Mo	Y	Ag	Sb	Ba	RE	Pb	
Clays				V	Cr		Ni	Cu	Zn											
Phosphates											Sr			Y			Ba	RE		U

Table 7.6 *Association of trace elements with landscape mineralogy*

	Tardy (1969)	Lelong *et al.* (1976)	Roquin *et al.* (1990)	Butt & Smith (1992)
Resistates			Si, Ti, Zr, Y, Ce	Cr, Ti, Sn, Au, REE, P
Bauxites and kaolinites	Ga, B	Al, Ga, B	K, Sr, Ba, Mg, Ni, Cu	
Ferruginous regolith	V, Cr	Fe, Ti, V, Cr, Mo	P, V, Cr, Nb, Cu, As, Mo	As, Bi, Co, Cu, Ni, Pb, Zn, Mn
Manganiferous regolith	Ba, Co, Pb	Mn, Fe, Co, Pb, Ba, Sn, Ni, Cu	Ba, Co, Ce	
Smectites	Ni, Ba, Pb, Zn	Si, Ti, Al, Fe, Mg, Ni, Zn, Pb, Cu		
Calcareous crusts	Ba, Sr	Ca, Sr, Ba, Zn		V, U

and is generally dispersed far from the ore-body and is not obviously enhanced in the vicinity of the ore.

Other processes besides pH change may promote hydromorphic dispersion, including the influence of organic compounds, variable salinity or change in redox state. Because different pathfinder elements have different regional abundances, reliance on a single pathfinder is unwise – that element may be regionally elevated. It is preferable to use a suite of pathfinders suited to the target commodity. An exploration programme then has a far greater certainty of detecting anomalous levels of As and other pathfinder elements because they have been dispersed some distance from the source without being too greatly diluted. Thus they are easily detected by routine analysis and their dispersion halo provides a broad exploration target surrounding a sulfide ore-body.

ENVIRONMENTS AND ELEMENT MOBILITY

From the earlier discussion of the chemistry of weathering, it is evident that the mobility of an element in regolith processes depends on its local environment, particularly on the eh–pH conditions and on the presence or absence of carbonate, sulfate or phosphate in the water. In order to compare mobility, Lelong *et al.* (1976), following Pédro & Delmas (1970), classified the elements according to their ionic potential (IP) (ratio of charge to ionic radius). In this classification, Group (a) ($IP < 2$), contained the alkali metals and alkaline earths excluding Mg, i.e. elements that form soluble ions or hydrates at all pH values. Group (b) ($2 < IP < 7$) comprised the metals which form insoluble oxides or oxyhydroxides in certain pH ranges, and Group (c) included elements which form other complexes (e.g. elements such as S, N, P, Si). Within the groups, Lelong *et al.* ranked the elements according to their solubility products, while pointing out that in the presence of carbonate, sulfide or sulfate, the formation of insoluble salts would change the mobility order based on hydroxide solubility (Table 7.7). Stumm & Morgan (1970, p. 269) give several examples (see also Chapter 1) of the changes in solubility of an element depending on the actual environment, and they '. . . underline the need to identify all pertinent species and to consider all reactions and equilibria related to the solubility of a mineral'. Garrels & Christ (1965) similarly exemplify this point for U^{6+} which, while soluble in the presence of carbonate, is precipitated as carnotite if V is also present. Knowing the solubility product of various U compounds may be irrelevant to mobility predictions if V is also in the solution.

As an example, unpublished analyses of parent rock and bauxite at Andoom, north Queensland, Australia, allow immobility figures to be estimated for several elements. These are determined as the ratio between each element's concentration in the bauxite to its concentration in the parent rock, and then scaled so that the 'immobile' elements Zr, Ti, Al and Nb have immobilities of 1, with a lower figure indicating greater

Table 7.7 *Data for Figure 7.3. Ionic radii (from Shannon 1976), ionic charge, ionic potential (IP, = charge/radius), and solubility products (from Lelong* et al. *1976, and calculated from tabulations of Gibbs'* free energy*)*

	IR (vi)	Charge	IP	SP hydroxides	SP carbonates
Na^{+}	1.1	1	0.9	−2.9	
Cs^{+}	1.78	1	0.6	−2.8	
K^{+}	1.46	1	0.7	−2.6	
Li^{+}	0.82	1	1.2	−1.4	
Ba^{2+}	1.44	2	1.4	2.3	8.8
Sr^{2+}	1.21	2	1.7	3.5	9.6
Ca^{2+}	1.08	2	1.9	5.3	8.4
Mg^{2+}	0.8	2	2.5	11.0	5.1
Mn^{2+}	0.92	2	2.2	12.7	10.2
Cd^{2+}	1.03	2	1.9	13.7	11.3
Ni^{2+}	0.68	2	2.9	14.7	6.9
Co^{2+}	0.83	2	2.4	14.8	12.8
Ce^{4+}	0.87	4	4.6	37.5	
Fe^{2+}	0.86	2	2.3	15.1	10.5
Pb^{2+}	1.26	2	1.6	15.3	13.1
Zn^{2+}	0.83	2	2.4	17.0	10.8
Y^{3+}	1.02	3	2.9	22.0	
Cu^{2+}	0.81	2	2.5	19.7	9.3
Be^{2+}	0.35	2	5.7	21.4	
La^{3+}	1.16	3	2.6	22.2	
Ce^{3+}	1.14	3	2.6	22.3	
Sc^{3+}	0.87	3	3.4	29.6	
Cr^{3+}	0.70	3	4.3	30.0	
Bi^{3+}	1.10	3	2.7	30.4	
Al^{3+}	0.61	3	4.9	32.5	
V^{3+}	0.72	3	4.2	34.4	
Ga^{3+}	0.70	3	4.3	35.0	
Fe^{3+}	0.73	3	4.1	38.0	
Ti^{4+}	0.69	4	5.8	40.0	
Th^{4+}	0.94	4	3.4	44.7	
Tl^{3+}	0.97	3	3.1	45.0	
U^{4+}	0.97	4	4.1	45.0	
Mn^{4+}	0.62	4	6.5	56.0	
Sn^{4+}	0.77	4	5.2	56.0	
Zr^{4+}	0.72	4	5.6	57.0	

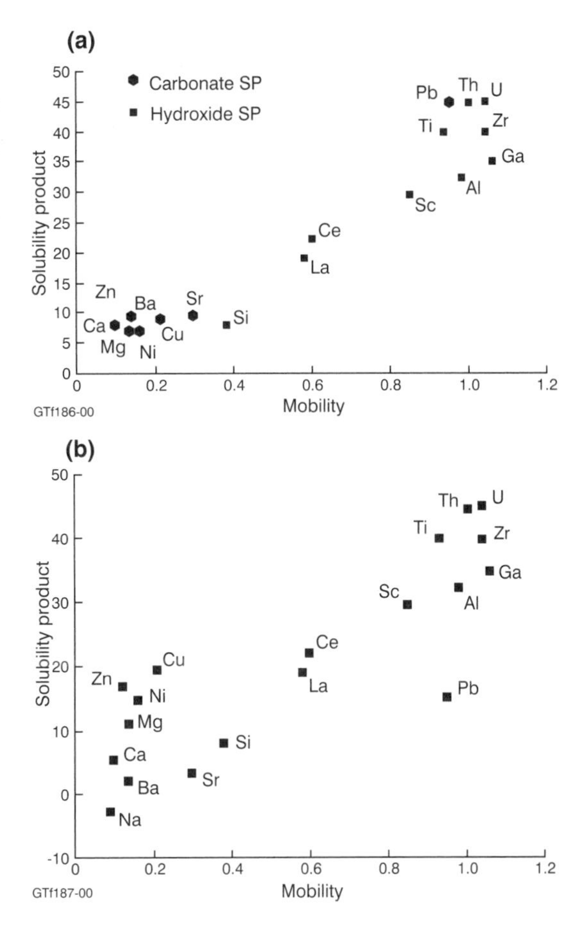

Figure 7.3 *(a) Mobility factors for Andoom bauxite (Weipa, north Queensland) compared to oxyhydroxide solubility products; (b) mobility of ions in Weipa bauxite versus solubility product (SP). Squares: SP for oxyhydroxides, circles: SP for carbonate*

mobility. In this instance, the four immobile elements have close to identical concentration factors throughout the profile, indicative of geochemical coherence, and interpreted as immobility. When the immobility index is plotted against the solubility product for oxides and oxyhydroxides (Figure 7.3a), a clear trend is evident, except for Mg, Ni, Zn and Cu, which are much more mobile than their SP would suggest and Pb, which is much less mobile. Using the SP for the carbonate brings these elements closer to the overall trend, and in each case, the carbonate is the theoretically stable species in the conditions at Andoom (Figure 7.3b). Pb may be present as hydrocerussite as suggested by its carbonate SP, but more probably it is trapped in zircon as the product of radioactive decay of U.

Geochemistry by element

Abundances are rounded and averaged, intended to give a broad indication only. Where possible, mafic igneous data are taken from continental basalts, felsic igneous from granitoids. Though all the data are 'variable', that is, elemental abundances in rocks vary according to the rock type, and in soils abundance depends heavily on local sources, an average figure is reported here. This represents the mean of a range that is mostly from about 25% of the average to about five times the average. Where the data are more widely ranging, 'variable' has been used. Refer to original sources for detail: W: Wedepohl (1969–78), K-P&P: Kabata-Pendias & Pendias (1984), U&B: Ure &

Berrow (1982). Solution species refers to the probable main dissolved species in natural aerobic waters (Stumm & Morgan 1981, 1996).

The elements considered are arranged by atomic number and identified by symbol, as in the alphabetical list below:

Element	Symbol	Atomic number
Aluminium	Al	13
Arsenic	As	33
Beryllium	Be	4
Boron	B	5
Calcium	Ca	20
Chromium	Cr	24
Cobalt	Co	27
Copper	Cu	29
Gallium	Ga	31
Gold	Au	79
Iron	Fe	26
Lead	Pb	82
Lithium	Li	3
Magnesium	Mg	12
Manganese	Mn	25
Molybdenum	Mo	42
Nickel	Ni	28
Phosphorus	P	15
Platinum Group	Pt, etc.	44–46, 75–78
Potassium	K	19
Rare earth elements (see Y)	REE	57–70
Scandium	Sc	21
Selenium	Se	34
Silicon	Si	14
Sodium	Na	11
Sulfur	S	16
Thorium	Th	90
Titanium	Ti	22
Uranium	U	92
Vanadium	V	23
Yttrium	Y	39
Zinc	Zn	30
Zirconium	Zr	40

LITHIUM

Abundance (g/t)	W	K-P&P	U&B
Mafic igneous	16		
Felsic igneous	45		
Shales	70		
Soils		25	31
Rivers	23 (median 1) μg/kg		
Solution species: Li^+			

Lithium is most abundant in felsic igneous rocks, reaching major element proportions in pegmatites where it occurs in the minerals lepidolite (mica), spodumene, petalite and several other Li-silicates.

During weathering, Li^+ is largely released to solution, however, Li has a high affinity for Mn, being an essential element in the regolith mineral lithiophorite ($Al_2Li(OH)_6[Mn^{2+}{}_{0.5}Mn^{4+}{}_{2.5}O_6]$). It is also found in the octahedral sheet of layer silicates, particularly in the smectite hectorite.

BERYLLIUM

Abundance (g/t)	W	K-P&P	U&B
Mafic igneous	1		
Felsic igneous	4–10		
Shales	2		
Rivers	10 μg/l–1 mg/l (Measures & Edmond 1983)		
Soils		2	1.5
Solution species: Be^{2+}, $Be(OH)^+$			

Beryllium in the crust occurs at low concentrations in many silicates, for example of the order of 10 g/t in feldspar and 30 g/t in muscovite. Its concentration is highest in granitic pegmatites where it may occur in beryl ($Be_3Al_2Si_6O_{18}$) or in several other Be-silicates.

Edmunds & Trafford (1993) measured the Be content of groundwater and river water in the UK, finding detectable Be (>0.05 μg/l) in only 12 of 924 samples. These were waters draining granites, and they contained levels of Be ranging from below the limit of detection up to 1 mg/l Be. At acid pH, Be occurs in solution as Be^{2+}, while from pH 5.5 to 8.5 $Be(OH)^+$ is the predominant species. Levels in the South American Amazon and Orinoco rivers range from 10 to 50 μg/l, being higher in the Orinoco, which drains more siliceous rocks than the Amazon.

According to laboratory experiments by You *et al.* (1989), Be has a five times stronger affinity for artificial δ-MnO_2, and for bauxitic soil, than for smectite, kaolinite or illite. They suggest Fe-oxides and oxyhydroxides may be the adsorbers in the bauxitic soil. Anderson *et al.* (1990) could not detect any water-soluble Be (<0.2 μg/l) in soils from south-eastern USA, and found of the order of 10 μg/kg exchangeable Be.

BORON

Abundance (g/t)	W	K-P&P	U&B
Mafic igneous	6		
Felsic igneous	20		
Shales	130		
Rivers	10 μg/kg		
Soils		35	38
Solution species: $B(OH)_3$, $B(OH)_4^-$			

The primary igneous source of B is from silicates, largely muscovite, where it may reach several hundred grams per tonne, and more prominently in the late-stage borosilicate mineral tourmaline. Because it is resistant to weathering, tourmaline survives into sediments and metamorphic rocks.

Below pH 9, the dominant species in solution is $B(OH)_3$, while at higher pH $B(OH)_4^-$ predominates.

The most immediate source of B for industry is the regolith, particularly evaporites of arid regions where B-rich lake waters precipitate a range of B-bearing minerals. Borax is a common precipitate when Na is high, but a complex assemblage and sequence can result from different water chemistries. Smith & Medrano (1996) provide detail of B-mineral preponderance fields at varying solution composition.

There are two stable isotopes of boron: ^{10}B and ^{11}B. In evaporite minerals, ^{11}B is fractionated into trigonally coordinated B in minerals (such as borax) rather than in tetrahedrally coordinated B-minerals (such as colemanite). Palmer & Swihart (1996) show that B-isotopes can be useful indicators of evaporitic environments, ^{11}B being enriched in marine evaporites compared to terrestrial evaporites.

SODIUM

Abundance (%)	W	K-P&P	U&B
Felsic igneous	3.3		
Mafic igneous	2.7		
Shales	1		
Rivers	Depends on distance from ocean		
Soils			1.1
Solution : Na^+			

One of the most abundant elements in the crust, Na occurs mainly in plagioclase feldspars, feldspathoids and zeolites. All these framework silicates are relatively susceptible to weathering (see Chapter 8), releasing Na to the regolith in solution. Except under the most arid conditions where minerals such as halite, natro-jarosite and borax precipitate, there are no stable Na-minerals in the regolith, and few regolith minerals contain more than trace amounts of sodium. Smectites retain Na in exchange sites provided the Ca content of the groundwater is not too high. The bulk of the Na released by weathering moves into streams and so to the ocean. Aerosol Na in the form of sea-salt is responsible for a high proportion of the salinity found in the regolith.

MAGNESIUM

Abundance (%)	W	K-P&P	U&B
Mafic igneous	6		
Felsic igneous	1.2		
Shales	1.5		
Rivers	4 mg/l		
Soils			0.8
Solution species: Mg^{2+}			

In mafic rocks, Mg is a major element, occurring in olivine and pyroxene. In more felsic rocks amphibole and pyroxene are the main hosts for Mg. Sedimentary rocks have Mg largely in chlorite, and low-grade metamorphic rocks hold Mg in amphibole, talc, chlorite, serpentine, garnet, cordierite and biotite.

Magnesium is highly soluble in water, ranking fourth among major elements (after Na, K and Ca) on the basis of its solubility product. The solubility of ideal Mg-silicates and magnesite are shown in Figure 7.4a, assuming $\log[H_4SiO_4]$ is set at −3.1 by 'soil silica' (Lindsay 1979), a value appropriate for the quartz-free environment of most high-Mg rocks, and at $\log[CO_2] = -3.5$. Magnesite is seen to be more soluble than talc or diopside and calcite (Figure 7.4b), and this is borne out in field observation that magnesite is a rare regolith mineral. At pH 8, under atmospheric CO_2 levels, Mg in solution in equilibrium with magnesite is about 600 mg/l. At higher CO_2 levels, as in organic-rich soils, the solubility of Mg drops to 20 mg/l. Where ultramafic rocks are weathering, Mg in solution may percolate into regions of high organic content and precipitate as magnesite nodules, as occurs at Kunwarara, north of Rockhampton, in Queensland. Here a broad valley filled with river sediments over granitic bedrock receives groundwater from flanking hills of ultramafic rocks,

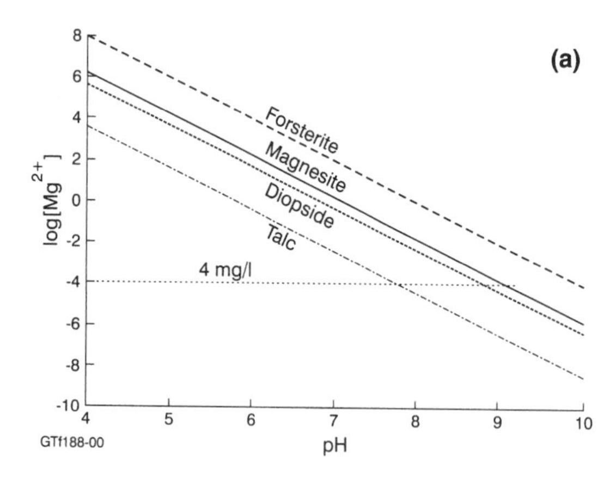

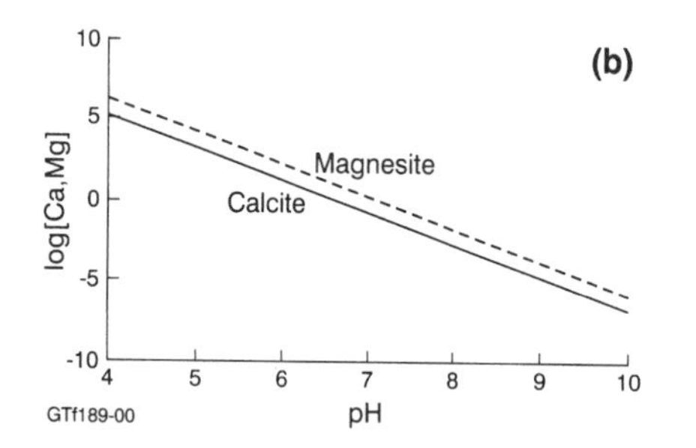

Figure 7.4 *(a) Solubility of various Mg-minerals at log[H_4SiO_4] = −3.1, log[CO_2] = −3.5 (data from Lindsay 1979). The average stream content of 4 mg/l is also shown; (b) comparison of the solubility of magnesite and calcite (data as in Figure 7.6).*

and magnesite nodules the size (and appearance) of cauliflowers have formed (Milburn & Wilcock 1994).

Under arid conditions, in closed basins where high-salinity groundwaters accumulate, the fibrous Mg-silicates sepiolite and palygorskite occur. A reaction between detrital clays, largely smectite, and high-Mg alkaline water derived from mafic rock weathering is generally envisaged (Weaver & Beck 1977). Sanchez & Galan (1995) describe palygorskite in a Spanish continental basin where Mg and clays were sourced from chlorite-rich shales, and Zhou *et al.* (1999) report palygorskite formed from basaltic ash in a fluvio-lacustrine environment in Anhui Province, People's Republic of China. Callan (1984) summarized the environments of sepiolite–palygorskite deposits and notes a marked relation between latitude (30°–40° N and S) and the occurrence of these minerals, interpreting this as a control by aridity on their continental formation. Palygorskite deposits are widespread in palaeochannels on the Eyre Peninsula, South Australia (Keeling and Self 1996) associated with dolomite and illite–smectite clays. Weaver (1984) shows that in the presence of silica, palygorskite is favoured over smectite by increase in both Mg and pH.

ALUMINIUM

Abundance (%)	W	K-P&P	U&B
Mafic igneous	7.5		
Felsic igneous	7		
Shales	4		
Rivers	0.02 mg/l		
Soils			6.7
Solution species: $Al(OH)_2^+$, $Al(OH)_4^-$			

Aluminium in the regolith is controlled first by pH, and second by organic complexing, as discussed in Chapter 6. At acid pH, Al is soluble as Al^{3+}, at neutral pH as $Al(OH)_3$, and at alkaline pH largely as $Al(OH)_4^-$. Solubility in the pH range of most regolith (6–8) is very low, as the solubility product (32.5) indicates. Analyses of Al in groundwater are fraught with difficulties, the major one being that colloidal clay particles may be finer than the filters used prior to analysis (typically 0.5 μm), so reported Al may include both dissolved and finely particulate species.

Many measurements of acid stream and groundwater dissolved Al yield results close to the theoretical solubility figures, in the range 0.1–1 mg/l. There is, however, much evidence, particularly from soil research, indicating that Al moves in solution through the weathering profile to a greater extent than simple solubility calculations would allow. McFarlane & Bowden (1992) present field evidence, in the form of amorphous aluminous encrustations on grass at the bottom of seasonally waterlogged valleys, for the migration and precipitation of alumina through a weathering profile in Malawi. Analyses of Al in groundwater (pH 6) ranged from about 0.5 mg/l unfiltered to about 0.15 mg/l after acidification and filtering (0.45 μm). Since the equilibrium solubility of Al at pH 6 is only about 0.002 mg/l, they conclude that the dissolved Al is organically bound, where it is isolated from the Si also in solution and thus does not react to form kaolinite.

As the most insoluble of the major elements in the crust, Al minerals dominate the regolith. As well as the earlier section of this chapter, Chapters 3, 6, 8 and 9 devote considerable space to the varieties of and transformations between Al-bearing minerals.

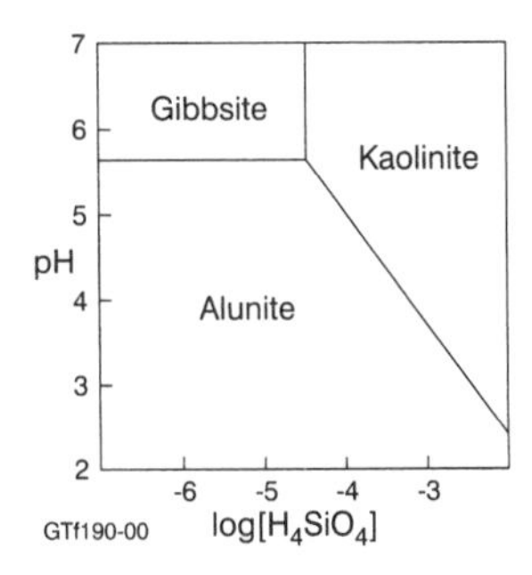

Figure 7.5 *Stability fields for gibbsite, kaolinite and alunite in the presence of K ($10^{-3.2}$ M) and SO_4 ($10^{-2.8}$ M) (after Raymahashay 1968)*

The stability fields for the three Al-minerals gibbsite, kaolinite and alunite have been mapped out by Raymahashay (1968) and by Nordstrom (1982) (Figure 7.5). At low pH, alunite will crystallize in preference to gibbsite at low silica levels, and alunite–kaolinite–quartz can coexist at pH 5 or thereabouts ($\log[H_4SiO_4] = -4$) provided sulfate and K are available.

SILICON

Abundance (%)	W	K-P&P	U&B
Mafic igneous	23		
Felsic igneous	30		
Shales	Range	27 Turekian & Wedepohl	
Rivers	6 mg/l		
Soils		33	
Solution species: H_4SiO_4			

As the most abundant metal in the crust, Si in the regolith may be sourced from almost every rock type. Weathering of silicates contributes Si in the form of H_4SiO_4 to groundwater, at levels depending on the equilibrium species (see Chapter 6). Typical values for SiO_2 concentration in rivers range from 4 to 25 mg/l (2–12 mg/l Si) (Aston 1983). Loss of silica from weathering rocks is gradual and continuous, and provides a ready measure of the progress of chemical weathering (Table 7.8).

By far the commonest silicon mineral in the regolith is quartz, derived from granitic rocks or quartz-rich sediments and metamorphics. Because quartz dissolves congruently, no secondary minerals are produced during its weathering, unlike, say, feldspar. Once feldspars weather to clays, they break down by physical means quite rapidly and are easily eroded. Quartz, by contrast, remains entire and granular, and so is difficult to transport. Quartz grains (silt-size and larger) are residual in the regolith, and in most weathering profiles on quartz-bearing rocks quartz increases in abundance relative to other primary minerals (excluding zircon and similar resistates). Only under lateritic weathering is quartz entirely dissolved, leaving Al and Fe-oxides and hydroxides. Despite the longevity of quartz, silica is steadily lost from all weathering profiles. Commonly, the uppermost part of a profile is silica enriched, even over quartz-free bedrock. Aeolian deposition, movement of colluvium and vegetation additions in the form of phytoliths contribute to the silica content of the soil. Table 7.8 lists the silica content of progressively weathered dolerite from the Darling Range, Western Australia (Davy 1979). Although the dolerite, which when fresh has almost 50% silica, has been weathered to an almost silica-free bauxite, the overlying soil contains high silica, probably derived from adjacent bauxitized granites which retain appreciable quartz through to the soil zone. At the Weipa bauxite in north Queensland, where there is no outcropping siliceous rock, the upper 2 m of the deposit is enriched in silica (Table 7.9).

PHOSPHORUS

Abundance (%)	W	K-P&P	U&B
Mafic igneous	0.2		
Felsic igneous	0.1		
Shales	0.07		
Rivers	0.1 mg/l (very variable)		
Soils	Variable	0.8	
Solution species: HPO_4^{2-}			

Table 7.8 *Silica content from bedrock to soil over dolerite in the Darling Ranges, Western Australia (Davy 1979)*

	Dolerite →	Progressively weathered dolerite →				Bauxite	Caprock base	Caprock	Soil
SiO_2 (%)	48	30.5	21.3	10.7	2.3	2	6.8	3	36.8

Table 7.9 *SiO_2 content with depth in the Weipa bauxite (Jacaranda pit)*

Depth (cm)	% SiO_2
125	6.66
200	11.06
250	1.43
350	1.22
450	1.31
550	20.29
700	28.19
800	28.15
950	30.52
1100	31.62
1200	45.26
1525	52.6
1700	59.0

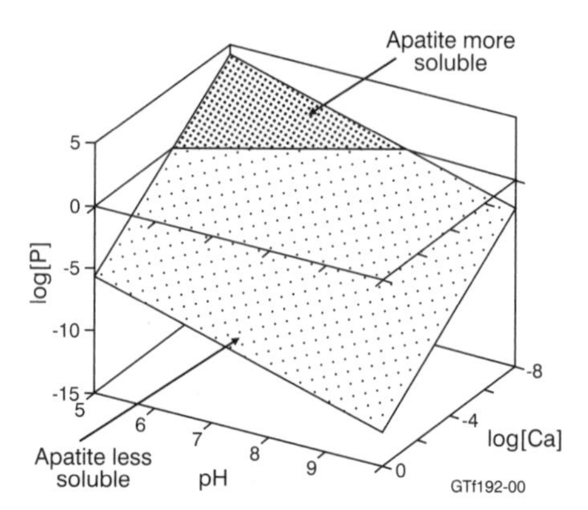

Figure 7.6 *Diagram showing the solubility of apatite with varying pH, [P] and [Ca]. If the solution composition lies above and in front of the inclined surface apatite will crystallize, below or behind the surface apatite dissolves*

The solubility of P in weathering solutions is determined for most common rock types by the solubility of apatite. According to Sposito (1989), in the presence of calcite at pH 7, the equilibrium phosphate in solution with apatite is about 10^{-10} M. Figure 7.6 shows the phosphate and calcium content of water in equilibrium with apatite at varying pH. Apatite becomes soluble in acid conditions or in the absence of calcium.

Lindsay (1979) devotes a chapter to P in soils and shows how the stable phosphate mineral depends on the major mineral assemblage. The minerals that determine the activity of Fe, Al and Ca ultimately control the phosphate levels; in high-Ca soils apatite precipitates, but in acid soils where apatite is soluble, the Al- and Fe-phosphates may control P levels. Because of its importance in agriculture, there is a vast literature on P in soils (see Dixon & Weed 1989).

Flicoteaux & Lucas (1984) describe in some detail the weathering of phosphate rock. In essence, primary carbonate apatite is altered to fluorapatite, then with further weathering millisite ($CaAl_3(PO_4)_2(O,OH)_4.2H_2O$) develops which gives way to crandallite ($CaAl_3(PO_4)_2(OH)_5.H_2O$) which in turn gives way to wavellite ($Al_3(PO_4)_2(OH)_3$) and augelite ($Al_2PO_4(OH)_3$). In this reaction, Ca is leached, Al becomes the dominant cation and the water content increases. Fe-phosphates may also form, but these are generally not prominent under oxidizing conditions where goethite and haematite become the major Fe-minerals.

Many phosphate minerals of the crandallite group occur in the regolith, becoming the host mineral for large cations such as Pb, Ba and REE (see earlier in this chapter under 'phosphates' and under the particular elements).

Organisms are particularly significant in the weathering of apatite, since they require P for metabolism. Banfield & Eggleton (1989) found that bacteria attacked apatites very early in the weathering of granite, dissolving the minerals from within biotite books and sequestering the P.

SULFUR

Abundance (%)	W	K-P&P	U&B
Mafic igneous	0.065		
Felsic igneous	0.33		
Shales	0.24		
Rivers	Highly variable		
Soils		0.04	
Solution species: SO_4^{2-}			

Sulfur reaches the regolith from the lithosphere by the weathering of sulfides, from the atmosphere following volcanic eruptions and industrial release, and as an aerosol from the ocean. Most of the S released to the regolith from bedrock comes from the mineral pyrite, FeS_2, which is a common minor mineral in igneous rocks, particularly volcanics, and is also significant in coal and various sedimentary rocks. Data summarized by Berner & Berner (1987) suggest that pyrite weathering alone releases of the order of 2×10^7 t of S to the regolith annually, much of which reaches the ocean as sulfate. Though minor in amount, the weathering of sulfide ore-bodies is important to the

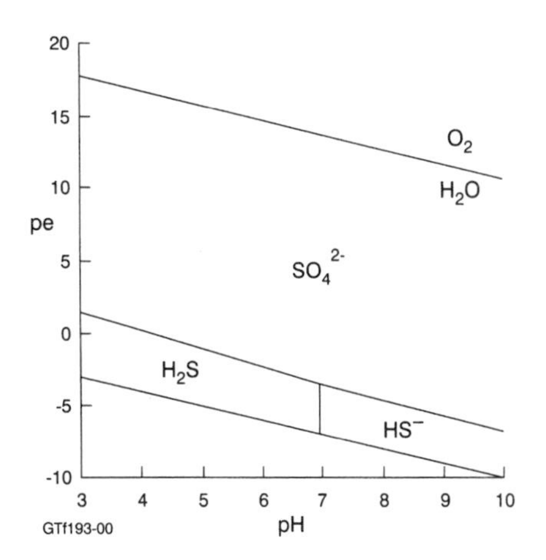

Figure 7.7 *Preponderance diagram for S species (data from Brookins 1988)*

mineral exploration industry, and this aspect is taken up in the section on gossans in Chapter 9.

During weathering, S in the form of metal sulfide, is oxidized to sulfate. Considered in isolation, the preponderance diagram for the major S-species in the regolith is shown in Figure 7.7. Over most pe/pH conditions of the regolith, sulfate is the predominant species. Thiosulfate (S_2O_3) may occur as a metastable species in groundwater to levels of the order of 1–5 mg/l. According to Williams (1990, p. 64), thiosulfate occurs in the early stages of pyrite oxidation at high pH, and is a 'key intermediate species formed during the oxidation of all sulfide minerals over a considerable pH range'.

The reaction of sulfide with air and water is a potent source of acid:

$$2S^{2-} + 2H_2O + O_2 \rightarrow 2H_2SO_4 + 4e^-$$

Where pyrite is the source of S, though there is a series of reaction steps (see section on Fe below), the overall reaction is

$$4FeS_2 + 15O_2 + 14H_2O \rightarrow 4Fe(OH)_3 + 8H_2SO_4$$

In this reaction acid is produced by the oxidation of S and of Fe. Acid mine drainage and the development of acid sulfate soils are two results of this reaction.

Acid sulfate soils develop whenever pyritic rocks or sediments are exposed to an oxidizing environment. Many coastal muds, particularly those of mangrove swamps such as fringe much of the east coast of Australia, contain pyrite at depth as a result of the biogenic reduction of ferric iron in the sediment and sulfate in sea water. Using CH_2O to represent the bacterially derived organic matter, the beginning and end states can be written as (Pons *et al.* 1982)

$$Fe_2O_3 + 4SO_4^{2-} + 8CH_2O + O \rightarrow 2FeS_2 + 8HCO_3^- + 4H_2O$$

Draining the swamps allows air to reach the pyrite, causing oxidation and the production of sulfuric acid. The acid released when the pyrite is oxidized may be neutralized by alkalis in clay exchange sites or calcite, by the Mg in saponite (van Breeman 1982) or by reaction with feldspars. Quartz–kaolin sediments, however, neutralize little of the acid, which then is released into waterways with disastrous consequences for most wetland life.

By far the largest store of S in the regolith is the mineral gypsum ($CaSO_4.2H_2O$). Throughout arid Australia, and in arid terrains around the world, evaporite lacustrine deposits contain large amounts of gypsum. Gypsum is also found in the regolith over weathering sulfides, mainly in areas of low rainfall, such as at the Goonumbla copper deposit near Parkes in central New South Wales, Australia. Here, Ca released during the weathering of the host monzonites has reacted with sulfate from the oxidation of the sulfide ore minerals to produce centimetre-sized gypsum crystals in clay-rich soils.

POTASSIUM

Abundance (%)	W	K-P&P	U&B
Mafic igneous	0.5		
Felsic igneous	3.2		
Shales	2		
Rivers	6.5 mg/l		
Soils		1.5	1.8
Solution species: K^+			

Potassium is a major element in the crust, found largely in alkali feldspar and mica. The K-feldspar and muscovite are relatively slow to weather, hence K is retained in the regolith longer than Na or the alkaline earth elements, which occur in the more weatherable minerals plagioclase (Na, Ca), olivine (Mg) and pyroxene (Ca, Mg). Some of the K released during weathering is taken up by neo-formed illite, in the exchange sites of smectites, or by plants. Manganese oxides also adsorb or include K, particularly the minerals cryptomelane, birnessite, todorokite and romanchéite. In arid environments K may precipitate as jarosite or alunite. Ultimately, though, the bulk of crustal K is lost in solution.

CALCIUM

Abundance (%)	W	K-P&P	U&B
Mafic igneous	7		
Felsic igneous	2		
Shales	3		
Rivers	15 mg/l		
Soils	Variable	2	
Solution species: Ca^{2+}			

Calcium has a crustal abundance of about 4%, thus it is a major element in many rocks. In igneous and mafic metamorphic rocks, plagioclase is the commonest Ca-mineral, followed by pyroxenes and amphiboles, with calcite in limestones and marls being the host for calcium in sedimentary rocks.

Calcium is quite soluble in water, and all Ca-minerals are readily weathered at acid pH. Compilations of the order of element mobility during weathering place Ca in the first four of the major elements, and on the basis of solubility product; only Na and K are more soluble. Under alkaline conditions in the presence of carbonate, calcite is stable (Figure 7.8).

Figure 7.8 indicates that anorthite is more soluble than diopside, and that both are unstable with respect to calcite in atmospheric equilibrated water. As was shown in Chapter 6, the equilibrium between water, atmospheric CO_2 and calcite occurs at pH 8.4. The average Ca concentration in rivers of 15 mg/l is plotted in Figure 7.8, indicating that most natural waters are undersaturated in Ca with respect to the common Ca-silicates, but are in equilibrium with calcite.

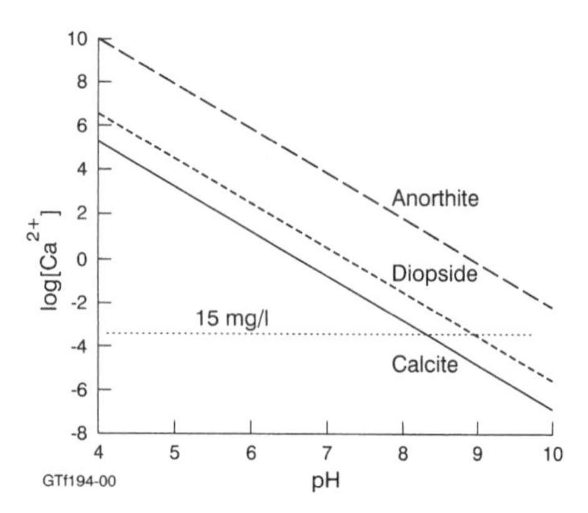

Figure 7.8 *Ca^{2+} concentrations in equilibrium with anorthite ($CaAl_2Si_2O_8$), diopside ($CaMgSi_2O_6$, assuming $[Ca^{2+}] = [Mg^{2+}]$) and calcite ($CaCO_3$), in the presence of kaolinite and quartz, $\log[CO_2] = -3.5$ (data from Lindsay 1979)*

SCANDIUM

Abundance (g/t)	W	K-P&P	U&B Horovitz (1975)
Mafic igneous	30		
Felsic igneous	7		
Shales	12		
Rivers			<0.3 μg/l
Soils		5–10	10
Solution species: Sc^{3+}, $Sc(OH)^{2+}$, $Sc(OH)_3$			

Scandium is widely distributed in rock-forming minerals, substituting for Fe^{3+}, Y^{3+} and less commonly Al^{3+}. Levels rarely exceed 1000 g/t, and are generally much less than this. Crandallite group phosphates may contain almost 1% Sc_2O_3. Scandium is highly insoluble in water, and is among the more immobile elements during weathering.

Scandium is concentrated in residual regolith, such as bauxites, where it reaches levels of 100 g/t (Jamaica, Wagh & Pinnock 1987), or 60 g/t at Weipa, Queensland, Australia, a threefold increase relative to fresh rock (unpublished date of the authors). Concentration may be through incorporation in crandallite group minerals or in anatase (Wagh & Pinnock 1987), adsorbed on Fe-oxyhydroxides (Vlasov 1968). da Costa & Araújo (1996) report a mean Sc content of 117 g/t in 27 samples of lateritic Fe crust from Brazil. The crusts are highly phosphatic (mean P_2O_5 = 11.85%); however, statistical analysis of the data revealed the geochemical association of Sc with Fe and other transition elements, not with P_2O_5 or with anatase.

TITANIUM

Abundance (%)	W	K-P&P	U&B
Mafic igneous	1.9		
Felsic igneous	0.5		
Shales	1		
Rivers	~1 μg/kg, but no reliable data		
Soils			0.5
Bauxite/laterite	3		
Solution species: $Ti(OH)_4$. ?$HTiO_3^-$			

Figure 7.9 *Submicron-sized anatase crystals set in a web of smectite after the weathering of titanite, viewed by scanning electron microscopy (photo David Tilley)*

Titanium is a common minor or trace element in many silicates, as well as occurring in the accessory minerals rutile, ilmenite and sphene. Weathering of silicates and oxides containing Ti, such as sphene biotite, pyroxene, garnet, amphibole or ilmenite leads to the precipitation of Ti as submicron-sized crystals of anatase (Figure 7.9) within the pseudomorph of the parent crystal (Anand & Gilkes 1984). Rutile is highly resistant to weathering and is a common resistate mineral in the regolith. Ilmenite, while sufficiently resistant to weathering to become an important ingredient in beach sands and other resistate mineral accumulations, alters through a sequence of phases toward haematite or goethite plus anatase.

Van Baalen (1993) has summarized the scant literature on the concentration of Ti in natural waters and its hydrolysis in laboratory experiments. Titanium solubility at typical groundwater pH is so low that measurement is extremely difficult. Van Baalen's collected observations hint at a U- or V-shaped hydrolysis diagram, with total dissolved Ti at pH 3 of the order of 50 μg/l, falling below 0.01 μg/l at pH 7, then increasing to about 50 μg/l at pH 11 or thereabouts.

The apparent short travel distance for the Ti is clear evidence for its insolubility in groundwater. Under extreme weathering anatase may be mobilized, possibly as colloidal particles rather than in solution, and move down through the weathering profile. It is a common mineral in silcretes where it forms caps on the upper surfaces of quartz grains.

VANADIUM

Abundance (g/t)	W	K-P&P	U&B
Mafic igneous	240		
Felsic igneous	70		
Shales	150		
Rivers	~1 μg/l, but no reliable data		
Soils		60	108
Solution species (oxidized): $H_2VO_4^-$, HVO_4^{2-}			

In igneous minerals, V substitutes as V^{3+} for Al and Fe^{3+}, and is present in greatest amounts in pyroxenes and micas, magnetite, haematite and titanite.

With progressive oxidation, V in solution changes from trivalent to tetravalent, and eventually becomes quinquevalent. The chemistry is not simple, as Garrels & Christ (1965) show, and the stable phases depend, as always, on the other ions in solution. In the case of V, there has been much investigation into the mineralogy of alteration of U–V ores (see Environments and element mobility, p. 122).

During weathering, trivalent V maintains an association with Al, occurring in dioctahedral smectites, where it can reach major element abundance (30% V_2O_3 in volkonskoite) (Guven & Hower 1979). Norrish (1975) reported up to 7 wt% V_2O_5 in an interstratified mica-montmorillonite formed from the weathering of shale. In goethite, V substitutes for ferric iron in reducing environments, and where the V

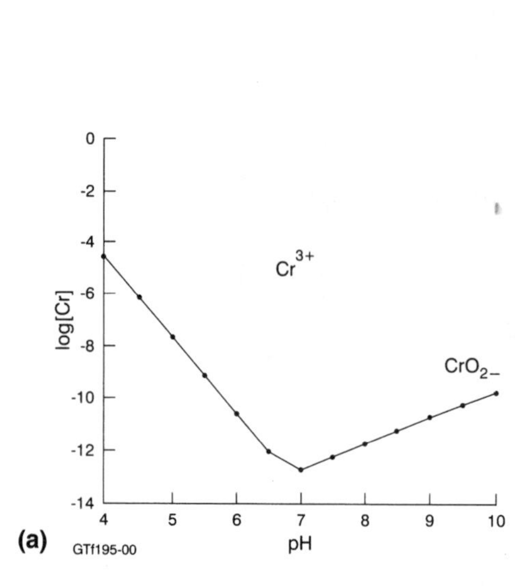

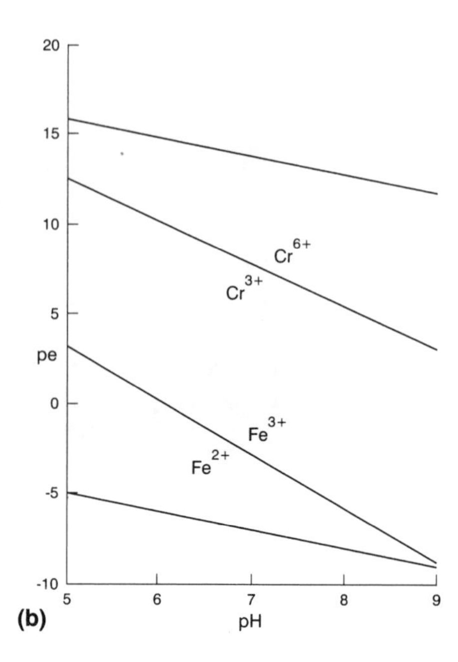

Figure 7.10 *(a) Solubility of Cr in reducing solution in equilibrium with Cr_2O_3, expressed as moles/litre of the sum of Cr^{3+} and CrO_2^-. (b) pe/pH relationship for Cr^{3+} to Cr^{6+} compared with the ferrous/ferric transition (upper and lower lines mark the stability limits of water as in Figure 6.17)*

content is high the mineral montroseite (VO(OH)), which is isostructural with goethite, may form.

Under oxidizing conditions vanadates form, with a variety of minerals resulting, but these only become significant in the oxidized zone of ore-bodies rich in V.

CHROMIUM

Abundance (g/t)	W	K-P&P	U&B
Mafic igneous	300		
Felsic igneous	10		
Shales	80		
Rivers	1 μg/l	5 μg/l (McFarlane *et al.* 1994)	
Soils	60		
Solution species: $Cr^{iii}(OH)_3$, $Cr^{vi}O_4^{2-}$			

Chromium is found primarily in the oxide chromite, and substituting for Fe^{3+} in some silicates. Chromium reaches major element proportions in some silicates: garnet (uvarovite, $Ca_3Cr_2[SiO_4]_3$), pyroxenes (cosmochlor, $NaCr[Si_2O_6]$) and the Cr-muscovite fuchsite, and these mineral groups commonly contain trace amounts of Cr.

In regolith minerals chromium (Cr^{3+}) can be incorporated into the octahedral sheet of dioctahedral smectites (e.g. volkonskoite or chromian nontronite), as Paquet *et al.* (1987) describe from the weathering of the Cr-mineral stichtite or Cr-chlorite from Campo Formoso, Brazil.

Depending on pe and pH, Cr may exist in reducing regolith waters as Cr^{3+} below pH 7 or CrO_2^- above that pH, and under oxidizing conditions below pH 6 as $Cr_2O_7^{2-}$ or CrO_4^{2-} above pH 6. The solubility under reducing conditions has a minimum around pH 6–7, and shows that Cr^{3+} is less soluble (by a factor of about 250) than ferric iron (see Figure 7.10a).

Manceau & Charlet (1992) present evidence that Cr^{3+} is oxidized to Cr^{6+} by Mn^{4+} in birnessite and by Mn^{3+} in other Mn-oxides and oxyhydroxides. They describe the process as inner-sphere adsorption to the Mn-oxide/oxyhydroxide surface, oxidation, and then release into solution. Thus Mn-oxides and oxyhydroxides would not be expected to be hosts for Cr^{3+} in the weathering environment.

McFarlane *et al.* (1994) found that Cr was leached from lateritic weathering profiles in Malawi to the extent of about 70%. Following experimental work, they interpreted the leaching, and the relatively high levels of Cr in groundwater samples (10 μg/l), to be the result of mobilization of Cr by micro-organisms. Bartlett & Kimble (1976), who were unable to oxidize chromic solutions in experiments lasting 40 days, under varying pH and aeration conditions, found that chromate solutions were quickly reduced to chromic in the presence of organic matter. It would appear that if Cr is mobilized biotically, this is not achieved via oxidation to the more soluble chromate.

Although McFarlane *et al.* (1994) conclude that biological solubilization of Cr is too severe to permit its use as an immobile element to estimate degree of weathering, many other workers have found or assumed Cr to be immobile under weathering.

Examination of the pe/pH relationships for Cr show that it is probably relatively immobile in weakly oxidizing environments (at or above the pe at which the ferrous/ferric transition occurs), but becomes quite mobile at higher oxidation potential when chromate becomes the dominant species (Figure 7.10b).

MANGANESE

Abundance (%)	W	K-P&P	U&BT
Mafic igneous	0.1		
Felsic igneous	0.035		
Shales	0.06		
Soils	0.05	0.076	0.078
Solution species: Mn^{2+}			

Manganese occurs widely in the minerals of fresh rocks, largely camouflaged in octahedral sites of silicates. In igneous rocks it is generally in the divalent (reduced) state, though in some metamorphic rocks it may be trivalent (in epidote) or tetravalent (braunite). On weathering, Mn is oxidized to the tri- and tetravalent states according to the reactions:

$$4Mn^{2+} + O_2 + 4H_2O \rightarrow 2Mn_2O_3 + 8H^+$$
$$2Mn^{2+} + O_2 + 2H_2O \rightarrow 2MnO_2 + 4H^+$$
$$2Mn_2O_3 + O_2 \rightarrow 4MnO_2$$

Garrels & Christ (1965) describe the stability relations among several Mn-species, and one such diagram is shown in Figure 7.11 where the stability fields of Mn-oxides and carbonate at $P_{CO2} = 10^{-4}$ at 25°C, are compared with similar data for Fe. The diagram shows that Mn^{2+} remains stable to higher oxidation potential than Fe^{2+}, as was shown in Figure

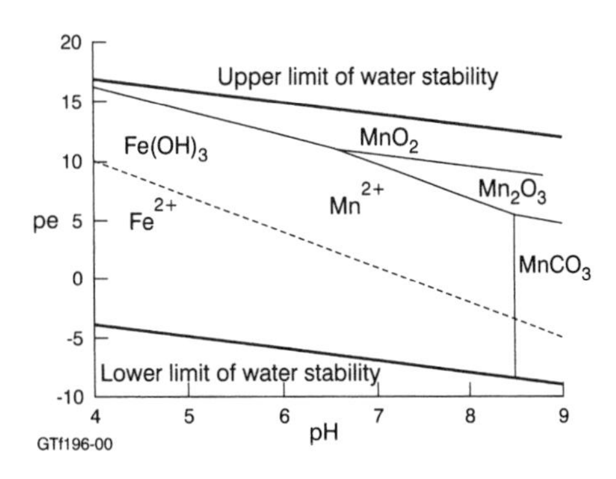

Figure 7.11 *pe:pH diagram for Mn at $P_{CO2} = 10^{-4}$ at 25°C, $[Mn^{2+}] = 10^{-6}$ (after Garrels & Christ 1965). The dashed line marks the ferrous–ferric transition*

6.16, and that siderite is not stable at this low P_{CO2}. In solution, only Mn^{2+} and its hydrolysis products occur to any significant extent. Mn^{3+} is unstable with respect to Mn^{2+} and MnO_2, and the solubility product of Mn^{4+} is so high (56, compare Fe^{3+}, 38) that Mn^{4+} is below all limits of detection in solution.

Manganese remains divalent and soluble to much higher pe values than does Fe^{2+}; however, in the upper part of the regolith where oxidation potential is high, Mn precipitates as the oxide or oxyhydroxide. Most Mn-minerals in the regolith are very poorly crystallized (see Chapter 3), have high surface areas and readily adsorb other ions.

While the weathering environment remains reducing, Mn^{2+} can be incorporated into the octahedral sheet of smectites, as Paquet *et al.* (1987) describe for a weathering tephroite (Mn_2SiO_4) in a dunite from the Ivory Coast, Africa. Up to 80% of the octahedral cations in the smectite were Mn^{2+}.

IRON

Abundance (%)	W	K-P&P	U&B
Mafic igneous			
Felsic igneous			
Shales	4.7		
Rivers			
Soils			3
Solution species: several, see Figure 6.7			

Because of its ability to undergo oxidation and reduction (ferric to ferrous), and its abundance in the crust, Fe can be deemed the most important element in regolith geochemistry. Depending on the conditions, Fe may be mobilized as various hydrolysis species or sequestered in solution by chelating compounds, or it may be fixed as oxide, oxyhydroxide, carbonate, sulfide, phosphate or silicate.

The inorganic behaviour of Fe is documented in detail by, among others, Garrels & Christ (1965), Stumm & Morgan (1970) and Drever (1988).

In igneous rocks, Fe is largely present in the ferrous state, although the common accessory mineral magnetite ($FeFe_2O_4$) has one ferrous and two ferric iron ions, and some ferric iron is found in many amphiboles and in biotite. Oxidative weathering quickly produces ferric hydroxide (ferrihydrite, or amorphous Fe-hydroxide). To examine the weathering of ferrous-iron-bearing minerals, the behaviour of Fe under various solution conditions can be seen with the help of pe–pH diagrams (see Chapter 6).

Figure 7.12a shows that amorphous ferric oxyhydroxide, or ferrihydrite, is stable over a wide pH range under oxidizing conditions. However, in swamps and other regolith where organic matter maintains a reducing environment, ferrous iron is the stable species.

If the partial pressure of CO_2 is high (say $P_{CO2} = 10^{-2}$, typical of soil CO_2), siderite becomes a stable phase at high pH in a reducing environment (Figure 7.12b). Many rocks contain minor pyrite, and the weathering of this mineral may convert sulfide to elemental S or to sulfate, and ferrous iron to ferric. The oxidation of pyrite is complex; the reaction sequence may be written as (Stumm & Morgan 1970)

$$2FeS_2 + 7O_2 + 2H_2O \rightarrow 2Fe^{2+} + 4SO_4^{2-} + 4H^+$$

This reaction releases ferrous iron to solution, and acidifies the water.

With further oxidation, the ferrous iron is converted to ferric iron, which may then hydrolyse and precipitate as amorphous ferric hydroxide:

$$4Fe^{2+} + O_2 + 10H_2O \rightarrow 4Fe(OH)_3 + 8H^+$$

This reaction is not limited to pyritic weathering environments; the acidification of weathering solutions by Fe hydrolysis is an important feature of the weathering of all ferrous minerals, and a process of major importance at the water-table, which is commonly the oxidation–reduction boundary.

Returning to pyritic weathering, if the ferric iron remains in solution (for example the weathering solution is not diluted by water of higher pH so that hydrolysis is limited), then the ferric ions can themselves oxidize pyrite,

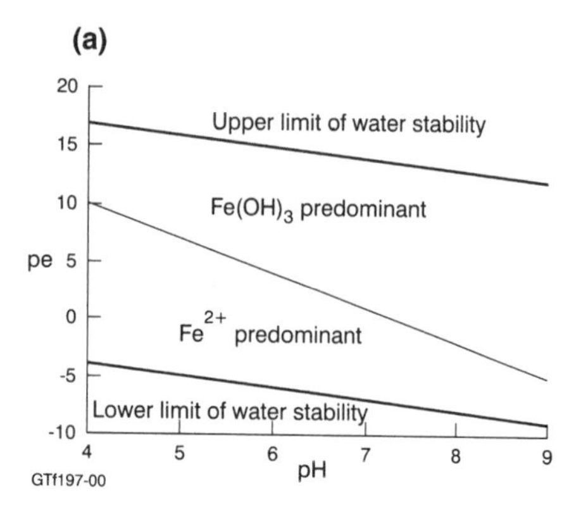

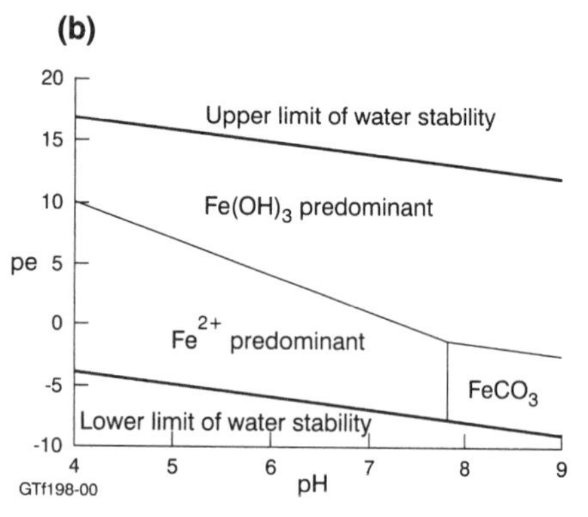

Figure 7.12 *(a) pe–pH diagram for Fe–O–H_2O, at log[Fe] = −6, assuming amorphous $Fe(OH)_3$ is the ferric hydroxide. At pH < 3 Fe^{3+} becomes significant, at pH > 9 $Fe(OH)_2$ is stable at pe less than about −4. (b) pe–pH diagram for Fe–O–H_2O–CO_2 at log [Fe^{2+}] = −6, $P_{CO2} = 10^{-2}$ (after Drever 1988)*

$$FeS_2 + 14Fe^{3+} + 8H_2O \rightarrow 15Fe^{2+} + 2SO_4^{2-} + 16H^+$$

further acidifying the water. Thus ferric iron and pyrite do not coexist.

Garrels & Christ (1965) show that at the levels of S in normal groundwater (of the order of 10^{-4}–10^{-6}), pyrite has only a small stability field, restricted to pH between about 5 and 8, and pe values around −5.

Taylor (1988) proposed a mechanism for the migration of ferric iron in soils, following experiments in which he added hydrolysed (flocculated) $Fe(OH)_3$ at pH 4 to a solution of Al-hydroxy species. This induced further hydrolysis with consequent reduction of pH, but by maintaining a pH of 4, the precipitated ferric hydroxide returned to solution as an Al–Fe hydroxy complex. Such a complex, he suggested, would be mobile in the regolith, and allow Fe (and Al) to migrate.

It is the Fe-oxides and oxyhydroxides that dominate the weathering of Fe under most regolith conditions. The early precipitate, ferrihydrite, is a highly reactive mineral because of its huge surface area, and is capable of adsorbing a wide range of trace elements. Ferrihydrite slowly converts to goethite or haematite, and many of its adsorbed ions remain trapped in the better crystalline mineral, thus Fe becomes an important host for many elements in the regolith.

COBALT

Abundance (g/t)	W	K-P&P	U&BT
Mafic igneous	45		
Felsic igneous	5		
Shales	20		
Rivers	0.2 μg/l		
Soils		8	1211
Solution species: Co^{2+}			

Cobalt occurs in a range of primary silicates substituting for Fe, in olivine of the order of 200 g/t, in pyroxene half that amount (Wedepohl 1969–78).

In natural waters, up to about pH 9, the stable species in solution is Co^{2+}. Above pH 9 $Co(OH)_2$ is stable except in highly oxidizing environments where $Co(OH)_3$ can exist. However, according to Garrels & Christ (1965), carbonate displaces the stability of the trivalent hydroxide to pH > 9 and expands the stability field of $CoCO_3$ so that it occupies all of the pe/pH diagram above pH 6.

The solubility of Co is considerably lowered in the presence of ferric oxides and hydroxides, by over five orders of magnitude at pH 7, according to Thornber & Wildman (1984). In soils Co has been shown to be strongly associated with Mn-oxides by Taylor & Mackenzie (1966), and Loganathan *et al.* (1977), in a study of the relation between pH and Zn/Co uptake on δ-MnO_2, show a fourfold increase in Co sorption between pH 6 and 7. Grana Gomez *et al.* (1990) estimated the fractionation of Co between various soil components, and concluded that the Fe and Mn-oxyhydroxides were the dominant hosts for Co, accounting for over 60% of the total Co. The mobility of Co during weathering would appear to be controlled more by its adsorption to Fe and Mn minerals than by other chemical factors.

Cobalt is particularly abundant in the regolith over intensely weathered ultramafic rocks. Manganese wad, a field term for a black, fine-grained aggregate of Mn-oxides and oxyhydroxides, is common in the oxidized zone over such rocks, and Mn-minerals such as lithiophorite and asbolane may contain 10–20% cobalt oxide, as well as appreciable Ni (~10%). The Ni-laterite deposits over ultramafic rocks near Marlborough, Queensland, contain 0.06% Co in 1% Ni-ore, with values reaching 1.5% Co locally.

NICKEL

Abundance (g/t)	W	K-P&P	U&B
Mafic igneous	120		
Felsic igneous	10		
Shales	50		
Rivers	Insufficient data		
Soils		20	34
Solution species: Ni^{2+}			

Nickel in igneous minerals occurs as Ni^{2+}, where it substitutes in the octahedral sites of silicates. It is also concentrated in spinels and much more in the sulfides niccolite and pyrrhotite. Nickel is much more abundant in mafic igneous rocks than in felsic, and Ni-ores occur either as sulfide deposits within the mafic rocks or in the weathering profiles of ultramafics.

In solution at pH below 7, Ni^{2+} is the stable species, and as its solubility product of 15.7 indicates, Ni^{2+} is readily soluble in water. The weathering profiles of mafic rocks tend to be alkaline at depth, under which condition Ni-rich clays are stable. According to theory (Garrels & Christ 1965), at pH of about 9, $Ni(OH)_2$ precipitates. Only rarely has this mineral (theophrastite) been reported. Silica is generally high in weathering ultramafics (in the presence of amorphous silica, Figure 6.3), presumably shifting the equilibrium towards silicates, but experimental data are lacking.

Where Ni in weathering bedrock is high, Ni-silicates are formed in the regolith. Garnierite is a field term for green, non-swelling, Ni-rich clays, and garnierite is generally a mixture of 1 : 1 (serpentine group) and 2 : 1 (talc-like) clays. The Ni-dominated end-member clays are

- 1 : 1 nepouite, brindleyite and pecoraite
- 2 : 1 willemsite (talc analogue) and pimelite (variable, poorly defined)
- 2 : 2 nimite (chlorite), stevensite (smectite)

In the early stages of the weathering of ultramafics, the primary silicates alter to various FeMg-layer silicates, and these are commonly Ni-rich. Paquet *et al.* (1987) describe trioctahedral smectites containing up to 30% NiO forming close to parent pyroxenes in a weathering ultramafic from Jacuba, Brazil. Further up the profile, the NiO content decreases, and the smectites change to dioctahedral, containing 10% NiO. The final stage of weathering leads to the formation of Ni-rich goethite. Similarly Trescases *et al.* (1987a) describe Ni-smectite in the saprolite of a weathered serpentinite from Sao Joao, Brazil. The fresh serpentinite contains only 0.5% NiO, while the saprolite contains 2% NiO. Trescases *et al.* (1987a) attribute the increase in Ni in the saprolite to lateral migration of this element from a Ni-rich laterite higher in the toposequence.

Semi-lucent emerald green Ni-talc (pimelite), with some expanding character, has been described by Wiewióra *et al.* (1982) from the weathering profile of ultramafic rocks in Szklary, Poland. The Ni-clays form veinlets in the nontronite-rich zone of the weathering profile. The most Ni-rich sample has almost 33 wt% NiO, occupying two of the three octahedral sites in the structure.

Schellmann (1983) examined a number of Ni-rich residual ores on serpentinite, and concluded that the Ni was largely contained within weathered, Mg-depleted serpentine, where it appeared to form an amorphous Ni-silicate. The source of the Ni, though ultimately from the ultramafic below the residual ore, was interpreted as having been leached from goethite containing up to 1.5% NiO in the overlying laterite and fixed in the slightly weathered serpentinite.

At the Brolga nickel mine, Central Queensland, Ni and Co occur in lateritic alteration of sepentinized peridotites (Parianos 1994). The mineralogy is dominated by goethite, chalcedony and quartz, and Fe–Ni smectite (nontronite–pimelite).

Brand *et al.* (1998) classify regolith Ni deposits according to their Ni mineralogy as

Type A: dominated by garnierite, generally deep in the saprolite, mostly found in humid tropical regions.

Type B: dominated by Ni-bearing smectite, generally in the upper saprolite, mostly found in semi-arid regions (e.g Western Australia).

Type C oxide deposits, occurring at the pedolith–saprolith boundary, found in all climatic environments.

COPPER

Abundance (g/t)	W	K-P&P	U&BT
Mafic igneous	50		
Felsic igneous	10		
Shales	35		
Rivers	3 μg/l		
Soils		25	2622
Solution species: $CuCO_3$, $CuOH^+$			

In the reducing environment of igneous rocks, Cu occurs as Cu^+. Copper has a higher affinity for S than for silicates, thus in the presence of S, Cu-sulfides are formed, mainly chalcopyrite. In mafic silicate minerals, Cu concentration averages 50 g/t, falling to 2–10 g/t in quartz and K-feldspar.

On weathering, Cu combines with oxygen to form first cuprite (Cu_2O) and at higher oxidation potential tenorite (CuO). Garrels & Christ (1965) indicate that with increasing oxidation at acid pH, chalcopyrite oxidizes to release Cu^{2+} to solution and Fe-oxide precipitates. At more alkaline pH, cuprite or tenorite are formed with Fe-oxides.

In the presence of carbonate, Stumm & Morgan (1981) show that below pH 7, malachite is the stable Cu-mineral, and tenorite above pH 7. Precipitation of malachite at pH 6 requires a Cu concentration in excess of 10^{-4} M (6 mg/l), whereas at pH 8, tenorite maintains equilibrium with a total dissolved Cu content about 1000 times less, close to the average of groundwater Cu concentration.

Mosser & Zeegers (1988) found that during the weathering of disseminated Cu mineralization in a granodiorite, the Cu-content fell from 2600 to 1000 g/t. Throughout the 4 m weathering profile, more than half the Cu was found in the $<2\ \mu m$ fraction, leading Mosser & Zeegers to conclude that Cu was enriched in the unaltered mica, vermiculite and smectite (8000 g/t), and kaolinite (2000 g/t). Although Mn-oxides were not major, their Cu content was high (6%). Haematite and goethite had 2300 and 3000 g/t Cu respectively. In a later publication, Mosser *et al.* (1990) give evidence that up to 1500 g/t Cu can be incorporated into the octahedral sheet of kaolinite and smectite.

Copper is preferentially adsorbed on Fe-oxides above pH 5, and because of the economic significance of ironstone gossans, there have been many studies which have shown a strong correlation between Cu and Fe content of weathered rocks. At Johnson Creek, Queensland, Australia, weathering of sideritic rocks containing minor chalcopyrite has led to the formation of extensive strata-bound ironstones with up to 4000 g/t Cu (Taylor & Thornber 1992), with a strong correlation between Fe and Cu (Britt 1993). Near by, at Paradise Valley, an ironstone containing up to 3% Cu occurs along a fault line. There is no immediate source of Cu, and scavenging of the metal by precipitating Fe in the fault is thought to have given rise to the high Cu content. At Mt Isa, also in Queensland, the Cu content of the gossans reaches nearly 6%, the values enhanced by the neutralization of acid weathering solutions by carbonates. By contrast, barren gossans associated only with pyritic rocks, typically contain only tens to hundreds of grams per tonne Cu. The low values are partly the consequence of low pH and high mobility of Cu following the oxidation of pyrite during weathering.

Copper may be strongly associated with Mn-oxides (Mosser & Zeegers 1988), where it may be co-precipitated or adsorbed. The pH_{pzc} for Mn-oxides are much lower than for Fe-oxides (see Table 7.3), thus Mn-oxides will adsorb at lower pH than Fe oxides. Figure 7.13 shows the predominant areas for Cu and its oxides, with the areas for Mn superimposed. At all acid pH and in oxidizing conditions, the dominant copper species is Cu^{2+}, whereas Mn-oxides precipitate at high values of pe. Copper is therefore unlikely to co-precipitate with Mn in acid environments, but may well be adsorbed on to Mn-oxides. Under a change from acid to alkaline solutions at high pe, the two metals may co-precipitate.

The weathering of Cu–Fe-sulfides such as chalcopyrite leads to a variety of products, depending on the pe and pH conditions. At pH > 6 (Thornber 1985), ferric hydroxide and chalcocite precipitate according to the reaction

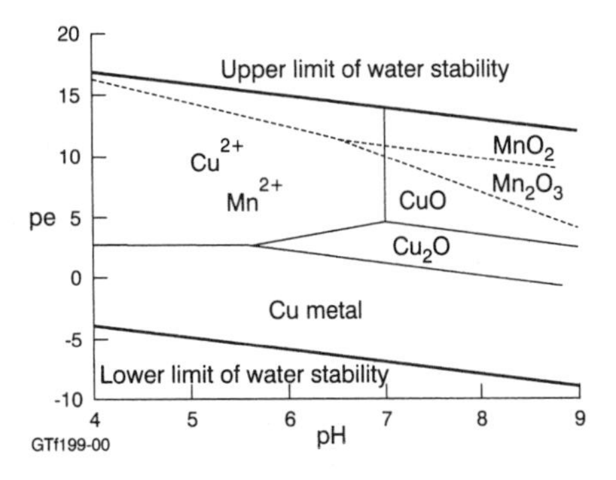

Figure 7.13 *pe–pH diagram for Cu–H_2O–O_2 with the stability fields for Mn superimposed (after Garrels & Christ 1965)*

$$2Fe^{2+} + 2Cu^{2+} + 7H_2O \rightarrow 2Fe(OH)_3 + Cu_2O + 8H^+$$

Below pH 6 native Cu may form from Cu in solution: $2Cu^+ \rightarrow Cu^{2+} + Cu^0$. Where copper is a minor constituent of the sulfide, the Fe-oxyhydroxides formed during the oxidation of Fe-sulfides such as pyrite provide a ready adsorption site for Cu.

ZINC

Abundance (g/t)	W	K-P&P	U&BT
Mafic igneous	100		
Felsic igneous	50		
Shales	100		
Rivers	10		
Soils		60	6034
Solution species: $Zn(OH)^+$, Zn^{2+}, $ZnCO_3$			

In igneous rocks, Zn substitutes for Fe and Mg in magnetite and in most of the common silicates, at levels of the order of 100–1000 g/t. Sphalerite (ZnS) occurs as an accessory mineral in some mafic rocks.

Since zinc only occurs in the +2 valence state, redox conditions do not affect its behaviour during weathering, although as it is adsorbed on to Fe-oxyhydroxides, redox affects the availability of adsorption media. In the vicinity of a major Zn source, phosphates (e.g. scholzite, hopeite), carbonates (smithsonite under high CO_2 levels, hydrozincite at atmospheric CO_2), and silicates (hemimorphite, Zn-smectite or sauconite) may form during weathering. In gossans, goethite has been reported with 2.5% Zn (Nickel & Daniels 1985). Thornber (1985) considered that sphalerite was slow to weather and that as a result the release of Zn to solution was relatively slow, inhibiting its precipitation and allowing time for adsorption on prior minerals.

Thornber & Wildman (1984) showed from laboratory experiments that co-precipitation of Zn with Fe reduces the equilibrium concentration of Zn in solution by several orders of magnitude when Fe >> Zn. At pH between 5 and 9, only Fe-hydroxides appeared as the precipitate, but these contained all the precipitated Zn. Loganathan *et al.* (1977) report that five times as much Zn is sorbed on δ-MnO_2 at pH 7 as at pH 6. This experimental conclusion is amply reflected in many geochemical studies of weathering profiles over ore-bodies on the Yilgarn Craton of Western

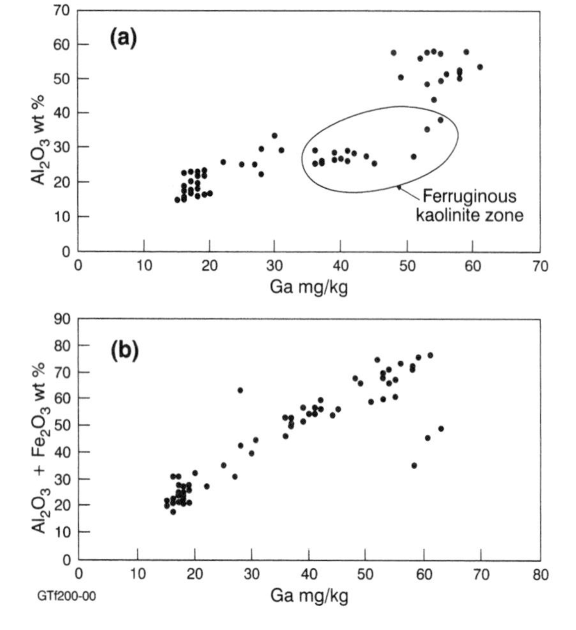

Figure 7.14 *(a) Ga versus Al_2O_3 at Andoom, north Queensland, Australia. (b) Ga versus Al_2O_3 + Fe_2O_3 at Andoom, north Queensland, Australia*

Australia, where Zn is found to occur with Fe and Mn-oxides and hydroxides.

Paquet *et al.* (1987) found sauconite (Zn-smectite) as the first stage of weathering of willemite (Zn_2SiO_4) at Niari, Congo, Africa, where the Zn-content of the clay reached 2.1 per three octahedral sites. Further weathering at this location led to the formation of zincite (ZnO).

Kabata-Pendias & Pendias (1984) in summarizing the hosts for Zn in soils, suggest that most is associated with the hydrous oxides and with poorly crystalline clays such as allophane. In such sites an increase in acidity leads to mobilization of Zn into solution.

GALLIUM

Abundance (g/t)	W	K-P&P	U&B
Mafic igneous	20		
Felsic igneous	20		
Shales			
Rivers			
Soils		25	21
Solution species: $Ga(OH)_4^-$			

Geochemically, Ga is similar to Al and ferric iron in charge and ionic radius, and consequently is found camouflaged in aluminosilicates and Fe-oxides/oxyhydroxides. Feldspars have from 10 to 100 g/t Ga, micas up to 300 g/t. Gallium is slightly less soluble in water than Al (SP Ga = 35, Al = 32.5), hence it is slightly enriched relative to Al during weathering.

At Weipa, in north Queensland, Australia, where shales have weathered to bauxites, Ga shows close coherence with Al, and is slightly enriched (by 5%) in the bauxite relative to Al. Below the bauxite, at a depth between 6 and 12 m where there is a zone of Fe enrichment, the Ga : Al ratio increases by about 30% (Figure 7.14a). Preferential incorporation of Ga in the Fe-oxides and oxyhydroxides is consistent with the crystallochemical character of Ga; the ionic radius of Ga is between those of Al and Fe^{3+}, hence Ga correlates closely with (Al_2O_3 + Fe_2O_3) (Figure 7.14b).

Hieronymus *et al.* (1990) examined Ga/Al/Fe relationships in several bauxitic and lateritic weathering profiles. They found that there was the expected correlation between Ga and Al in 'classical' deeply weathered profiles, where Al and Fe are residual and all other major elements depleted. However, where Al has been mobilized from the surface lateritic crust, Ga may be retained in the Fe-oxides/hydroxides, to the extent that there is a negative correlation between Al and Ga. They attribute this to more intense leaching of Al than Ga under acid conditions.

ARSENIC

Abundance (g/t)	W	K-P&P	U&B
Mafic igneous	1.5		
Felsic igneous	1.5		
Shales	13		
Rivers	1 μg/l		
Soils		8	11
Solution species: $HAsO_4^{2-}$, $H_2AsO_4^-$			

Arsenic is a chalcophile element, occurring most commonly as arsenopyrite (FeAsS); however, in most igneous rocks the bulk of the As is present in magnetite and ilmenite.

Cherry *et al.* (1979) present the preponderance diagram for As-species in solution at the eh/pH range of natural waters (Figure 7.15). The transition from As^{3+} to As^{4+} occurs in the same eh region as the oxidation of ferrous iron. At pH < 7, Cherry *et al.* show that Fe^{3+} rapidly oxidizes As^{3+}, thus As in the

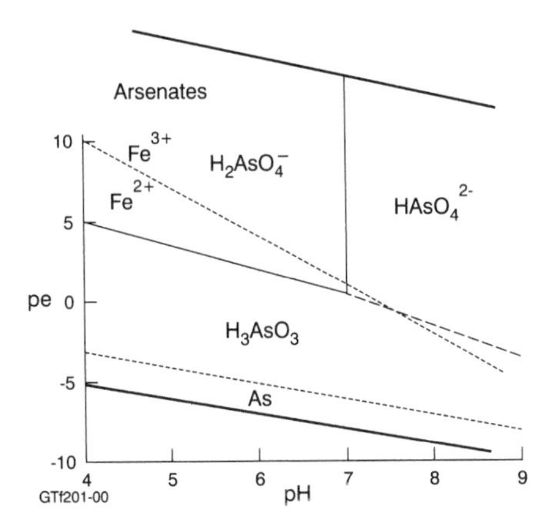

Figure 7.15 *Preponderance diagram for arsenic (total dissolved As = 50 µg/l) drawn from the thermodynamic data of Cherry* et al. *1979)*

oxidized zone of the regolith can be expected to be entirely arsenate.

The affinity of ferric hydroxide for arsenate is so high that it was used as an antidote for arsenic poisoning at least up until the 1950s (Sollmann 1957). A dilute solution of ferric sulfate mixed on demand with magnesia was recommended. These medications would have the effect of shifting the stomach pH to higher values, ensuring that the arsenic species became the less soluble arsenate, precipitating ferrihydrite, and enhancing the adsorption of As. This affinity is strongly reflected in the As content of gossans, where the dominant mineralogy is generally goethite, and As levels may exceed 1000 g/t. This association can result either from adsorption or co-precipitation. If other metals are present, a wide range of arsenate minerals may form, including beudantite ($PbFe_3[AsO_4][SO_4](OH)_6$).

SELENIUM

Abundance (g/t)	W	K-P&P	U&B
Mafic igneous	0.1	0.3	
Felsic igneous	0.1	0.3	
Shales	1	0.6	
Rivers		0.1 µg/l	
Soils	0.4		0.4
Solution species: SeO_3^{2-}, $HSeO_3^-$, SeO_4^{2-}			

Selenium is a chalcophile element, found in sulfide minerals generally in the range 10–50 g/t. At a given pH, Se oxidizes at higher oxidation potential than does S, initially to the less soluble selenite ion (SeO_3^{2-}), then with further oxidation to the more mobile selenate (SeO_4^{2-}). Both ions are adsorbed by Fe-oxides and hydroxides, selenate more so than selenite. Frost & Griffin (1977) found the adsorption of Se^{4+} in the form of $HSeO_3^-$ to kaolinite and smectite was pH dependent, being maximized at pH 2–3, and falling by a factor of about 3 at pH 7. There is a considerable agricultural literature on Se because of its biological significance (see for example Ihnat 1989; Jacobs 1989). Selenium may serve as a pathfinder element for Cu–Pb–Zn sulfide ores because of its tendency to be fixed in Fe-oxide-oxyhydroxides during weathering of the sulfide ore-body.

YTTRIUM AND THE RARE EARTH ELEMENTS (REE)

Solution species $Y(OH)_3$
REE^{3+}, $REE(OH)^{2+}$

Yttrium and the rare earths are moderately large trivalent ions, found largely in igneous rocks as phosphates, either as allanite and monazite, or in apatite. Normally all are trivalent in minerals and solution, but under oxidizing conditions Ce^{4+} is formed, and under extremely reducing conditions Sm^{2+}, Eu^{2+} and Yb^{2+} (Wood 1990). According to Wood, in typical acid groundwater, REE^{3+} is the dominant species, $REESO_4^+$ sub-dominant. Above pH 6, carbonate species will dominate under typical carbonate concentration of 10^{-4} M. Only at very high levels of P will phosphate species be significant. Many investigators have found that there is REE fractionation during weathering, with preferential retention of the light REE in the saprolite, and cerium being concentrated (as Ce^{4+}) in the upper more oxidized parts of a weathering profile. These observations are supported by Elderfield *et al.* (1990) who demonstrated that the heavy REE are enriched over the light in river waters, implying that 'the HREE's will be preferentially released to solution during the weathering of source rocks'.

During the weathering of igneous rocks there is much evidence for the truth of Nesbitt's (1979) suggestion that rare earths weathered from the primary minerals enter solution and then are recycled in the weathering profile. He reports that the outer weathering rinds of spheroidally weathered Torrongo granite (east central Victoria, Australia) are enriched more than threefold in heavy rare earths, and up to twice in the lighter rare earths. In contrast, the kaolinized saprolite between the corestones is depleted in all rare earths.

Banfield & Eggleton (1989) found that the REE excluding Ce were significantly enriched (about five-fold) in weathered granite. At the first stages of weathering of the granite, the REE phosphates rhabdophane and florencite occurred as precipitated crystals and in doughnut-shaped forms suggestive of fossilized bacteria, inside the shells of weathering apatite crystals in biotite. They concluded that the REE released to the groundwater from the weathering of apatite and allanite elsewhere in the profile, became fixed by the P in the organic matter *in situ*, but that Ce became oxidized and immobilized immediately on release from weathering allanite.

This general behaviour of the REE during weathering has been confirmed by a number of studies, including those of Middleburg *et al.* (1988), Price *et al.* (1991), Mongelli (1993) and Braun *et al.* (1993). Soubies *et al.* (1990) found a two-stage process in which apatite first loses REE and perovskite is dissolved, with Ca-rhabdophane crystals and anatase formed in the solution cavity, then when apatite dissolution is complete, the rhabdophane is partly dissolved and replaced by REE-florencite. Angélica & da Costa (1993) studied the REE distribution in lateritic rocks from Brazil. They found that the crandallite group of phosphates, particularly florencite, were important scavengers of REE in the Fe crust. High levels of Ce^{4+} and anatase in the Fe crust were interpreted as Ce substituting for Ti in the anatase.

Trescases *et al.* (1987b) show clear evidence for the change in REE solubility with pH. In the upper, acidic parts of a weathering sequence established on REE-rich argillaceous sediments in Brazil, rare earths were leached, except for Ce. The lighter rare earths became deposited in fissures as lanthanite ($REE_2(CO_3)_3.8H_2O$) immediately above a calcrete horizon, and the heavier rare earths were leached from the profile. The change in pH at the calcrete horizon caused the precipitation of the REE as carbonate.

It has been commonly reported that rare earths are associated with clays in the regolith (for example Marker & De Oliveira (1990) report REE and Ba bound to hydrobiotite and vermiculite in the saprolite of a weathered syenite), and the interpretation is made that the REE are present in exchange sites. REE phosphates described from weathering profiles are typically in the size range 1–5 μm, with many being <2 μm. The clay fraction (<2 μm) separated from weathered rocks is likely to contain a concentration of REE-phosphates, and this could lead to the assumption that the REE are chemically associated with the major clay minerals when in fact they may be only physically associated with that size fraction. Retention of REE in the weathering profile appears to depend on the presence of phosphate, which is largely sourced from apatite. The common association of apatite with biotite in igneous rocks may lead to an association of REE-phosphates with mica weathering products.

ZIRCONIUM

Abundance (mg/l)	W	K-P&P	U&BP&B-J
Mafic igneous	200		
Felsic igneous	200		
Shales	200		
Rivers			2.6 μg/l
Soils	250	345	250
Solution species: $Zr(OH)_3^+$, $Zr(OH)_4$, $Zr(OH)_5^-$			

Zirconium occurs in rocks predominantly as zircon. Zircon is highly resistant to weathering and is a common resistate mineral of the regolith. Hafnium and U/Th radioactive decay series elements are commonly retained in residual zircons, though the more soluble products such as Ra and Pb may be leached from metamict zircons during weathering.

Zirconium is highly insoluble in water, and data on its hydrolysis are few. According to Baes & Mesmer (1976) total Zr in solution shows a U-shaped relation with pH; in very acid solution (pH 2) Zr solubility is about 10^{-6} M, the minimum ($10^{-11.4}$ M) is at pH 5.4, and by pH 12 the solubility has returned to 10^{-6}.

Bruno & Sellin (1992) reported $Zr(OH)_5^-$ as the dominant species in granitic groundwater, and estimated total dissolved Zr to be in the range 10^{-10}–10^{-11} M.

Zirconium is very widely used as an immobile reference element in rock-weathering studies, because of its very low solubility and because it occurs in most rocks only in the mineral zircon, which is very resistant to weathering. Provided there is no mechanical movement of zircon in or out of the profile, Zr is probably the least mobile of the common elements. Nonetheless, there is a suggestion in the literature that Zr is mobile at high pH. Carroll (1953) suggested that Ca- and Na-bicarbonate-rich waters may be able to dissolve zircon, Moore (1996a) found evidence for Zr mobility during the alkaline weathering of basalt, and

Tejan-Kella *et al.* (1991) note greater etching on zircons from a young alkaline soil than from older acidic soils. Braun *et al.* (1993) show clear evidence for zircon dissolution during the weathering of a syenite from Cameroon. Increased weatherability of zircon in alkaline groundwater is consistent with the limited data on Zr solubility.

MOLYBDENUM

Abundance (ppm)	W	K-P&P	U&BT
Mafic igneous		1	
Felsic igneous		2	
Shales		1	
Rivers			
Soils		2	3.22
Solution species: MoO_4^-			

Because of its biological importance, there is a considerable agriculture and soils literature on Mo (see, for example, Kabata-Pendias & Pendias 1984; Nicholas & Egan 1975). In sulfide ore bodies, Mo occurs as molybdenite, MoS_2. Under oxidizing weathering, Mo is mobilized largely as $HMoO_4^-$.

Molybdenum is considered to be a pathfinder element for Au. Gossans over auriferous lodes in arid Australia typically carry of the order of 10 g/t Mo (Butt & Zeegers 1992a), appreciably higher concentrations than in average rocks. At low pH (below 6) under oxidizing conditions, the oxyanions MoO_4^{2-} and $HMoO_4^-$ are very sparingly soluble, of the order of 10^{-7} at pH 9 (Vlek & Lindsay 1977). Thus under the acid conditions of sulfide gossan development Mo should be immobile. Adsorption to Fe-oxides and oxyhydroxides is thought to fix Mo in the regolith, and it also precipitates as ferrimolybdite ($Fe_2(MoO_4)_3 .8H_2O$) (Nickel & Daniels 1985) and wulfenite ($PbMoO_4$, Vlek & Lindsay 1977).

PLATINUM GROUP ELEMENTS (RUTHENIUM, RHODIUM, PALLADIUM, RHENIUM, OSMIUM, IRIDIUM, PLATINUM)

The platinum group elements (PGEs) occur in mafic igneous rocks either as sulfides (except Rh) or metallic alloyed with Fe (Buchanan 1988). Oxidation of PGE sulfides is achieved by the reduction of the PGE to metal as the S is oxidized, thus both metallic and sulfide sources of PGEs release metal to the weathering environment.

Generally, PGEs are released as native metals from weathering ultramafics, to concentrate as placers because of their high density. They are essentially unreactive under normal weathering conditions, though Os and Ru oxidize at lower pe than the other PGEs, and so gain some relative mobility during weathering (Brookins 1988). Gray *et al.* (1996) found a three- to fivefold increase in Pt and Pd in the ferruginous zone overlying the Ora Banda Sill in Western Australia, with these PGEs being incorporated in the regolith Fe–Al-oxides and oxyhydroxides. The extent of element concentration was of the same order as that for Cr, Zr and Cu, indicating immobility of Pd and Pt during weathering at that site.

Fuchs & Rose (1974) found that Pd was depleted relative to Pt in surface and soil horizons over the Stillwater Complex, Montana, USA. On the basis of thermodynamic calculations, they conclude that chloride promotes the solubility of both Pt and Pd (to solubilities of the order of 10^{-12} M).

GOLD

Abundance (μg/kg)	W	K-P&P	U&B
Mafic igneous	4		
Felsic igneous	2		
Shales	3		
Rivers	0.1		
Soils		Varies	
Solution species: $AuCl_2^-$, $AuCl(OH)^-$, $Au(S_2O_3)_2^{3-}$			

The literature on Au in the regolith is large, summarized for the regolith of arid terrains by Gray *et al.* (1992). In essence, Au is immobile under almost all regolith conditions, and its concentration is most commonly by physical translocation of particulate gold to a sediment trap. Nonetheless, Au is detected in regolith positions that suggest migration in solution, particularly associated with Fe-oxides and oxyhydroxides or with pedogenic calcrete. Bowell *et al.* (1996) examined four lateritic weathering profiles over auriferous rocks in tropical West Africa. They concluded that Au showed a sigmoidal distribution, being enriched in the lower saprolite depleted in the upper saprolite, and enriched again but not as strongly in the mottled clay just below the ferruginous crust (cuirasse) boundary. The upper enrichment was interpreted as hydromorphically dispersed Au precipitated at the water-table.

Highly saline groundwaters (chloride concentrations of the order of 50 000 mg/l) in an oxidizing environment can dissolve Au to form one of the chloride complexes $AuCl_2^-$, $AuCl(OH)^-$. If such solutions should encounter ferrous or metallic iron, precipitation of Au follows the equation

$$AuCl_2^- + Fe^{2+} + 3H_2O \rightarrow Au + Fe(OH)_3 + 3H^+$$

Co-precipitation of Au with Fe appears to explain Au enrichment at the oxidation front of a weathering profile.

Webster (1986) examined the mobility of Au during the weathering of sulfide–carbonate ore, where the conditions are alkaline, and concluded that both Au and Ag could transport as soluble thiosulfate ($Au(S_2O_3)_2^{3-}$, $Ag(S_2O_3)_2^{3-}$) in oxidizing environments. Williams (1990) notes that the metastable occurrence of thiosulfate during the oxidation of sulfide minerals could allow the solution and transport of Au in this form. Figure 7.16 shows the pH and redox conditions where various Au-species are stable.

Gold may also be solubilized by cyanide to form $Au(CN)_2^-$, an important reaction used in the extraction of Au from ore. Some plants and bacteria produce cyanide, and though so far unproven, these may play a part in the regolith mobilization of Au. The influence of cyanide-producing organisms may be clouded by the added influence of other organic substances that may affect Au solubility.

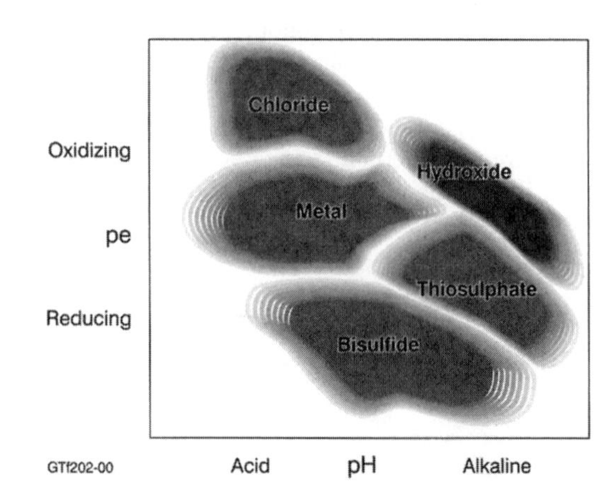

Figure 7.16 *pe/pH diagram showing the conditions for various forms of soluble Au*

LEAD

Abundance (g/t)	W	K-P&P	U&B
Mafic igneous	4		
Felsic igneous	20		
Shales	25		
Rivers	3 μg/l		
Soils		25	29

Lead substitutes for K in silicates, and so is highest in micas and feldspars (of the order of 50 g/t), reaching percentage levels in amazonstone (green K-feldspar) such as at Broken Hill, Australia. In sulfide ore-bodies, Pb is largely present as galena (PbS).

The weathering of galena is rapid in an oxidizing environment (Thornber 1985) releasing Pb^{2+} to solution. Mann & Deutscher (1980) investigated the solubility controls on Pb under various regolith solution chemistries assuming atmospheric levels of CO_2. Between pH 8 and 10, cerussite ($PbCO_3$) controls the concentration of Pb in solution at about 0.35 mg/l, rising sharply at pH < 7 or > 10. At acid pH, sulfate causes the precipitation of anglesite ($PbSO_4$), limiting the solubility of Pb to about 1.4 mg/l. High levels of chloride in groundwater considerably raise the concentration of Pb in solutions in equilibrium with the chlorides laurionite ($Pb_2Cl(OH)_3$) or cotunnite ($PbCl_2$). Anglesite and cerussite are common secondary Pb-minerals.

Clays are known to be able to adsorb lead, smectites particularly (Helios-Rybicka *et al.* 1995; Sikora & Budek 1994). Lead is strongly adsorbed on to Fe-oxides/hydroxides, and in the presence of sulfate, plumbojarosite ($PbFe_3[SO_4]_2(OH)_6$) can form from the resulting Fe–Pb–sulfate association. Arsenate is commonly present where sulfides are weathering, in which case beudantite ($PbFe_3[AsO_4][SO_4](OH)_6$) may form. Where P levels are high, plumbogummite ($PbAl_3[PO_4]_2(OH)_5.H_2O$) or pyromorphite ($Pb_5(PO_4)_3Cl$) may precipitate. Manganese oxides also are strong Pb adsorbers, where Pb occurs in tunnels in their structure, particularly in coronadite, or as an interlayer cation in the phyllomanganates.

THORIUM

Abundance (g/t)	W	K-P&P	U&B
Mafic igneous	1		
Felsic igneous	20		
Shales	10		
Rivers	0.01		
Soils			14
Solution species: $Th(OH)_4$			

Thorium occurs in igneous rocks mainly in allanite and monazite, as well as in U-bearing minerals as a product of radioactive decay of the U. Thorianite (ThO_2) and thorite ($ThSiO_4$) occur in pegmatites and as detrital grains in sediments.

The dominant Th-species in solution is $Th(OH)_4$. Bruno & Sellin (1992) report the solubility of Th in groundwater to be 2×10^{-10} mol/l in equilibrium with crystalline ThO_2, the same solubility as given by Baes & Mesmer (1976). Ryan & Rai (1984) suggest the solubility product K_{sp} for the reaction

$$ThO_{2\ (amorphous)} = Th^{4+} + 4(OH)^{-} \text{ is } K_{sp} = 45.9$$

compared with the value of 44.7 in the compilation of Lelong *et al.* (1976) shown in Table 7.5. Ryan & Rai conclude that the solubility of $Th(OH)_4$ is lower than 10^{-18} M. Measurements at such low levels are extremely difficult, but the data clearly show that Th is highly insoluble in the weathering environment with a solubility at least 10 000 times lower than Al, hence Th is generally classed with Zr and Ti as an immobile element.

Östhols *et al.* (1994) examined the influence of carbonate on Th solubility, and found that at high CO_2 partial pressures (e.g. 0.1 or greater), Th hydroxides dominated the solution species below pH 5. However, from pH 5.5 to 7.5 the dominant species in solution was $Th(OH)_3CO_3^-$, while above pH 7.5 $Th(CO_3)_5^{6-}$ became dominant. The experimentally determined total dissolved Th in the presence of 10% CO_2 at pH 7 is of the order of 10^{-5} M, rising with increasing pH. Data from alkaline lakes show Th concentrations of the order of 10^{-7} M, suggesting compounds other than the carbonates control Th solubility in natural systems.

URANIUM

Abundance (g/t)	W	K-P&P	U&B
Mafic igneous	0.5		
Felsic igneous	15		
Shales	4		
Rivers	~1 ppb (varies)		1 μg/l Palmer & Edmond 1993
Soils	1–10 ppm		2

Solution species: reducing: $U(OH)_4$, oxidizing: $UO_2(CO_3)_2^{2-}$, $UO_2(CO_3)_3^{4-}$

Uranium in the crust occurs as U^{4+}, where it is strongly partitioned into felsic igneous rocks because of its large size and high charge. Where U concentrations are high, uraninite (UO_2) may be an accessory mineral in granite or a minor mineral in pegmatites and hydrothermal veins. More commonly, U occurs camouflaged in other accessory minerals such as apatite, sphene and zircon.

In solution under reducing conditions U^{4+} is highly insoluble, whereas U^{6+} is readily soluble in oxidizing groundwaters (2×10^{-7} cf. 3×10^{-4} mol/l, Bruno & Sellin 1992). Frick (1986) reported that under arid climate weathering of a South African granite, U levels in the rock below the water-table (reducing) ranged between 10 and 30 mg/l, whereas above the water-table (oxidizing) concentrations were generally <5 mg/l, the difference attributed to dissolution during weathering. The U in this granite occurred largely in apatite, monazite and biotite, and on weathering these minerals released U, which then became concentrated in groundwater carbonates downstream.

Bicarbonate in solution increases the solubility of U, leading to the formation of carbonate-hosted U deposits when CO_2 loss causes precipitation of insoluble U-carbonates.

Studies of U mobility at the Ranger and Koongarra deposits of the Northern Territory, Australia (von Gunter *et al.* 1999) showed that 80–90% of the U released by weathering was quickly adsorbed on to Fe-oxides and oxyhydroxides, initially ferrihydrite, remaining fixed as that mineral transformed to goethite and haematite. Uranium levels in the groundwater were very low, of the order of 5 μg/l.

8
Mineral weathering

Introduction

In Chapters 3–6 we have developed two interwoven strands – the minerals of the regolith and the environmental forces that affect them. We have shown the structural similarities and differences of the minerals, the weathering agents that act on them, how those agents, particularly water, behave in the landscape, and how minerals and water react. In Chapter 7 we emphasized the regolith mineral hosts for the elements that are released and redistributed during chemical weathering. We are now in a position to look in some detail at what happens to the minerals as they weather; at the sequences of mineral change and the cumulative effect of these changes on weathering rocks.

The weathering of the silicates can be described ultimately in terms of three processes:

- replacement of more soluble ions by protons;
- change of Al coordination from tetrahedral to octahedral; and,
- oxidation of ferrous iron.

These processes lead to changes in the bulk chemistry, generally quantified by a 'chemical index of weathering'. There is a consequent change in volume of the weathered rock through increased porosity, quantified by density measurement as long as the rock fabric is preserved, but unquantifiable beyond that stage of weathering because there is no longer any reference point. Once the rock fabric has been lost, or possibly even before, introduced materials or loss of macroscopic material such as clays, can change the bulk chemistry in a way unrelated to that rock's chemical weathering.

REPLACEMENT OF MORE SOLUBLE IONS BY PROTONS

Consider first the starting and final mineralogy and composition of a weathered intermediate igneous rock such as a syenite (Table 8.1). The simplified and idealized weathering reactions are:

$K[Si_3Al]O_8 + 8H_2O \rightarrow Al(OH)_3 + KOH + 3Si(OH)_4$
K-feldspar + water → gibbsite + solubles
oxygen loss: 5

$Ca_{0.5}Na_{0.5}[Si_{2.5}Al_{1.5}]O_8 + 35/4H_2O \rightarrow 1.5Al(OH)_3 + 0.5Ca(OH)_2 + 0.5NaOH + 2.5Si(OH)_4$
intermediate plagioclase + water → gibbsite + solubles
oxygen loss: 3.5

$Ca_2(Mg_2Fe_2Al)[Si_7Al]O_{22}(OH)_2 + 20H_2O \rightarrow Al(OH)_3 + FeO(OH) + 2Ca(OH)_2 + 2Mg(OH)_2 + 7Si(OH)_4$
hornblende + water → gibbsite + goethite + solubles
oxygen loss: 19

$K(Mg_2Fe)[Si_3Al]O_{10}(OH)_2 + 10H_2O \rightarrow Al(OH)_3 + FeO(OH) + 2Mg(OH)_2 + KOH + 3Si(OH)_4$
biotite + water → gibbsite + goethite + solubles
oxygen loss: 7

Comparing the original syenite and final bauxite separately, Table 8.1 shows the differences in composition expressed as atom %. This very clearly indicates that Fe and Al are unaffected, and that H is hugely increased, at the expense of all other cations. However, this example compares 100% of syenite with 100% of bauxite, whereas in reality 100 g of syenite only produce 36 g of bauxite. Figure 8.1a shows the changes caused by weathering on a mass basis, starting from 100 g of syenite. It is again evident from this figure that the dominant chemical change is the replacement of Si, Mg, Ca, Na and K by H.

Each weathering reaction involves the addition of water and the loss of ions in solution. Overall there is a loss of oxygen from the weathered rock, for with each cation leached from the rock goes a charge equivalent

Table 8.1 *Comparison in atom percentage between syenite and its weathered equivalent bauxite*

	Syenite atom%	Bauxite atom%
Si	19.7	0
Al	8.0	13.0
Fe	1.4	2.3
Mg	1.9	0
Ca	2.1	0
Na	1.2	0
K	3.5	0
H	1.9	41.2
O	60.2	43.5

amount of oxygen. In the end, only the oxygen needed to balance the charges on Al and Fe, plus the additional protons, remain. Figure 8.1b shows the chemical changes per 100 atoms in the original syenite.

Figure 8.2 compares the atomic constitution of the Earth's crust with that of intensely weathered regolith. Assuming a global rate of weathering of 1 mm/1000 years (see Chapter 10), of the order of 40 million t of water are fixed annually by rock weathering. Compared to the amount of water available in the ocean (10^{18} t) or falling annually on the Australian continent's surface (10^{12} t), the amount fixed by weathering is small.

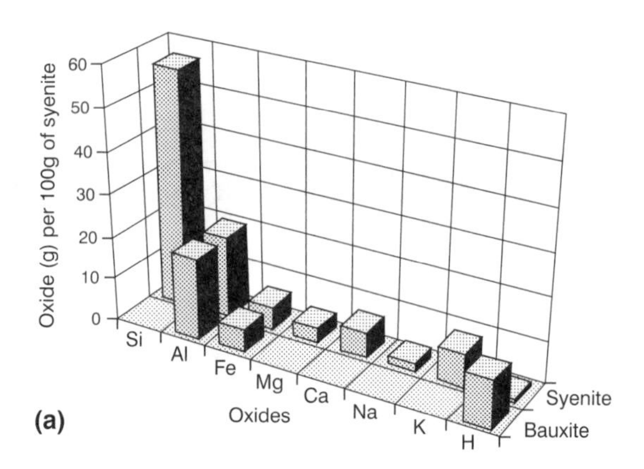

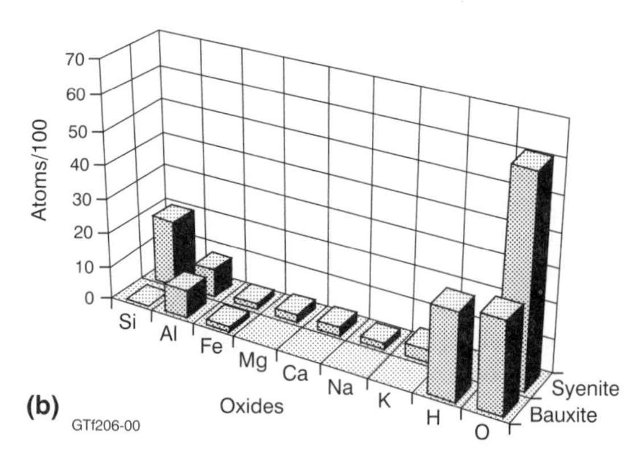

Figure 8.1 *(a) Change in major element content resulting from the weathering of a syenite to bauxite, assuming complete immobility of Al and Fe and complete loss of all other oxides. (b) Change in major element proportion as syenite weathers to bauxite. Over half the original oxygen atoms are lost*

CHANGE IN ALUMINIUM COORDINATION TO OXYGEN

Because Al has an ionic radius between those for ideal tetrahedral and octahedral coordination to oxygen (see Chapter 3), in more open structures (higher temperature) it occurs in tetrahedral coordination whereas at low temperature (or at high pressure) it is in octahedral sites.

The most abundant igneous minerals are the feldspars, and in these Al is in tetrahedral coordination to oxygen. In pyroxenes there is little Al except in the rare mineral jadeite; what little there is substitutes for Si in the tetrahedral chains. In amphiboles there is both tetrahedral and octahedral Al, in roughly equal

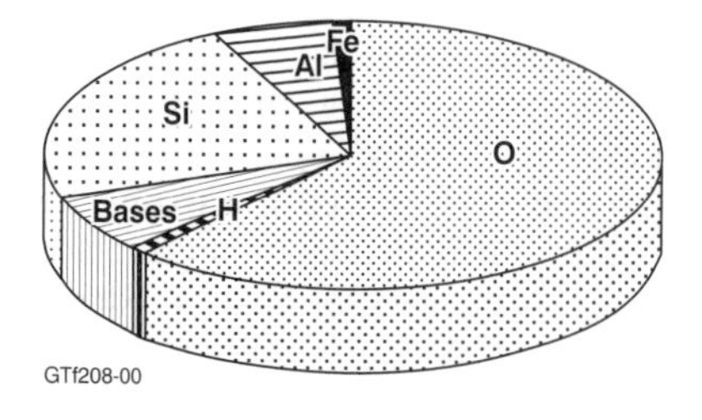

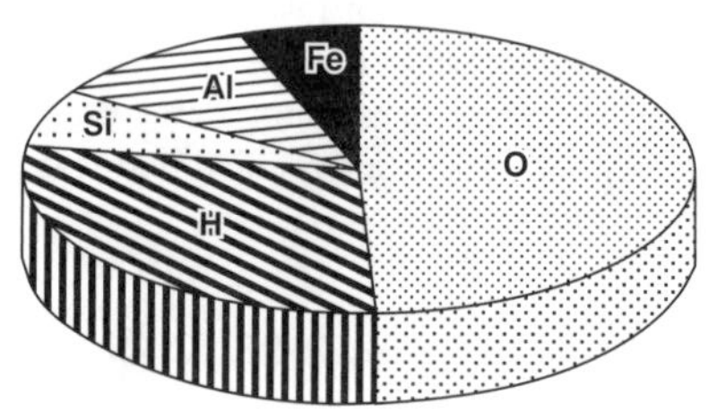

Figure 8.2 *Major atom percentage in (a) crustal rocks, and (b) deeply weathered rocks*

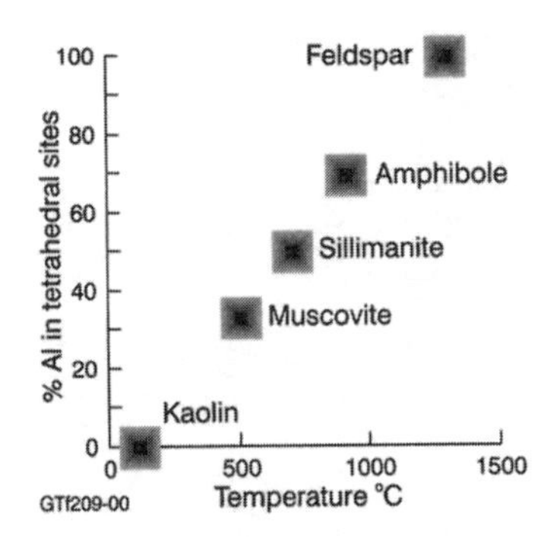

Figure 8.3 *Tetrahedral Al as a ratio of total Al plotted against approximate temperature of mineral formation for a selection of silicates*

proportions, in muscovite the octahedral to tetrahedral Al ratio is two to one. Biotite has Al in the tetrahedral sheet and generally a lesser amount in the octahedral sheet. Thus the dominant coordination for Al in igneous rocks is fourfold.

Metamorphic rocks, in so far as they are composed of micas, amphiboles and feldspars, have similar Al coordination. Metamorphic rocks derived from pelitic sediments ultimately yield garnets, aluminosilicates such as andalusite, cordierite and chlorite. In these minerals, Al coordination varies: octahedral in garnets, the aluminosilicates and many chlorites, and also tetrahedral in sillimanite and chlorite.

The mineral products of weathering are few, but in none is there a major amount of tetrahedrally coordinated Al. Kaolinite has only octahedrally coordinated Al, as do gibbsite and boehmite. Aluminium in goethite can reach 30 mol%, all in octahedral sites substituting for Fe^{3+}. The aluminous smectite beidellite has an octahedral : tetrahedral ratio of about 6 : 1, and although the trioctahedral smectites and vermiculite have in general only tetrahedral Al, they have no more than do the aluminous smectites.

The general trend from tetrahedral coordination at high temperature to octahedral coordination at low temperature is exemplified by the common aluminous minerals (Figure 8.3). In this figure the temperature of formation has been broadly approximated, and the data restricted to low-pressure environments.

OXIDATION OF FERROUS IRON

The oxidation of Fe has a very marked and visible effect on a weathering rock. Most obvious is the change in colour from the pale greens or browns of ferrous minerals olivine, pyroxene, amphiboles and biotites to the strongly pigmenting red and brown of haematite and goethite. Oxidation destroys the crystal structure of almost all silicates as a result of the valence change from Fe^{2+} to Fe^{3+}, demanding a concomitant change in the anion charge. While this is possible for a few minerals through a coupled substitution of $(OH)^-$ for O^{2-}, only stilpnomelane, and to a lesser extent biotite, ever undergo this reaction. All other minerals undergo structural breakdown, leading to a weakening of the rock fabric. A strong feedback mechanism accentuates this fabric change, because the structural collapse that follows oxidation opens the weathering rock, giving easier access to oxygen and water, causing further oxidation and dissolution. By this process minute cracks enlarge, and the oxidation 'front' migrates into the weathering rock.

Mineral weathering

While it is evident that the weathering of a rock can be expressed only in terms of the changes in its chemistry, the chemicals are always present in the form of an association of minerals. Knowing the processes and stages of mineral weathering provides the basis for understanding both the major and trace element geochemistry of weathering.

Mineral weathering reactions are relatively easy to write, but not so easy to get right. The equation stated earlier: feldspar + water = gibbsite + solubles is a fair description of both the starting and finishing mineralogy and the starting and finishing chemistry, but gives no indication of the complex steps that connect parent and offspring.

Whenever dissolution is occurring, the rate of that process is temperature dependent. In theory:

$$\text{rate} = K \exp(-E/RT)$$

where K and E are reaction-dependent constants, R is the gas constant and T the temperature (°kelvin). Lasaga (1995) concluded that the average activation energy for silicate dissolution was of the order of 15 kcal/mol. Using this value, dissolution would occur ten times faster at 25°C than at 0°C. In experiment, Grandstaff (1986) found forsteritic olivine dissolved five times as fast at 26°C as it did at 1°C, and 12 times faster at 49°C. Arctic weathering rinds are certainly thinner than those of temperate regions, but this may result from water availability and organisms as much as from temperature.

A most fundamental control on the rate of mineral dissolution, and hence on the rate of chemical weathering, is the composition of the weathering solution. As long as the concentration of the dissolved species is in equilibrium with the solid species, no dissolution occurs and the rate of weathering is zero. There are many reported instances of the way the solution composition controls the weathering products. For example, regions of a catena, where groundwater ponds become saturated in silica, so that though kaolin might be the stable phase in the higher, well-drained parts of the profile, smectites dominate the hydromorphic regions. Norrish (1972) summarized the effect of K concentration on the weathering of mica, and found that as little as 1 mg/l K in the weathering solution was enough to stop K loss from illite. Schmitt (1983) concluded that complete albitization of Triassic sediments had occurred as the result of continental deep weathering beneath a Jurassic land surface during a period when the groundwater had a high Na content.

The nature of the weathering solutions (temperature, pH, chemistry) affects not only which minerals weather first and their weathering rate, but the way in which they do it. Chapter 6 (Chemistry of weathering) described the difference between congruent and incongruent weathering, and indicated that under equilibrium conditions, some minerals dissolve congruently while others pass through a sequence of neoformed minerals, that is, they dissolve incongruently. Chemical equilibrium, however, is not always attained during weathering. The introduction of fresh rain to a weathering profile that is in chemical equilibrium with its existing weathering solutions rapidly changes the solution chemistry much faster than mineral weathering can restore equilibrium. This new, dilute water will be able to dissolve any components of the primary minerals, and those of the secondary products, until the least soluble element reaches saturation. During this initial attack, dissolution will be congruent, only returning to incongruent dissolution when saturation with one or more elements is reached.

Mineral weathering begins by chemical attack at points of weakness in the mineral structure. All minerals have defects, and these provide the points for the entry of water to the crystal structure. Grain boundaries are the initial avenues for water penetration, then dissolution may begin at edge and screw dislocations, at exsolution lamellae or along twin planes. Cracks, caused by shrinking during cooling, expansion from unloading by erosion or external stresses such as diurnal temperature change or shearing, provide another weakness along which water can penetrate.

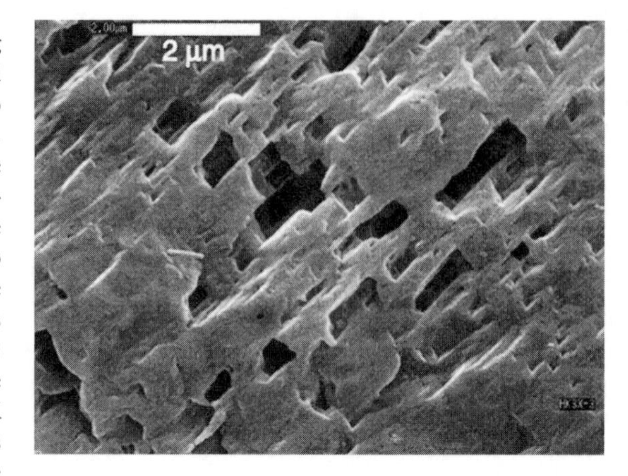

Figure 8.4 *Etch pits in feldspar (photo R. A. Eggleton)*

Studies of silicate mineral weathering have repeatedly shown that dissolution causes the formation of etch pits (Figure 8.4); pits on the mineral surface whose sides parallel crystallographic directions of the mineral (Parham 1969; Berner & Holdren 1979; Berner *et al.* 1980). As etch pits enlarge, they coalesce into larger cavities, until the crystal becomes a honeycomb and is eventually completely dissolved. Velbel (1983) described how this process can lead to the formation of pseudomorphs of secondary minerals after the primary mineral. When the weathering solutions are undersaturated with respect to all the elements of the mineral, congruent dissolution occurs, focused at dislocations and defects with the formation of etch pits. Changes in water chemistry with time or space may lead to an environment where a secondary mineral, such as kaolinite or gibbsite, starts to precipitate in any available cavity, including the etch pits left by the earlier congruent dissolution. A return to undersaturation will cause more rapid dissolution of the primary mineral than the secondary product, former etch pits enlarge and new ones develop, which in turn become filled with clay when the solution becomes sufficiently saturated. Eventually the primary mineral is wholly replaced by the clay in such a way that even subtle crystallographic properties of the parent, such as twin planes, may be reflected in the pseudomorph.

SPECIFIC MINERALS

Glass

Glass is found in quickly cooled volcanic rocks. It is extremely reactive, and soon forms a weathered rind. The first product of basaltic glass is commonly a yellowish smectite called *palagonite*. Andesitic glass (containing more Si and less Fe) weathers to a whiter aluminous smectite, generally referred to as *bentonite*. The alteration of volcanic glass to imogolite and allophane is commonly reported (see e.g. Wada 1989).

Olivine

Olivine is one of the most reactive minerals. It contains ferrous iron and Mg, both react quickly with air and water. The common weathering products are smectite and goethite (Delvigne *et al.* 1979), producing accurate pseudomorphs known as iddingsite (Figure 8.5a). Eggleton (1984) and Smith *et al.* (1987) showed that the oxidative alteration of olivine to iddingsite begins with the development of comb- or tooth-like etch channels spaced at about 200 Å (see Figure 11.4b). Within these channels, precursor clay minerals grow, forming slim bridges across the channels. With increased alteration, these precursor clays enlarge to become recognizable as saponite. At the same time, the Fe in the olivine oxidizes and minute (10–20 nm wide) goethite crystals grow, oriented in crystallographic continuity with the olivine, presumably inheriting their orientation from the oxygen framework of the parent crystals of olivine (Figure 8.5b). Some Mg is dissolved out and the rest remains as Mg-smectite (incongruent solution of olivine). At this stage the altered olivine is called *iddingsite*.

The opening of the olivine crystal through the loss of Mg, according to the reaction

$$8(MgFe)_2SiO_4 + 16H^+ + O \rightarrow 2Mg_3Si_4O_{10}(OH)_2$$

olivine — 'saponite' + neglecting tetrahedral Al and interlayer cations

$$- 2FeO(OH) + 8Mg^{2+} + 5H_2O$$

allows Al to diffuse into the crystal, following weathering of surrounding glass or feldspar. This promotes the formation of aluminous smectite, and with further weathering, spherical halloysite. Details of the later stages are less clear, but the final product of mild weathering is a goethite + kaolinite/halloysite pseudomorph:

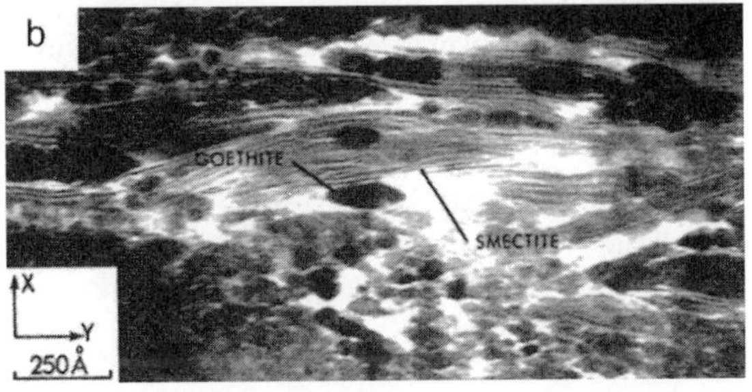

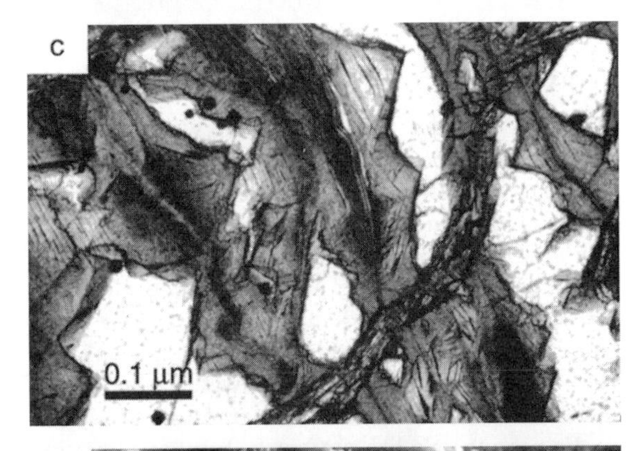

Figure 8.5 *Weathering of olivine. (a) Iddingsite rim on olivine seen in optical thin section; (b) TEM of goethite and smectite making up iddingsite; (c) optical micrograph of bowlingite alteration of olivine (clear); (d) SEM image of random smectite on etched olivine (photos a, b: R. A. Eggleton; c, d: K. L. Smith)*

$$8Mg_{.75}Fe_{.25}SiO_4 + 2H_2O + O_2 + 0.5Ca + Al \rightarrow$$
olivine
$$2Ca_{.25}Mg_3[Si_{3.5}Al_{.5}]O_{10}(OH)_2.H_2O + 2FeO(OH)$$
Mg-smectite goethite
$$+ SiO_2$$

If the conditions of weathering preferentially remove Al over Fe, the iddingsite pseudomorphs may be entirely goethite.

If the olivine weathers below the water-table where air may be excluded, the Fe is not oxidized, and a pseudomorph of smectite (saponitic) or vermiculite forms called *bowlingite* (Figure 8.5c, d) (Delvigne *et al.* 1979). When this clay is exposed to air, the Fe in it quickly oxidizes and either nontronite or goethite results:

$$8(MgFe)_2SiO_4 + 2H_2O + O_2 + 0.5Ca + Al \rightarrow$$
olivine
$$2Ca_{.25}(MgFe)_3[Si_{3.5}Al_{.5}]O_{10}(OH)_2.H_2O$$
Mg-smectite
$$+Mg(OH)_2 + SiO_2$$

Pyroxenes

Pyroxenes weather to clay minerals whose ultimate composition depends on the pyroxene composition. The initial reaction appears to depend more on the pyroxene structure, and to involve the production of complex biopyriboles. Eggleton & Boland (1982) examined the weathering of orthopyroxene from an igneous rock composed dominantly of enstatite. They found a sequence of reactions, beginning with the development of 2-, 3-, 4- and wider chain-width structures (Figure 3.15) and their gradual evolution to an Fe-bearing talc-like structure. Further alteration led to the ejection of Fe from the talc and the production of discrete Fe-oxides and talc.

Wang (1988) examined the weathering of augite from a syenite, and found the same initial reactions (see Figure 11.4a), leading to the formation of complex chain-width pyriboles. These in turn altered to smectites, initially maintaining structural coherence with the parent augite, but later collapsing into randomly oriented clays before changing to kaolinite/halloysite.

Amphibole

Amphiboles have a good cleavage that allows water and air relatively easy access to the mineral. Alteration begins along cleavage cracks with the production of an 'intergrade mineral' (Proust & Velde 1978) or a talc-like mineral, as it does in Mg-rich pyroxenes (Wang 1988), or smectite (Jeong 1992). Iron-rich amphiboles on oxidation of Fe produce a ferric-iron rich smectite (nontronite), and those poorer in Fe, beidellite, both with dissolution of Ca, Mg and some Si. Early in weathering, the layer silicates form in structural continuity with the parent amphibole (Figure 8.6a). Once weathering has proceeded far enough to open the amphibole structure, etch pits form, in which goethite and kaolinite precipitate (Figure 8.6b) (Velbel 1989; Jeong 1992). Iron and Al are largely conserved, and the other elements are variably lost.

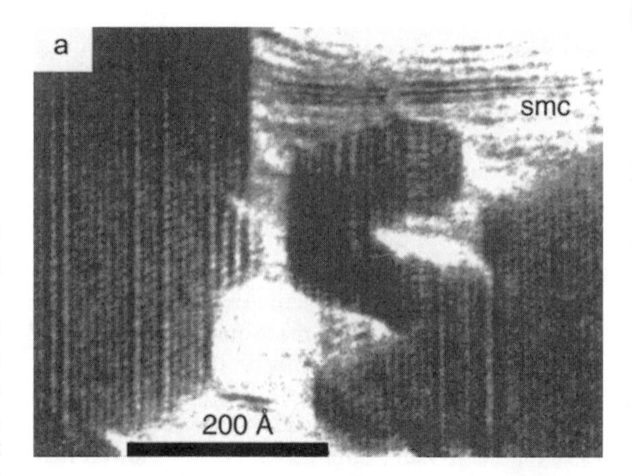

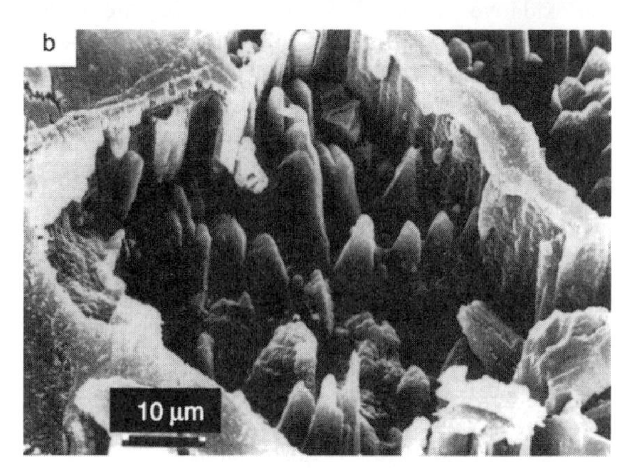

Figure 8.6 *Weathering of amphibole. (a) 3- and wider chain width pyriboles and smectite (SMC) formed from amphibole (cf. Figure 3.15). (b) Box-work formed by clays and Fe-oxyhydroxides precipitated along cleavages early in weathering, with later clays replacing amphibole in the boxes (photo Q-M. Wang)*

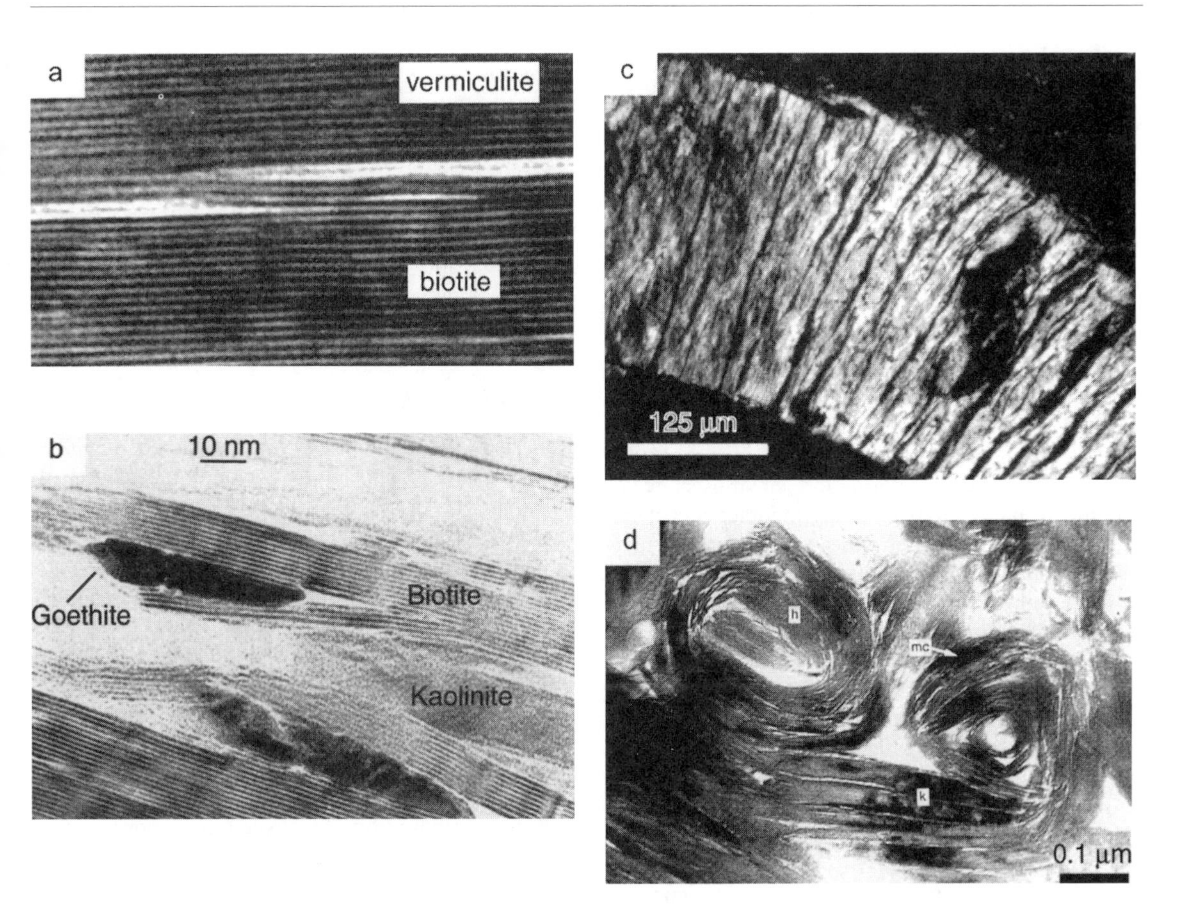

Figure 8.7 *Layer silicate weathering. (a) TEM image of biotite altered to vermiculite in structural continuity. (b) Later alteration of biotite showing remnant biotite (10 Å fringes), goethite (dark pods) and kaolinite (featureless regions, photos M. Aspandiar 1998). (c) 'Accordion' of kaolinite weathered from muscovite (optical thin section). (d) TEM image of kaolin after muscovite (k) curling to form halloysite (h). Remnants of unaltered mica (mc) have rolled with the kaolinite (c, d: from Robertson & Eggleton 1991)*

Biotite

Many studies of biotite weathering have demonstrated the initial formation of 'hydrobiotite' (biotite with K lost from alternate layers), followed by interstratified biotite vermiculite, then by vermiculite (e.g. Coleman *et al.* 1963; Norrish 1972; Graham *et al.* 1989). In hand specimen, biotite flakes become gold-coloured, particularly when released from saprolite and seen in hill slope wash or creeks. Banfield & Eggleton (1988) and Wang (1988) examined the weathering of granitic and monzonitic biotite respectively, reaching essentially the same conclusions. In the initial stages, K is leached from alternate interlayers, leading to an unexpanded random interstratification of biotite and K-depleted biotite (all layers still 10 Å). Next, brucite-like layers occupy the K-depleted interlayers, producing short sequences of regularly interstratified biotite–vermiculite interspersed with biotite which gradually evolves to vermiculite (Figure 8.7a).

The Mg required for this appears to come from biotite layers that have been more completely dissolved by weathering. Consequently, although there is an increase in thickness of the 'hydrobiotite' from 2×10 Å to $10 + 14$ Å, there is no overall volume increase, because each 14 Å layer is produced at the expense of two 10 Å layers (one dissolved and one expanded). Overall, there is a volume decrease.

Later in biotite weathering, this decrease is accompanied by the replacement of vermiculite by kaolinite and goethite growing in crystallographic orientation to the biotite, probably by epitaxy (Figure 8.7b) as well as by halloysite in open spaces left by dissolution.

Chlorite

Chlorite initially weathers to interstratified chlorite–vermiculite (Proust 1982; Jeong 1992; Aspandiar 1992), and then with further weathering to dioctahedral vermiculite.

Muscovite

The initial breakdown of muscovite yields dioctahedral vermiculite and kaolinite (Meunier & Velde 1979). Transmission electron microscopy (TEM) by Banfield & Eggleton (1990), Robertson & Eggleton (1991), Singh & Gilkes (1991) and Jeong (1992), showed that the transformation of muscovite to kaolinite or halloysite is probably topotactic, evolving from a 2 : 1 layer silicate (either muscovite or a first-stage alteration illite–smectite or vermiculite), so that kaolinite inherits much of its structure from the muscovite (Figure 8.7c). Because muscovite has one out of three tetrahedral sites occupied by Al, and kaolinite has no Al in its tetrahedral sheet, some reconstitution of the structure must occur. Singh & Gilkes (1991) showed a high degree of inheritance of the octahedral sheet in the conversion of a Cr-muscovite to a Cr-kaolinite.

The reaction may be written as

$$\underset{\text{muscovite}}{2KAl_2[Si_3Al]O10(OH)_2} + 4H_2O \rightarrow \underset{\text{kaolinite}}{3Al_2Si_2O_5(OH)_4} + K_2O$$

Vermiculite

Itself a product of weathering rather than a primary mineral, vermiculite is an intermediate stage in the weathering sequence for pyroxenes, amphiboles, muscovite and chlorite, but most commonly vermiculite is formed from biotite. Most vermiculite weathers to a mixture of kaolinite and goethite (e.g. Jeong 1992), with the release of Mg and Si. For example,

$$2Mg_{0.5}.6H_2O.(Fe^{3+}Mg_5)[Si_6Al_2]O_{20}(OH)_4 + 16H_2O \rightarrow 2Al_2Si_2O_5(OH)_4 + 2FeO(OH) + 11Mg(OH)_2 + 8H_4SiO_4$$

According to Buurman *et al.* (1988), a trioctahedral vermiculite similar in composition to that above, weathered to nontronite by loss of Mg from the vermiculite with retention and oxidation of the octahedral iron.

Smectite

Though themselves commonly the product of weathering (bisiallitisation), smectites also weather, mainly by loss of silica to kaolinite (monosiallitisation). For beidellite, the chemistry of this process is described in Chapter 7. Saponite releases its major constituents (magnesia and silica) to solution, though the silica may precipitate near by as opaline silica. Nontronite weathers to goethite, also with the release of silica.

Kaolinite and halloysite

Robertson & Eggleton (1991) and Singh & Gilkes (1992a) have proposed that kaolinite can alter to halloysite (Figure 8.7d). Jeong (1992) demurs, giving evidence suggesting that kaolinite and halloysite alteration products of mica have formed independently, and also showing an example of kaolinite forming from halloysite (Jeong 1998). Singh & Mackinnon (1996) showed experimentally that kaolinite can be persuaded to roll by repeated intercalation. This issue of the early alteration of kaolinite is unresolved, but there is ample evidence that thereafter, both kaolinite and halloysite weather following the path indicated in Chapter 6, with the loss of silica and the crystallization of gibbsite (allitisation):

$$\underset{\text{kaolin}}{Al_2Si_2O_5(OH)_4} + H_2O \rightarrow \underset{\text{gibbsite}}{2Al(OH)_3} + 2SiO_2$$

This is a very common weathering reaction in deeply weathered terrains, where laterite and bauxite are prominent. The upper parts of such profiles are rich in gibbsite, formed by leaching of kaolinite.

Feldspars and feldspathoids

Clay minerals and muscovite have Al in octahedral coordination to oxygen (Al surrounded by six oxygens). In feldspar the Al is in tetrahedral coordination (surrounded by four oxygens). To produce a clay from feldspar requires Al to change coordination, and that involves the entire breakdown of the structure. Feldspars initially weather by breakage of an Si–O–Si bond in the [AlSi] tetrahedra framework, under the influence of an H^+ ion. The H^+ replaces Ca^{2+} or Na^+ or K^+, and the structure is quickly destabilized. Submicron-sized etch pits created by dissolution are detectable in optical thin section because they impart a turbid appearance to the feldspar (Figure 8.8a). Reports of synthetic weathering commonly refer to a cation-depleted coat that may or may not impede diffusion as dissolution proceeds. The literature on this subject is extensive (see e.g. Wollast & Chou 1985; Banfield & Eggleton 1990). Other evidence, both from observations of naturally weathered plagioclase and from experimental dissolution, suggests plagioclase weathering can be congruent, that is, the mineral dissolves without the formation of an intermediate phase.

Most of the artificial weathering studies drew conclusions from rate of dissolution and rate of cation addition to the solvent, or by indirect physical methods (e.g. X-ray photoelectron spectroscopy). Using SEM, Holdren & Berner (1979) were able to show both the existence and nature of clay on the feldspar surface. They concluded that it was patchy and highly permeable.

Tazaki & Fyfe (1987) showed very clear evidence for an intermediate stage between feldspar and its weathered product halloysite. They called this material 'primitive clay precursors', and showed it to be an Fe aluminosilicate with a circular or spheroidal habit, 150–200 Å in diameter. Delvigne (1998, p. 294) shows micrographs of plagioclase extending from fresh rock to a weathered crust. Between the plagioclase and its alteration product gibbsite is a zone of isotropic material, interpreted as the gibbsite precursor.

The existence of a thin (0.05–0.25 μm) non-crystalline coat, grading into smectite in the case of K-spar and Na-plagioclase, or to halloysite in the case of Ca-plagioclase, has been shown by Banfield & Eggleton (1990) and Wang (1988). Furthermore, this non-crystalline coat is seen (by microprobe analysis) to be relatively enriched in Fe (Figure 8.8b), as Tazaki & Fyfe (1987) had found, thus providing a clear view of the reason for the yellow staining that is seen to penetrate weathering feldspars; the Fe released from the weathering of biotite or olivine diffuses along cracks in the feldspar, and combines with the non-crystalline breakdown product of feldspar weathering. Crystallization of this material leads to an Fe-bearing smectite (but not nontronite), and ultimately to a mixture of halloysite and goethite (Figure 8.8c)

In contrast to the research giving evidence for incongruent dissolution of plagioclase and the direct formation of intermediate phases, there is a body of work that supports congruent dissolution. Velbel (1983) used light microscope and SEM images of the early stages of plagioclase weathering from granites at Coweeta, SC, USA, to conclude that the feldspar

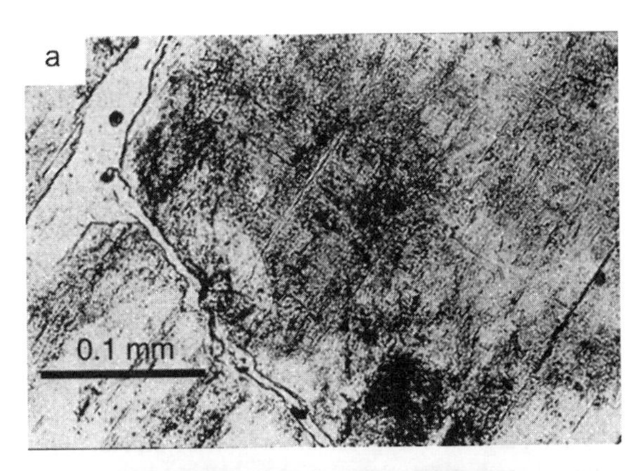

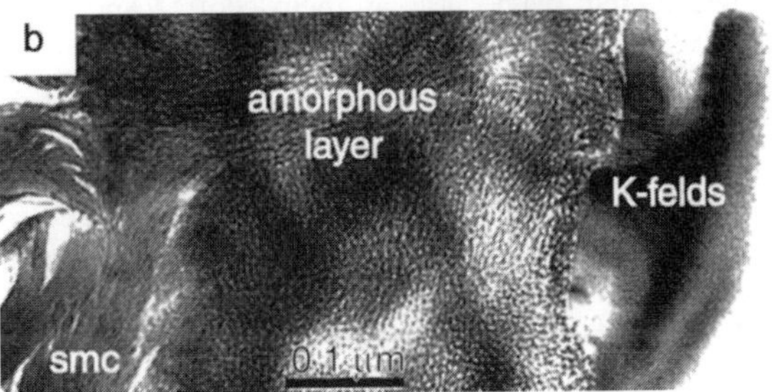

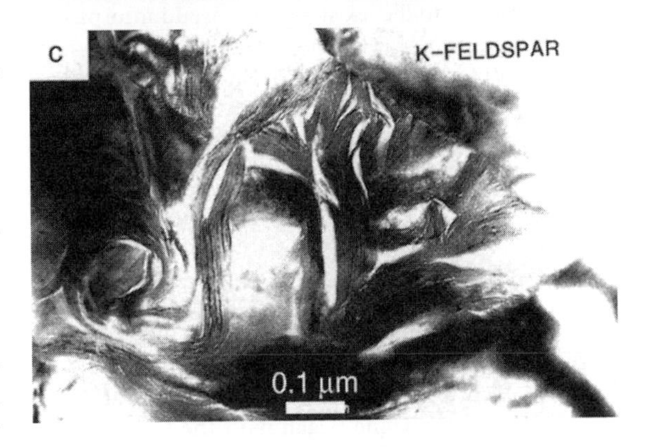

Figure 8.8 *Feldspar weathering. (a) Turbid slightly weathered feldspar seen in optical thin section. (b) Fe-rich aluminosilicate coat of weathered K-feldspar (Banfield & Eggleton 1990). (c) Etch pit in feldspar filled with halloysite (photo Q-M. Wang)*

weathered by a dissolution–precipitation process with no intermediate stage. At the early stages, nearest to unaltered plagioclase, etch pits are completely devoid of any weathering product or alteration coating, suggesting congruent dissolution. Precipitation of kaolinite in regions of the regolith away from the initial dissolution sites was observed to be identical in feldspar pseudomorphs and cracks in other minerals such as quartz. Boulet (1974) showed that plagioclase weathering transformation began with embayments around the margins of grains where they were in contact with other minerals like quartz and K-feldspar. The embayments were filled by an amorphous aluminosilicate gel (similar in composition to kaolinite in wetter climates and smectite in drier climates). Okumura (1986) described the weathering of plagioclase (An_{90}) from a gabbro, and concluded that the dissolution mechanism depended on the 'surrounding aquatic condition'. Deeper in the profile, plagioclase alters directly to kaolin minerals, whereas at intermediate depths congruent dissolution takes place giving rise to dissolution voids in the plagioclase. In the highest parts of the profile, corestones still contain unweathered plagioclase at their centre, and their weathering yields skeletal gibbsite separated from the plagioclase by an 0.2 mm thick film of allophane. Jeong (1992) describes the weathering of labradorite to kaolin minerals as a congruent solution–precipitation process, but with an amorphous precursor to halloysite very early in the weathering. The differences between the various studies may be one of scale, with TEM revealing early, transient, incongruence that is followed by congruent loss of Si and Al. These studies demonstrate that the existence of pseudomorphs of clay replacing feldspar is not evidence for a solid-state replacement process. The form of the feldspar, including the twin lamellae, can be preserved during a weathering process in which dissolution and precipitation occur repeatedly, gradually replacing the host mineral.

Nepheline

Wang (1988) examined the weathering of nepheline from a nepheline syenite of the Mt Dromedary complex, New South Wales, Australia, and found that the etching of nepheline began with the formation of tiny etch pits (Figure 8.9). Coalescence of the etch pits produced bigger corrosion holes where halloysite was found. Halloysite was the only secondary product of nepheline seen in this study. By contrast, Schumann (1993) found that nepheline from a syenite in Brazil weathered directly to gibbsite with no intervening clay. The difference between these observations probably results from climatic differences, as there is twice the rainfall (1700 mm) at the Brazilian site than on the New South Wales coast (900 mm).

The process of feldspar and feldspathoid weathering appears to be independent of rock type, except in so far as the feldspar composition is related to rock type. The main difference between feldspar weathering in mafic and felsic rocks is that the more

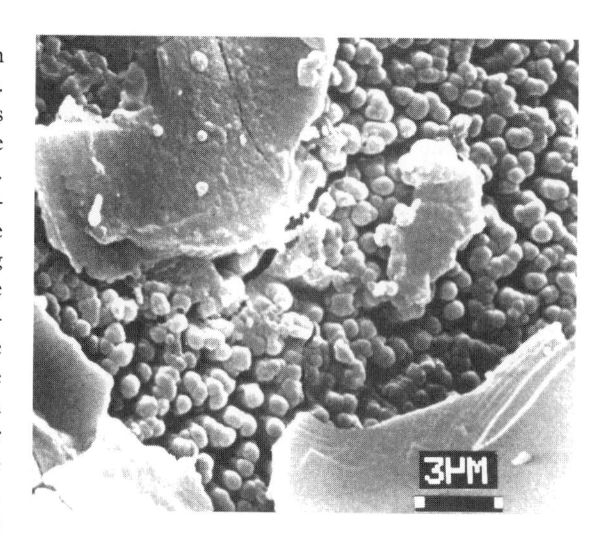

Figure 8.9 *Weathering of nepheline showing halloysite-filled etch pits (Wang 1988)*

aluminous anorthite (and nepheline) of mafic rocks have a composition similar to kaolinite, whereas albite and orthoclase have compositions closer to smectite. These similarities lead to a more direct conversion of anorthite and nepheline to kaolinite/halloysite than of albite or orthoclase. Ignoring the involvement of Fe in the early stages, the chemical and mineralogical steps are:

$$\underset{\text{analcime}}{2NaAlSiO_4} + 2H_2O \rightarrow \underset{\text{kaolinite}}{Al_2Si_2O_5(OH)_4} + Na_2O$$

$$\underset{\text{anorthite}}{CaAl_2Si_2O_8} + 2H_2O \rightarrow \text{non-crystalline} + CaO \rightarrow \underset{\text{kaolinite}}{Al_2Si_2O_5(OH)_4}$$

The overall reaction for an alkali feldspar might be:

$$\underset{\text{alkali feldspar}}{2(K,Na,Ca)[Al_{1.15}Si_{2.85}]O_8} + 2H_2O \rightarrow \underset{\text{smectite}}{Ca_{0.15}Al_2[Al_{0.3}Si_{3.7}]O_{10}(OH)_2.H_2O} + \underset{\text{in solution}}{2.2SiO_2} + (K,Na)_2O$$

$$\underset{\text{smectite}}{0.87Ca_{0.15}Al_2[Al_{0.3}Si_{3.7}]O_{10}(OH)_2.H_2O} \rightarrow \underset{\text{kaolinite}}{Al_2[Si_2]O_5(OH)_4} + \underset{\text{in solution}}{1.2SiO_2 + 0.13CaO}$$

Quartz

Quartz, while not changed to a new-formed mineral, is commonly dissolved during weathering, and close

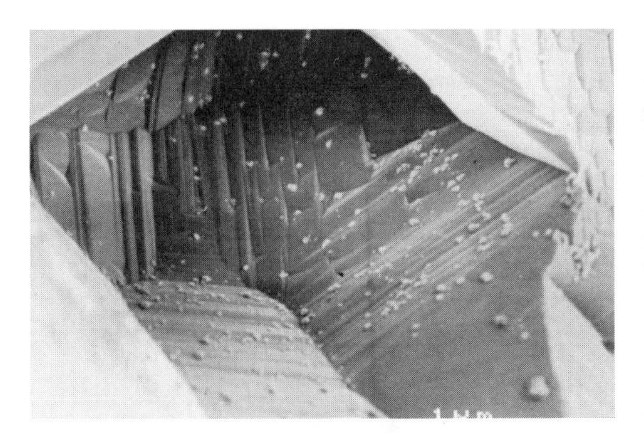

Figure 8.10 *An etch pit in quartz, approximately 10 μm wide (photo David B. Tilley)*

examination usually reveals crystallographically oriented triangular etch pits and preferential solution along fractures and crystal dislocation sites (Pye & Mazzullo 1994) (Figure 8.10). The size of the residual quartz is generally less than that of the grains in the fresh rock as a result of solution of fractures (Pye & Mazzullo 1994), stress release (Moss & Green 1975) and solution. Crook (1968) reports quartz rounding occurring as a result of solution under acid soil conditions, while Pye & Mazzullo (1994) show clearly that over about 8000 years, weathering of dune sands actually increases the angularity due to solution and secondary quartz precipitation, but that the grain form remains. This effect was due to the high concentrations of organic acids leached from the soils on the dunes. Brantley *et al.* (1986) showed that in a 1 m thick saprolitic weathering profile on Proterozoic granite in Venezuela, quartz was densely etched at the surface and down to about 60 cm and they suggested the degree of etching was directly related to the concentration of silica in the weathering solutions. At 60 cm the concentration reached saturation for 25°C and below that concentrations were above critical so no pitting occurred.

Ilmenite

Ilmenite weathers by oxidation of the ferrous iron to pseudorutile (~$Fe_2Ti_3O_9$), which in turn weathers to rutile (Grey & Reid 1975) or anatase (Anand & Gilkes 1984) (Figure 8.11). All three minerals (ilmenite, pseudorutile and rutile) are important constituents of heavy mineral beach sands. Rutile itself is highly resistant to weathering and is a common resistate mineral in the regolith.

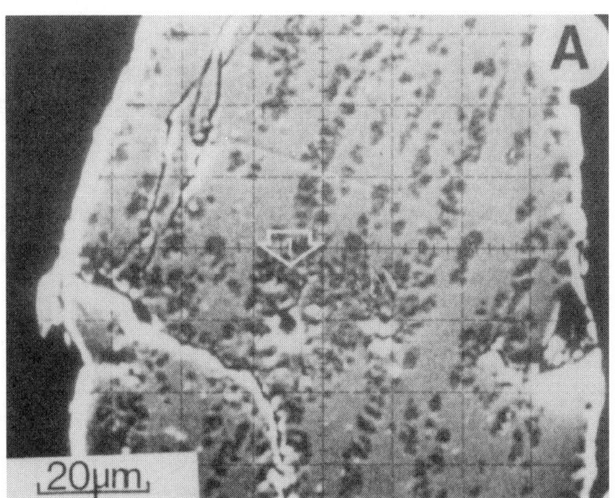

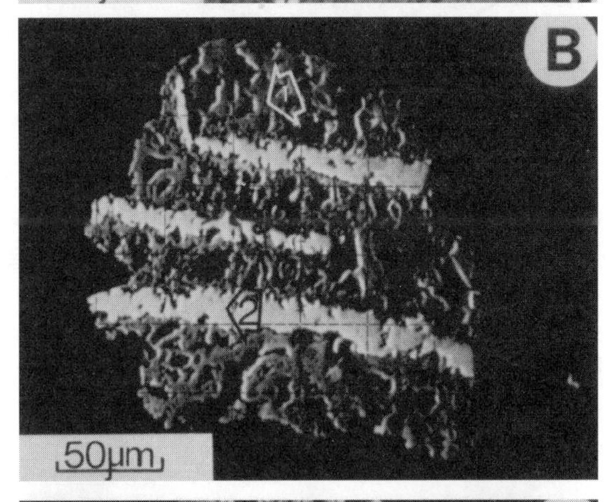

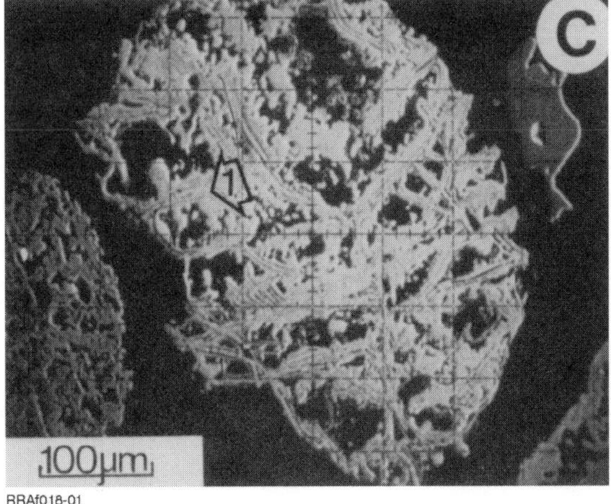

RRAf018-01

Figure 8.11 *Backscattered electron images of polished sections of altered ilmenite grains. (a) Patchy pseudorutile (1) within ilmenite. (b) Porous framework of anatase (1) separated by approximately parallel zones of unweathered ilmenite and/or pseudorutile (2). (c) A porous framework consisting solely of anatase (1) (Anand & Gilkes 1984)*

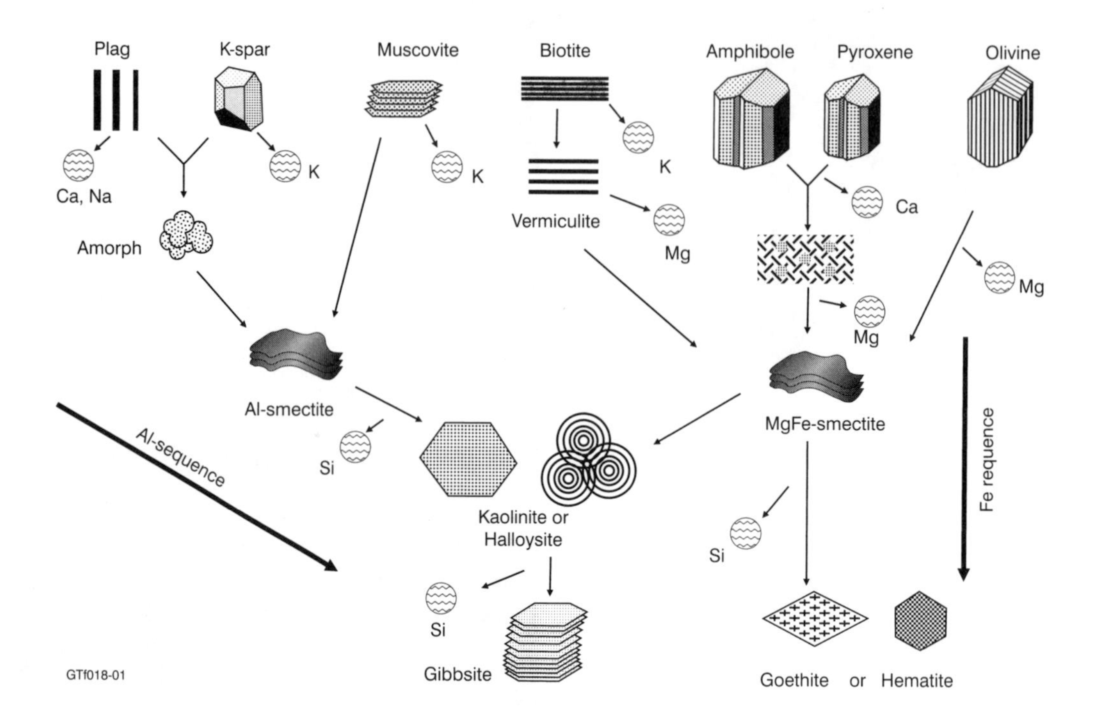

Figure 8.12 *Summary of silicate mineral weathering*

OTHER MINERALS

Mineral	Common weathering products
Garnet:	
Pyralspite	Smectite and goethite or Mn-oxides and hydroxides
Grandite	Silica and goethite
Titanite	Beidellite and anatase
Kyanite	Kaolinite
Glaucophane	Na-saponite
Cordierite	Halloysite
Glauconite	Smectite → kaolinite + goethite
Stilpnomelane	Vermiculite → nontronite → goethite + opal
Talc	Saponite
Serpentine	Magnesite and opal
Magnetite	Maghemite, haematite

Conclusion

In summary, chemical weathering is a sequential process. The larger, lower-charged cations are most soluble. In the early stages of weathering of olivine, chain-silicates and micas, loss of alkalis and alkaline earths causes minimal disruption to the structure of the mineral or the fabric of the rock. The early replacement minerals occupy the original mineral volume with a topotactic or epitactic relation to the parent, and have a composition dependent on that of the original mineral. By contrast, framework silicates weather by a complete loss of structure immediately after the replacement of Na, Ca or K by H. The resulting non-crystalline aluminosilicate is transformed to a clay, whose composition reflects the parent silicate – smectite from high silica feldspar, halloysite from intermediate plagioclases or nepheline. Later in weathering, when diffusion avenues have been opened by volume loss, small-scale mobility of Al and Fe leads to homogenization of the mineralogy as the early clay products (smectite and ferrihydrite) are replaced by the intermediate products kaolinite, halloysite and goethite, and ultimately by gibbsite and haematite or goethite. A summary of mineral weathering pathways and products is shown in Figure 8.12.

9
Rock weathering

Rocks weather when they interact with air, water and biota in the near-surface environment. The interaction is greatest near the surface, and declines with depth. There is therefore a progressive change in weathering from surface to interior, and the physical extent of that region of change is called the *weathering profile*. A weathering profile is a one-dimensional section through weathered mineral or rock material. Although it is part of a larger three-dimensional picture, the profile is a site-specific concept. It is independent of scale and may equally apply to a weathered crust on a pebble (commonly called a 'weathering rind') or to a section several hundreds of metres thick. Weathering profiles may be made up of any of the products formed by mineral and rock weathering. This includes material chemically or physically weathered.

The variations observed through a profile are variously called zones (e.g. Ruxton & Berry 1957; Ollier 1969), horizons (e.g. Tardy 1997; Birkeland 1984; Nahon 1991), facies (e.g. Boulangé *et al.* 1997) or layers (e.g. Boulangé *et al.* 1997). All of these names are reasonable, but for many people each has specific meanings other than those related to weathering profiles. 'Horizons' specifically refer to variations in soil profiles. In a geological context a horizon refers to a thin layer or time-plane with a characteristic litho- or bio-facies or a surface separating two beds of rock. 'Layers', while commonly called beds in sedimentary geology, has a sedimentary connotation and is not a preferred term for this reason. 'Facies' is a word simply meaning the appearance or aspect of any rock or the sum total of its characteristics. Many, including Boulangé *et al.* (1997), seem to use it to refer to mineralogical aspects of the profile ('kaolinitic facies'). 'Zone' has many uses in geological literature, but in particular it refers to a particular unit of rock defined by the characteristics of the fossils it contains. In plain English, horizon means the line which marks the joining of the Earth and sky, layer means a thickness of material, facies has no plain meaning, and zone commonly means a continuous area which differs in some respect from adjacent areas. We prefer to use the term 'zone' for cognate parts of weathering profile, but facies would be equally acceptable.

By *in situ* we mean materials that have not been transported but are in the place they were originally formed by the weathering process that formed the profile. This concept of weathering profiles being *in situ* has many complications, as the processes of weathering themselves require the movement of materials at various scales to form the profile. Let us explain.

Consider a weathering profile developed on granite. The processes of weathering may initially be physical, such as the development of physical openings along joints and stress fractures as the granite is progressively uncovered by erosion. These processes may lead to small blocks and fragments of granite moving locally. As chemical weathering occurs solutes are moved by water passing through the granite. The new-formed minerals are often very fine-grained (from nanometres to micrometres in diameter) and are consequently capable of being physically moved in the solutions moving through the granite. Plants and animals living in the regolith create voids through which larger and smaller particles (sand-sized) may move. None of these movements negate the profile being considered *in situ*.

The reverse of *in situ* is transported regolith, which may still form part of a seemingly continuous weathering profile. By transported we mean material that has been moved by surface processes usually associated with physical erosion. Such materials may be moved by gravity (hill-creep, colluvial processes), water (sheet wash, alluvial processes), wind or ice. These materials may deposit over an *in situ* weathering profile and later be incorporated into it, sometimes to such a degree that it is indistinguishable from the *in situ* material. Such profiles we call polygenetic

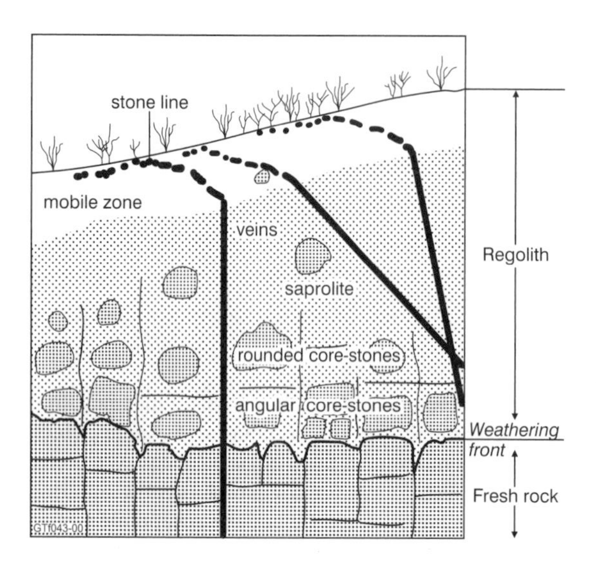

Figure 9.1 *A typical* in situ *weathering developed on granite with aplite or quartz veining and joints*

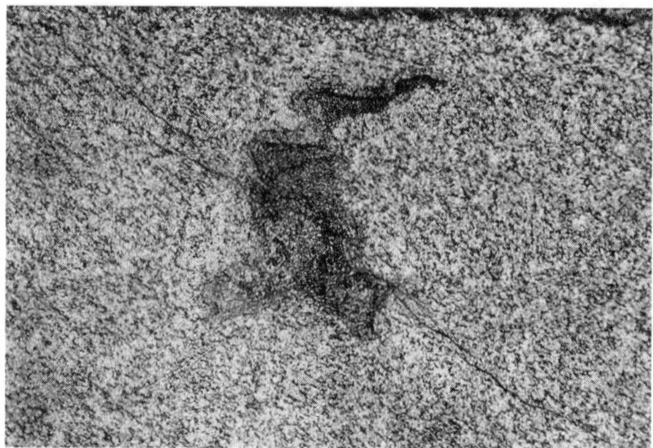

Figure 9.2 *Weathering in a granite obelisk (Adelaide) marked by oxidation of Fe^{2+} (from biotite) to Fe^{3+} (red-brown coloured goethite) (photo GT)*

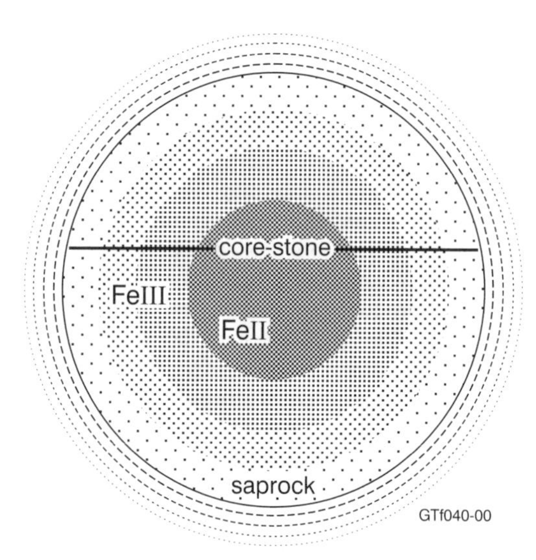

Figure 9.3 *Nature of the weathering changes across weathering fronts illustrated by considering a weathering rind on a corestone*

weathering profiles. In other cases it is possible to separate the transported from the *in situ* materials, but if weathering subsequent to the deposition of another layer has formed what appears to be a single profile they would still be considered a polygenetic profile. It may happen that the layer of sediment added to the *in situ* profile has been thick enough to cause weathering to be stopped in the covered profile so we have two weathering profiles, one sitting upon and isolated from the other.

A typical *in situ* weathering profile

The various zones of an *in situ* profile (Figure 9.1) from the fresh rock up through the profile consist of the following.

FRESH ROCK

Fresh rock is rock unaltered by the processes of weathering. In cases where the rock contains Fe^{2+} the most obvious signs of weathering are the appearance of red or yellow Fe^{3+} oxides or oxyhydroxides (Figure 9.2). In other cases the boundary between fresh and weathered rock may not be clear. Minerals typical of chemical weathering products can be part of the original rock, particularly in the case of sedimentary rocks or those formed by hydrothermal alteration.

WEATHERING FRONT

The weathering front is the boundary between fresh and weathered rock. In most cases the boundary is sharp and clear when observed in the field. At smaller scales however, the boundary is more diffuse. Figure 9.3 illustrates the gradational change across the weathering front. The weathering of homogeneous rocks like granite or basalt proceeds down joints in the rock so the weathering front is usually irregular and deeply indented. But in such rocks it is common for

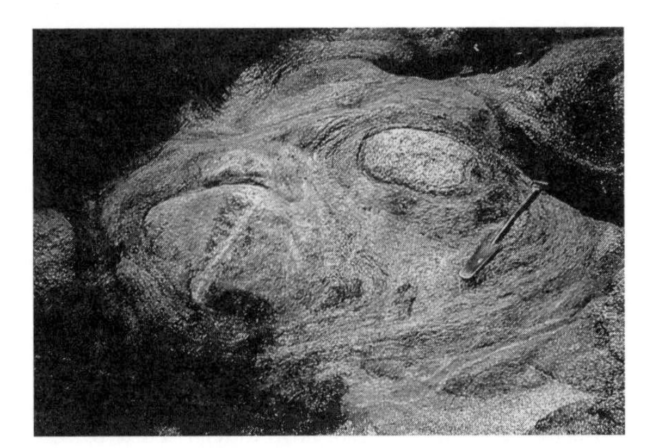

Figure 9.4 *Spheroidally weathered granite showing the gradational nature of the weathering front in the corestone (photo GT)*

lumps of fresh rock with spherical or ellipsoidal shapes (corestones) to occur detached from the fresh rock below the front. In this case isolated corestones of fresh rock are encased in sheets of progressively more weathered rock which give way to unstructured saprolite (Figure 9.4). In other cases the corestone may pass more rapidly outwards into relatively unstructured saprolite. In less homogeneous rocks like sedimentary or metamorphic rocks the small-scale weathering front is usually irregular, as the different lithologies making up the rock mass weather at different rates (Figure 9.5).

Despite these irregularities, with the passage of time the weathering front moves further and further into the rock, creating a thicker and thicker regolith. It therefore follows that the regolith closest to the surface has been weathering longer than material deeper in the regolith. Thus the upper parts of a weathering profile, unless some exceptional circumstances obtain, are older and more weathered than those in deeper parts of the profile.

SAPROCK

Saprock is the first stage of weathering. It consists of partially weathered minerals and as yet unweathered minerals (e.g. feldspars have begun to alter to clay minerals or olivine has altered to iddingsite). Saprock maintains all the fabric and structural features of the fresh rock. It is distinguished from the more weathered 'saprolite' by being physically strong compared with it, usually requiring a sharp hammer blow to break it. Saprock may or may not contain corestones, or the outer rim of corestones may be saprock.

Figure 9.5 *An irregular weathering front on Proterozoic sedimentary rocks at Fowler's Gap, north of Broken Hill, New South Wales (photo Steve Hill)*

SAPROLITE

Saprolite is more altered than saprock, but still retains the fabric and most of the structural characteristics of the parent rock (Figure 9.6). Saprolite is a term coined by Becker (1895) derived from the Greek σαπροσ (sapros) meaning 'rotten'.[1] This, like saprock, implies there has been little or no bulk volume change during the weathering processes. In saprolite, weathering-resistant minerals remain more or less as they occurred in the parent and weatherable minerals are wholly or partially pseudomorphed by clays and/or oxides and oxyhydroxides so that rock fabric is maintained. Saprolite may or may not contain corestones.

Although saprolite retains rock fabrics at a field observation scale, closer examination reveals that considerable proportions of the material present may

1. Merrill (1897) makes some interesting observations on the use of this term. Saprolite fits his category of regolith (Chapter 1) called residual and he considered it impossible to include all residual materials under one term. He thought Becker's use of saprolite as the term for such materials objectionable as it conveys the idea of putridity. He also comments on 'geest', a provincial term to describe such materials because it had a multiplicity of meanings in contemporary dictionaries. Similarly, he criticized the old German word *Gruss* for the same reasons. Gruss (or grus) is used now to mean granitic saprolite by many researchers.

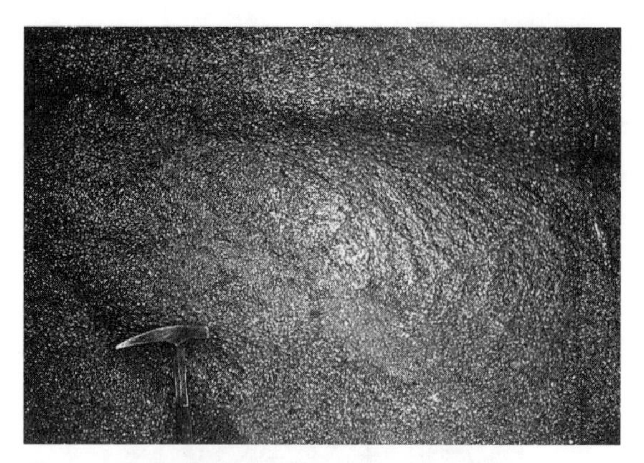

Figure 9.6 *Detail of saprolite showing rock fabrics and joints preserved (photo GT)*

be locally transported. The major regolith components to move are fine-grained new-formed mineral particles and sands as well as chemically transported and deposited materials.

ABOVE THE SAPROLITE

Upward from the saprolite, the processes of weathering wreak changes to the rock fabric to the point where the parent is no longer distinguishable. These changes are gradual, so that despite the loss of fabric, the material can still be inferred to be derived from the parent rock, and indeed chemical analyses show smooth trends in element composition throughout the profile from fresh rock into and commonly above this region of profound change.

The causes of change are many. Simple gravitational movement, particularly in downward-percolating water, moves the finer clay particles, and as they are moved, the coarser mineral grains such as quartz can themselves settle. Plant roots can penetrate this clay-bearing region, and as they grow they push the mineral grains around. Invertebrates tunnel and excavate, displacing grains, and the seasonal wetting and drying of the profile cause smectites to swell and shrink, working the regolith into ever-changing fabrics.

The changes that transform the saprolite are largely the same as those that form soil. In some profiles this region where the parent fabric begins to be lost can be viewed as the base of the B-horizon. Consequently, for some authors, everything above saprolite is termed the 'pedolith', while the saprock and saprolite are collectively termed 'saprolith'. Because the changes in this region are the result of movement, largely of particles, but also of chemicals, the term 'mobile zone' is used by others. In the French literature, the saprock is referred to as the zone where the fabric of the parent rock is conserved, and this is contrasted with the zone above, where the fabric has been lost.

There appears to be no word in the literature that uniquely and unequivocally can be applied to this part of a weathering profile. 'Mobile zone' implies a process, and certainly in many instances mobility would seem to be self-evident. The 'zone of lost fabric' implies the presence of a former fabric; in general we try to avoid defining things on the basis of features that are not evident, because such an approach colours observation with conceptual models. Part of the problem stems from the fact that there is no simple boundary between weathered rock below and reorganized regolith above; the boundary is invariably undulating, and there may be patches of saprolite (e.g. corestones) entirely surrounded by the reorganized material. Nonetheless, in any profile one can usually put a hand on the place where pedogenic processes are more apparent above, and rock fabric more evident below. We will refer to the upper region as the pedolith.

Within the pedolith there may be a variety of subregions, or zones. If these are arranged horizontally, and particularly if they are part of the solum, they may be called 'horizons'. Where clays are dominant, 'plasmic zone' is used, because 'plasma' is defined as any soil material of diameter <2 μm (clay sized). Where Fe has been lost so that the colour of the region is white compared to the material above and below, 'pallid zone' is commonly used. The region where quartz grains dominate above granite saprolite has been referred to as the 'arenose zone'. If the region shows evidence of being the remnants of a former saprolite but having undergone compaction by gravitational settling, 'collapsed saprolite' has been used, though this is a somewhat self-contradictory term, for 'saprolite' is defined as retaining the fabric of the parent. A part of the pedolith may be colour-mottled (generally in red and white) and so may be called the mottled zone. However, once the word 'zone' has been introduced it is difficult not to see the profile in terms of some conceptual model in which a sequence of 'zones' are the consequence of some perceived process.

Robertson *et al.* (1998) report that in the plasmic or arenose zone, resistant minerals like quartz are physically fragmented to produce angular shard-like grains. The secondary growth of kaolinite aggregates, or blasts, also causes the quartz to be segregated from the growing kaolinite aggregates to form a sandy material (Figure 9.7).

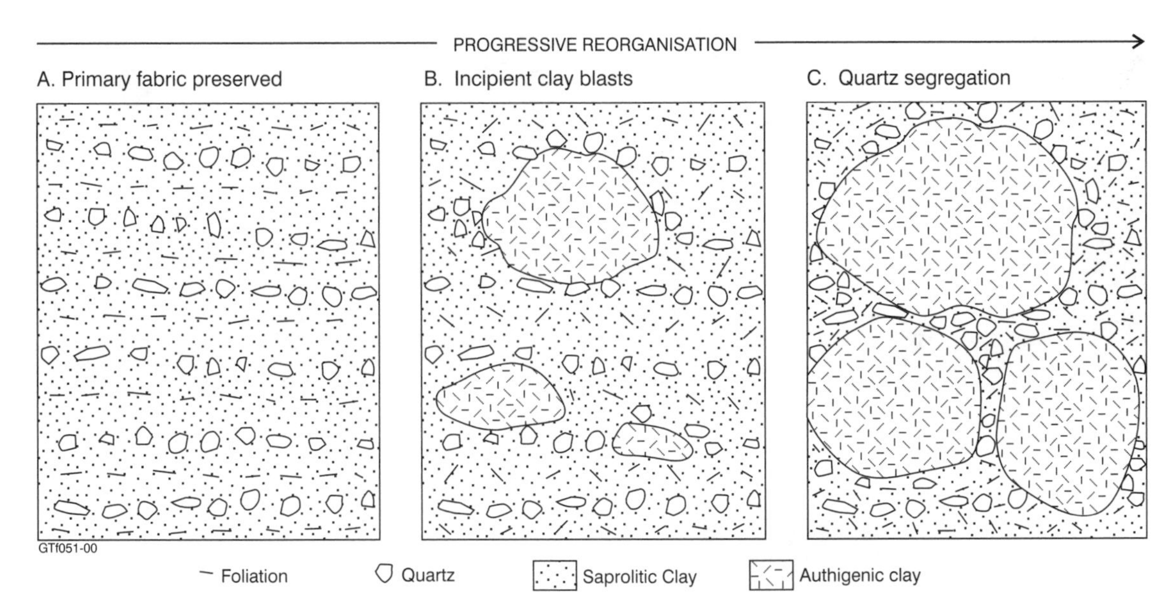

Figure 9.7 *Loss of a schistose fabric by progressive development of clots of secondary kaolinite around which remnant quartz becomes segregated (from Robertson* et al. *1998)*

Figure 9.8 *Hill creep photo showing cleavage planes in Ordovician sediments bending downhill (photo GT)*

Mobility of part of a weathering profile is most easily recognized where veins and dykes cut the rock and the movement of the upper part of the profile by hill-creep is demonstrated by a breakage of the vein and its development into a stone line migrating downslope (Figure 9.1). Another indicator of movement may be the tilting of steeply dipping strata or cleavage downslope due to drag caused by regolith movement (Figure 9.8).

Bioturbation as a cause of saprolite collapse is common, and because the main biological activity is near the surface this is why it occurs in the upper parts of profiles. Common forms of bioturbation include the activity of fauna that inhabit the regolith such as termites, ants, worms and other similar creatures. Milne (1947) reports termite activity produced the upper layer of regolith on slopes in Tanzania. In Queensland Watson & Gay (1970) report 20% of the ground covered by 2–3 m diameter termite nests indicating a considerable biological turnover of regolith in its upper parts. There are similar reports by Wheeler (1910) from the central USA, Glover *et al.* (1964) from Kenya and many others. Paton *et al.* (1995) report termite turnover as ranging from 0.02 to 5.9 t/ha per year. Holt & Greenslade (1979) report turnover rates of 0.4–0.8 t/ha per year from Charters Towers, Queensland. At rates such as these every part of the pedolith can be disturbed over a period of 50 000 years. At Coober Pedy in South Australia termite galleries are observed up to 30 m below the surface, and across northern Australia they are common down to the dry-season water-table which can vary in depth to about 30 m (Figure 9.9).

As well as animals, plants have a significant effect in bioturbation of the regolith, in the upper part in particular. Tree growth displaces regolith materials as the bowl and root system expands. On the death of the trees the roots rot and cavities are left into which the regolith can collapse (Figure 9.10). Similarly, material from higher up the profile may collapse down the

Figure 9.9 *Termite activity at ~25 m depth in an opal open pit at Coober Pedy. Note darker material that has entered deep into the profile through termite galleries (photo GT)*

Figure 9.10 *Silicified rhizomorphs in clay-rich pallid zone filled with sand from higher in the profile, Innaminka, South Australia (photo GT)*

voids causing regolith mixing. Tree fall is another form of plant bioturbation that causes significant disruption to the uppermost parts of the profile. Tree fall generally only disturbs the upper 0.5–1.5 m at most of a profile (Figure 9.11).

These upper regions of the regolith merge imperceptibly with the soil. This book does not concentrate on soil science per se, but we do summarize soil classification systems in Chapter 11, where the more 'bedrock-independent' aspects of the regolith profile are considered.

Figure 9.11 *Tree fall at Weipa after a cyclone has passed through. Note the top metre or so of regolith is exposed and will gradually mix as it is washed from the root bowl (photo GT)*

Profile environment

Through any weathered profile, the environment of weathering and the changes that occur are largely independent of rock type. The saprock is likely to be wet, at least for some part of the year, the water : rock ratio is low, the water is silica saturated, and oxidation is slow or absent. Silicates alter to 2 : 1 clays (bisialitization) and sulfides oxidize to sulfate.

The saprolite is likely to be intermittently wet and dry as the water-table fluctuates through the year. The groundwater is continually freshened by rain and so is undersaturated with silica. Unweathered silicates and 2 : 1 clays alter to kaolin minerals (monosiallitisation); halloysite may dominate deeper, kaolinite higher up. Ferrous iron and Mn become oxidized, colouring the saprolite in browns, reds and black depending on local concentration of the oxidized products.

If erosion is slow relative to the weathering rate, the saprolite evolves towards a lateritic profile. Prolonged regular seasonal rise and fall of the water-table lead to the gradual removal of Fe from the zone of intermittent saturation. When saturated, reducing conditions develop, ferrous iron dissolves. The Fe diffuses through the profile, precipitating on oxidation at the water-table. By repeated dissolution and precipitation, Fe is moved from this mobile zone and is left at the top of the wet-season water-table. Simultaneously, the silica-undersaturated water dissolves any primary quartz, the two processes yielding a kaolin-rich zone in the profile (referred to as the plasmic zone or, because of its white colour, the pallid zone) (Figure 9.12).

At the level of Fe precipitation, the regolith is also heavily colonized by plant roots and organisms. These

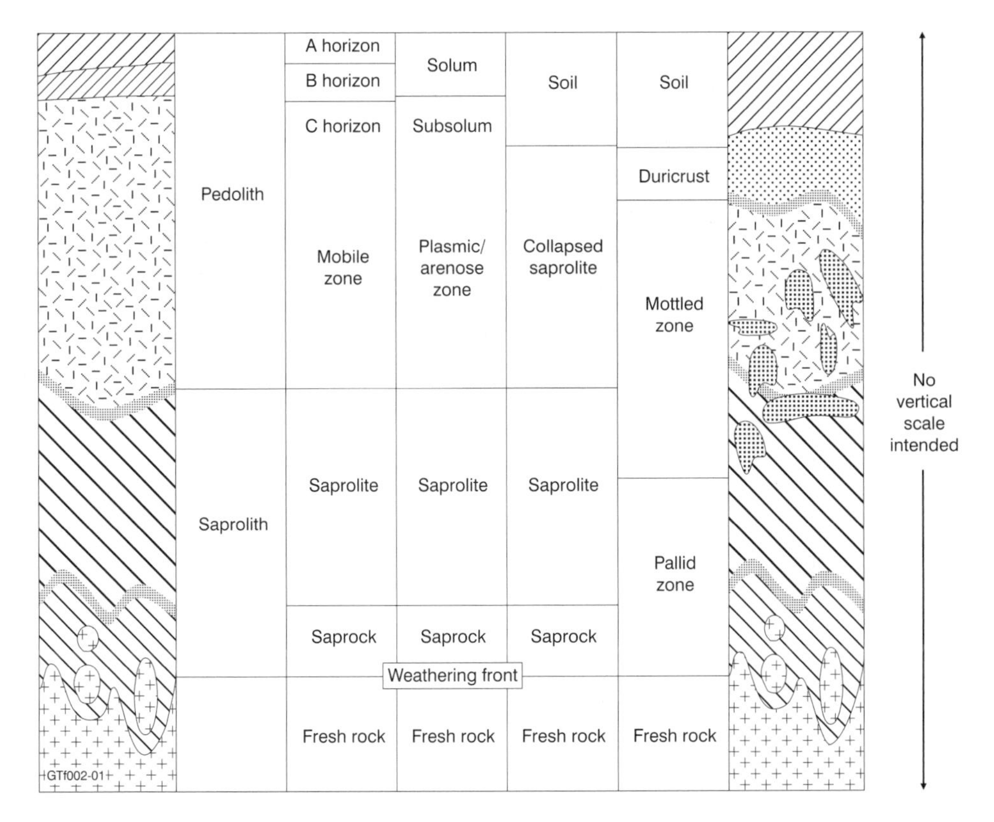

Figure 9.12 *A schematic showing the variety of terms used to refer to parts of a regolith profile. Differences may reflect real variations in the nature of the profile or author preference*

set up local regions of sharply varying redox state, reducing in the vicinity of decaying organic matter, oxidizing where cracks and channels give direct access to the atmosphere. Iron mobility is therefore very variable locally, becoming apparent as blotches or patches of red and white, in a region known as the mottled zone. Mottling is not confined to intensely weathered (lateritic) profiles, but occurs wherever organisms provide variable redox.

Above the wet-season water-table in regions of low erosion, kaolin is weathered to gibbsite (allitisation), yielding either bauxite if the rock is relatively low in Fe, or the ferruginous carapace of a lateritic profile.

MICROENVIRONMENTS

Since rocks are aggregates of minerals, it is natural to assume that rock weathering might be described as the sum of the processes of individual mineral's weathering. In terms of the overall changes in mineralogy and chemistry caused by the weathering of a rock, this

Table 9.1 *Microenvironments of weathering*

Rock dominated	Contact microsystem	Saprock
Rock solution	Plasmic microsystem	Lower saprolite
Solution dominated	Fissural microsystem	Upper saprolite

is largely correct. However, when rock weathering is examined in more detail it becomes apparent that the processes that occur, and the changes that take place, result from microenvironments in the weathering rock. So what occurs at, say, a feldspar/biotite interface, may differ from what is happening inside either of these minerals.

Much of the research into the microenvironments of rock weathering took place in France, and the ideas and results are summarized in Proust & Meunier (1989) and Nahon (1991). The first stages of rock weathering occur at depth, under conditions where the groundwater reaching the rock is essentially not oxidizing, where organic activity is slight but not non-existent and where the rock dominates the environment. This weathering regime is called 'rock dominated'.

Figure 9.13 *The microsystems of rock weathering. (a) Fresh rock: (1) widening of grain boundaries; (2) microfissures. (b) Contact microsystem: (1) new mica formed at grain boundaries; (2) inert contact between quartz and plagioclase; (3) initial internal mineral breakdown. (c) Development of the plasmic system: (1) further internal breakdown; (2) plasma; (3) clay along grain boundary; (4) fissure. (d) Development of fissural system: (1) plasma; (2) fissure; (3) clay concentration; (4) recent clay deposit; (5) Fe-oxyhydroxide band (modified from Meunier & Velde 1979)*

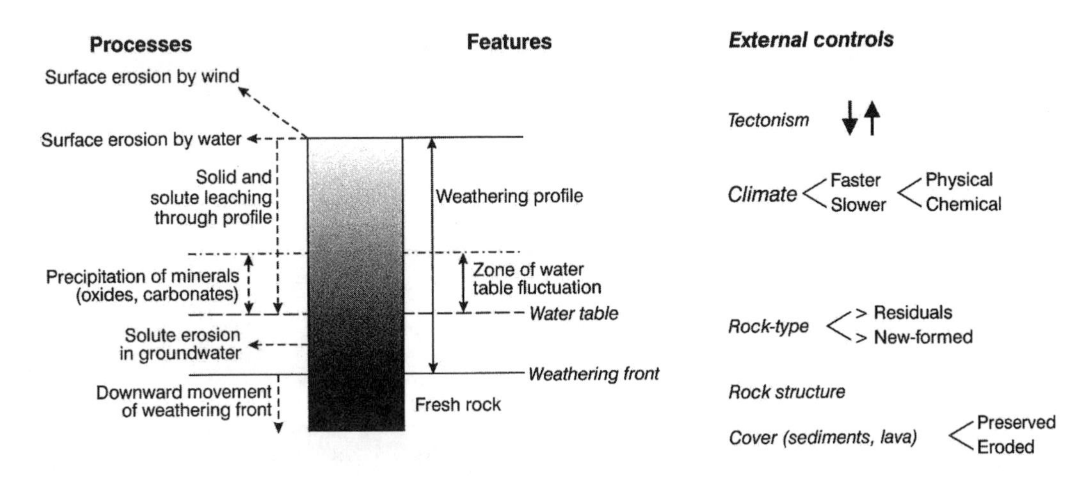

Figure 9.14 *(a) Illustrates the dynamics of weathering products in a profile through which water percolates and some of the factors controlling these processes. (b) Movement of components of weathering profiles and the factors that control the long-term evolution of the profile*

In a study of granite weathering, Meunier & Velde (1979) distinguished three types of rock alteration zones: coherent rock, friable rock and fissural (Table 9.1). In the coherent rock, new minerals are found where two primary minerals react to form a third (e.g. kaolinite at a plagioclase–muscovite boundary). It is notable that no new phases form against quartz crystals. This region is called the *contact microsystem* (Figure 9.13b). In this zone water moves along grain boundaries. In profile terms, the material is saprock.

In the friable zone, minerals break down internally, water pathways extend into the weathering crystals and the alteration chemistry is close to equilibrium. Transformation in this zone leads to the formation of heterogeneous plasma, and the region is called the *plasmic microsystem* (Figure 9.13c). The parent rock fabric is clearly evident and there is no significant volume change, even though little if any of the non-quartz minerals remain unaltered. The plasmic microsystem dominates in the saprolite. This stage of weathering is also called the 'rock-solution interactive microsystem'.

The *fissural system*, as the name implies, cuts through the other two systems (Figure 9.13d). It is the region of highest water and air throughflow, and is generally the site of Fe-oxyhydroxides and kaolin. No primary minerals remain, nor are their pseudomorphs evident. This stage of weathering is also called 'solution dominated'.

Pereira *et al.* (1993) emphasize the importance of microenvironment in their paper on the weathering of biotite from two granites. For example, where K-feldspar is more abundant, the K and Si concentrations in weathering solutions are elevated, creating a chemical potential which prevents the loss of these elements from adjacent biotite, leading to the formation of illite at the common boundary.

MOVEMENT THROUGH THE PROFILE

Within a weathering profile the products of weathering do not necessarily remain where they are formed (Figure 9.14). The changing microenvironments in a weathering profile lead to changes in the way the products of weathering move. In the earliest stages of rock weathering, only fluids, gases and dissolved species move through a weathering profile, thus nearest to the fresh rock all transport is by solution. As pathways widen within and between mineral grains (development of the fissural stage), colloidal and particulate material is translocated. The eluviation of clays from the A-horizon of the soil to the B-horizon is the most prominent type of this movement, and there is also physical migration of particles throughout the upper saprolite. Nesbitt & Markovics (1997) note the accumulation of clay in fracture fills below granite corestones, indicating a region of accumulation of material moved down the profile where it may have further weathered. Smaller accumulations of clay or Fe-oxyhydroxide as coatings around individual grains, or on crack and void walls, known as cutans, are ubiquitous once the profile has opened sufficiently to allow particle movement. Once these processes begin, or bioturbation, or shrinking and swelling occur, the original rock fabric starts to be destroyed and all the solid components of weathering are subject to movement. Nahon (1991) has a particularly lengthy and detailed description of these movements of materials within the profile.

Gravity is the main agent of translocation, so there is always a general downward movement of all mobile products. On slopes, there is also lateral migration, particularly of solutions and by mass movement. Particulate material appears to largely move vertically, as evidenced by the position of clay cutans on the upper surfaces of larger grains.

Many rocks undergo hydrothermal alteration, and the changes induced by this are difficult to distinguish from changes caused by weathering, particularly in minerals such as feldspar, which are not susceptible to oxidation. During hydrothermal alteration, feldspars may be altered to sericite (ultra-finely divided muscovite) or kaolinite, biotite to chlorite and hornblende to actinolite. Of these reactions, only the alteration of feldspar to kaolinite mimics weathering. Dixon & Young (1981) interpreted variations in the topographic expression of granites of the Bega Batholith of south-eastern New South Wales, Australia, as being a consequence of variable 'preconditioning' by hydrothermal activity. While this probably is a factor, there is a greater effect on topography from mineralogical variation among the many plutons comprising the batholith. Broadly, quartz and K-feldspar-rich granites are more resistant to weathering than quartz-poor plagioclase-rich diorites, and this variation in modal mineralogy has a major influence on topography (excluding other geological factors such as tectonics).

Figure 9.15 *(a) Joint pattern preserved in granite saprolite with corestones at Island Bend, south-eastern New South Wales and (b) exfoliation rind (photo GT)*

Weathering of different rocks

PLUTONIC ROCKS

Granites (including granite, adamellite, granodiorite, tonalite, diorite)

Granitic rocks (granitoids) are characterized by the presence of quartz, K-feldspar and plagioclase, and varying amounts of amphibole, biotite and muscovite, with accessory apatite, zircon and sphene. They have a coarse grain size (centimetric) and have widely spaced (1 m or more) joints, leading to the formation of large corestones in a matrix of weathered granite (saprolite, or grus), which may eventually appear at the surface as tors. I-type granitoids, those of igneous source, tend to have higher Ca, contain amphibole and biotite and are unfoliated (Hine *et al.* 1978). S-type granitoids have a sedimentary source, are lower in Ca and contain biotite but little, if any, amphibole. High-level or late-stage granites may carry muscovite.

Granites exhibit onion skin weathering or exfoliation, terms used for the weathering pattern of corestones surrounded by concentric weathering rinds (Figure 9.15a, b). Outward from the corestone, successive rinds are, on average, more intensely weathered; however, the centre of any rind may be less weathered than the margin of the preceding rind, leading to an oscillation of weathering intensity outward from the corestone. Rinds are thought to be the consequence of expansion of the outer surface of the granitic boulder in response to unloading or thermal change, causing cracks parallel to the boulder surface. The cracks provide avenues of entry for water and air, accelerating weathering, particularly oxidation.

In the non-oxidizing earliest stages of weathering (contact microsystem), both I- and S-type granites are affected through loss of K from biotite and the development of vermiculite (see biotite weathering, Chapter 8). Apatite, commonly enclosed by biotite, starts to dissolve, a process possibly mediated by micro-organisms (Banfield & Eggleton 1989). At this early stage of weathering the rock is still hard and solid, requiring a sledgehammer to break it.

At a slightly more advanced stage of weathering, plagioclase feldspar becomes cloudy when viewed in optical thin section, the result of the formation of solution pits and vacuoles. Amphibole hydrates by the production of a series of biopyroboles (see amphibole weathering in Chapter 8) which also makes it cloudy or turbid. Dissolution begins to open up pathways into the weathering rock, particularly via cleavages and fractures. K-feldspar and quartz remain little altered except for dissolution creating etch pits. If the rock remains below the region of oxidation, which is essentially below the permanent water-table, weathering continues as dissolution via etch pits and fracture pathways. The development of the plasmic microsystem is shown by the conversion of most of the granitic minerals to smectites whose composition reflects that of each parent mineral – saponite from the mafics, beidellite from the feldspars. Quartz is essentially insoluble in this weathering environment if the SiO_2 content of the water remains saturated with respect to quartz (see Figure 6.16, the mineral stability field for granitic rocks under concentrated ground waters ('dry')).

If the environment permits the free passage of weathering solutions, desilication will be progressive and follow the path described in Chapter 6. During this, the plasmic microsystem stage, the 2 : 1 layer silicates such as chlorite, illite and smectite lose silica, alkalis and alkaline earth cations to become 1 : 1 kaolinite or halloysite (monosiallitisation, Pedro 1966). Further leaching and silica loss converts the kaolin minerals to gibbsite (allitisation). At the same time oxidation begins to affect the weathering granite. Biotite reacts, breaking down to goethite, kaolinite and anatase with the loss of K, Mg some silica in solution. The total volume of mineral product left by biotite oxidation and leaching can be of the order of 80% of the parent mineral. Mongelli & Moresi (1990) provide data on the weathering of granitic biotite to kaolinite. Assuming no loss of Al, Fe or Ti, the parent biotite $(K_2(Fe_{3.00}Mg_{1.61}Al_{0.68}Ti_{0.35})[Si_{5.29}Al_{2.71}]O_{20}(OH)_4)$ occupies a volume of 490 $Å^3$, the weathering products kaolinite, goethite and anatase occupy 405 $Å^3$. As plagioclase weathers, assuming constant Al, beidellite replacement of plagioclase occupies slightly more volume than granitic andesine, but evidence suggests that there is some loss of all constituents. Kaolinite after andesine occupies only about 60% of the parent volume. This volume loss is sufficient to open pathways for the infiltration of weathering fluids, and during this stage of oxidative weathering, fissural or solution-dominated weathering is entered.

The opening of wider fissures accelerates the access of water and air to the weathered granite which, by this stage, has little if any primary mafic minerals or plagioclase. It may, however, still retain K-feldspar, much of its original quartz, and all of the original resistant accessory minerals such as zircon or ilmenite. Fissures may be filled by gibbsite, kaolinite and silica (Calvo *et al.* 1983). The retention of quartz and K-feldspar, and the almost volume-for-volume replacement of plagioclase and biotite by kaolinite, smectite and goethite, maintains the volumetric relationship of the granite fabric. That is, plagioclase and biotite pseudomorphs are recognizable as such, and the weathered rock is termed 'saprolite'.

The next stage of granite weathering is associated with severe physical movement of the material. Bioturbation combined with the shrink–swell behaviour of smectite and the simple gravitational movement of quartz grains is responsible for destruction of the saprolite fabric. Within the profile this leads to formation of the pedolith, which is commonly pale coloured in which case it may be referred to as the pallid zone (because the removal of Fe by organically mediated reduction leaves this region dominated by kaolin and quartz). Nearer the surface, the eluviation of clay leads to a zone dominated by quartz grains, referred to by some as collapsed saprolite. Fractures in the saprolite become filled with eluviated clays, precipitated Fe-oxyhydroxides and silica. Small dykes and veins cutting the granite may retain their form throughout these zones, testifying that the main movements have been vertical. Not all profiles continue to weather long enough to develop a thick pedolith; on slopes particularly, erosion keeps pace with the production of saprolite, and the soil horizons lie directly on saprolite.

Nesbitt & Young (1989) and Nesbitt & Markovics (1997) have summarized two studies on the weathering of granite. For the Stone Mountain Granite, they show that on a volume percentage basis, quartz and muscovite are essentially constant at about 30 and 10% respectively during weathering to saprolite, plagioclase is entirely lost and K-feldspar decreases by about 60%, while kaolinite and illite increase to account for approximately 55%

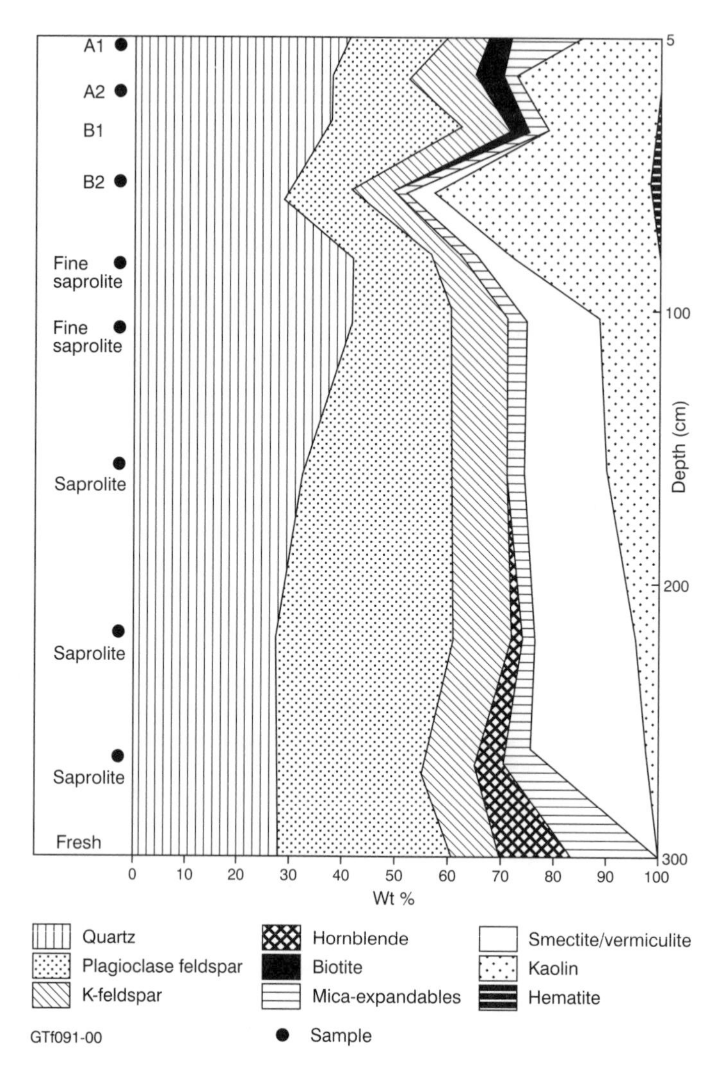

Figure 9.16 *Quantitative mineralogy of a regolith profile developed on the Broughton River Granodiorite from the Charters Towers region of Queensland. The granodiorite is overlain by 25 cm of transported regolith (from Aspandiar 1998)*

of the weathered rock. These mineralogical changes are paralleled by chemical changes, which show that of the major oxides ($CaO + Na_2O$) decreases most rapidly with weathering, reflecting the greater susceptibility to weathering of plagioclase compared to K-feldspar.

For the Toorongo Granodiorite of Victoria, Australia, Nesbitt & Markovics (1997) examined chemical and mineralogical changes from the fresh rock through saprolite to a clay-filled fracture zone. They found that plagioclase cores and biotite were the first minerals to weather, followed by amphibole and K-feldspar. Plagioclase was replaced by kaolinite, and K-feldspar by kaolinite and illite. Biotite altered to vermiculite and Fe-oxyhydroxides. Quartz remained ubiquitous and little altered except for grain size reduction by fracturing. In the most weathered samples of the clay-filled fracture zone, illite, kaolinite and quartz dominated the alteration assemblage.

The resulting granitic weathering profile (Figure 9.16) comes about from this complex interplay of mineral weathering, physical disruption of the rock fabric and translocation of some minerals. Trends are reflected in the density and the fabrics of the regolith materials. Quartz is concentrated in the upper, mobile, part of the weathering profile relative to the lower profile zones because of the removal of new-formed minerals by erosion and downward leaching or eluviation. Across much of the Yilgarn in Western Australia, and Broken Hill, New South Wales, the surface is covered by a quartz sand lag as a result of these processes (Craig 1993; Ollier *et al.* 1988; Butt & Smith 1992). As Chesworth (1979) pointed out, granite weathering leads to residua dominated by kaolinite, quartz, gibbsite and goethite or haematite.

The accumulation of quartz at the surface as the end product of weathering has been used to explain these quartz sand sheets on the Yilgarn Craton of Western Australia, and its origin has been attributed to deflation of finer particles, but it is possible that they are the result of leaching of Al^{3+} and Fe^{3+} under conditions of podzolization. In parts of Amazonia (Bleackley & Kahn 1963; Heyligers 1963; Ab'Saber 1982; Lucas *et al.* 1987, 1988; Stallard & Edmond 1981, 1983, 1987) and in Borneo (Thorp *et al.* 1990) white sands accumulate as Fe and Al are leached under conditions of podzolization. Such conditions require abundant water, ample organic complexing agents (vegetation residues) and high water-tables. As it is known that during much of Australia's Palaeogene and Neogene climates were wet enough to support rainforests, podzolization should be considered as a possible origin for these sand sheets. Podzolization currently occurs in the tropics of Australia, particularly along the wet eastern coast as well as in cool, wet climates along the west coast of Tasmania. The common factor in this is an abundance of water and vegetation.

Further than this, De Ploey (1965) suggested that during earlier wet climates this process may have led to the formation of the sand now forming the dunes of the Kalahari. It is interesting to speculate whether much of the sand in the Australian dune systems which covers 40% of the continent may owe its origin, at least in part, to podzolization during the Tertiary. Another similar situation exists with the formation of silcrete where significant dealuminification (see Chapter 13) must have occurred. It is possible that much of the sand and quartzose regolith in Australia, and possibly other similar landscape features, were formed by podzolization, even though much of it may have been redistributed by aeolian processes after continental drying at about 5 Ma.

Chemical changes

The changes in mineralogy during granite weathering combine to produce a change in the composition of the weathered rock. Calcium and Na are lost early, along with Rb and Sr, reflecting early weathering of plagioclase. As biotite and amphibole weather, Mg, Ca and K are lost and ferrous iron is oxidized. Production of clay minerals adds water to the weathered rock. Most studies have found or concluded that Zr and Ti are immobile (e.g. Nesbitt & Markovics 1997).

Gardner *et al.* (1978) have assessed the rate at which major elements are lost from a weathering granite. They found that relative to silica loss, which was linear with density decrease, Na and Ca were lost faster, so that almost 90% of the original Ca and Na were lost by the time 50% of the silica had been leached. Very little K, by contrast, had been lost by this stage. These results are consistent with the greater resistance to weathering of K-feldspar than plagioclase. Banfield (1985) found similar chemical changes; Figure 9.17 shows the progression of element loss from a weathering granite. In this figure, the oxide is expressed as weight per unit volume, to show the actual loss, and is plotted against Nesbitt & Young's (1982) CIA index:

$$CIA = [Al_2O_3/(Al_2O_3 + CaO + Na_2O + K_2O)] \times 100$$

which charts the course of weathering. Some correlation between CaO, Na_2O and K_2O and CIA is to be expected, since these are the essential oxides that determine the index, but beyond that trend, the shape of the data trends reflect the progression of the mineral weathering. Both Ca and Na, largely in plagioclase, are lost more rapidly early in the weathering, whereas K, present in K-feldspar and biotite, and Mg, present in biotite and amphibole, show a more linear trend. Silicon (Figure 9.17b), which is leached from all the major minerals, also shows a linear trend.

Chemical change is commonly represented on a triangular (ternary) diagram. Three components may be selected from the bulk chemical analysis of samples from a weathering profile and their total recalculated to 100%. The trend such data points make on a ternary diagram indicates chemical change. Nesbitt & Markovics (1997), for example, select molar Al_2O_3 for one apex, $CaO + Na_2O$ for the second and K_2O for the third (Figure 9.18). The resulting diagram reveals progressive loss of $CaO + Na_2O$ at constant K_2O through most of the profile, with K_2O only leaching out in the most weathered samples.

An alternative representation is to select SiO_2 at one apex, $Fe_2O_3 + Al_2O_3$ (sesquioxides) at the second, and $CaO + MgO + Na_2O + K_2O$ (alkalis) at the third. On such a diagram, progressive weathering shows a trend towards the sesquioxide corner of the triangular

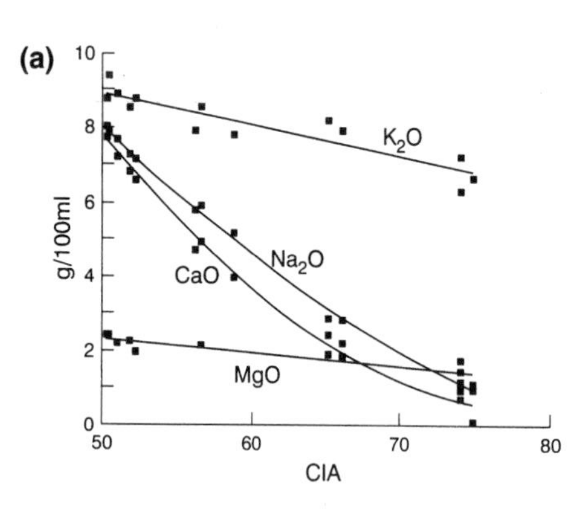

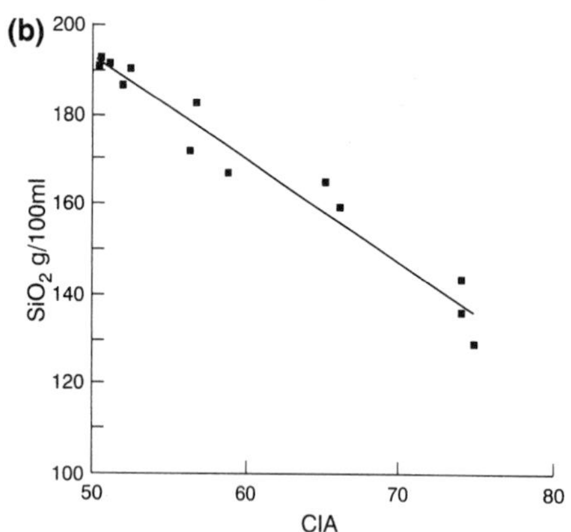

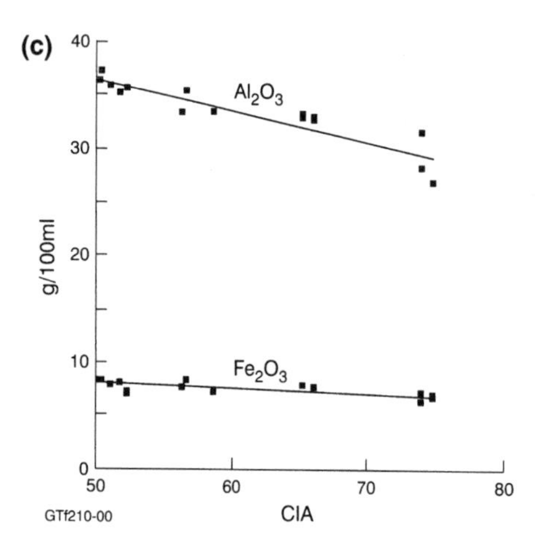

Figure 9.17 *Trend of element loss during granitic rock weathering. (a) alkalis and alkaline earths from weathering Jindabyne Tonalite; (b) silica from Bemboka Granodiorite; (c) sesquioxide from Bemboka Granodiorite. CIA = [Al_2O_3/(Al_2O_3 + CaO + Na_2O + K_2O)] × 100*

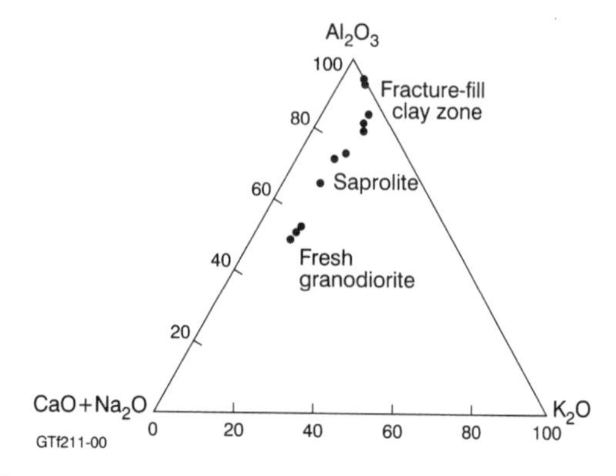

Figure 9.18 *Toorongo Granodiorite chemical changes with progressive weathering (data from Nesbitt & Markovics 1997)*

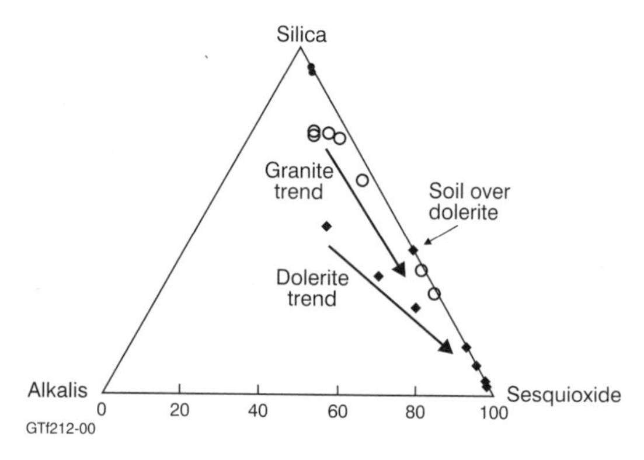

Figure 9.19 *Compositional changes with weathering of granite (circles) and dolerite (squares) to bauxite in the Darling Range of Western Australia (data from Davy 1979)*

diagram, as weathering of silicates releases their more soluble ions, as is shown for granite and intruded dolerite dykes in the Darling Ranges of Western Australia in Figure 9.19. With extreme weathering, the assemblage reaches the sesquioxide corner. It is suggested from its position on the diagram, that the soil over dolerite is derived from granite weathering. In the Darling Ranges the dolerites are only a minor part of the bedrock, and granitic detritus has apparently covered the dolerite weathered profile.

Monzonite–syenite

Monzonites and syenites weather in much the same way as granites, with the establishment of corestone and rind morphology in the saprock and saprolite. Because these rocks have little or no quartz, there is no relatively insoluble major phase to help retain the parent rock fabric, with the result that the saprolite readily collapses. When there are feldspathoids in the rock, weathering proceeds even more quickly, and the almost unweathered corestones may have a very thin (10 mm) weathering rind surrounded by clay-rich regolith.

In the Little Rock area of Arkansas, Gordon & Tracey (1952) found that nepheline syenite produced a deep profile in which fresh rock grades upward into a completely kaolinized zone, but still retains, in part, the original texture of the parent rock. The profile then grades into a zone of compact kaolinitic clays that are in between the kaolinized nepheline syenite and an overlying massive bauxite. In the Poços de Caldas, Brazil (Harder 1952; Schumann 1993), a massive bauxite deposit is situated directly on a nepheline-rich syenite. The boundary is so sharp that a single nepheline crystal may be completely fresh at one end below the contact, while altered to gibbsite at the other above the contact. In the transition zone, nepheline breaks into a mosaic of fragments, apparently converting directly to gibbsite, as neither smectite nor kaolinite was observed. Ferromagnesian minerals in this zone alter to aluminous goethite. Orthoclase is the most resistant mineral to weathering, eventually breaking down to gibbsite.

Wang (1988) found that the weathering susceptibility of individual minerals in the Mt Dromedary monzonites in south-eastern Australia can be summarized as:

nepheline > plagioclase > biotite > hornblende > augite > K-feldspar > opaque minerals

The order of augite and hornblende susceptibility was found to be reversed in this study compared to that commonly observed (e.g. Loughnan 1969). This is partly because augite crystals are relatively well protected by the rims of hornblende and, more importantly, hornblende formed by uralitization of augite has many ‹110› direction diffusion channels arranged in a diamond shape, thus the weathering rate of hornblende is dramatically accelerated.

Marker & De Oliveira (1990) found the weathering sequence of minerals in a sodalite–nepheline syenite to be:

sodalite = calcite = apatite > nepheline > microcline

Gabbro

Like granite, gabbro weathers along joints, producing sub-spherical corestones with 'onion-skin' weathering rinds. Fritz (1988) noted that whereas granite weathering produced much clay (gibbsite–kaolinite), an adjacent gabbro weathered to a mixture of unweathered primary minerals and Fe-oxyhydroxides with only a little vermiculite as a clay product. At the base of a weathered gabbro in Japan, smectites, kaolinite and unweathered primary minerals dominate (Okumura 1988). Higher in the profile the smectites give way to kaolinite and goethite, and locally gibbsite. Both Fritz and Okumura conclude that the gabbro saprolites they studied were susceptible to erosion, thus limiting the opportunity to develop a very clay-rich later stage assemblage.

In a summary of mafic rock weathering, Proust & Meunier (1989) recognized four weathering levels. At the base of the profile in the slightly weathered coherent rock, weathering occurs along mineral fractures (contact microsystem). In the crumbly saprock above this, weathering occurs in plasmic microsystems, and the individual minerals are highly altered, being largely pseudomorphs of clay minerals. Higher in the profile the primary minerals are no longer recognizable, the rock fabric is lost, and the mineralogy is dominated by secondary clays surrounding primary mineral relicts. At the top of the profile lies completely reorganized argillaceous regolith.

Ildefonse (1978) studied the weathering of a gabbro from the Massif du Pallette, France, composed of two amphiboles, cummingtonite and actinolite, and plagioclase (An_{52}). The weathering reactions were found to depend on the microsystem and on the mineralogical microenvironment. In the lower part of the profile where the original rock fabric was preserved (saprock) both amphiboles altered to nontronite + talc + an amorphous ferric hydroxide, the plagioclase to dioctahedral vermiculite and beidellite occurred along the actinolite–feldspar boundary. Higher in the profile where the rock fabric was lost, cummingtonite had all altered, actinolite altered to vermiculite + beidellite + amorphous ferric hydroxide, and plagioclase to vermiculite + ferric beidellite. The major chemical changes, based on weight %, were a loss of Ca and Mg, which is consistent with the mineralogical changes, and a marked increase in K_2O (from 0.3% in the fresh rock to over 1% in the most weathered samples).

Ultramafics: pyroxenites, peridotites, lamprophyres

Ultramafic igneous rocks have a low alumina content, and this makes their weathering process distinct from those of feldspathic rocks. The clay minerals that form during the development of saprolite are dominated by talc, vermiculite and Fe–Mg smectites, reflecting the high Mg content of the dominant parent minerals, olivine and pyroxene. With further weathering these clays break down, yielding Mg in solution which in arid climates commonly precipitates along fractures as magnesite. According to Proust & Meunier (1989), a lherzolite containing chrysotile, orthopyroxene and clinopyroxene weathered initially (weathered coherent rock) by alteration of orthopyroxene to talc and chrysotile to saponite. In the saprolite, the talc altered to saponite + magnesian gel. In the plasmic zone above this the saponite had weathered to a mixture of nontronite and Fe-oxides and oxyhydroxides. The upper regions of ultramafic weathering profiles are typically dominated by nontronite and goethite, while the silica released from the breakdown of the Mg–Fe clays precipitates as quartz, opal or chalcedony. Duparc *et al.* (1927) referred to this weathering assemblage as birbirite. Nickel, which is relatively abundant in ultramafics, becomes concentrated in the clay minerals, particularly in pimelite and garnierite, and in goethite (Schellman 1981; Wiewióra *et al.* 1982; Colin *et al.* 1990; Paquet *et al.* 1987; Trescases *et al.* 1987a) (see also the section on nickel in Chapter 7).

Figure 9.20 *Distribution of corestones in basalt profiles from temperate to semi-arid weathering environments; (a) photograph of a basalt weathering profile from the Monaro, south-eastern New South Wales (photo GT); (b) photo of basalt weathering profiles from tropical weathering environments from near Cairns, northern Queensland (photo GT)*

EXTRUSIVE IGNEOUS ROCKS

Basalt–andesite

Basalts, unlike granitic rocks, have few minerals that are highly stable in the weathering environment, and most of the parent minerals will eventually alter to new-formed minerals with time. However, in the same way as granitic rocks, basalt weathering profiles can retain some parent minerals at or near the top. Under conditions of intense weathering basalt rarely retains any parent minerals, but when weathering is less intense or erosion of weathered products is rapid, residual materials are often retained in the profile.

The most common parent materials are rock fragments of basalt. These vary considerably in size from large corestones littering the top of profile to sand-sized fragments in the upper portions of the profile. In areas where climates are relatively dry (e.g. the Monaro region south of Canberra, Australia, where rainfall is about 400 mm) corestones of fresh basalt up to 0.75 m are preserved within the upper part of the profile and on the surface. This occurs because moisture rapidly drains into the profile, leaving the upper portions relatively dry and thus weathering is relatively ineffective. Below this layer is an accumulation of new-formed minerals and virtually nothing else. Further down the profile fragments and corestones of basalt begin to appear again, being progressively less weathered down the profile (Figure 9.20).

Macaire *et al.* (1994) report weathering of basalts from the Middle Atlas Mountains in Morocco where rainfalls are between 570 and 1100 mm with average temperatures in the temperate range. It is significant that the region experiences heavy frosts during winter. They show that sediments derived from the basalts weathering in two small catchments were predominantly sand-sized. Table 9.2 summarizes their results.

This study clearly shows that it is possible not only to obtain basalt rock fragments but also discrete grains of basaltic minerals released from the weathering of

Table 9.2 *Grain-size characteristics of sediments derived from basaltic weathering profiles in Morocco (from Macaire* et al. *1994)*

	Basalt with no glass (%)	Basalt with glass (%)
% catchment basalt covered	18	9.5
Grain size		
Sand	75[a]	72[b]
Silt	20	27
Clay	5	2
% basalt derived sandy detritus	2	21
Sand mineralogy		
Olivine	3	18
Augite	<1	2
Quartz	3	51
Chalcedony	<1	11
Feldspar	0	1
Sandstone	0	3
Carbonates (organic)	92	36
Opaques	3	5
<10 μm mineralogy		
Kaolinite	43	33
Illite	45	37
14 Å clay	12	31
Quartz/feldspar	>0	>0

[a] Averages to nearest whole number ($N = 6$).
[b] Averages to nearest whole number ($N = 4$).

basalts. This is more noticeable in the case of glassy basalt where microfractures (?cooling) allow the release of individual grains of minerals such as olivine and augite. In the case of holocrystalline basalts the finer crystalline ground mass must first be chemically weathered and hence the release of sand-sized basaltic mineral grains is diminished, due to their reduction in size within the profile prior to release.

In an extensive study by Colman & Pierce (1981), the initial stages of weathering of basalts and andesites were investigated. Within coherent corestones they found the development of new minerals of indeterminable character, and established an order of susceptibility to weathering of the primary minerals as

glass > olivine > plagioclase ≈ pyroxene > K-feldspar >> titanomagnetite

Colman & Pierce assessed absolute changes in major element chemistry by reference to immobile Ti and at constant volume, and found the order to be

$$Ca \geq Na > Mg > Si > Al \geq K > Fe > Ti$$

The position of K in this sequence differs from its normally found solubility, possibly because there is generally so little in these rocks that any released by weathering can be taken up in the neo-formed clays.

Eggleton *et al.* (1987) found the same sequence of mineral susceptibility, and also assessed the relative loss rate of the major elements from three basalt profiles. They noted that although plagioclase began to alter before pyroxene, pyroxene weathering, once initiated, was more rapid, so that in more intensely weathered samples all the pyroxene was gone while some plagioclase remained. Within corestones and their weathering rinds, Fe, Ti, Cr, Nb, Zr, V and Ni were effectively immobile. For the other major elements, at the point where one-third of the original mass had been lost by dissolution, the amounts lost were:

Ca 90%, Mg 80%, Na 70%, K 65%, P 55%,
Si 45%, Mn ~40%, Al 5%

Trace element loss was found to be highly variable, though loss of Rb and Sr was consistently high (≈90%), and Cu and Zn low (~20%).

Moore (1996a) studied basalt weathering in eastern Australia, investigating the changes from fresh rock through the saprolite into the soil. She found the same sequence of mineralogical change as described above, as well as documenting the progress of clay mineral development. Initially, saponite, Fe-saponite and goethite form from weathering glass and olivine, which subsequently alter to dioctahedral nontronite and Fe-beidellite, and in some instances hisingerite. As feldspar begins to weather, Al-smectite forms and later breaks down to kaolinite; however, in areas of high water throughflow, kaolinite may be the first detectable clay mineral. Clinopyroxene, though slow to begin weathering, breaks down to nontronite, beidellite and goethite; vermiculite was found in weathered orthopyroxene-bearing basalts. Random interstratification of smectite/kaolinite developed later in weathering, gradually giving way to kaolin or halloysite and in some profiles, gibbsite. Figure 9.21 summarizes the changes that occur during basalt weathering.

Felsic volcanics

Volcanic ash weathers extremely quickly, typically to a mixture of allophane, halloysite, kaolinite and smectite. Gomes & Massa (1992) describe halloysite-dominated profiles formed by the weathering of trachytic pyroclastics. Kirkham (1975) described a weathering

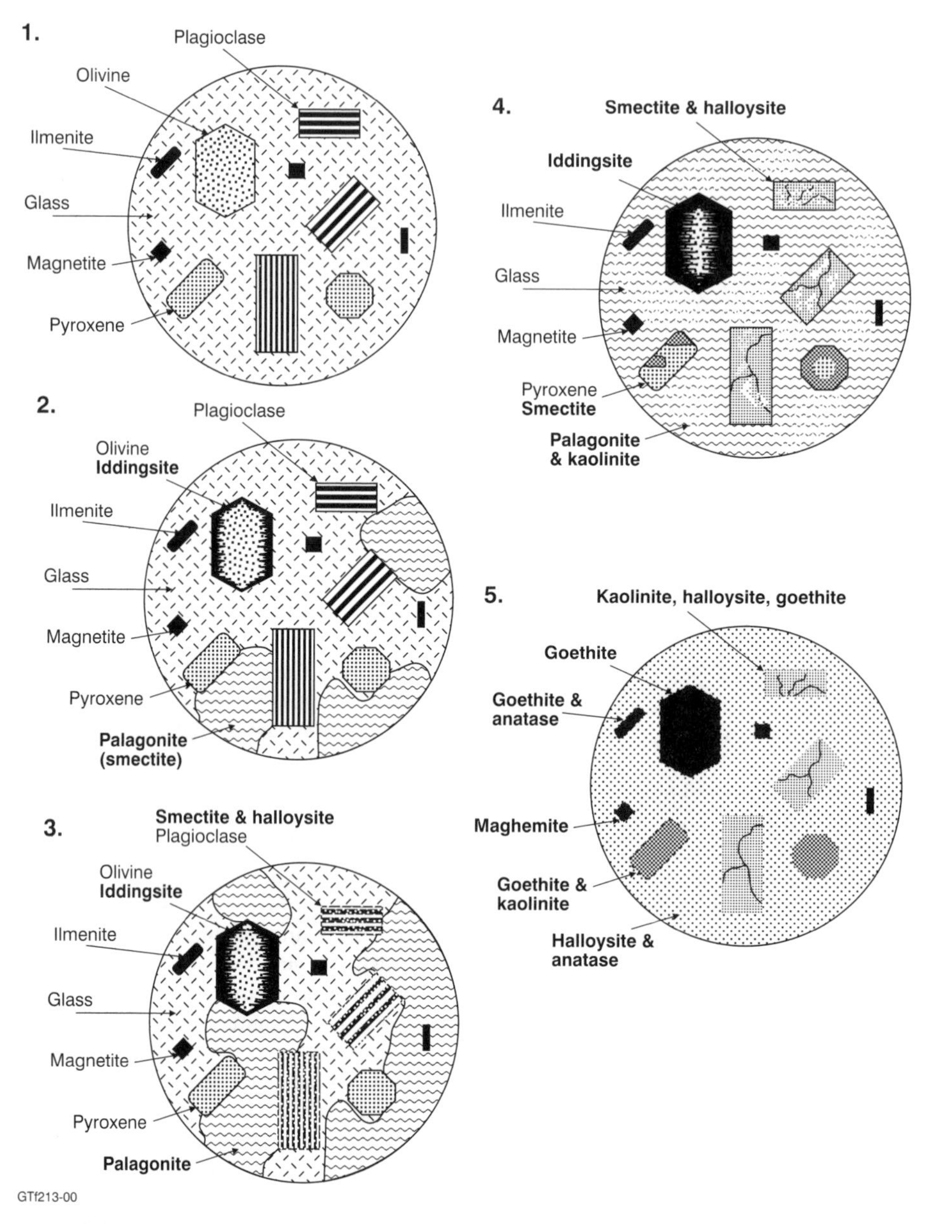

Figure 9.21 *Diagram showing the changes to minerals during basalt weathering. (1) Fresh basalt. (2 and 3) Glass alters to palagonite (smectite), olivine margins become etched with saponite in the etch channels, and plagioclase breaks down along cracks to a non-crystalline material. (4 and 5) Olivine is iddingsitized, converted largely to goethite. Plagioclase has altered to a smectite-halloysite mixture, and pyroxene has altered via a talc-like phase to smectite. Further weathering may lead to gibbsite and goethite*

sequence for New Zealand rhyolitic tephra of allophane–halloysite (10 Å)–metahalloysite (7.5 Å). Parfitt *et al.* (1983, 1984) noted that the alteration products of rhyolitic glass depended on the degree of leaching, and by implication on the silica content of the leaching solutions. Where rainfall and leaching were higher, allophane with a high Al : Si ratio (2 : 1) dominated, whereas under lower rainfall or more stagnant conditions halloysite (Al : Si = 1 : 1) was the more abundant. More andesitic ash, having a lower Si content, weathered only to allophane. The common occurrence of allophane and imogolite in weathering profiles on volcanic ash and pumice, and their transition to halloysite, was also noted by Wada (1989).

SEDIMENTARY ROCKS

The detrital components of sedimentary rocks have already been through one weathering cycle, so they are closer to equilibrium with the regolith environment than igneous rocks. Studies of significantly weathered sedimentary rocks are not common, with the exception of limestone/dolomite where very large numbers of studies have been conducted. Glenn *et al.* (1960) studied weathering of loess-derived silt loams composed of a variety of clay minerals including montmorillonite, vermiculite, chlorite, illite and kaolinite with about 10% amorphous and other material at depths of 4 m which changed very little between there and the surface. Surface and near-surface differences include an increase in chlorite, illite and amorphous material at the expense of montmorillonite and vermiculite. In this case the residual minerals are very similar to those of the original rocks, with the exception of the reduction in smectitic minerals. In eastern Australia, a typical Ordovician shale contains quartz, muscovite and chlorite. Loughnan (1960) and Bayliss & Loughnan (1964) described a number of shales used for structural clay extraction, as well as more extensively weathered shales used as a kaolin resource. Comparison of their mineralogical and chemical make-up (Table 9.3) shows how shales are affected by weathering.

Weathering of muscovite to kaolinite or halloysite, some dissolution of quartz, and the breakdown of chlorite lead to a kaolin–quartz assemblage, usually with a small amount of goethite and anatase. With continued or intense weathering, a shale profile can readily grade upward to a bauxite, as at Weipa, Queensland (Figure 9.22). Here the mineral assemblage changes from quartz + glauconite to quartz + glauconite + smectite to quartz + kaolinite + goethite to gibbsite + goethite.

In the arid Yilgarn region of Western Australia (Panglo Deposit), Scott & Dotter (1990) describe the weathering of shales in an environment that is now highly saline. The fresh rocks contain quartz, white mica, chlorite, pyrite, ± albite, talc, dolomite and siderite. Albite, talc, pyrite and carbonates are lost early in weathering, to produce a yellow–brown saprolite dominated by quartz, kaolinite, mica, chlorite, goethite and halite. Higher in the profile, in the paler-coloured leached saprolite, chlorite is lost and alunite and smectite appear. Both the alunite and the mica in the saprolite are sodic, possibly related to the salinity of the groundwater.

Table 9.3 *Approximate mineralogical and chemical compositions (weight %) of fresh and weathered shales (from Loughnan 1960; Bayliss & Loughnan 1964)*

	Average fresh shale	Structural clay shale	Kaolinized shale
Quartz	50	45	30
Chlorite	20	15	0
Mica	30	25	10
Kaolin	0	15	60
SiO_2	70	67	55
TiO_2	1	1	1
Al_2O_3	14	18	28
Fe_2O_3	4	4	2
K_2O	4	3	1
H_2O	5	6	12

SULFIDE WEATHERING

The weathering of sulfides differs very greatly from silicate weathering because of the production of acid as S is oxidized to sulfate. Whereas feldspar weathering, for example, increases the solution pH through the release of alkali cations, the reaction of sulfides with oxygenated water lowers pH. For example, the reaction

$$PbS + 4H_2O \rightarrow PbSO_4 + 8H^+ + 8e^-$$

releases hydrogen ions, whereas

$$CaAl_2Si_2O_8 + 3H_2O \rightarrow Al_2Si_2O_5(OH)_4 + Ca(OH)_2$$

produces limewater, which is alkaline. Sulfide weathering therefore becomes itself an important agent of silicate weathering, by acidifying the groundwater and making it more aggressive to the surrounding rocks.

As with all solution-dominated processes, the composition of the solution directs the course of the reaction. For example, carbonate-rich waters cause rapid precipitation of Pb as cerussite, whereas chloride-rich solutions allow some Pb mobilization (Thornber & Taylor 1992). A wide variety of metal-bearing minerals may be precipitated following the weathering of the sulfide, depending on the least soluble species and on the competition by highly adsorptive major minerals such as Fe- and Mn-oxyhydroxides. Sulfates are more likely low in the weathering profile, replaced by carbonates further up and oxides nearer the surface.

While it is not possible to rank the sulfide minerals in order of susceptibility to weathering because the

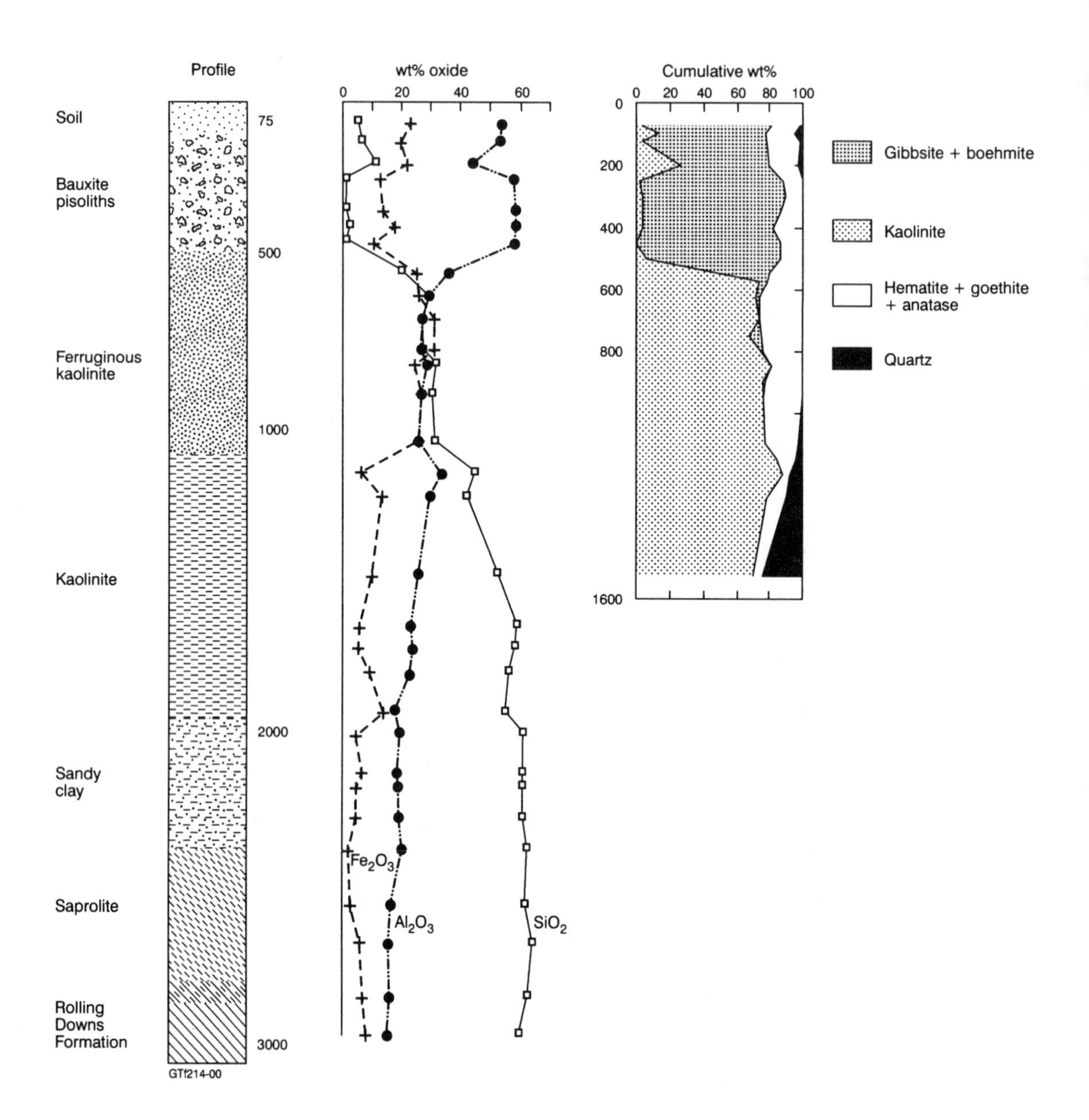

Figure 9.22 *Profile, geochemistry and mineralogy of the weathered Cretaceous Rolling Downs Group from the Jacaranda drill hole at Andoom near Weipa, northern Queensland*

order depends on the specific environment, Thornber & Taylor suggest the following sequence of sulfide mineral weathering:

> Galena > pentlandite > pyrrhotite
> > chalcopyrite > pyrite > Cu-sulfides

The mechanism of sulfide weathering is one of some debate. Simple solution and oxidation chemistry describes the reacting species, but other influences may affect the rate and pathway of the reactions. Thornber & Taylor (1992) emphasize the electrochemical processes that occur during sulfide weathering, regarding sulfide ore-bodies as a large assemblage of small galvanic cells. Adjacent sulfides of different reactivity immersed in the same weathering solution establish a voltage differential which leads to migration

Table 9.4 *Some sulfide weathering products*

Sulfide	Sulfate	Carbonate	Oxyhydroxide	Chloride	Silicate	Phosphate
Pyrite, Marcasite, Pyrrhotite			Goethite			
Galena	Anglesite, plumbo-jarosite	Cerussite		Cotunnite		Plumbo-gummite
Chalcopyrite, bornite, etc.	Chalcanthite	Malachite, azurite	Tenorite, cuprite goethite		Chrysocolla	Turquoise
Sphalerite		Smithsonite, hydrozincite			Hemimorphite	

of cations and anions between them. On a larger scale, the near-surface oxidative environment becomes anodic, the base of the water-table cathodic, so that anions move down and cations up.

According to Nickel & Daniels (1985), the weathering of galena given above takes place at the anode of a galvanic cell, with the balancing cathodic reaction being

$$4H_2O + 2O_2 + 8e^- \rightarrow 8(OH)^-$$

Nordstrom & Southam (1997) treat sulfide weathering as a bacterially mediated process, where bacteria such as *Thiobacillus* spp., which gain energy from the oxidation of S, catalyse the reaction. The data they summarize indicate that the rate of sulfide oxidation is limited by the bacterial requirements, and that bacteria can increase the rate by several times compared to inorganic oxidation. These observations do not negate the significance of galvanic processes, rather the bacterial cell is seen as the conductor of electrons from the mineral to oxygen.

Sulfate is by far the most abundant anion formed during sulfide weathering, leading to the production of a large range of metal sulfates. There is too wide a variety of oxidation products of sulfide mineral weathering to attempt to cover this topic here; Nickel & Daniels (1985) and Williams (1990) provide details. The commoner products, each depending on the dominant anion at particular levels in the sulfide weathering profile, are listed in Table 9.4.

During oxidative sulfide weathering, the most abundant ion released is ferric iron, originally contained in pyrite, marcasite, pyrrhotite, aresenopyrite, chalcopyrite, tetrehedrite–tennantite and stannite. Ferric iron is highly insoluble compared to the other metal ions released during weathering (Table 9.5).

Table 9.5 *Solubility products of some metal hydroxides*

Ion	SP
Cd^{2+}	13.7
Ni^{2+}	14.7
Co^{2+}	14.8
Pb^{2+}	15.3
Zn^{2+}	17.0
Cu^{2+}	19.7
Fe^{3+}	38.0

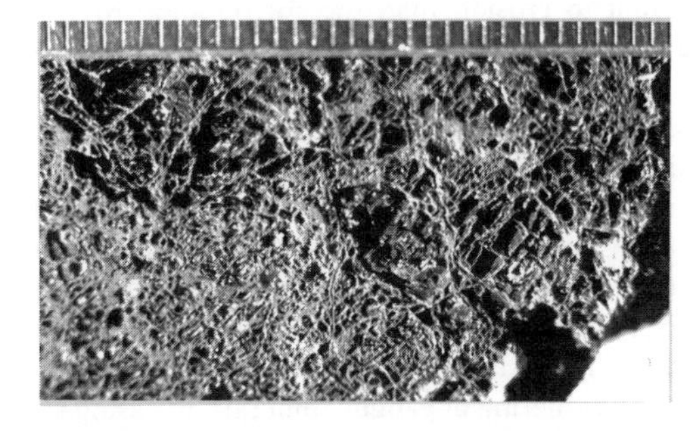

Figure 9.23 *Box-work in gossan from Mt Morgan, Queensland, Australia. Scale bars are 1 mm (photo R. A. Eggleton)*

Sulfide weathering takes the same physical pathways as silicate weathering, with dissolution following cleavage cracks and twin planes, i.e. initial weathering opens channels along crystallographic directions. Most of the liberated metals move in the groundwater and are either lost to the weathering ore-body or precipitate as sulfates, carbonates, etc. Because of its lower solubility, Fe commonly precipitates directly in the dissolution channels, leading to a pattern of intersecting planar goethite deposits known as 'box-work' (Figure 9.23).

Intersecting planes define cells, or boxes. The cells may be empty, or contain unweathered sulfide, or be filled with secondary ore minerals or other Fe-oxyhydroxides. The pattern of the box-work texture depends on the crystallography of the parent sulfide, and this relation allows the box-work to be a useful aid in identifying the original sulfide. Nickel & Daniels (1985) figure many box-work patterns, and indicate certain diagnostic patterns:

Chalcopyrite	Orthogonal and triangular box-work cells, trellised cell walls often with totally leached cell centres
Galena	Orthogonal and triangular box-work cells with straight walls
Pentlandite	Octahedral cleavage
Pyrite	Cubic pseudomorphs, thick cell walls
Pyrrhotite	Lamellar and curvi-lamellar banding or spongy texture
Sphalerite	No pattern known to be conclusive

Gossans

A gossan is the Fe-oxide-rich outcrop expression of a weathered sulfide body. Many gossans are resistant to erosion and so form hills or ridges, though some are less resistant to erosion than the surrounding rocks and are not prominent. Gossans are dominantly quartz–goethite rocks, and many show box-work fabric. Gossans may show pseudomorphs of prior sulfides, and careful examination of the fabric can help to identify the original mineralogy. The high Fe content of a gossan results from the dominance of pyrite, marcasite or pyrrhotite in sulfide deposits. Weathering of iron sulfides produces goethite (as described in Chapter 7), and the consequent acidity increases the rate of weathering of gangue minerals, releasing silica.

$$4FeS_2 + 10H_2O + 15O_2 \rightarrow 4FeO(OH) + 8H_2SO_4$$

$$2KAlSi_3O_8 + H_2SO_4 + 9H_2O \rightarrow Al_2Si_2O_5(OH)_4 + 4H_4SiO_4 + K_2SO_4$$

Since lowering pH does not increase silica solubility, the large quantities of silica liberated during rapid silicate weathering may precipitate with the goethite, leading to the typical gossan mineralogy.

Oxygen is consumed at the reaction front (water-table) and solutes produced move downwards within the profile, reacting with the ore and releasing cations which migrate to the oxidation front causing enrichment in the ore elements (supergene enrichment) (Figure 9.24a).

Gossans are favoured targets for mineral exploration, since sulfide bodies may host economically significant metals. Taylor & Thornber (1992) suggest certain pathfinder elements (see Chapter 7) of gossans to be used in reconnaissance surveys (Table 9.6). The near-surface parts of gossans are liable to be depleted of all but the most insoluble components as a result of their longer exposure to rain and dilute groundwaters. Quartz of the original rock may survive, and it may protect small sulfide grains from weathering. Beneath this leached zone, but still above the water-table, the gossan may contain secondary minerals of low solubility such as azurite, plumbogummite (Pb–Al-phosphate) and hydrozincite, or slightly less insoluble minerals such as malachite, cerussite or hemimorphite (Nickel & Daniels 1985). Below the water-table the weathering conditions are liable to be reducing, or at least not as strongly oxidizing, leading to a region of supergene enrichment.

The position of the water-table is important in controlling the nature of a gossan profile. A rising water-table will limit the rate of gossan formation by restricting O_2 access and weathering will stop. A static water-table will have a similar effect, but weathering will continue until O_2 is depleted and cations can no longer move out of the ore minerals. Falling water-tables promote weathering to greater depths by allowing O_2 penetration to greater depths. Thus maximal gossan profiles develop during phases of falling water-tables (Figure 9.24b). Examples of the types of gossans formed are given in Figure 9.24b.

Because of the high weatherability of sulfides, there can often be very deep weathering over them. At Kambalda, Western Australia, the Ni-sulfides are weathered to depths of 100 m or more below the oxidation front at the water-table. The weathering zone over sulfide deposits can vary from a few metres to many hundreds. Ollier & Pain (1996) quote depths of weathering of 600 m at Kennecott, Alaska, and Tintic, Utah, 900 m in Zimbabwe and 1000 m at Tsumeb, Namibia. Gossans and supergene caps are one type of representation of an ore deposit. Like Fe-gossan material, other Fe-oxides in the regolith also scavenge trace elements indicative of ore minerals below the surface, either within the regolith or below it.

OTHER IRON-RICH REGOLITH MATERIALS

False gossans

Not all quartz–goethite/haematite outcrops are gossans. Shear zones, which are easy conduits for

(a)

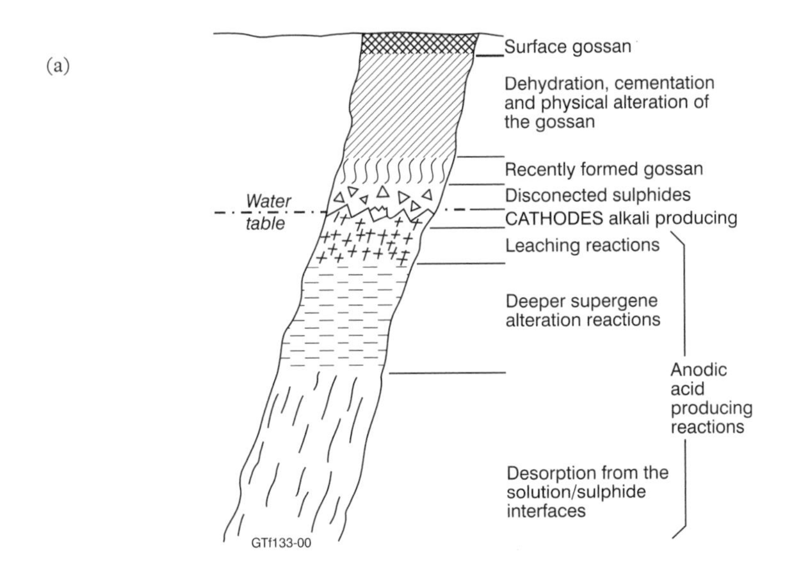

(b)

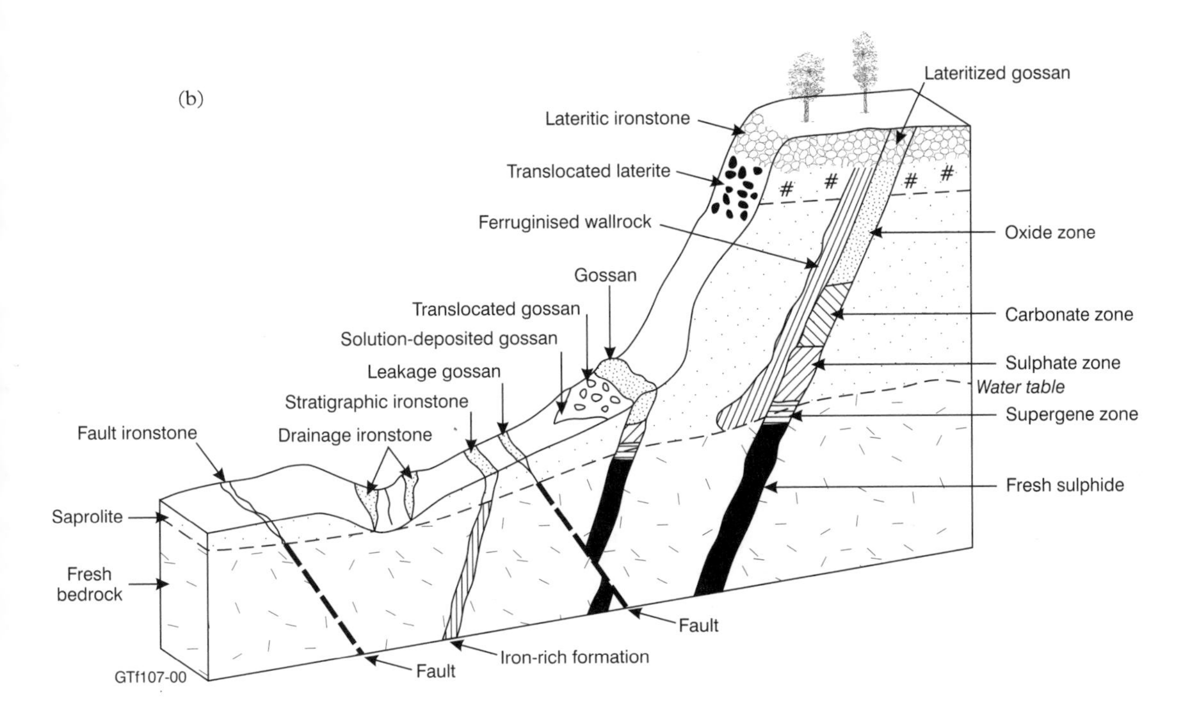

Figure 9.24 *(a) Diagram showing the development of gossans and supergene ores from primary sulfides (Butt & Zeegers 1992) (b) Types of ironstone (ferruginization from translocated Fe) and gossans (ferrunginization of sulfide mineralization* in situ*) (from Butt & Zeegers 1992)*

Table 9.6 *Suggested suites of target pathfinder and lithophile elements to be analysed during reconnaissance surveys of gossans (elements in parentheses are negative indicator) (from Taylor & Thornber 1992)*

Host rocks	Expected mineralization	Elements
Mafic–ultramafic volcanics	Ni–Cu	Ni, Cu, Co, Pt, Pd, Ir, Te (Cr), (Mn), (Zn)
	Au	Au, Pb, As, Sb, Se, Te, Bi, W
Felsic volcanics	VMS	Cu, Pb, Ag, Au, As, Sb, Bi, Se, Hg, Sn, Ba
	Cu–Mo–Au	Cu, Mo, Au, Re
Sediments	Pb–Zn–Ag	Pb, Zn, Ag, Cu, As, Hg, Sb (Mn), (Ba), (Co), (Ni)
	Cu	Cu, As, Pb, Sb, Ag, Hg (Mn), (Zn), (Co), (Ni)
Carbonates	Pb–Zn	Pb, Zn, Cd
Lateritic ironstones	–	Al, Si, Cr, V, Ti, Mn, P
Skarns	Cu–Zn, Pb–Zn Sn, W, Au–Ag	Cu, Pb, Zn, Sn, W, Au, Ag, Mn, Ca

VMS = volcanic massive sulfides

groundwater, or weathered Fe-rich strata (stratigraphic ironstones), may contain accumulated Fe-minerals and silica and form a resistant linear outcrop (Figure 9.24b). Such Fe-rich rocks may also contain anomalous quantities of Cu concentrated by adsorption from normal levels in regional groundwaters (e.g. Johnson Creek ironstone, Taylor & Thornber 1992). Geochemical analysis for an appropriate range of elements can provide a distinction between such ferruginized rocks and gossans.

Barren gossans

A gossan developed from the weathering of pyritic rocks lacking significant economic minerals is termed a barren gossan, a term which some contrast with fertile gossan, one hosting economically significant elements. In a barren gossan, systematic geochemical analysis will show a lack of ore-associated elements.

Transported gossans

Physical or chemical movement of the Fe of a gossan may leave parts of it shifted downslope relative to the gossan outcrop. The resulting ironstone mass may still contain ore-associated elements, but the translocated gossan will no longer be a direct guide to sulfides at depth.

Ironstones and 'laterites'

A wide variety of regolith Fe accumulations occur, and these are dealt with in Chapter 11. Valley-flank and valley-bottom ironstone accumulations (ferricretes) may mimic gossans, being roughly linear bodies rich in goethite or haematite. They do not generally have the box-work structure typical of gossans, and again trace element analysis may help distinguish these from gossans. So-called laterites, the ferruginous upper part of many deep-weathering profiles, can generally be distinguished from gossans by their fabric, relation to other parts of a weathering profile and landscape position (see Chapter 11).

GEOCHEMICAL DISPERSION AROUND SULFIDE BODIES

Because base metals occur as sulfides, and much Au is associated with sulfides, the detection of buried, or very extensively weathered sulfide bodies, is a major concern of mineral exploration. Sulfides do not always yield obvious gossans, or the gossan may be buried by fluvial or colluvial deposits. Provided the burial is not too deep (<100 m), and the groundwater regime has been suitable, trace elements from the weathered or buried ore-body may disperse into the overlying regolith. Butt & Zeegers (1992a) provide details of the nature of this dispersion. Butt *et al.* (1993) suggest that dispersion of 'pathfinder' elements associated with Au mineralization on the Yilgarn of Western Australia occurs in the order (most dispersed-least dispersed):

$$S > Ag > Te \gg As > Sb > Au > W > Ba, K, Rb$$

The greater mobility of the first three elements can lead to their being reduced to the regional or background levels, which reduces their usefulness, whereas the last three alkaline elements may reflect the broader alteration zone around the sulfide body.

Variation in weathering process

DISSOLUTION RATES OF MINERALS

Comparisons of the relative solubility, or rate of weathering, of the silicate minerals have been based on three kinds of approach: thermodynamics, laboratory experiments and studies of naturally weathered profiles. All approaches give the same general sequence of mineral susceptibility to weathering, though differences in detail are common. A reason for the differences in observations of natural weathering is that one mineral may begin to weather before another, but persist longer, thus their relative susceptibility depends on the criterion used. A second is that the relative rates may change throughout the profile in response to changes in water chemistry, mineral grain size and biological activity. Laboratory studies have tended to use simplified systems (e.g. Fe-free minerals to avoid redox effects, end-member mineral compositions, simple starting solution compositions), and the results may not be directly applicable to natural systems. Similarly, it is difficult to compare rankings because of the profound effect Fe-oxidation has on the integrity of a silicate mineral.

Drever & Clow (1995) list common minerals in order of their relative rates of dissolution based on laboratory measurements, and Sverdrup & Warfvinge (1995) estimate weathering rates of minerals from a catchment based on calculations (Table 9.7). These can be compared to several field observations, and to Reiche's (1943) weathering potential indices (Table 9.8).

Wasklewicz *et al.* (1993) noted that olivine in basalts on the rain-shadow side of Hawaii did not begin to weather until after plagioclase and pyroxene. They suggested this might result from the greater integrity of the olivine crystals compared to the more fractured plagioclase, or from reduced organic weathering agents in the arid local environment.

Table 9.7 *Relative rates of dissolution of different minerals in laboratory experiments at pH 5 far from equilibrium (Drever & Clow 1995)*

Mineral	Relative dissolution rate
Quartz	0.02
Muscovite	0.22
Biotite	0.6
Microcline	0.6
Sanidine	2
Albite	1
Oligoclase	1
Andesine	7
Bytownite	15
Enstatite	57
Diopside	85
Forsterite	250
Dolomite	360 000
Calcite	6 000 000

DISSOLUTION RATES OF ELEMENTS

The rate at which different elements are dissolved from a weathering profile can be compared from the slopes of element concentration with progressive weathering (e.g. see Figure 9.17). There is no independent measure of time in such an analysis of rate, thus any relation must be between successive amounts of an element and some necessarily dependent variable. For example, density (which changes partly because the element in question is being lost), or the amount of immobile element (which increases relatively partly because the element in question is being lost). Nonetheless, provided there is no addition of material to the profile, relative solubilities can be assessed from the slope of concentration versus a selected weathering index. Within the saprolite, addition of material is generally thought to be slight, and studies have established a general sequence of element loss. Such sequences may differ from the theoretical solubility order, because they depend on the association of minerals undergoing weathering, on specific features of each mineral such as defect density or compositional variation, and in some instances are affected by additions to the profile or particle movement through it.

Table 9.9 summarizes several studies of relative loss of elements from a profile, in terms of the proportion of the amount originally present. The differences between observed orders and the theoretical order based on solubility products reflects the difference in weatherability of the host minerals. In a more general way, Butt *et al.* (1993) show the extent of release and retention of a variety of elements during humid weathering of a weathering profile (Table 9.10).

CHEMICAL FRACTIONATION BY WEATHERING

The processes of rock weathering inevitably lead to a separation of the constituents of the crust. Both the particulate and the dissolved products of weathering move towards the lowest parts of the landscape and ultimately to the oceans. Eroded particles become the

Table 9.8 *Mineral resistance to weathering, observed, measured and calculated*

Reiche	Granite	Granite	Granite	Granite till	Monzonite	Gabbro	Gabbro	Basalt	Basalt	Laboratory	Computed
Quartz	Quartz	Quartz	Quartz	Quartz	Quartz	Quartz		Ilmenite	Titanomagnetite	Quartz	
K-feldspar	K-feldspar	K-feldspar	K-feldspar	K-feldspar	K-feldspar				K-feldspar		K-feldspar
Albite											
Biotite		Hornblende		Hornblende		Hornblende				Biotite	
Anorthite	Plagioclase	Biotite	Plagioclase	Plagioclase	Augite	Plagioclase	Plagioclase			K-feldspar	
Nepheline	Hornblende	Plagioclase			Hornblende	Pyroxene	Hornblende				
Hornblende	Biotite		Biotite	Biotite	Biotite	Biotite	Biotite				Biotite
Augite					Plagioclase		Augite	Pyroxene	Pyroxene	Oligoclase	
					Nepheline						Plagioclase
								Plagioclase	Plagioclase	Bytownite	Pyroxene
										Diopside	Hornblende
Olivine								Olivine	Olivine	Forsterite	
								Glass	Glass		
Reiche (1943)	Aspandiar (1998)	Markovics (1977)	Fritz (1988)	Law *et al.* (1991)	Wang (1988)	Creasey *et al.* (1986)	Fritz (1988)	Eggleton *et al.* (1987)	Colman & Pierce (1981)	Drever & Clow (1995)	Sverdrup & Warfvinge (1995)

Table 9.9 *Relative solubility of major elements from several profiles*

Author	Rock	Order of decreasing loss							
Law *et al.* (1991)	Till	Na	Al	K	Si	Ca	Fe	Mg	
Colman & Pierce (1981)	Basalt	Ca	Na	Mg	Si	Al	K	Fe	Ti
Eggleton *et al.* (1987)	Basalt	Ca	Mg	Na	K	Si	Al	Fe	Ti
Weipa unpublished	Shale	Na	Ca	Mg	K	Si	Ti	Al	Fe
Banfield (1985)	Granite	Ca	Na	Mg	Si	K	Al	Fe	Ti
Nesbitt & Markovics (1997)	Granite	Ca	Na	Mg	Fe	K	Si	Al	Ti
Davy (1979)	Granite	Ca	Na	Mg	K	Si	Fe	Al	Ti
Ildefonse (1978)	Gabbro	Ca	Mg	Fe	Si=	Al=	Na=	Ti	K
Solubility product		Na	K	Ca	Mg	Si	Al	Fe	Ti

Table 9.10 *Element mobility during weathering (after Butt* et al. *1993)*

Released from host mineral		Partly retained in secondary mineral	
Released at weathering front			
Sulfides	As, Cd, Co, Cu, Mo, Ni, Zn, S	Fe-oxyhydroxides	As, Cu, Ni, Pb, Sb, Zn
Carbonates	Ca, Mg, Mn, Sr		
Released in the lower saprolite			
Aluminosilicates	Ca, Cs, K, Na, Rb	Kaolin	Si, Al
		Barite	Ba
Ferromagnesian minerals	Ca, Mg	Fe- and Mn-oxyhydroxides	Ti, V, Cr, Mn, Fe, Co, Ni, Ga
Released in the upper saprolite			
Aluminosilicates (muscovite, smectite)	Na, Mg, K, Ca, Rb, Cs	Kaolin	Si, Al
Secondary Ferromagnesian minerals	Mg, Li	Fe-oxyhydroxides	Ti, V, Cr, Mn, Fe, Co, Ni, Ga
Released in the mottled and ferruginous zones			
Aluminosilicates	K, Rb, Cs, Si	Kaolin, gibbsite	Si, Al
Fe-oxyhydroxide	trace elements	Goethite, haematite	Fe
Retained in stable minerals			
B, Cr, Fe, Hf, K, Nb, Rb, REE, Th, V, W, Zr			

raw material of sediments and sedimentary rocks, and these are largely quartz and clay minerals. Through this process, alumina and silica become separated from the other components of crustal rocks. Zircon, Ti, Au, Sn and rare earth elements (REE) (in monazite), are released largely unweathered, removed by erosion, and concentrated in placers or beach deposits. The soluble ions – Ca, Na, Mg, Rb, Sr and K – are carried to the ocean where they ultimately form the ingredients for clastic and biogenic sediments and evaporites, a process that further fractionates: K and Rb into clays, Ca and Sr into organisms, Na into solution or evaporites and Mg into sediments or evaporites. Such fractionation explains why some granites – those formed from the melting of sediments – are Ca-poor, for the source rocks of such granites have been through a weathering cycle and lost Ca. By contrast, granites melted from prior igneous rocks are richer in Ca.

The use of geochemical signatures for provenance studies of altered rocks must take into account the possibility of fractionation during weathering. Nesbitt & Markovics (1997) assessed the extent of fractionation for the REE during granite weathering, and concluded that, except for Ce, the long-term average REE content of sediments reflected their source, even though some fractionation might occur within the weathering profile.

MINERALOGICAL AND CHEMICAL CONVERGENCE

It is apparent from the descriptions of mineral and rock weathering that if the process continues long enough, only the highly insoluble Al- and Fe-oxides and oxyhydroxides will be left as major components of the weathering profile, along with minor quantities of Ti, Zr, Cr and, perhaps, some less abundant low-solubility trace elements. This means that the mineralogy of all weathering profiles tends towards the same end products: gibbsite or boehmite, goethite or haematite and minor anatase, zircon and Cr-spinel. In some parts, perhaps as the result of fire, alumina and maghemite (Fe_2O_3) may also be present.

The upper saprolite, by definition, retains the fabric of the parent rock, but few primary minerals remain. Quartz grains, which weather little, are separated by feldspar and ferromagnesian pseudomorphs, in which the dominant minerals are kaolin and goethite. Above this is a region where micro-movement of the profile leads to the destruction of the fabric. Gravitational settling as mass is lost by dissolution is aggravated by bioturbation and shrink–swell behaviour of smectitic clays. This process marks the conversion to pedogenesis, and in many profiles the zone above the saprolite is the soil B-horizon. If the weathering process has not been prolonged or intense, the mineralogy of the B-horizon will include unweathered primary minerals as well as goethite, kaolinite and smectite. In more deeply weathered profiles, the process of alteration continues with the destruction of all primary minerals except quartz and the resistates (zircon, spinel, ilmenite, rutile, etc.), desilication of smectite to kaolin minerals, and possible further desilication to gibbsite. Sometimes over granite, the clays are eluviated so effectively that in the upper region, the so-called 'collapsed saprolite', only self-supporting quartz grains with minor clays and Fe-oxides remain.

VARIATION IN WEATHERING

We tend to think of weathering as a gradual, steady, change from fresh rock to weathered product. The weathering front moves inexorably down, rinds on corestones thicken over time, profiles deepen unless erosion overtakes them, but in reality weathering is a rock's response to its environment, and that environment changes in both space and time.

Spatially, we saw earlier in this chapter that the initial weathering processes in a rock occur at mineral interfaces, and that they differ from one pair of minerals to the next. At the scale of metres, only a small proportion of the regolith is water saturated at any time. Water finds pathways into the regolith, along which it penetrates, leaving the adjacent regions relatively dry. Weathering and dissolution are restricted to these small and irregularly shaped volumes. At the landscape scale, the changes in regolith environment from hill crest to valley bottom are marked. This is discussed further in Chapter 11 in the section on weathering catenas.

Probably the most important variable affecting weathering is climate, and climate varies at every scale from daily to geological eras. The diurnal changes in temperature, and the weekly changes in air pressure, humidity and temperature might seem to be trivial on the timescale of weathering. But weathering *is* a daily event. A weathering rate of 1 mm a millennium is 3 nm, or four kaolinite layers, a day. Many kaolinite crystals are less than one week thick!

At a longer timescale, groundwater levels can vary quite markedly over periods of years. Coventry (1982) made detailed measurements of water-tables and regolith in the Charters Towers area of Queensland. Failure of the annual monsoonal rains for many years after that study saw the water-tables drop to such depths that Coventry's work could never have been done in the years since.

But it is not simply the rate that changes, the very direction of weathering can change. In bauxitic profiles, resilication from gibbsite to kaolinite has been recorded, in saline waters clay can be converted to feldspar. Reversals of weathering direction can be extensive and long-lived, or local and ephemeral. It is the latter kind that have been suggested as a factor in the evolution of nodules and pisoliths. Rather than imagining weathering as continuous, uniform, monotonic change, we should think of it as erratic, uncertain, chaotic, but with an overall direction.

It is this essential character to weathering that challenges us in the interpretation of any regolith feature. Did it form directly from a precursor, or did it pass through a series of changes? If we return next year will it be the same?

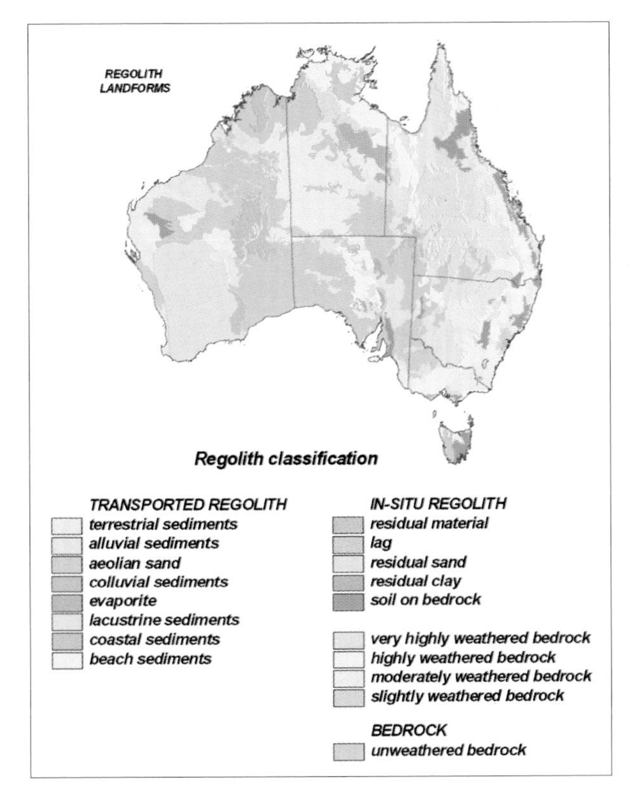

Figure 1.7 *1 : 10 000 000 map of Australian regolith (from AGSO)*

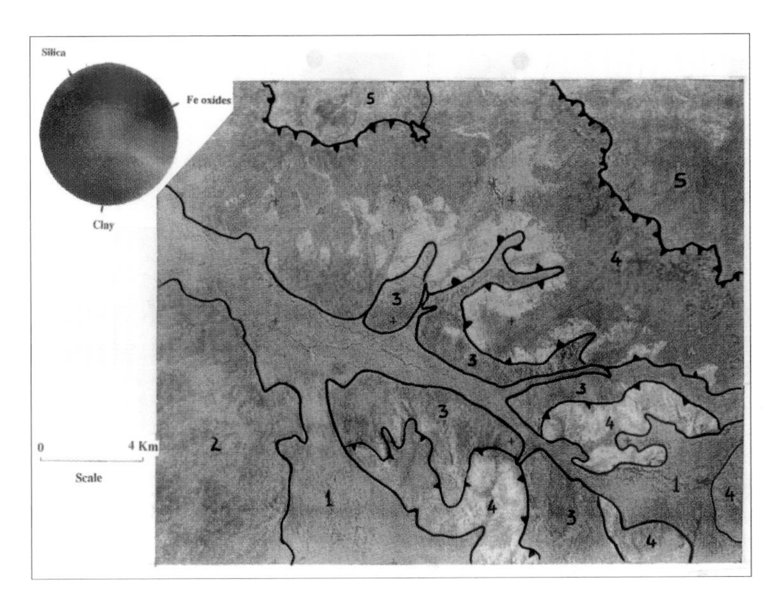

Figure 14.15 *Landsat thematic mapper false colour image from the Broken Hill region. It has been processed to highlight clay minerals (reds), Fe-oxides (greens) and silica (blues). The polygons shown on the image refer to landforms: (1) is a broad alluvial plain, (2) is an alluvial plain modified by aeolian dunes, (3) is erosional plains with <9 m relief bounded by minor escarpments, (4) are rises with 9–30 m relief, and (5) are erosional plains with <9 m relief bounded by major escarpments. The composition of the surface materials is easily identified on this image (image processed and compiled by John Wilford and David Gibson)*

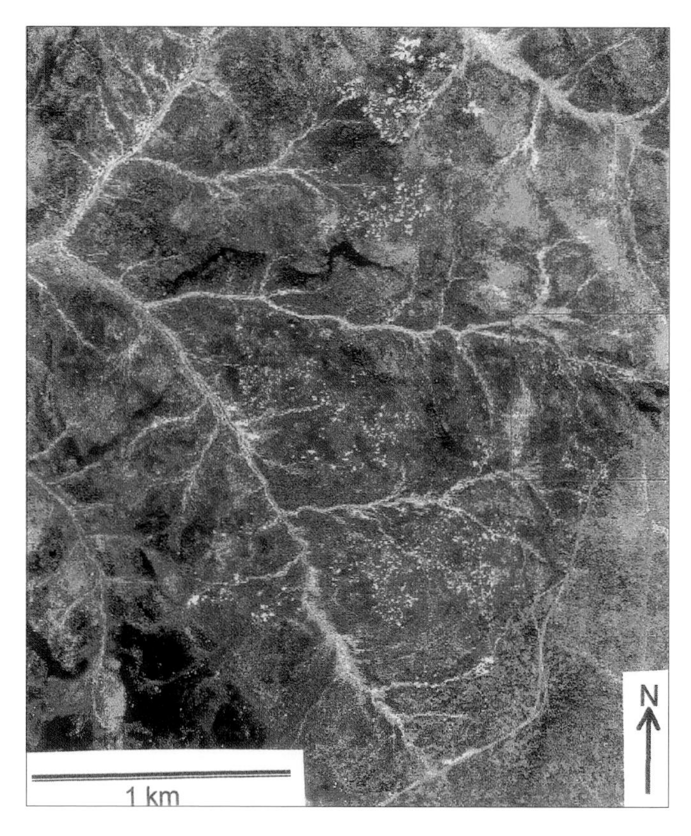

Figure 14.16 *HyMap™ image of an area in central western New South Wales showing colour differences corresponding to different spectral reflectories depending mainly on surface mineralogy and vegetation. Quartzose stream sediments are white, ferruginous surface lags are red-yellow, green is kaolinite-rich and blue tones are vegetation*

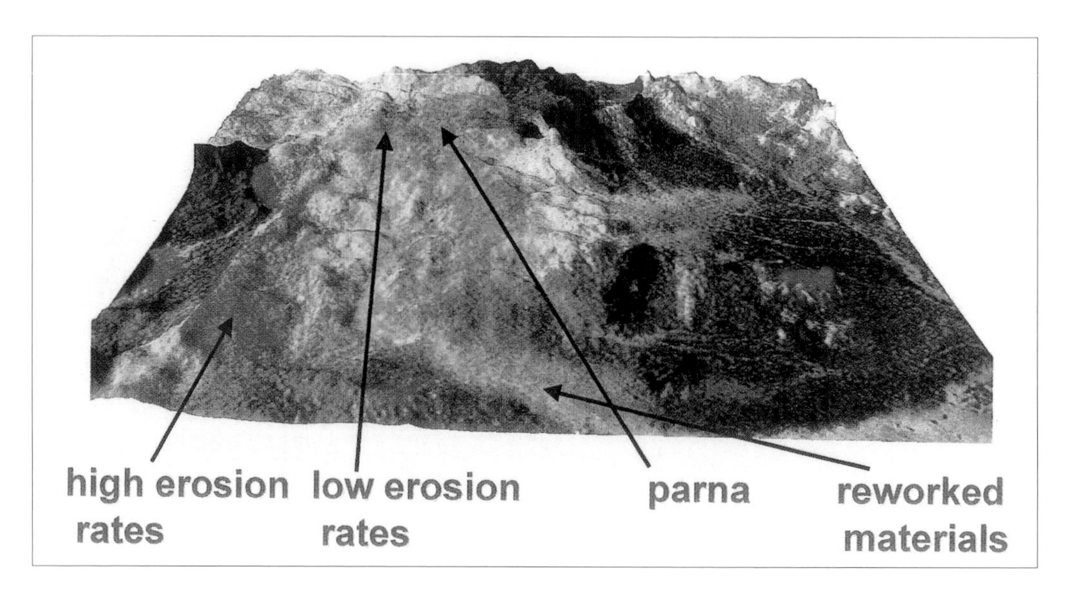

Figure 14.18 *A radiometric image of the West Wyalong region of New South Wales draped over a digital elevation model. The red colours are K response, green Th, and blue U. White areas have a high radiometric response in all channels and black areas have no response. A brief interpretation of landscape processes is shown (image John Wilford)*

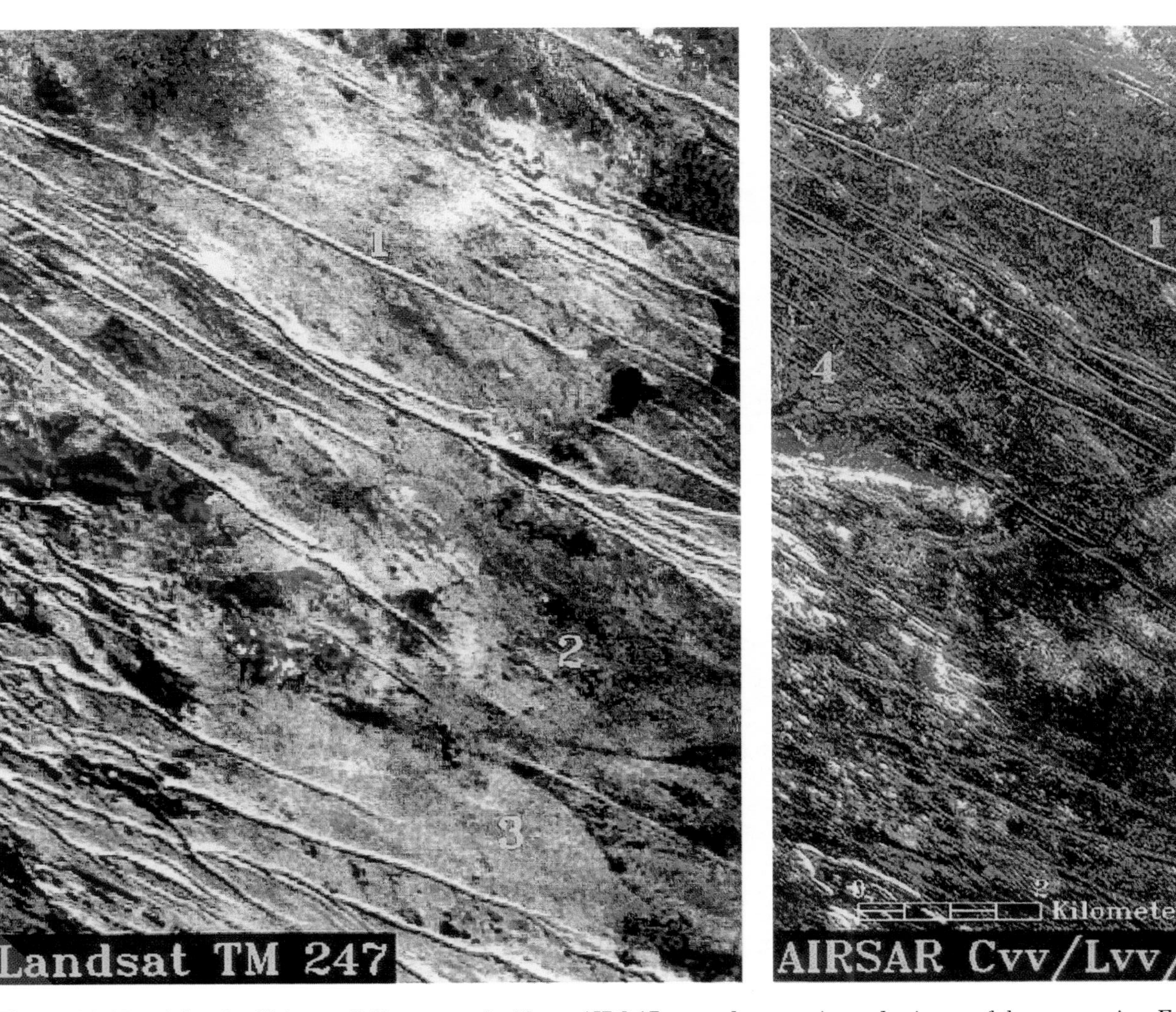

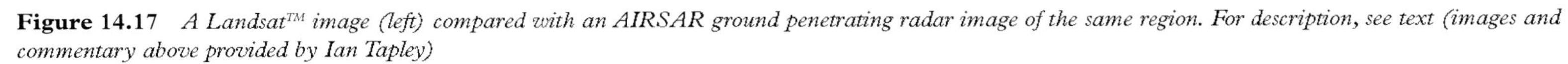

Figure 14.17 *A Landsat™ image (left) compared with an AIRSAR ground penetrating radar image of the same region. For description, see text (images and commentary above provided by Ian Tapley)*

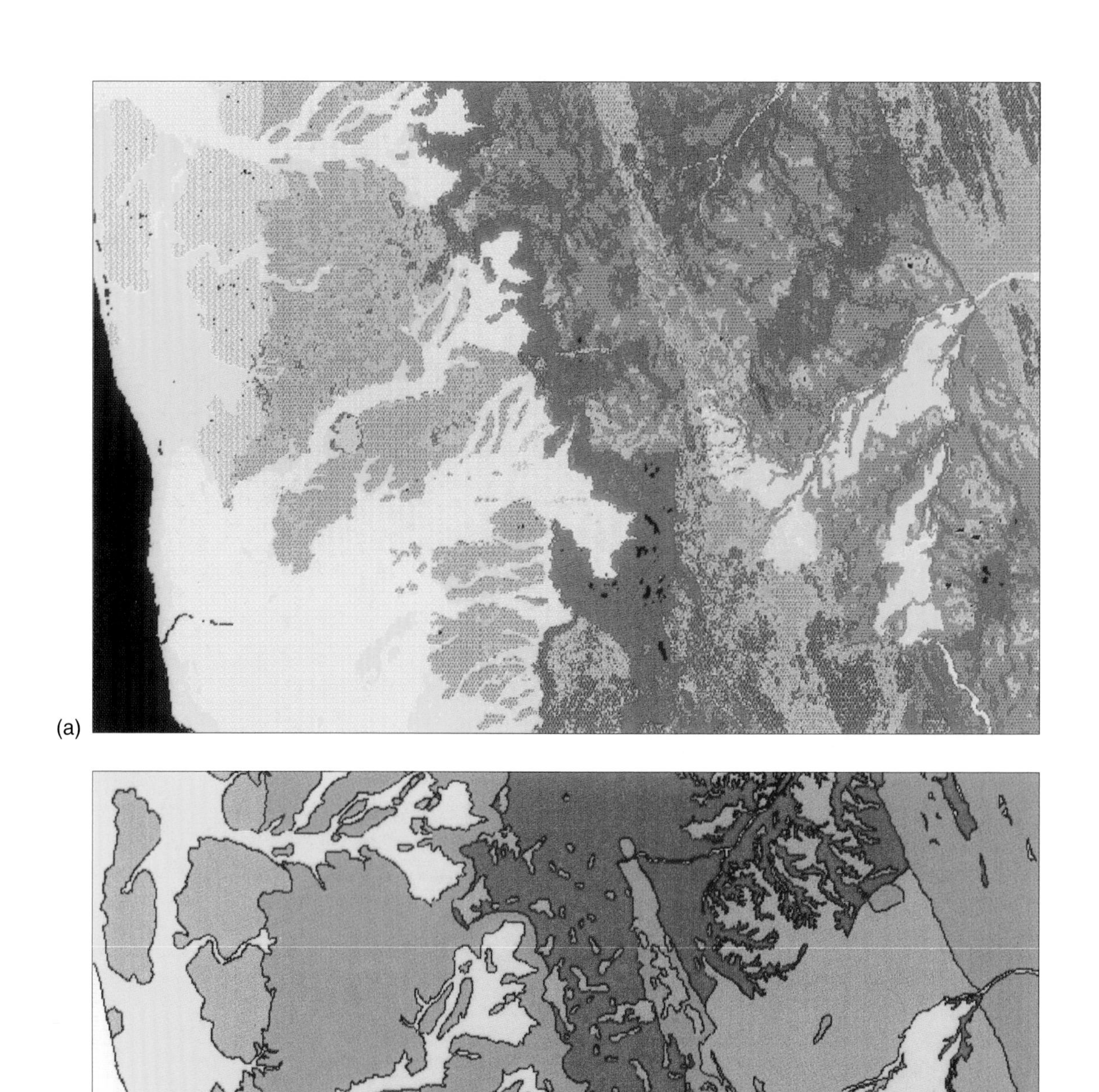

Figure 14.21 *(a) Map of part of the Ebagoola 1 : 250 000 regolith terrain map produced using an intelligent systems approach, compared to the original (b) made by conventional mapping*

10
Products of weathering

Introduction

Regolith processes lead to a range of products. Some materials from the parent rock remain at the site of weathering and are referred to as residual. Some of the minerals formed during weathering and some of the elements released may also remain in the place they formed, particularly clays and Fe- and Al-oxyhydroxides as well as the trace elements that adsorb to them. Much of the primary minerals and of the weathered products are removed as particulates during erosion. Chemicals in solution are largely carried into streams, and so into the sea or inland lakes, but some will precipitate if the conditions change. Sulfur, for example, is carried to the sea if there is enough rainfall, but will be reprecipitated as gypsum in the profile under more arid conditions. Similarly Ca is lost in a humid climate, but may form pedogenic or groundwater calcrete under more arid conditions. Ferrous iron precipitates on oxidation, and this process is probably responsible for most Fe concentrations in the regolith. However, ferric iron has some solubility, particularly at acid pH, and can move in the weathering profile in solution, and be precipitated by evaporation.

The nature of the materials left to observe at any particular place depends on a number of factors, paramount among which is the degree of erosion that occurred, and the amount of water moving through the weathered materials. These in turn depend primarily on local (depending on scale of observation) relief and climate as well as on the type of weathering. Chemical weathering will result, to a large degree, in a mixture of residual chemically inert minerals, new-formed minerals and solutes. A dominance of physical weathering will open up the rock and produce rock fragments. While that enhances the opportunity for chemical processes by exposing fresh rock to further chemical attack, the overall result is a regolith dominated by comminuted fragments of the original rock. A mixture of the two is common, but the proportions depend on the age of the regolith and the various processes acting on it through time. Table 10.1a, b summarizes the general fate of minerals and elements as a result of weathering.

Some indication of the total amount of weathered product can be gained from looking at the amount of sediment and solutes being transferred from the continents to the ocean (Figures 2.22, 10.1 and Table 10.2). These figures are not the whole story as there are the influences of climate (runoff) and tectonic relief to be accounted for. In Figure 10.1 it is clear that the major oceanic sediments are in areas of high rainfall and/or high relief, for example Asia and the Americas. It is equally notable that Australia contributes virtually no sediments to the oceans, yet it has huge reserves of weathered material stored in the regolith. This is mainly because of the general aridity and low relief of the continent. In Chapter 1 we mentioned maps showing the thickness of regolith in Australia, and these form a better estimate, but such maps are not available for other continents so sediment yield is the only realistic surrogate.

It is also worth keeping in mind that if one believes in the concepts of dynamic equilibrium, then weathering products are themselves subject to change even after weathering has been 'completed'. This means the materials produced are continually undergoing change, particularly as the environments in which they find themselves change through time and place.

A theoretical example of this would be the new-formed mineral, kaolinite. At the point of its formation it is a residual weathering product residing in the weathering profile. From this point many routes are possible. If it remains *in situ* then it may weather over time to gibbsite. It may remain in the local environment but, because of water percolating through the weathering profile, be removed in suspension downward through the weathering profile and deposited as kaolinite skins (cutans or argillans) lower in the same

Table 10.1a *Fate of some elements by weathering*

Insoluble; residual	Zr, Ti, Al, Fe, Th, Nb, Sn
Soluble	Na, K, Mg, Ca, Cl
Precipitated by evaporation	Ca, Si, Na, [CO_3], [SO_4], Cl
Precipitated by oxidation	Fe, Mn, Cu, S
Precipitated by reduction	Au
Precipitated by pH decrease	Si
Precipitated by pH increase	Ni, Co, Cu, Zn
Precipitated by CO_2 loss	Mg, Ca
Precipitated as insoluble salt	Ba, P, V, As
Adsorbed	Li, Cu, Zn, P

Table 10.1b *Fate of minerals by weathering*

	Residual	Changed	Eroded	Dissolved
From the primary rock	Zircon, spinels, rutile, ilmenite, tourmaline, garnet, corundum, quartz, Au	Olivine, pyroxene, amphibole, mica, feldspar (most silicates)	Quartz, clays	Calcite, dolomite, quartz
From the weathered rock	Gibbsite, haematite, goethite, Mn-oxides, anatase, crandallite group, barite	Vermiculite, smectite, kaolinite	Clays	Calcite, gypsum, halite

Table 10.2 *Estimates of total transport by rivers of solids and solutes to the oceans and the estimated equivalent denudation rates (from Summerfield 1991)*

Author	Mean load		Equivalent denudation rate
	10^9 t/yr	t/km^2 per yr	(mm/10^3 yr)
Solid load			*Mechanical*
Fournier (1960)	58.1	392.6	145.4
Jansen & Painter (1974)	26.7	180.4	66.8
Schumm (1963)	20.5	138.5	51.3
Holeman (1968)	18.3	123.6	45.8
Milliman & Meade (1983)	13.5	91.2	33.8
Lopatin (1952)	12.7	85.8	31.8
Solute load			*Chemical*
Goldberg (1978)	3.9	26.4	5.9
Livingstone (1963)	3.8	25.7	5.7
Meybeck (1979)	3.7	25.0	5.6
Meybeck (1976)	3.3	22.3	5.0
Alekin & Brazhnikova (1968)	3.2	21.6	4.8

profile. If erosion is significant then the kaolinite may be eroded and transported by surface water to a place of deposition. Environmental conditions at the depositional site may covert the kaolinite to gibbsite or resilicify it to form smectite. Equally the kaolinite may remain as a residual new-formed mineral in the weathering profile for very long periods of time as changes in the environmental conditions may be within the range of kaolinite stability.

Rates of weathering

The concept of weathering rate depends to some extent on the context. In Chapter 8, the rates of dissolution of various minerals are compared, and this is one aspect of weathering rate. The rate at which the landscape is lowered (or the rate of denudation) might be assessed, and, on the assumption that only weathered material is eroded, interpreted as the weathering

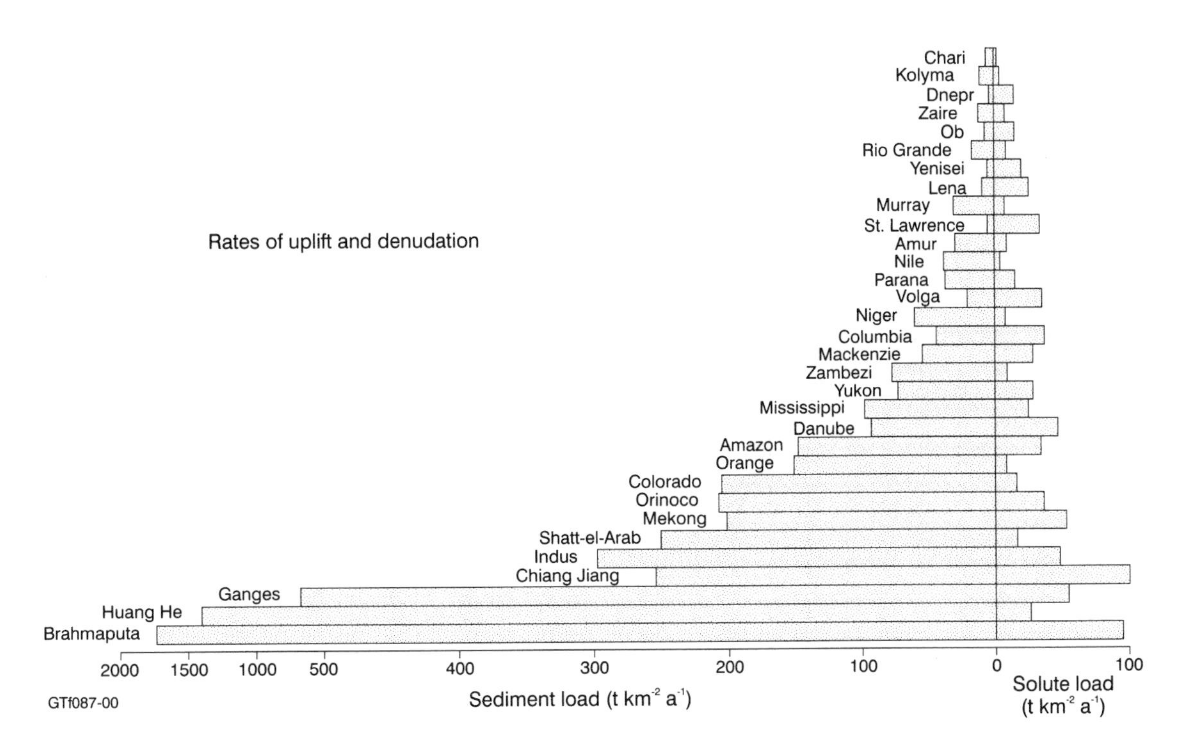

Figure 10.1 *Sediment and solute loads for the world's largest drainage basins. The solute load is represented by the estimated denudational component only and non-denudational components have been removed (from Summerfield 1991). The estimates are for the basin loads and do not necessarily represent the fluxes to the oceans of these drainage basins*

rate. This is clearly not always the case as weathering profiles are preserved in the landscape and geological record. It is possible to estimate chemical erosion (weathering) rates from catchment mass balance studies (e.g. Velbel 1985) if it is assumed that all material going into solution from weathering reactions exits the catchment or that it is assumed that alkalis leaving are all sourced from one or several minerals. It is possible that this occurs in some cases, but in many cases significant solutes remain in the regolith within new-formed minerals and the solutes so trapped are not reflected in the catchment solute load and thus mass balance calculations do not reflect true chemical weathering rates. Another approach might be to examine the rate of migration of the weathering front into a rock. A fourth is to determine the rate of conversion of saprolite to soil.

Colman & Dethier (1986) provide a number of detailed studies of weathering rates. Broadly, they show that surface lowering rates range from about 2 to 30 mm/ka, whereas rind formation on rocks is much slower, with values given between 0.005 and 1 mm/ka. Pillans (1997) showed that many studies of weathering rates and soil development rates on bedrock fell in the range 1 mm/ka, or slightly less, whereas soil formation rates on transported regolith ranged from a high of as much as 1 mm/ka to 10 mm/ka. Such determinations are highly dependent on climate, topography and rock type. Pillans found that on basalts in semi-arid central Queensland, soil formation rates were only 0.3 mm/ka.

It is possible to estimate erosion rates, at least locally, from the topography and age of igneous rocks. On the south coast of New South Wales the Mt Dromedary igneous complex reveals the presence of a volcanic edifice that erupted 95 Ma ago (Smith *et al.* 1988). On the presumption that the volcano was of the order of 1 km higher than the coarse-grained syenite stock now exposed, the erosion rate over the Tertiary can be estimated as at least 10 mm/ka. Apatite fission track dates suggest that of the order of 2 km of rock has been eroded from the region in the past 100 Ma, indicating an erosion rate of the order of 20 mm/ka. While this gives an erosion rate, weathering rates may have been somewhat different. The volcanic edifice may well have contained ashes that are comparatively easily eroded without being significantly weathered first. On the Monaro of New South Wales, 60–35 Ma old basalts have valleys cut into them that are 50–100 m or so deep, requiring a local erosion rate of at least 1–4 mm/ka. In this case the basaltic pile filling the valleys contained little ash and the basalt would have to be appreciably weathered before it could erode. Bauxite and lateritic bauxite up to 3 m thick have developed on lavas in the pile and palaeomagnetic studies have shown that these took between 0.5 and 5 Ma to form, suggesting a weathering rate of the basalts of between 0.6 and 6 mm/ka.

Palaeomagnetic investigations of lateritic profiles in French Guiana (Théveniaut & Freyssenet 1999) demonstrate that over the last 780 ka the profile has deepened at a rate of 11.3 ± 0.5 mm/ka. The upper parts of this 54 m thick profile are also shown to be >10 Ma old, suggesting weathering rates in excess of 5.4 mm/ka.

Residual materials

This group of minerals includes all those that remain resident in the weathering profile for a period. It includes those original rock minerals little or unaffected by weathering and all the new-formed minerals not removed. They can generally be considered in three major groups: residual major rock-forming minerals, new-formed minerals and resistate minor minerals. New-formed minerals are discussed in Chapter 8.

RESIDUAL ROCK-FORMING MINERALS

Included here are all those minerals not completely destroyed by the weathering process. Their nature depends to a large degree on the type of parent material being weathered and the position in the weathering profile.

Rock type obviously plays an important part in determining the residual minerals. Granite-like rocks will mostly weather leaving residual quartz, possibly muscovite, and commonly K-feldspars, as well as some minor relatively unweatherable minerals (resistate minerals) like tourmaline, zircon, monazite, titanite, spinels and rutile.

Quartz is probably the most abundant residual mineral simply because it is an abundant rock-forming mineral and it is the least soluble of the rock-forming minerals (Table 9.7). Quartz does weather by dissolution, as is described in Chapter 8.

In less intensely weathered examples than occur on the Yilgarn, where many granitic profiles are capped by silcrete, it is usual to find residual feldspars and micas in the upper parts of the weathering zone. At Charters Towers, Aspandiar (1998) records feldspars at the top of the *in situ* weathered profiles of granodiorite (see Figure 9.16). Biotite is rapidly transformed to vermiculite by 175 cm, while hornblende continues to within 100 cm of the surface of the *in situ* profile. Deeper in the weathering profile residual minerals generally consist of a progressively increasing proportion of the original minerals plus their alteration products until fresh rock is reached. The nature of the materials depends to a large degree on the position of the permanent water-table, and if it occurs within the profile then the minerals below it are likely to be reduced and richer in alkaline earths than the minerals above.

Basaltic rocks, on the other hand, tend to have very few residual minerals except for those extremely resistant to weathering such as chromite, ilmenite, magnetite and titanite. These trace minerals may be altered to other Fe^{3+} and Ti^{4+} minerals, for example pseudo-rutile, which themselves tend to remain as residual components. It is common, particularly in cool temperate and arid climates, for profiles derived from basalt weathering to contain residual rock fragments, particularly near the upper parts of the profile where water drains rapidly leaving little time for the rock fragments to weather.

Metamorphosed pelitic rocks tend to leave muscovite and quartz as the major resistates, with garnet, tourmaline, andalusite, kyanite or sapphirine as characteristic minor resistates. Calc-silicate metamorphics may leave only garnet as a resistate, the other component minerals such as actinolite, diopside, epidote and wollastonite being fairly susceptible to weathering. Gniesses react like granitic rocks, having similar mineralogy, although their metamorphic fabric provides such ready access to solutions that they weather more readily. Typical resistates from gneisses are quartz and muscovite, as well as trace minerals such as garnet and zircon.

Sedimentary rocks are themselves the product of at least one cycle of weathering and as a consequence detrital sedimentary rocks contain minerals already relatively resistant to weathering. Thus sediment and sedimentary rock weathering leaves similar residual minerals to the original rock – quartz, clay minerals and possibly feldspars with a small concentration of resistate minerals, commonly including zircon, tourmaline, rutile, ilmenite, monazite and other heavy minerals. In these rocks weathering may do no more than dissolve a more soluble cement (e.g. calcite) or slightly dissolve framework grains of quartz and so provide space for interstitial clays to be washed away.

Chemically or biochemically formed sediments, on the other hand, are mostly composed of minerals which dissolve very easily, e.g. calcite, dolomite, gypsum, halite. Thus they weather by congruent dissolution and leave almost no resistates. Minor quantities of clays in the chemical sediment may become concentrated as the host rock is dissolved, and wash into low points (solution caves, sink-holes, etc. in limestone karst) where they can form residual deposits, or be weathered to gibbsite as deposits of karst bauxite. Even silcretes and cherts dissolve, leaving only the insoluble anatase as a residue.

The nature of the residual minerals present in the weathering profile depends also on the position in the profile being considered. It is generally true to view a weathering profile as 'younging downwards', although many weathering profiles contain unconformities (see Chapter 2). That is, the profile was formed by a downward-moving weathering front progressively converting fresh rock to regolith. The consequence of this, ignoring complications discussed elsewhere, is that the residual weathering products contain an increasingly large proportion of each of the parent rock minerals deeper in the profile. Because in most landscapes there is some erosion at the surface, the absolute amount of any particular residual mineral may increase up or down the profile, but the variety of those residual from the parent generally decreases up-profile.

Resistate minerals

Important components of the residual weathering products are the mineral group commonly called 'resistates' or 'resistant minerals'. These are minerals not significantly affected by the weathering process. Quartz has been considered in some detail above, and this is probably the most common resistate under most weathering conditions, but it will not be discussed again here.

Other common resistates include the minerals that occur as trace components in parent rock, but which are concentrated by depletion during the process of weathering. The most common are zircon, rutile, ilmenite, anatase, magnetite (and other spinels), garnet, tourmaline and monazite. Some of these minerals weather to secondary phases which are more stable in the weathering environment and also remain as resistates. In some cases these resistates also include Au, diamonds and cassiterite.

There is considerable argument about the geochemical stability of many of the resistates in various weathering environments. This is an important topic as their stability is related to the calculation of most 'weathering indices', either using the relative concentration of these stable minerals in the weathering profile or doing the same using their geochemistry. For example, the degree of weathering is often related to the increase in zircon or Zr, or Ti, relative to its content in the parent rock. If these minerals or elements are mobile in the weathering profile then these indices of weathering will be incorrect.

It is important to draw a distinction between resistate minerals and insoluble (or sparingly soluble) elements. Zircon is one of the least soluble elements in regolith water (solubility of about 10^{-11} M/l, Chapter 7). Zircon is by far the commonest Zr-bearing mineral in rocks and regolith, and it is highly weathering resistant. Thus since zircon tends to be concentrated in residual regolith, so must Zr. By contrast, Ti, while also a highly insoluble element (solubility about 10^{-9} M/l, Chapter 7), occurs in several minerals, in rutile it is as resistant to weathering as zircon. In ilmenite, upon weathering Ti precipitates as anatase along with goethite in a pseudomorph. In titanite or pyroxene Ti released by weathering is also precipitated as anatase within the periphery of the original mineral. While the anatase particles are highly insoluble, they are easily transported in percolating water when the other components of the pseudomorph are leached (see Figure 7.9). Where weathering is aggressive enough to rapidly break down silicates (say at low pH), the released anatase may be flushed out of the profile and give the appearance of Ti solubility from one region and precipitation in another.

Thorium is another essentially immobile element (solubility in groundwater $\approx 10^{-10}$, Chapter 7). It occurs in rocks primarily in zircon, so it has two reasons to be resistant to weathering: it is in a resistant mineral, and even should that mineral be open to weathering solutions (as are metamict zircons), the low solubility of Th ensures it travels no distance.

Additionally most of these resistate minerals are more dense than the majority of rock-forming silicates and most new-formed minerals. Those which do have a higher specific gravity, are collectively known as heavy minerals. Because they are heavier they tend to concentrate as lenses and beds as they are eroded, transported, winnowed and deposited. They form a significant ore in many unconsolidated sedimentary sequences. These deposits of heavy minerals are discussed in Chapter 15.

Continuous weathering

The weathering products themselves are exposed to continued weathering after they are released from the parent rock. Materials in an *in situ* profile continue to be altered as long as the profile is at or near the Earth's surface, and while weathering agents are able to affect the materials. Equally, the products eroded are subject to continual physical and chemical weathering until they are removed from weathering influences, or until the products are entirely composed of minerals which are stable under the prevailing weathering conditions.

The end products of this process are generally minerals like quartz, haematite/goethite, and kaolinite or gibbsite. It is worth noting, however, that quartz will still dissolve, even if very slowly, in most aqueous environments unless the water is saturated with respect to quartz. Haematite will, in dry environments be stable, but in wet ones it may hydrate to goethite and *vice versa*. Gibbsite may under conditions of elevated silica activity in the weathering solutions undergo resilication and form kaolinite, and Danjic (1983) showed that silica-rich solutions percolating into bauxite from overlying limestone had resilicated boehmite to kaolinite. It is also possible for kaolinite to be resilicified under these conditions to form smectite. Thus even the 'normal' end products of weathering react to changed environments. These changes are enhanced when the geochemical and physical conditions change, such as occurs with a climate change.

11
In situ weathering profiles

Introduction

The concept of the weathering profile was introduced in Chapter 9, and examples of profiles on several different rock types were given there. Towards the end of that chapter we pointed out that there was some degree of convergence towards a common suite of minerals as weathering progresses, with lateritic and bauxitic profiles being the typical end product of prolonged chemical weathering, almost independent of parent. In this chapter we address the weathering profile itself, with less regard to its evolution from whatever parent. Most weathering profiles exhibit changes in the nature of materials from top to bottom, and we begin here at the top.

Soil

Soils form in the uppermost part of a regolith profile, usually within the collapsed saprolite or a transported veneer. Soils form at the Earth's surface in response to many factors, many similar to those controlling regolith formation in general, which produce layers or *horizons* parallel or sub-parallel to the surface. The distinction from regolith is, in general, subtle, but the relationship with the surface is distinctive. To amplify the difference it may help to suggest that soils develop horizons from a regolith parent, while regolith develops from *in situ* or transported weathered rock material to form zones. There are many books that summarize the importance of soils as regolith materials and their influence on landscape evolution, including Ollier & Pain (1996), Hunt (1986), Fookes (1997), Paton *et al.* (1995), CSIRO (1983) and Birkeland (1984). Figure 9.14 shows the relationships of soil to regolith profiles.

SOIL HORIZONS

Major soil horizons are indicated in Figure 11.1, and are summarized here.

Solum horizons

A1 mineral horizon at or near the surface with organic matter accumulation

A2 mineral horizon with less organic matter than either the A1 or B-horizons, and generally of paler colour

A3 transitional horizon between A and B, but having properties more like the A

B1 transitional horizon between A and B, but having properties more like the B

B2 horizon with relatively high contents of one or more of clay, Fe, humus; there may be a maximum of soil structure or an intensification of colour. Sub-horizons may be defined B21, B22, etc.

B3 transitional between B and C-horizons but has properties more like the B2

Sub-solum horizons

C horizon below the solum (AB profile), partially weathered, little affected by pedogenesis

D horizon of soil material unlike the overlying solum (buried soil); and

R bedrock

Suffixes can be used to indicate the presence of particular features (e.g. B2k indicates the presence of carbonates in the B2). Check McDonald *et al.* (1990) for further details.

Soil classification

The only data on which a soil classification system is based are those obtained from the soil profile. There are, however, numerous (>10) major classification systems used around the world. The reason that there are so many is because soils are very complex entities.

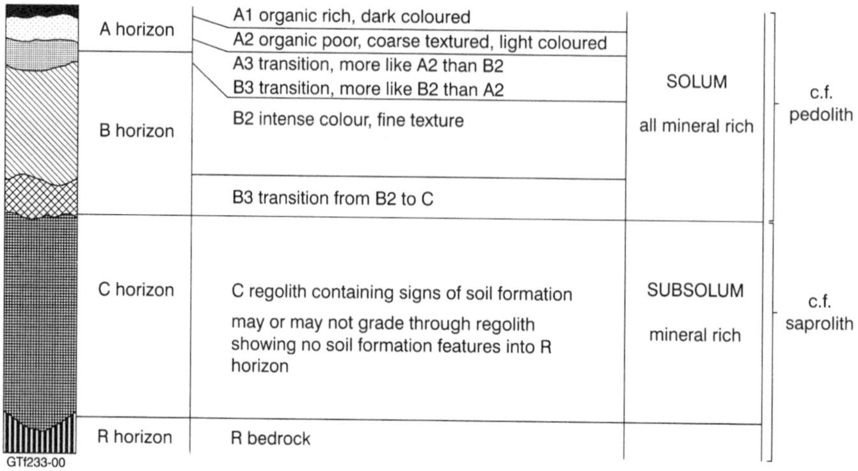

A horizon	A1 organic rich, dark coloured A2 organic poor, coarse textured, light coloured	SOLUM all mineral rich	c.f. pedolith
B horizon	A3 transition, more like A2 than B2 B3 transition, more like B2 than A2 B2 intense colour, fine texture B3 transition from B2 to C		
C horizon	C regolith containing signs of soil formation may or may not grade through regolith showing no soil formation features into R horizon	SUBSOLUM mineral rich	c.f. saprolith
R horizon	R bedrock		

GT1233-00

Figure 11.1 *Main mineral horizons of a soil profile. It is important to note that although all these horizons can be defined, not all may occur in any one profile*

There are basically three major forms of soil classification being used:

1. Hierarchical classifications consist of various levels of classification. The upper level is the universe of soils and the lowest is an individual entity (profile). The soil properties used have less priority at successive levels down the classification. In practice most widely used soil classifications are hierarchical. Hierarchical classifications are probably easier to comprehend because they only require the manipulation of a limited data set at one level.
2. Non-hierarchical classifications simply divide all soils into groupings at the one level on the basis of various soil properties. No grouping or soil property has priority over another.
3. Disjoint, conjoint and fuzzy classes. Many natural features fall into the disjoint classes (e.g. sand and mud) that have no overlap; rather they are divided into mutually exclusive classes. If the class boundaries are fuzzy or show overlap (conjoint) then differentiation between classes is difficult (e.g. sandy mud and muddy sand).

SOIL CLASSIFICATION SYSTEMS IN AUSTRALIA

There are two systems commonly used in Australia, the *Northcote Key* and the *Great Soil Group* (GSG). The former is an example of a hierarchical classification and the other of a non-hierarchical, fuzzy system. Both are widely used and are increasingly being used in conjunction with an overseas system, the *US Soil Taxonomy*, another example of a strictly hierarchical system and a new Australian Soil Classification system (Isbell 1996). Descriptions of each system now follow.

Great Soil Groups of Australia

This system was developed during a long period since the 1920s and 1930s and the standard version was published in 1968 as *The Handbook of Australian Soils* (Stace *et al.* 1968).

A. No-profile differentiation

1. *Solonchaks.* Saline soils with little profile development in 30 cm. Developed on a range of regolith.
2. *Alluvial soils.* Soils developed in alluvium with some development of the A-horizon. Sedimentary character of the regolith is preserved.
3. *Lithosols.* Thin stony soils on a range of bedrocks with some development of the A-horizon.

B. Minimal profile development

8. *Desert loams.* Thin, loamy A-horizon over a structured, clayey B of red to brown colour. Carbonates and gypsum occur through the profile.
9. *Red and brown hardpan soils.* Red to brown loamy soils over a silica cemented hardpan.

10. *Grey, brown and red clays.* Uniformly coloured soils of high clay content throughout; self-mulching surface; carbonates common at depth; crack deeply on drying.

C. Dark soils

11. *Black earths.* Uniformly very dark-brown to black colour, high clay content throughout; self-mulching surface; carbonates common at depth; crack deeply on drying.
12. *Rendzinas.* Shallow, dark-brown to black clay loams to light clays; strong crumb structure; carbonate nodules may occur with the limestone parent rock at depth.
13. *Chernozems.* Uniformly very dark-brown to black colour; usually a clay loam surface grading to a clay at depth; minor carbonates may occur at depth.
14. *Prairie soils.* Slightly acid soils with dark A-horizons with moderate crumb structure over clayey B-horizons; carbonates usually absent.

D. Mildly leached soils

16. *Solonetz.* Soils with prominent texture contrast between A and B-horizons; B-horizons are alkaline and have a coarse prismatic structure.
17. *Solodized solonetz/solodic soil.* Soils with very prominent texture contrast between A and B-horizons; A-horizon usually bleached; B-horizon alkaline with coarse columnar structure.
18. *Soloths.* Similar to solodized solonetz, but have slightly acid to neutral profiles with less prominent structure in the B-horizon.
20. *Red-brown earths.* Brownish, loamy A-horizons (usually with A2) abruptly overlying reddish clayey B-horizons of blocky to prismatic structure; carbonates are common at depth.
21. *Non-calcic brown soils.* Brownish, loamy A-horizons (no A2) abruptly overlying reddish clayey B horizons with blocky to prismatic structure; no carbonates at depth.
22. *Chocolate soils.* Dark-brown, loamy crumb-structured A-horizon grading into brown to red clayey B-horizon; carbonates absent; associated with basaltic rocks.
23. *Brown earths.* Dark A-horizon over yellowish, brownish or reddish B-horizon with little texture change at depth; profile acid throughout.

E. Soils dominated by sesquioxides

25. *Red earths.* Gradational, apedal or weakly structured soils with dark, sandy to loamy A-horizon changing gradually to a reddish, more clayey B; typically porous; sesquioxide nodules common in the B.
26. *Yellow earths.* Gradational profiles similar to red earths except for a yellow coloured B.
27. *Terra rossa soils.* Thin reddish soils over limestone or other highly calcareous rocks; brown loamy A-horizon with gradational change to reddish clayey B-horizon.
28. *Euchrozems.* Strongly structured, reddish clayey soils; similar to krasnozems except for their neutral reaction; on basalt or other basic rocks.
29. *Xanthozems.* Strongly structured, yellowish clayey soils; similar to krasnozems except for their yellow B-horizon colour; acid profiles on basaltic rocks.
30. *Krasnozems.* Strongly structured soils with dark A-horizon grading into clayey red B-horizon; acid reaction; formed on basaltic rocks.

F. Mildly to strongly leached soils

31. *Grey brown podzolic soils.* Have similar profile forms to 32, 33; they differ in their B-horizon colour.
32. *Red podzolic soils.* Loamy A-horizon abruptly overlying a red clayey B-horizon; a distinct A2 is present and the profile is acid throughout.
33. *Yellow podzolic soils.* As for red podzolic soils except in having a yellowish B-horizon.
34. *Brown podzolic soils.* Have similar profile forms to 32, 33; they differ in their B-horizon colour.
35. *Lateritic podzolic soils.* Have sesquioxide nodules in the B-horizon and usually a thick, pale-grey weathered zone at depth.
37. *Podzols.* Sandy texture throughout but strong differentiation into a greyish to pale A-horizon over a dark-brown to black organic and iron-enriched B. Australian podzols rarely have a decomposed litter horizon (A0) like their Northern Hemisphere counterparts.

G. Organic soils

These are usually peaty soils (>20% carbon) with high water-tables; the exceptions are the alpine humus soils which have highly organic A-horizons over yellowish

Table 11.1 *A brief summary of the Northcote Factual Key for the classification of Australian Soils*

Primary profile form		Subdivision		Section/class	Principal profile form examples
Organic soils	(O)	(0–30 cm >20%organic carbon)			
Uniform	(U)	Coarse textured	(Uc)	A-horizon features	Uc 2.33
		Medium textured	(Um)	A-horizon features	
		Fine textured, no cracking	(Uf)	A-horizon features	
		Fine textured, cracking	(Ug)	A-horizon features	Ug 5.16
Gradational	(G)	Calcareous throughout	(Gc)	B-horizon structured	
		Not calcareous throughout	(Gn)	B-horizon structured/A2/pH	Gn 2.14
Duplex	(D)	Red clay B-horizon	(Dr)	A-horizon/B mottle/pH	Dr 2.33
		Brown clay B-horizon	(Db)	A-horizon/B mottle/pH	
		Yellow/grey B-horizon	(Dy)	A-horizon/B mottle/pH	
		Dark B-horizon	(Dd)	A-horizon/B mottle/pH	
		Grey B-horizon	(Dg)	A-horizon/B mottle/pH	

or brownish B-horizons without an increase in clay at depth; they are strongly acid soils.

The main advantages of this system are that it is widely used in Australia and its terms have wide currency there and, to some extent, overseas. The terms are hence very familiar. Consequently they convey a picture of a particular type of soil profile (central concept).

Some disadvantages are that there is no key or schedule for this classification and it lacks clear definition of the range of properties in each group. This promotes subjectivity in its use and makes it difficult to modify to include new ideas. For this reason, as more work has been done on soils in Australia, the classification has not evolved to take account of the new information; it is becoming obsolete.

Factual Key of Northcote

The Key owes much to about 500 soil profile descriptions from across Australia. This is a hierarchical classification that is widely used in Australia and was devised to help in the production of a national soil map. The hierarchy has five levels:

- Divisions;
- Sub-divisions;
- Sections;
- Classes; and,
- Principal profile forms.

A brief summary is given in Table 11.1.

Some advantages of the Key are that it is simple and unambiguous to use, in fact a reasonably trained layperson could operate it. The keying attributes are all easily determined in the field, so it is particularly useful for mapping. Most of the attributes are physical and do not require sophisticated tests to determine.

Some disadvantages of the Key are that it is difficult to establish new soil classes as the knowledge base expands. The alpha/numeric nomenclature is confusing and is a poor means of communication to those not using the Key regularly. The criteria used to divide soils at the various levels of the Key are often contrived and seemingly rather abstract. Two soils which appear very similar in the field, but which key out differently on just one attribute, may end up being classes a great distance apart in the Key. The Key leaves no room for user discretion if it is applied with rigour.

Overall the Key is an artificial system which has its place where it is required that every profile shall be unambiguously allocated to a class on the basis of a limited number of properties. This is often a benefit where the system is being used by non-professionals; however, such systems are inherently unsatisfactory in the process of ordering knowledge. They are unpleasantly like filing systems in government departments, that is, once filed the job is complete and so may be forgotten.

Australian Soil Classification

A new classification system for Australian soils has been developed (Isbell 1996) which combines the best aspects of the GSG and the Northcote Key. This new scheme is a hierarchical system with five levels:

- Order;
- Suborder;
- Great group;
- Subgroup; and,
- Family.

Table 11.2 *A basic outline of the Australian Soil Classification (after Isbell 1996)*

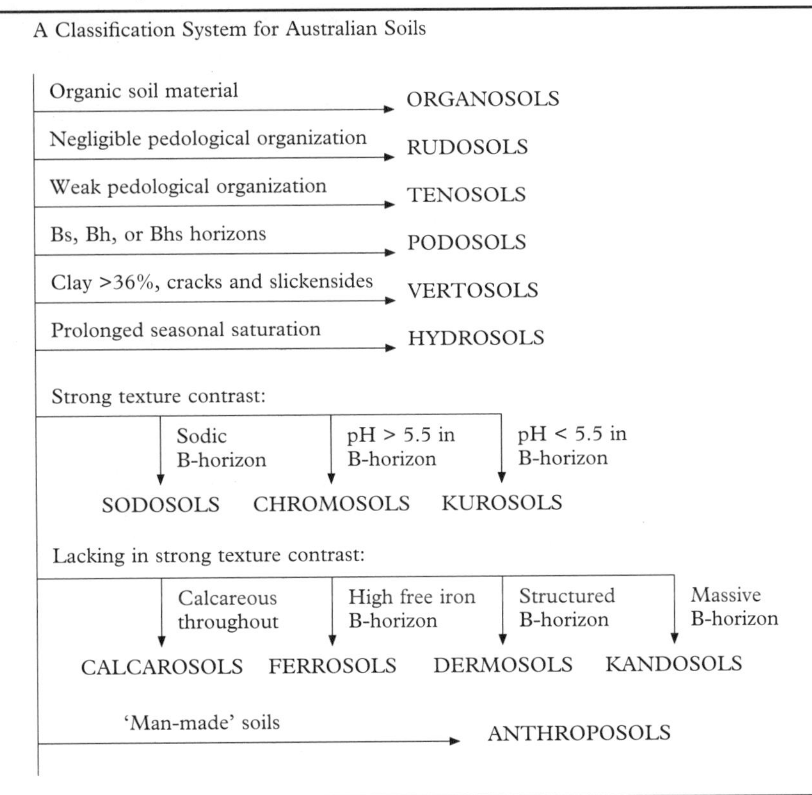

The brief features of the total 14 orders are outlined in Table 11.2. Some new terms are used in this system which are well defined to avoid some long-used but, to some degree, abused terms.

US SOIL TAXONOMY

An example of a globally used soil classification system is the US Soil Taxonomy (Soil Survey Staff 1975). This is a general purpose classification of a hierarchical type which was developed, like the Northcote Key, for use in soil survey work. The Taxonomy is based on about 10 000 profile descriptions and is more representative of the soils in the USA than of the world.

The Taxonomy operates at six hierarchical levels:

- Order;
- Suborder;
- Great Group;
- Subgroup;
- Family; and,
- Series.

The Taxonomy has a completely new language, based on Greek and Latin roots, and the names are intended to indicate the main properties of the particular class (Table 11.3).

The Taxonomy has become a de facto international classification by increased usage in the major international and national soil journals. The main subdivisions are orders and the 10 orders are summarized in Table 11.3 with the equivalent GSG terminology.

Some advantages of the US Soil Taxonomy are that it attempts to be a more comprehensive and straightforward classification system than any other. It is a valuable classification from the point of view of communication hence its use in the journals, and its new nomenclature avoids the use of long-used/misused terms with which most soil professionals are familiar, but which have different meaning to different users.

Some disadvantages of the Taxonomy are that information is lost about the similarity of soils close to the boundary of two classes because of the exclusivity of the units in the hierarchy (in a similar way to the Northcote Key). It does not handle intergrades well.

Table 11.3 *Major soil orders of the US Soil Taxonomy and an example of how the terms are used to form a soil name*

Soil orders	Equivalent GSG	Etymology
Entisols	Alluvial soils	(recent)
Inceptisols	Alpine humus soils	(inception)
Mollisols	Prairie soils	(L. mollis = soft)
Alfisols	Red-brown earths	(pedalfer = Al, Fe)
Ultisols	Red-podzolic soils	(L. ultimus = last)
Oxisols	Krasnozems	(oxides)
Spodosols	Podzols	(G. spodos = wood)
Vertisols	Black earths	(L. verto = turn)
Aridisols	Desert loams	(L. aridus = dry)
Histosols	Peats	(G. histos = tissue)

Example: subgroup: Natric Palexeralf is a soil with strong textural differentiation (alf) and a thick (pale), sodic (natr) B-horizon occurring in a dry (xer) environment. This is equivalent to a solodic soil.

Because all the possible categories at different levels have not been defined it is possible that some soils may not be recognized. Moreover, to define many of the classes it is necessary to have laboratory data so the system is difficult to use with accuracy in the field.

Figure 11.2 *Weipa weathering profile showing the much eroded bauxite above a ferricrete above the pallid zone of the deep weathering profile. Photo at Hay Point, north Queensland (photo GT)*

DURICRUST

Duricrusts cap many *in situ* weathering profiles. At Weipa, for example, the weathering profile (Figure 11.2) is capped by a ferricrete formed from the concentration and hardening of Fe-mottles and this is in turn overlain by a 2–5 m section of pisolitic bauxite (Eggleton & Taylor 1999). Ferricrete, sometimes referred to as 'laterite', overlies significant proportions of weathering profiles on the Yilgarn Craton in Western Australia (Butt & Smith 1992). Silcrete overlies deep weathering profiles in much of the Eromanga Basin of Queensland, South Australia and New South Wales (Senior 1978; Taylor 1978). Calcretes form in the upper parts of weathering profiles in much of arid central Australia (Milnes 1983), in weathered sediments and bedrock regolith at Broken Hill (Hill 1996) and in the Cobar region of central New South Wales (Leah 1996). Taylor *et al.* (1990b) report bauxitic duricrusts developed on weathered basalts from the Monaro in south-eastern New South Wales, and Ruxton & Taylor (1982) and Taylor & Ruxton (1987) report similar situations from the Shoalhaven catchment, also in southern New South Wales. Similar duricrust cappings are reported from throughout central and western Africa (McFarlane 1976; Thomas 1994; Goudie 1973; Summerfield, 1983), from equatorial South America (Boulet *et al.* 1997) and Europe (Thiry 1997).

As duricrusts are discussed in more detail in Chapter 13 we will not continue here.

Physical characteristics of profiles

The physical, as well as the chemical and biological, properties of regolith profiles are important as they control many of the processes which act in the regolith and these are often very different from those which act in the parent rock. While some of the characteristics noted below are not strictly physical properties their nature affects the physical nature of the regolith.

MINERALOGY

The mineralogy of *in situ* weathering profiles is predominantly affected by the nature of the parent rock and its weathering products. Mostly the physico-chemical properties of the clay minerals influence physical behaviour. They can shrink and swell on drying and wetting, they can become oriented under the same conditions. As they swell aggregates may rub

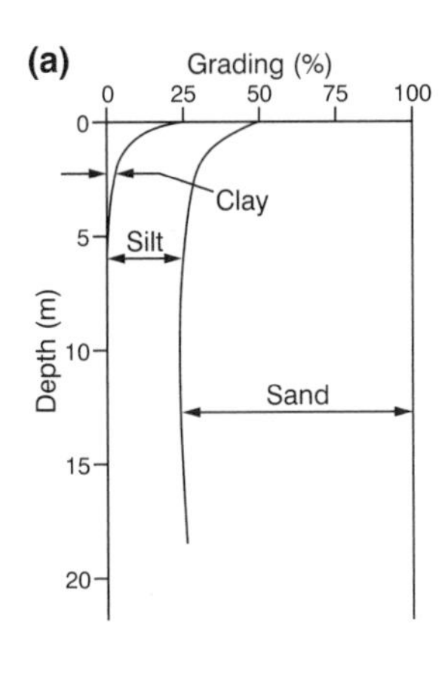

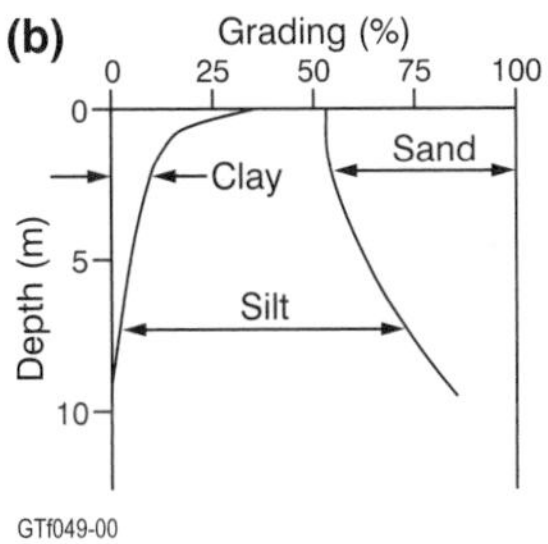

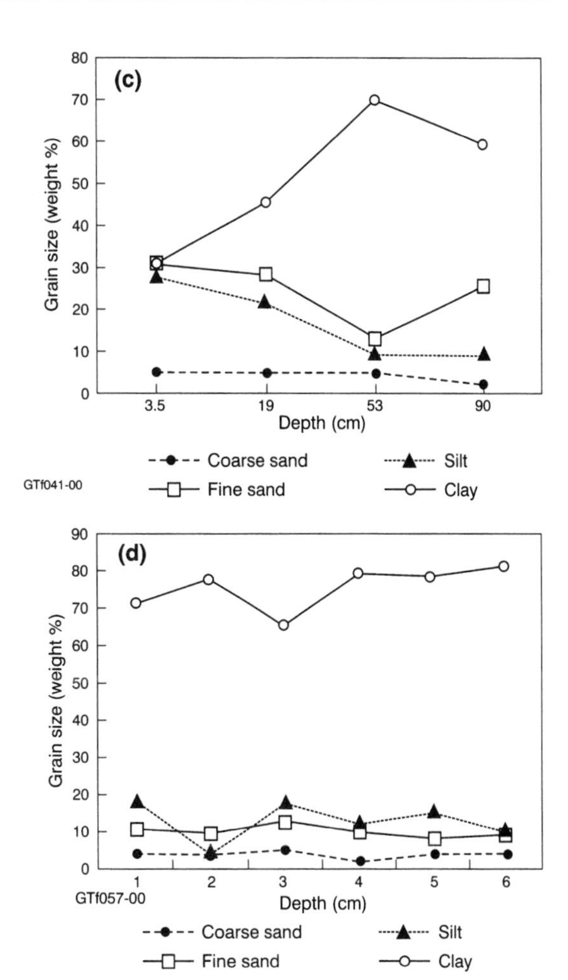

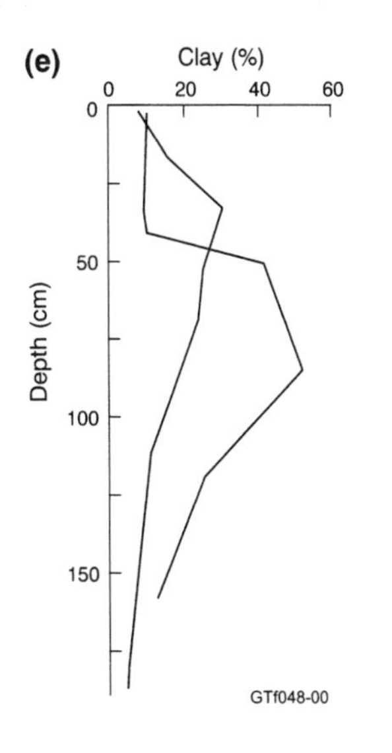

Figure 11.3 *Cumulative % grain-size of weathering profiles through: (a) granite; (b) acid volcanic rocks from Hong Kong (from Lumb 1975); (c and d) weathered basalt from the Armidale area of New South Wales (from Schafer & McGarity 1980); (e) granite saprolite (grus) from the New South Wales south coast (from Hough 1983)*

against one another, forming polished surfaces or neocutans (Brewer 1964). Each lump so formed forms a separate hydrological domain within the regolith, as the polished lumps are less permeable than the voids between them. This also means each lump forms a separate hydrochemical domain and may have entirely different composition solutions compared with the surrounding voids or other more permeable material.

Any stable minerals resistant to weathering will remain in the profile and affect the grain size as well as other physical parameters such as porosity and permeability.

TEXTURE

Grain size of regolith zones is again dependent on the nature of the parent rock. Take for example the weathering of granite as opposed to that on a basalt or gabbro (Figure 11.3).

Before discussion of the results shown in Figure 11.3 it is necessary to point out that the results of such analyses are dependent on the methods used to analyse grain size. Simple techniques such as used in sedimentology will not necessarily work. Clay minerals formed as weathering progresses are initially attached to larger grains of parent rock minerals and are often difficult to separate from them so the proportion of clay-sized material may be less than expected. It should also be noted that many other minerals including Fe-oxides and quartz occur as clay-sized fragments. As the clay mineral proportion increases it is also possible that the clays will be flocced (present as clumps or microaggregates which may get as large as silt-sized grains), particularly in the case of a basalt or gabbro where they are likely to be Ca^{2+} saturated. Cementation by new-formed Fe^{3+}-oxide minerals may also increase apparent grain size.

Weathered granites contain a mixture of sand-, silt- and clay-sized grains. The majority of the sand is composed of quartz when weathering is relatively intense, but in the lower saprolite zone quartz, feldspar and mica may make up the sand-sized component. Data from Hong Kong show that weathered granites there are comparatively clay poor below about 5 m and that sand and silt dominate the saprolite. This may also suggest that erosion is working ahead of weathering, so clay minerals do not develop before the profile is eroded. On the other hand, data from the Appalachians (Pavich 1985) clearly demonstrate clay-sized grains are present at considerable depth where hill slopes are less steep than in Hong Kong. Similarly, on the Yilgarn Craton many of the gneissic weathering profiles have zones at varying depths that are almost entirely kaolinitic (Butt & Zeegers 1992a). Data from Banfield (see Figure 11.6b) show that in granitoid rock weathering from southern New South Wales clay-sized contents do not necessarily increase up-profile. Figure 11.6 also clearly shows there are relatively high proportions of fine-clay-sized material making up the total clay-sized component of weathering.

Robertson *et al.* (1998) demonstrate petrographically that sutured quartz from metamorphic rocks and granites of the Yilgarn Craton in Western Australia are progressively desegregated along the suture boundaries as weathering proceeds, creating a finer modal size for the former framework grains. This is very similar to the findings of Moss (1972) and Moss *et al.* (1973) who showed unsound quartz crystals comminute rapidly on their release from granite in the Canberra region.

Weathered mafic and ultramafic rocks yield only clay minerals and Fe-oxides and oxyhydroxides as new-formed minerals, and the majority of these are clay sized except where Fe-nodules or vermiculite grows. This produces weathering profiles very different in grain-size character from those of granites because of the absence of weathering-resistant minerals in mafics. In some cases when local conditions are suitable, quartz may precipitate from the weathered solutes in the regolith as nodules varying in size from huge (>5 m) to silt sized, but this is a rare occurrence. The Ni-laterites of the Marlborough area in Queensland contain a high proportion of secondary chalcedony and chrysoprase.

In other rock weathering profiles the grain size of the weathered products follows similar lines as those in granites and basalts; the product depends on the original rock type, the new-formed minerals and precipitates.

Grain size has a major impact on weathering rate, as it is this that determines the potential surface area of minerals available for chemical reaction. In coarser-grained materials the available surface area is small compared to finer-grained materials. Thus the finer the mineral grains the more rapidly they are likely to react and come to equilibrium with their environment.

POROSITY, VOID RATIO, PERMEABILITY AND HYDRAULIC CONDUCTIVITY

Porosity and void ratio are similar measures of the relative space taken up by solids and voids in a material. In the case of weathering profiles this is important because it controls such features as the surface area available for chemical reaction, the potential amount of water a profile zone can hold and therefore the potential for chemical reaction. Permeability or hydraulic conductivity measures the ability of a fluid to move through a medium, in this case a weathering profile. Permeability is a measure of the degree of connection between voids in the material. Materials without any water could not weather chemically, similarly if the weathering solution is not removed prior to saturation with respect to the minerals with which it is in contact reactions will stop. Thus a hydraulic conductivity sufficient to allow solutions to move through the rock or regolith at a rate greater than the rates of chemical reaction is necessary for weathering to proceed and profiles to deepen. In most weathering situations the major part of hydraulic conductivity is provided by macropores such as joints

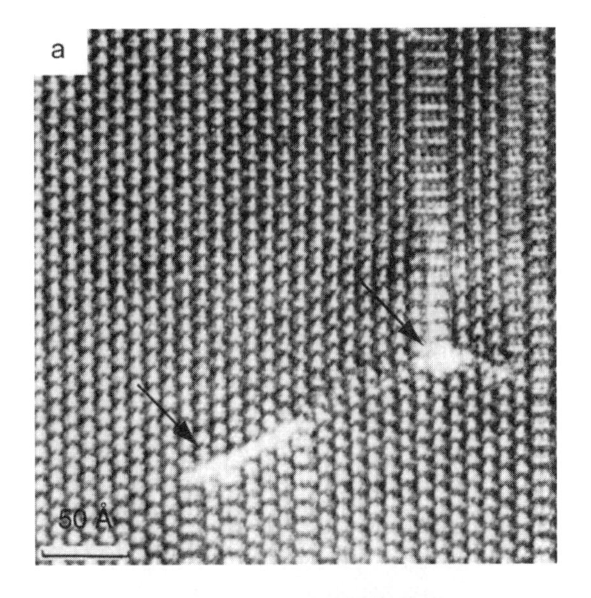

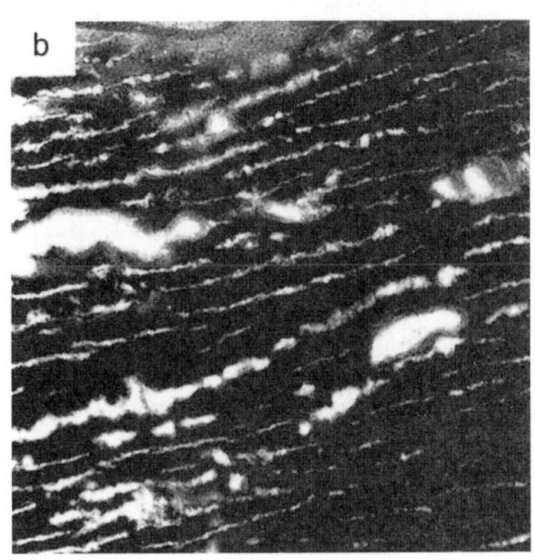

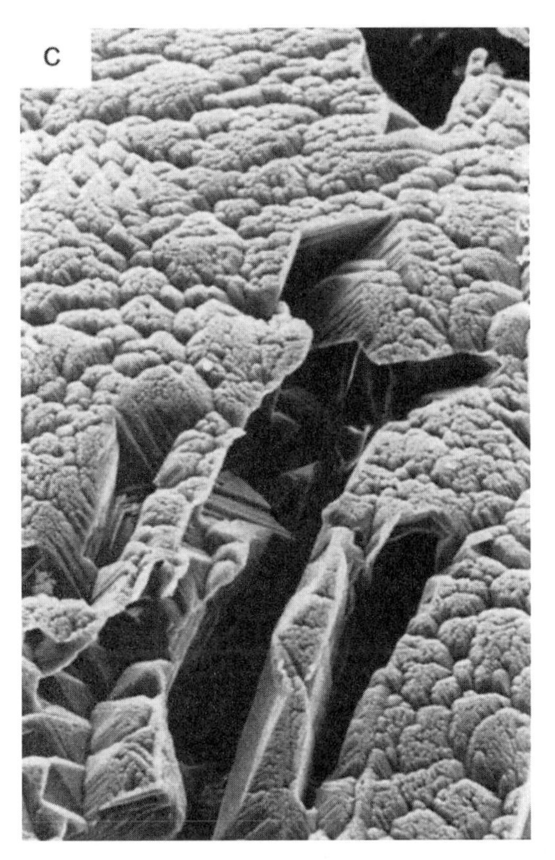

Figure 11.4 *(a) Molecule-sized holes (arrowed) in weathering amphibole viewed down the Z-axis. The 'knitted' fabric shows the TOT double chains of amphibole, the ladder-like strips are of wider TOT chains formed by coalescence of double chains during weathering (photo Q-M Wang). (b) Dissolution channels (white) in weathering olivine, 5–20 nm wide (photo RAE). (c) 10-micron wide solution channels in weathering quartz (photo D. B. Tilley)*

and fractures (or remnants of them in saprolite), biotic tubules or holes, and inter-grain boundaries after weathering has begun to remove or replace some minerals. While the bulk of fluids move through such pores, much of the weathering occurs at the edge and within mineral grains (Figure 11.4) accessible only via micropores. These solutions move much more slowly than do those in macropores.

Typically, porosity or void ratios in a weathering profile are high at the surface, low in the pedolith and high again in the saprolite and decrease again into fresh rock. This is well demonstrated by data from McNally (1992), Lumb (1975, from Thomas 1994) and Thomas (1994). McNally shows that in a granite weathering profile, saturated bulk density progressively decreases in the more highly weathered materials while porosity progressively increases (Figure 11.5a). He also illustrates similar relationships for quartzose sandstones. Similarly Lumb shows that void ratio increases up-profile with progressive weathering of granite and acid volcanic rocks from Hong Kong (Figure 11.5b, c). Results for profiles developed on weathered 'basement rocks' in Africa show more complex results than those from Hong Kong and Australia (McNally 1992) (Figure 11.5). The data of Thomas show relative permeability is high at the surface in residual materials from which many fine-

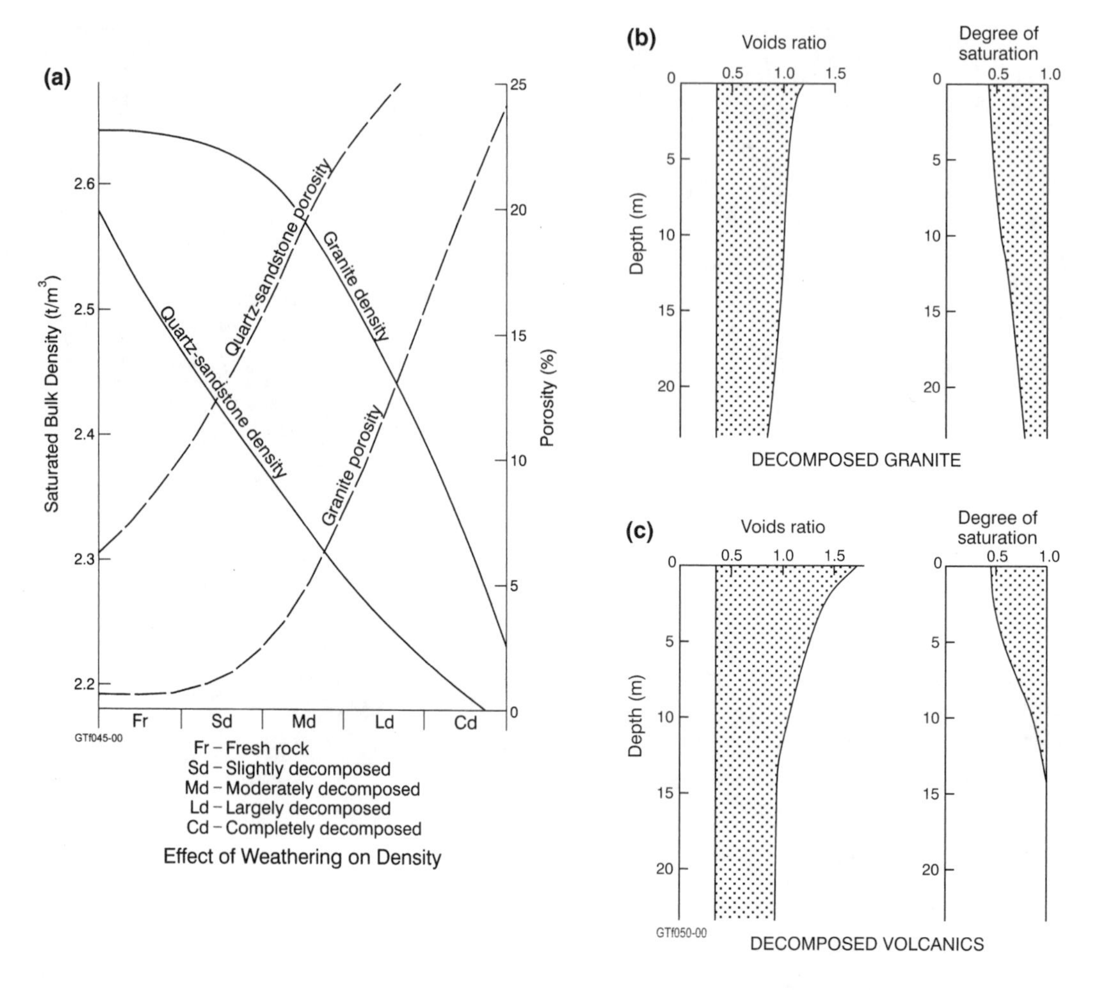

Figure 11.5 *(a) The effect of weathering on porosity and density (from McNally 1992), and (b) and (c) the void ratio (the proportion of voids in a material) and degree of saturation for weathered granite from Hong Kong (from Thomas 1994 after Lumb 1975)*

grained components have been winnowed or eluviated. It decreases rapidly in mobile and saprolite zones where clay-sized materials have not been removed or have been added to by illuviation. The fact that porosity is relatively high in this zone while permeability is low, suggests the clay minerals (and perhaps Fe-oxides) maintain the porosity but block fluids passing through the material. This relates to the flow domains mentioned above. Once the zones in which corestones occur is reached, permeability increases only to fall off again in the fresh basement. Hydraulic conductivity behaves similarly for similar reasons. Porosity behaves similarly to permeability in these profiles, with a progressive decrease downward except in the mobile zone, probably due to compaction and/or illuviation of clay particles.

As well as these considerations hydraulic conductivity also varies with weathering profile zones. The soil may have highly variable properties depending on the type of soil formed. The lower pedolith has lower conductivity than the more open saprolite and saprock, and all weathering profile zones generally have a higher conductivity than the fresh rock (Figure 11.6).

REGOLITH STRUCTURES

Regolith structures include a host of features which all affect its physical nature and behaviour. They include those described under the headings below.

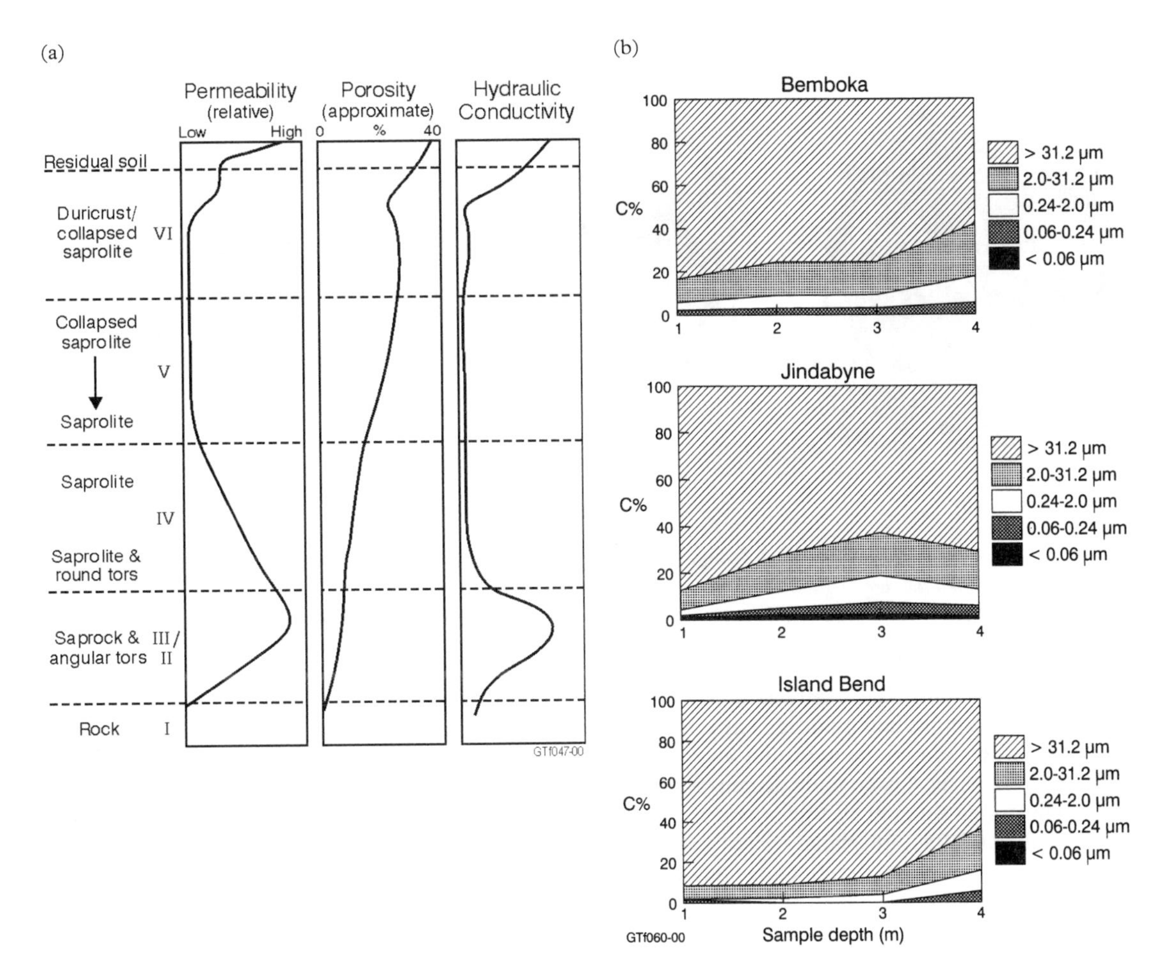

Figure 11.6 *(a) Typical curves for saprolite properties and hydraulic behaviour, based on experience with basement complex rocks in Africa (from Thomas 1994). (b) Grain size data from granitoid weathering in southern New South Wales (from Banfield 1985)*

Mottles, nodules and pisoliths

Mottles are weakly cemented accumulations of minerals coloured differently from the body of the regolith. They most commonly form by the precipitation of Fe^{3+} or less commonly Mn^{3+} in the regolith plasma (fine-grained phases like the clay minerals) (Brewer 1964). They initially develop without disruption to the fabric of the regolith plasma (Nahon 1991). As the process of mottle formation continues, mottles develop more defined outlines and instead of Fe^{3+} simply precipitating in the plasma it begins to replace the plasma as the Fe migrates by centripetal accumulation. This occurs by the replacement of kaolin with Al-rich haematite and the enclosure of any weathering-resistant minerals (quartz) present in the regolith. Similar processes occur with Mn, calcite and gypsum mottles.

Nodules and pisoliths are growths or glaebules of regolith minerals that form within it. *Nodules* are concentrations with no internal layering, but they may preserve the fabric of the original rock or regolith in which they form. *Pisoliths* are generally smaller than nodules, are rounder (pea-shaped, hence the name) and are concentrically layered. The layering or cortex may occur around a core nodule or a framework grain in the regolith.

By far the most common Fe segregations are Fe-oxide glaebules and mottles that form by the accumulation of Fe^{3+}-oxides to a point where they colour and/or cement the plasma in which they form. The processes of formation are controversial. Some clearly

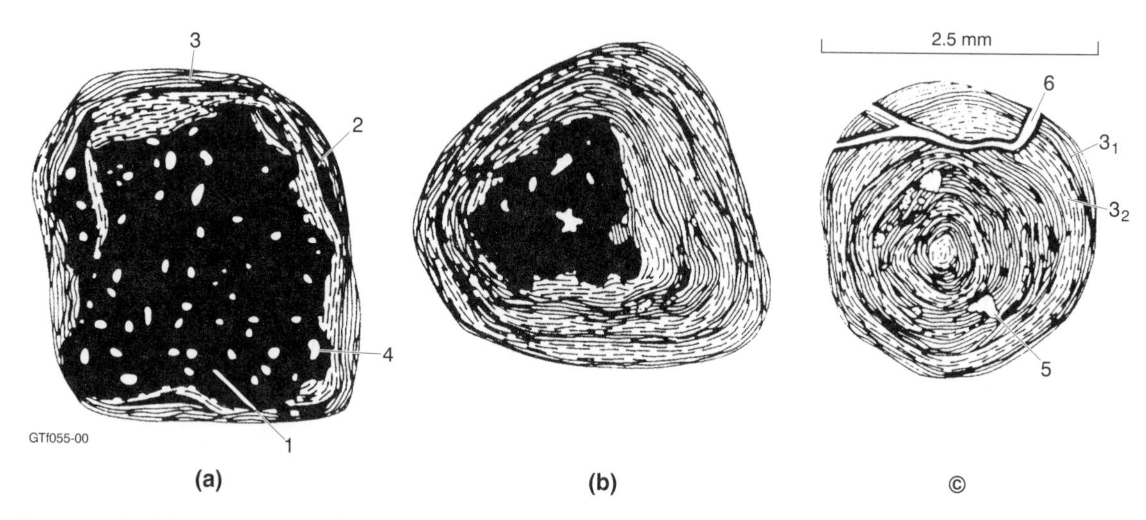

Figure 11.7 *Differentiation of geothitic banded cortex at the expense of haematitic nodules (from Nahon 1991). (a) Large pisolith with well-developed haematitic core (1) and thin cortex (3). (2) is a relic of haematite in the cortex and (4) are microvoids. (b) Intermediate pisolith between (a) and (c). (c) Small pisolith devoid of a haematitic core. Cortex (3) consists of alternately layered dark bands (3_1) and light-coloured bands (3_2). (5) are relics of quartz grains and (6) is a secondary fissure*

form by the migration of Fe^{2+} and its precipitation when the solutions move into oxidizing environments in the regolith, for example around, but away from, root tubules. This is most commonly demonstrated by Fe^{3+} accumulations along or in thin weathered rock adjoining joint and fracture planes where oxidizing solutions moving through the fractures encounter solutions containing Fe^{2+}.

Mottles that occur in the regolith begin as soft concentrations of Fe^{3+} that may harden by continued accumulation of Fe and replacement of the plasma (or through the progressive dehydration of hydrous forms of Fe-oxides). Such a process is obvious at the boundary between the mottled zone and ferricreted zone below the bauxite in the active weathering profiles at Weipa (Figure 11.2).

The process of pisolith formation is unclear. Most observers, seeing the regular concentric banding of a pisolith, and the not uncommon presence of an irregular nucleus, conclude that pisoliths grow by successive additions of surface material. However, Jones (1965) concluded that lateritic pisoliths evolve from nodules by a process of centripetal reorganization, that is, by chemical and mineralogical change inward from the outer rim (the cortex). Nahon (1991) also suggests that pisoliths develop from nodules. As nodules form in the regolith, wetting and drying cause the development of constraint cutans or plasma separations in which the clay minerals become oriented around the nodules. As these form, and wetting and drying continue, voids develop around the nodule. These voids cut the nodule off from its Fe supply and as a consequence the nodule begins to form a cortex. This process may continue until the whole nodule is converted to cortex with no original nodule visible as a core. These transitions from mottle to pisolith are illustrated in Figure 11.7.

Tilley (1994) found that Nahon's interpretation of the evolution of ferruginous pisoliths could be extended to the aluminous pisoliths of the Weipa bauxite deposit in northern Australia. In essence, Tilley recognized a constant interplay between bauxitization, kaolinization and Fe-oxide–hydroxide mineral precipitation and dissolution. As the conditions change, either from climatic change or position in the profile, nodules or earlier formed pisoliths may be resilicated by silica-bearing groundwater to yield a kaolinite coat, which in turn may be desilicated to gibbsite again. The redox cycle of Fe causes pH variation which promotes alternate dissolution and precipitation of Al species. Pore size within the pisolith dictates the chemical activity of water. In smaller pores (nanometric) the less hydrated species precipitate (boehmite or alumina), whereas in larger pores gibbsite is stable (Tardy & Nahon 1985). Thus within a nodule, successive chemical change smooths out shape differences, yielding a spherical pisolith. Movement may break pisoliths, and their fragments become the centre of new nodules by agglomeration of particles illuviated down-profile. Tilley recognized up to three pisolith-forming events at Weipa on the evidence of internal 'unconformities'.

Support for the concept of pisolith evolution from the rim to the centre, rather than by layer accretion, comes from a particular class of pisolith at Weipa having very hard, vitreous centres dominated by η-alumina and tohdite ($5Al_2O_3.H_2O$) and softer rims composed of gibbsite, boehmite and haematite. The anhydrous alumina species are found nowhere else in the deposit, and have only been reported elsewhere rarely, as the result of heat. If these pisoliths formed by accretion, then either at some time in the past, ooliths of anhydrous alumina minerals must have formed, to be later coated by the more hydrated species, or the cores of these pisoliths dehydrated *in situ*. The latter alternative is basically what the hypothesis of centripetal organization proposes anyway, that the changes leading to the production of a pisolith happen in that volume of material.

Nodules and pisoliths once formed make relatively large grains in the regolith. They act as foci for precipitation and may become cemented into a relatively impermeable layer. Such indurated regolith is shielded from groundwater interaction except at surfaces.

Biotubules

Biotubules are holes made in the regolith due to flora and fauna. Insects, other invertebrates and vertebrates make holes in regolith materials. Tree roots penetrate regolith deeply (20–30 m is common in semi-arid terrains). On the death of the root the space it occupied is left as a void. Ants, termites, worms and other invertebrates also make holes either from feeding or as living galleries in the regolith. Worm feeding trails tend to be filled by worm casts as the worms move through the regolith. Ants and termites, on the other hand, make open galleries, often lined with polished walls that tend to be stable even when moderately wet. These biotic voids provide much of the regolith's macroporosity through which fluids can freely move and which, when they collapse, cause mixing of the regolith.

Cutans

Cutans are skins deposited or precipitated on surfaces within the regolith. Cutans most commonly form from the deposition of clay minerals on void surfaces. When clays deposit from suspension in intra-regolith water they do so with their largest surface area parallel to the surface on which they deposit. This gradually causes the build-up of a coating of oriented clay platelets on the surface, like clays coating paper to provide its shiny surface. These are recognizable at field scale and are shown under the microscope to be composed of well-oriented clay mineral skins. These clay cutans (argillans) are relatively impervious and prevent or slow considerably the passage of fluids from the regolith plasma to the voids and from voids to the plasma. The clays that form argillans are generally eroded (eluviated or leached) by intra-regolith water from higher up in the profile or higher up the hydraulic gradient within the regolith to be deposited (illuviated) lower down in the profile.

Other cutans can be formed from Fe-oxides (ferrans), Fe-oxides and clay minerals (ferriargillans), calcite (calcitans) and any other minerals that may precipitate in weathering profiles.

Many researchers have discussed the nature, genesis and significance in more detail than we do here; they include, for example, Brewer (1964), Nahon (1991) and Retallack (1990).

Clay aggregates and plasmic orientations

Clay aggregates and plasmic orientations are common within fine-grained regolith materials. Above, we have described the formation of clay aggregates due to wetting and drying, causing swelling and contraction and associated compressive and tensional stresses in clay-rich material.

Wetting and drying stresses also produce domains of oriented clay minerals (sepic fabrics, Brewer 1964), in the regolith plasma. These domains are often related to framework grains and glaebules, but they may also form extensive fabric elements in clayey materials. The relationship of sepic fabrics to nodules produces the voids discussed above. These fabrics also affect significantly the regolith strength and may, although no data are available, affect fluid movement through regolith materials. In domains where the clays are very well oriented, fluid movement may well be impeded and diverted to areas of higher conductivity.

Clay aggregates may also be formed by flocculation, but aggregates so formed do not form oriented aggregates like those formed from illuviation.

Flocculation

Clays are very fine-grained layer silicates (see Chapter 3 for a discussion of clay minerals). Because they are small, generally less than 2 μm in diameter and a tenth that in thickness, they have a large surface to volume

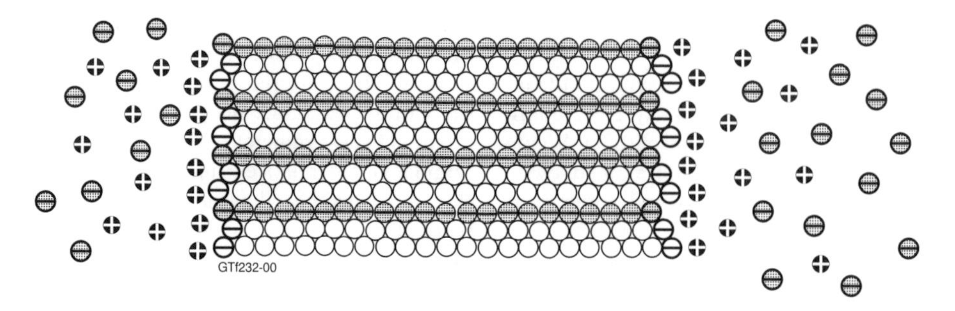

Figure 11.8 *Diagram of a small kaolinite crystal in solution. The anions at the sides of the crystal (dark-edged circles with a bar) are not charge satisfied by the structural cations, and are left with a negative charge. The grey circles represent hydroxyls, and although these form the upper surface, this surface is neutral, as is the basal plane of oxygens forming the lower surface (for details see Chapter 3). Cations in the solution (small circles marked +) are attracted towards the negative edges of the kaolinite in two parts, a closely adhering layer and a diffuse layer further out, leaving their balancing anions (grey circles marked −) locally in excess*

ratio, hence surface phenomena are important in their behaviour. A typical kaolinite crystal, for example, has 100 times the surface : volume ratio of a fine silt grain.

A clay crystal terminates with anions, either oxygens or hydroxyls, some of which are not charge satisfied by the cations within the crystal (see Figure 3.7). This leaves the crystal with a small net negative charge. When in water suspension, these negatively charged surface oxygens attract cations to themselves in two layers, one at the surface, the other more diffuse adjacent to the first. This arrangement of ions associated near the surface of an oppositely charged colloid is termed the electrical double layer. Overall the suspension must be charge neutral, but the charges are not uniformly distributed throughout (Figure 11.8). For more detail see for example van Olphen (1977).

When the clays are suspended in pure water or in water with a very low concentration of dissolved salts, the particles repel each other through their positive double layer, since they are all similarly charged. This keeps the clay in suspension. However, as well as the electrostatic repulsion, there is a weak van der Waals attraction between one particle and another because of uneven charge distribution. In the case of kaolinite, the hydroxyl surface has regions of weak positivity, since the proton of the (OH) group resides here, and this local positive may attract the negative edge of another crystal. But in most accidental close approaches between two crystals, the repulsive force is stronger than the attractive. Eventually, though, chance will lead to crystals achieving a close approach, and the attractive force at close range is enough to bind the crystals, leading to gradual agglomeration, or flocculation. For some undisturbed clay suspensions this may take months, and if the suspension is constantly agitated, as would happen in most natural environments, the clay may remain in suspension indefinitely. When clay crystals do self-aggregate, they are likely to do so in parallel arrangement, since their planar surfaces are neutral. This helps to develop oriented aggragates of clays such as form cutans (Figure 11.9a).

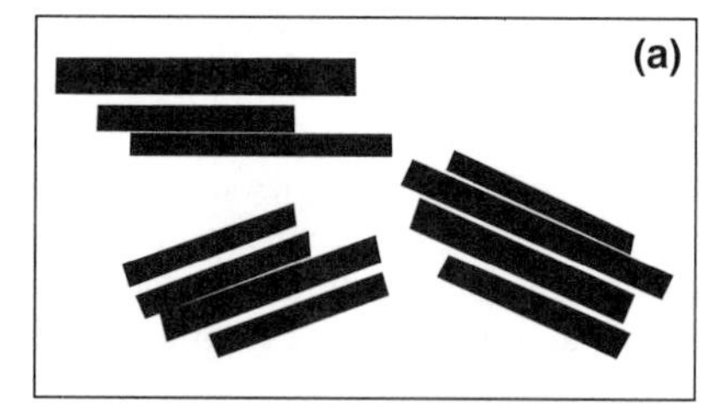

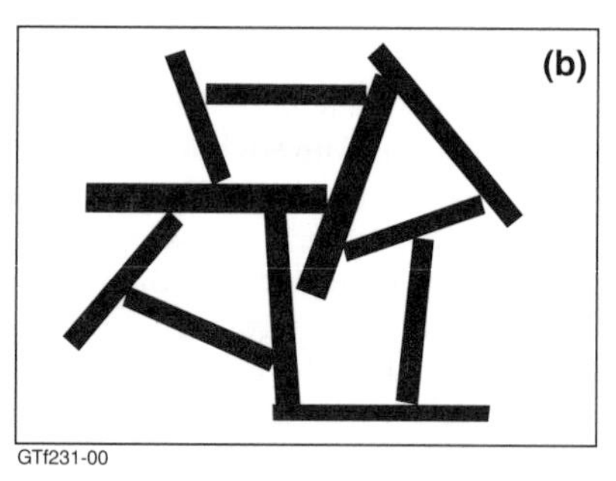

Figure 11.9 *(a) Oriented clay crystals in face-to-face aggregation. (b) 'House of cards' structure for a floc of clay with the constituent particles in edge-to-face association*

The addition of salts (electrolytes) to the suspension markedly changes the balance between the attractive and repulsive forces. The higher concentration of cations compresses the diffuse part of the double layer, so that close approach occurs more readily, allowing the attractive van der Waals forces to

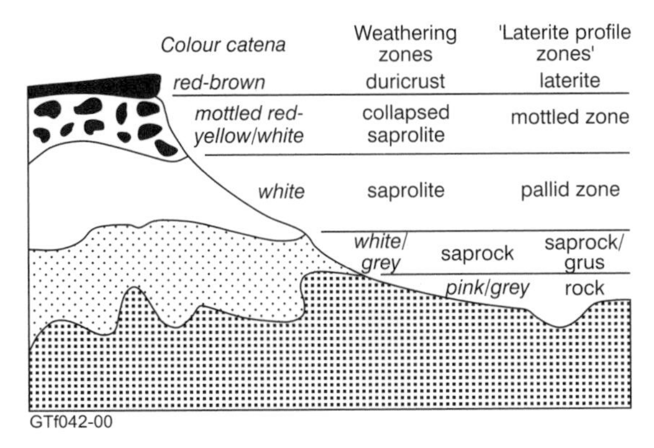

Figure 11.10 *Colour catena in a landscape*

overcome the electrostatic repulsion. This in turn increases the rate at which the clay particles come together, and flocculation occurs rapidly. Clay flocs form very open structures, called 'house of cards' structures, illustrated in Figure 11.9b.

Divalent or trivalent cations compress the diffuse layer more effectively than do monovalent cations, thus Ca^{2+} or Al^{3+} are very effective flocculating agents. Colloids in the water supply are flocculated by the addition of Al salts, and gardeners are familiar with the 'clay-breaking' power of limestone or gypsum. Undispersed clays can form face-to-face aggregations that prevent the inflow of water and air to the soil, giving poor tilth and making the soil difficult to work. The addition of a flocculant changes the clay structure to the more open edge-to-face 'house of cards' arrangement, allowing aeration and the penetration of water.

Weathering catenas

The term 'catena' comes from the shape of a catenary curve and is applied in soil science and geomorphology to systematic changes of regolith features with topography. Weathering catenas are the systematic variation in weathering in an area from the hilltops to valley bottoms. In weathering profiles this catenary variation can be due to many factors.

Colour weathering catenas are relatively common in weathering profiles. The major cause of colour variations is due to the interaction of water with new-formed minerals. In a landscape developed on a profile with a ferruginous duricrust (ferricrete) colour variations may often follow those described as 'lateritic profiles' and the various zones of the profiles will be exposed in different parts of the landscape. Figure 11.10 illustrates the case well.

These colour variations come from the variations in mineralogy of the weathering profile and their hydration state, which is in turn related to the hydrology of the site. Red colours in the duricrust and in the mottles are produced by haematite formed in the drier upper parts of the profile. Yellows can be common in the mottles and these result from hydrated Fe-oxides (goethite). The mottles themselves result from aggregation of Fe-oxides by the processes described above. This usually occurs close to, or in the zone of, fluctuating water-tables (e.g. Weipa). The white saprolite is the result of removal of all or nearly all Fe. Lower in the landscape colours tend to become darker greys, greens and blues as a result of permanent water in the profile which limits leaching and any Fe present remains in the ferrous state, often in the presence of smectitic clays.

Fabric varies in a catenary sequence as a result of variations in structural and textural patterns preserved or developed in the weathering profile. Such variations are important in the landscape context of a weathering profile as their variations can significantly control further weathering processes by influencing the movement of groundwater through the profile. This in turn influences the variation in mineralogy and chemistry of the profile.

Good examples of this are observed commonly where springs occur within weathering profiles. In some cases the plasmic zone forms a relatively impervious pan within the profile and water moves laterally through the weathering zone, leaching elements and physically removing finer particles. Geochemistry of spring waters points to the elements removed. On the Monaro of south-eastern New South Wales, H_4SiO_4 and alkalis are removed by lateral flow through fabric variations in the weathering profiles (Moore 1996b).

Volume changes with weathering

The crystalline products of weathering, predominantly clay silicates and Fe–Al oxyhydroxides, occupy considerably less volume than do the parent minerals. Velbel (1993) calculated that the volume of weathering product ranged from 0.99 of the volume of the parent in the case of anorthite weathering to kaolinite, 0.46 for K-feldspar to kaolinite, and only 0.11 for anthophyllite weathering to goethite, assuming Al and Fe were conserved. Of the common rock-forming silicates, only garnet appears to produce products whose volume exceeds that of the original garnet, and then only by about 10%. Velbel did not include smectites among the weathering products he considered. When andesine feldspar weathers to smectite with two water layers between the 2 : 1 layers (15.4 Å basal spacing), there is a theoretical volume increase of 5%:

$$4Ca_{0.17}Na_{0.83}[Al_{1.17}Si_{2.83}]O_8 + 10H_2O \rightarrow 2Ca_{0.17}Al_2[Si_{3.67}Al_{0.33})O_{10}(OH)_2.4H_2O$$

andesine smectite $+ 0.34CaO + 1.67Na_2O + 4SiO_2$
volume = 668 Å^3 volume = 704 Å^3

The common pseudomorphous replacement of pyrite by goethite, which produces 'Devil's dice', involves a theoretical volume loss of 11%.

$$4FeS_2 + 15O_2 + 10H_2O \rightarrow 4FeO(OH) + 8H_2SO_4$$

pyrite goethite
volume = 159 Å^3 volume = 142 Å^3

Some other volume changes upon weathering are tabulated in Table 11.4, assuming conservation of Al. Such calculations, and examination of natural weathering reactions, tend to support the contention that there is always a volume loss. As long as only a small fraction of the rock's minerals has weathered, the individual mineral grains remain coherent and strong enough to support the weathered rock body without any collapse, even if some minerals are partly or wholly pseudomorphed by weathering products. This part of the weathering profile is the saprock or saprolite. Eventually, however, the supporting grains lose sufficient volume that collapse is inevitable, and the original rock fabric is lost. Igneous rocks appear to be able to retain their fabric up until about one-third mass loss.

Despite these losses (and occasional gains), many studies of rock weathering have shown that at least as far as saprolitic weathering, the process occurs at constant volume. The presence of preserved fabric of the parent rock in the saprolite is used to infer that weathering has proceeded by solution loss only, with no collapse or with no introduction of external material (Nahon & Merino 1996). In such cases, although a large amount of void space is created in the process, much of this is between the sub-micron-sized newly formed clays and Fe-oxide–hydroxides, and these minerals are able to support the fabric of the weathered rock. Water may be retained within the voids in the saprock and saprolite.

Table 11.4 *Theoretical volume changes accompanying the weathering of some common rock-forming minerals*

Mineral transformation		Volume change (%)
From	*To*	
K-feldspar	Kaolin	–55
Andesine	Smectite	+5
Anorthite	Kaolin	–10
Chlorite	Smectite	–13
Muscovite	Kaolin	–28
Smectite	Kaolin	–42
Kaolin	Gibbsite	–39
Diopside	Talc	+4
Calcite	Solutes	–100

Classification of degree of weathering

There are many classifications of the degree of weathering of rocks. As mentioned above, the degree of weathering varies down-profile and this is reflected in the mineralogy and fabric of the profile compared to the parent rock. The degree of weathering is an important descriptor for field recording of the nature of weathering profiles, and very important in considering slope stability or the engineering response of regolith materials. Most schemes used to describe the degree of weathering are based on concepts derived from the 'ideal weathering profile' developed on a single parent rock with the assumption that the upper parts of the profile contain little if any transported debris, for example an *in situ* granite profile. Some examples of other parent rock profiles have been studied, but few are used as the basis for discussing the degree of weathering.

Perhaps the most quoted example of degrees of rock weathering is Ruxton & Berry's (1957) (Figure 11.11) study in Hong Kong. This study is descriptive and based on the needs in Hong Kong for classifications applicable to engineering undertakings.

More detailed and complex classifications of the degree of weathering have appeared over time, again based on rocks that are mineralogically homogeneous and simply jointed. An example of such a classification is the Geological Society's (1990) weathering profile and weathering grades (Figure 11.12). This essentially takes the Ruxton & Berry classification and adds a more complex, but still simple uniform parent rock weathering profile.

Pain *et al.* (1991) enunciated a schema for the classification of the degree of weathering of *in situ* rocks and transported materials for use in regolith mapping in Australia. Their categories and descriptions are as follows:

Properties of Rock Masses

Zone		Percentage solid rock
VI	Soil	0
V	Red-brown sandy clay	0
IV	Pallid silty sand with few rounded corestones	<50
III	Corestones still largely rectangular and interlocked	50-90
II	Minor weathering along joints	>90
I	Fresh bedrock	100

Weathering front

GTf044-00

Figure 11.11 *Features, weathering zones, composition and percentage of parent rock in a weathering profile from Hong Kong (modified from Ruxton & Berry 1957). Weathering 'zone' or grade as some engineers call it are shown in Roman numerals on the left. These weathering zones of Ruxton & Berry are the basis for most contemporary geotechnical classifications of weathering*

- *Unweathered*: regolith with no visible signs of weathering. Normally this class is confined to sedimentary regolith types because, by definition, fresh bedrock is not regolith
- *Slightly weathered*: rock has had traces of alteration, including weak Fe staining, and some earth material. Corestones, if present, are interlocking (*accommodated*) (our italics); there is slight decay of feldspar, and a few microfractures. Slightly weathered rock is easily broken with a hammer. Slightly weathered sediments have traces of alteration on the surface of the sedimentary particles, including weak Fe staining. Some earth material may be present, filling voids between coarse particles (cf. *saprock*) (our italics)
- *Moderately weathered*: rock has strong Fe staining, and up to 50% earth material. Corestones, if present, are rectangular and interlocked (*accommodated*) (our italics). Most feldspars have decayed, and there are microfractures throughout. Moderately weathered rock can be broken with a kick (with *steel-capped* (our italics) boots on), but not by

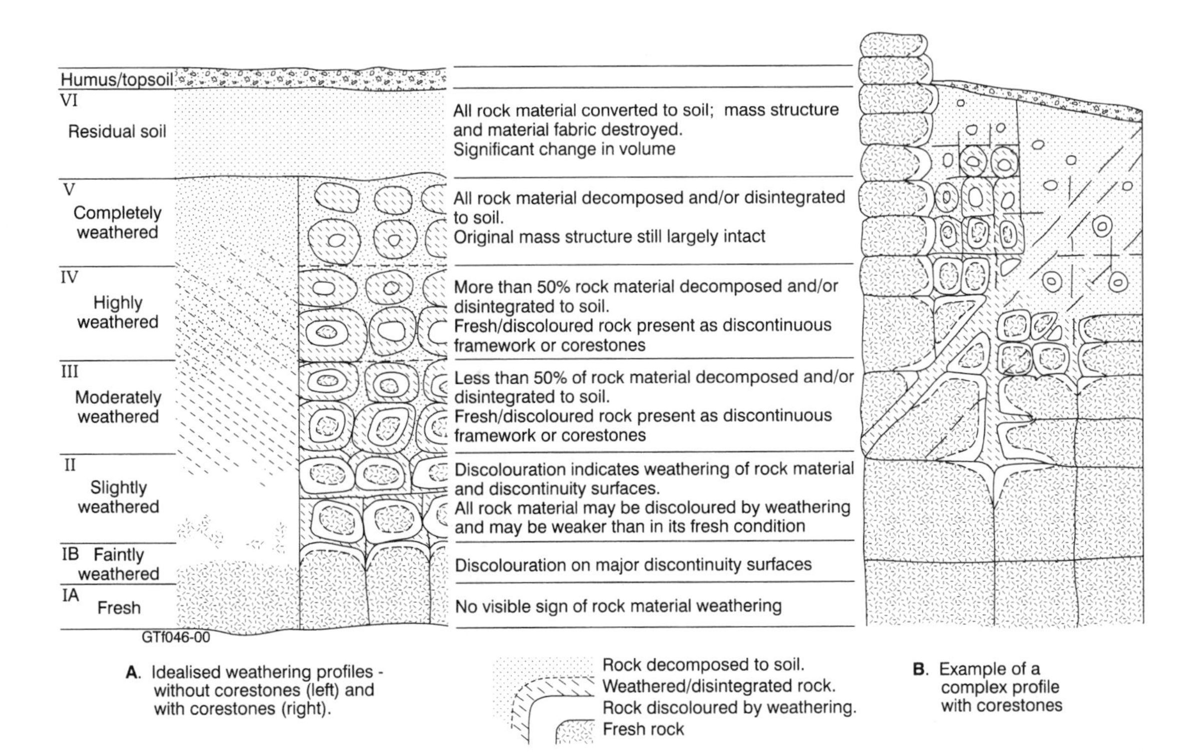

Figure 11.12 *Weathering profiles, weathering grades and weathering classification of rocks from an engineering perspective (from the Geological Society 1990)*

hand. Moderately weathered sediments have strong Fe staining, and up to 50% earth material. Labile particles up to gravel size are completely weathered. Larger particles have thick weathering skins (*rinds*) (our italics). Most feldspars in particles have decayed (cf. *saprock*) (our italics).

- *Highly weathered*: rock has strong Fe staining, and more than 50% earth material. Corestones, if present, are free (*separate from others and not accommodated*) (our italics) and rounded. Nearly all the feldspars are decayed, and there are numerous microfractures. The material can be broken apart in the hands with difficulty. Highly weathered sediment has strong Fe staining, and more than 50% earth material. All except the largest particles are weathered right through. Boulders have thick weathering skins (cf. *saprolite*) (our italics)
- *Very highly weathered*: rock is produced through decomposition of rock masses due to exposure to land surface processes. The material retains structures from the original rock. It may be pallid in colour, and is composed entirely of earth material. Corestones, if present, are rare and rounded. All feldspars are decayed. It can easily be broken by hand. Very highly weathered sediment is thoroughly decomposed, but still retains the shapes of the original sediment particles, as well as laminations and bedding. It is composed completely of earth material (cf. *saprolite or pallid zone*) (our italics)
- *Completely weathered*: rock retains no structures from the original rock. There are no corestones, but there may be mottling. It is composed completely of earth material. Completely weathered sediment retains no structures from the original sediment. It is composed completely of earth material. There may be mottling (cf. *pedolith mobile zone, etc.*) (our italics)

In the above descriptions we have inserted comparable names used elsewhere in this chapter; however, while there are similarities the degrees of weathering described by Pain *et al.* (1991) have no direct comparison with profile terminology.

None of the above classifications account for physical weathering and regolith composed almost entirely of physically formed debris is not considered. Such regolith types are very rare, but at field scale observation they do occur. In very cold and in very steep terrains they are common, but not of major importance elsewhere.

All the degrees of weathering classifications are qualitative and give no or little indication of the chemical changes which have occurred during weathering except by inference from mineralogical variations through the profile. More rigorous measures of the degree of weathering using quantitative measures are referred to as weathering indices.

WEATHERING INDICES

Those who have studied weathered rocks very often seek to quantify the extent of weathering exhibited by a specimen, and to this end several *weathering indices* have been proposed, based on a comparison between the chemistry of the fresh or parent rock, and that of the weathered rock or regolith. There are a number of considerations necessary before using such data. Abundances of the major elements are usually expressed as oxides because O is the main balancing anion, whereas trace elements are presented as grams per tonne (or ppm) of the element. Most analyses report H_2O^+ and H_2O^-. The former is crystalline water and forms part of the minerals, but the latter is adsorbed water released at about 105°C. A sensible comparison between fresh and weathered rock first requires that the analyses be recalculated to total 100% excluding H_2O^-.

To make use of chemical data for rocks and regolith there are a series of problems that need to be assessed before calculating indices of weathering. They include:

- Is the regolith uncontaminated, i.e., is it valid to presume the regolith to be directly derived from its underlying parent with no external additions? Many regolith profiles are capped by transported debris, which means the surface components may not be sourced from the same parent as the deeper parts of the profile. Aeolian additions to the regolith are probably ubiquitous, so it is almost certain that no upper regolith samples are free of external additions. Within the saprock and saprolite there is more likelihood that the assumption of uncontaminated parent holds
- Is the parent uniform? Sedimentary rock or sediment parent material is inherently inhomogeneous. For such parents, a representative analysis to compare with the regolith analysis is difficult to establish unless the parent is clearly uniform over a great enough depth or region to make presumption of homogeneity reasonable

- Has solid material been lost from the profile? There may be preferential particulate loss from the profile, more a certainty than a 'maybe' in the upper profile where clay moves quite freely through the more porous regolith. Addition of Al in the form of clays to the B-horizon of soils and loss from the A-horizon is one obvious place where a chemical analysis will tell less than all of the story

Analyses are expressed as weight per cent, thus both the parent and regolith analyses add up to 100. If, for example, 100 g of rock weathers, and in so doing loses 25 g of chemicals, the 75 g remaining constitutes 'the whole', and when that is analysed and totalled to 100, there is an apparent increase, even though they (the remaining elements) may not have changed at all. Table 11.5 exemplifies this problem known as 'closure' (adding up to 100%).

Several observations about chemical analyses of weathered materials compared to fresh rocks can be made:

- The amount of SiO_2 usually decreases in weathered materials because it is somewhat soluble, though less so than CaO, MgO, K_2O and Na_2O
- Alumina has a low solubility under normal weathering pH, so its relative proportion generally increases in weathered material
- Most fresh rock contains iron in the ferrous state. Ferrous iron is quite soluble, and would be leached if it stayed reduced, but weathering generally occurs in oxidizing conditions. Ferric iron is very insoluble so Fe tends to remain in weathered materials. In Table 11.5, no Fe has been lost or gained. The change from 4 g of FeO to 4.4 g Fe_2O_3 simply reflects the molecular weight ratio of Fe_2O_3 : 2FeO (159.7/143.7 = 1.111). Manganese behaves similarly
- Titania generally increases in relative proportion, as it is essentially insoluble in most weathering environments
- Zirconium, like Ti, is generally considered to be immobile
- The remaining major elements in fresh rocks (K, Na, Mg, Ca) decrease as weathering proceeds. Some of these anions are temporarily bound in clay minerals, but as the clay minerals themselves weather, the remaining alkalis and alkaline earths are progressively released

Weathering indices fall into three categories, based on the approach used to establish true gains and losses. One group of indices are based on a reference to the abundance of an assumed immobile element. A second group are based on the chemistry of the weathered rock as it is (not as it was), thus avoiding the problem of absolute changes by not referring to any parent. The third group make the assumption that the process was isovolumetric, thus the actual loss can be established by analysing a fixed *volume* of parent and an equal *volume* of regolith and comparing the mass of each element contained in that specified volume. Each is useful, and which type is applied depends on the nature of the samples, the reason for wanting a weathering index and probably the availability of appropriate analytical techniques.

Table 11.5 *Hypothetical rock analysis before and after change in the amount of certain components*

	100 g fresh rock, i.e wt% oxide	Change (g)	Weight after change (g)	Wt% after change
SiO_2	60	−10	50	63.0
TiO_2	1	0	1	1.3
Al_2O_3	15	0	15	18.9
FeO	4	−4	0	0.0
Fe_2O_3	0	+4.4	4.4	5.5
MgO	12	−10	2	2.5
CaO	6	−5	1	1.3
H_2O^+	2	+4	6	7.6
Total	100		79.4	100

Note: From Table 11.5, it would appear, looking at the original analysis (column 2, 100 g fresh rock) and column 5, % after change) that silica, titania and alumina had increased, whereas in fact silica has been lost and the other two were unaffected (immobile). Establishing a way of knowing what has happened to each element is critical to the application of weathering indices.

Immobile element indices

The immobile element approach assumes that one or more elements have been unaffected by weathering, undergoing neither dissolution nor physical translocation, or that their mobility is restricted to a region smaller than the sample size. With this assumption, the ratio of the immobile element content in a weathered sample to its content in the parent rock is a measure of the overall loss suffered by the rock during weathering. Solubility data (Table 11.6) suggest that Zr, Ti, Nb, Th and Al might be acceptably immobile, though the abundance of Nb is generally too low for accurate use. Zirconium is very widely used, mainly because it occurs in most rocks only in the mineral zircon, which is very resistant to weathering. Provided

Table 11.6 *Values of solubility products (SP) from Lelong* et al. *(1976) and calculated from tabulations of Gibbs free energy for the more abundant least soluble elements*

Element	SP hydroxides
Cr	30
Bi	30.4
Al	32.5
V	34.4
Ga	35
Ce	37.5
Fe	38
Ti	40
Th	44.7
Tl	45
U	45
Mn	56
Sn	56
Zr	57

there is no mechanical movement of zircon in or out of the profile, Zr is probably the least mobile of the common elements. Nonetheless, in some chemical environments Zr is thought to be mobilized (see Zr in Chapter 7), and even in seemingly totally *in situ* profiles zircon may be added from foreign sources. Colin *et al.* (1993) found that exotic zircons had been introduced to the depth of the root zone (3 m), that is, just above the gneissic saprolite, and Brimhall *et al.* (1988) concluded that zircons had been introduced by aeolian accession to the regolith of the Darling Ranges, Western Australia. Such additions would negate the use of Zr as an immobile element.

Titanium occurs in rocks largely as rutile, ilmenite and sphene, or in the structure of micas, amphiboles and pyroxenes. The susceptibility of these minerals to weathering increases in the order listed. Titanium is therefore released early in the weathering of igneous rocks, and continues to be released as weathering proceeds. Its mobility during weathering in solution is poorly understood because of its extreme insolubility at regolith pH. If Ti precipitates immediately on release from its parent mineral as anatase, it becomes fixed in the weathering profile and may be effectively immobile. Presumed physical migration of anatase particles has been documented (e.g. Milnes & Fitzpatrick 1989), particularly in silcretes.

Braun *et al.* (1993) assessed the mobility of Ti, Zr and Th in a weathered syenite in Cameroon, and concluded that Th was the least mobile of the three. The concentration of Th in the regolith profile was found to be controlled by breakdown of allanite, apatite, titanite and epidote, but the mineralogical host for Th could not be determined. The mobility of Ti was recognized by the presence of Ti-enriched cerianite in white clay seams.

Once an immobile element has been identified, determination of a weathering index usually follows Nesbitt's (1979) approach. He calculated the percentage change in any element during weathering by reference to an assumed immobile element of concentration I_p in the primary rock, I_s in the sample. If X_p is the concentration of any other element in the primary rock, and X_s its concentration in the sample, then

$$\% \text{ change} = 100\ [(X_s/I_s)/((X_p/I_p)-1)]$$

Chittleborough (1991) suggests a weathering ratio (WR) for soils based on the assumption that Zr is immobile, and that most of the Zr is contained in the 20–90 μm fraction of a soil:

$$\text{WR} = [(\text{CaO} + \text{MgO} + \text{Na}_2\text{O})/\text{ZrO}_2]_{20-90\mu\text{m}}$$

The isocon technique indices

Moore (1996a) introduced an extension of the immobile element method, the isocon technique. This approach to element mobility during weathering uses all analysed elements, and seeks evidence of immobility from within the data. The method relies on comparison of the chemistry of a weathered rock with that of its unweathered parent. Each analysed element is allotted a scaling number whose purpose is simply to ensure that all the elements are spread uniformly along the two axes of the diagram. On the horizontal axis, the scaled concentrations of elements in the parent rock are plotted, while the weathered composition is plotted on the vertical axis. Scaling numbers are selected so that all the expected mobile elements are spread across the diagram, rather than being grouped at one or other end. The concept is that truly immobile elements will all be equally relatively enriched or diluted in the weathered sample. If this happens, when the scaled amounts in the two samples are plotted, immobile elements, all having the same ratio, will plot on a straight line. Examination of the isocon diagram shows immediately if such a straight line relation can be found. If it can, then the elements on the line are immobile, those above have been enriched, those below depleted. Figure 11.13 shows an isocon plot for the Baynton weathered basalt sample B8 compared to fresh B1 (Eggleton *et al.*

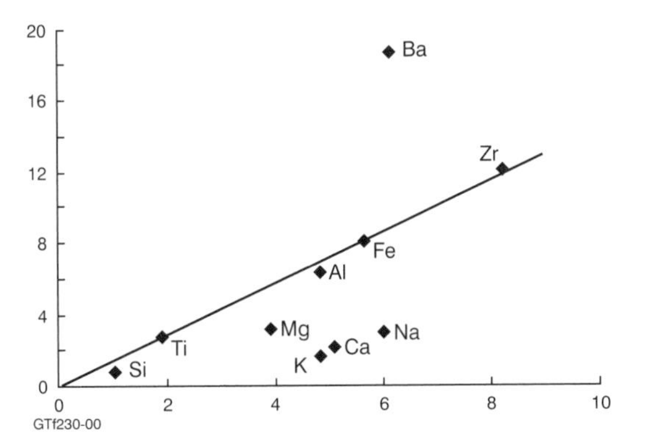

Figure 11.13 *Isocon diagram for Baynton Basalt samples B1 and B8 (Table 11.7). The Isocon is the line of best fit through the presumed immobile elements Ti, Al, Fe and Zr. Immobility is accepted because these elements, generally known to be highly insoluble,* do *fall on a single line*

1987). A line passes through Ti, Fe and Zr, with Al very close to that line. This line is then accepted as the isocon. If the isocon is the line indicating no loss or gain, all the elements below the line have been lost, any above gained. Thus in sample B8, there is a gain of Ba, minor loss of Si, more loss of Mg and greatest loss of K, Ca and Na compared to the parent B1.

The isocon technique provides a visual way of assessing which elements are immobile, gained or lost. The same information is available in the original data simply by ratioing weathered composition against parent. The same data as were used in Figure 11.13 are given in Table 11.7, but without the individual scaling factors used in the isocon diagram. Inspection of the ratios shows that Ti, Al, Fe and Zr all have the same ratio, Ba has a greater ratio and the other elements have smaller ratios. The equality of ratio for Ti, Al, Fe and Zr is strong evidence for their immobility, just as the isocon diagram portrayed.

Rock chemistry indices

Several weathering indices, or weatherability indices, have been devised based on the chemical composition of the weathered rock. These largely depend on establishing ratios between more soluble and less soluble constituents, such as Reiche's (1943, 1950) weathering potential index (WPI) and product index (PI):

$$WPI = 100(\sum \text{bases} - H_2O)/(\sum \text{bases} + SiO_2 + \sum R_2O_3)$$
$$PI = 100SiO_2/(SiO_2 + \sum R_2O_3)$$

Table 11.7 *Baynton basalt analytical data (from Eggleton* et al. *1987)*

	Fresh basalt (B1)	Weathered basalt (B8)	Weathered/Fresh
SiO_2%	52.2	42.7	0.82
TiO_2%	1.9	2.8	1.47
Al_2O_3%	14.4	19.0	1.32
Fe as Fe_2O_3%	11.2	16.0	1.43
MnO%	0.1	0.1	1.00
MgO%	5.9	4.8	0.82
CaO%	7.6	3.3	0.43
Na_2O%	3.0	1.5	0.50
K_2O%	1.2	0.4	0.33
Ba (g/t)	245.0	745.0	3.04
Zr (g/t)	189.0	275.0	1.46

Vogel (1975) modified Reiche's index to

$$MWPI = [(Na_2O + K_2O + CaO + MgO)/(SiO_2 + Al_2O_3 + Fe_2O_3)] \times 100$$

Harnois (1988) proposed the chemical index of weathering (CIW):

$$CIW = [Al_2O_3/(Al_2O_3 + CaO + Na_2O)] \times 100$$

Colman & Pierce (1981) studied weathering rinds on basaltic and andesitic corestones, and applied Reiche's WPI and PI indices. They found that this pair of indices gave a very clear discrimination between successive layers of a weathered corestone and so provided a useful measure of degree of chemical weathering. Chittleborough (1991) discusses several indices, arguing that most are flawed because they assume Al to be immobile. Inasmuch as Al and Fe are much less mobile than silica, the alkaline earths and the alkalis, these indices have considerable usefulness provided the mobility of Al and Fe is recognized.

Nonetheless, as a result of more recent research many of the above ratios and indices have been shown to be poor indicators of weathering. Thomas (1994) summarizes some of the more recent attempts to derive useful weathering indices. Fookes (1997) outlines a silica–sesquioxide molar ratio (K_r) to indicate the extent of weathering and to differentiate regolith types:

$$K_r = [Si]/([Al] + [Fe])$$

where square brackets denote mole fraction, namely

$$[Si] = \%SiO_2/60.09 \qquad [Al] = \%Al_2O_3/101.94$$
$$[Fe] = \%Fe_2O_3/159.7$$

Table 11.8 *CWI values for a weathering profile in granite*

CWI (%)	Profile zone (Figures 11.11 and 11.12)	Description (Figure 11.12)
60–90		Duricrust
40–60	VI	Residual soils
20–40	V	Completely weathered
	IV	Highly weathered
15–20	III	Moderately weathered
	II	Slightly weathered
13–15	I	Fresh rock

Ferruginous regolith has K_r values <2.0, clay and quartz rich materials K_r = >2 and ferricrete has generally K_r <1.33.

Sueoka (1988) proposed a chemical weathering index (CWI) based on the molecular percentages of sesquioxides plus water (loss on ignition) as a proportion of alkalis, alkaline earths and silica:

$$CWI = 100 \times ([Al] + Fe] + [Ti] + [H])/(\text{sum of mole fraction of all other components})$$

Sueoka claims this index corresponds closely with the physical properties of weathered granite with CWI = 13–15% for fresh granite, CWI = 15–20% (weathered granite in his terms), and CWI = 20–60% for collapsed saprolite (residual soil) and >60% for ferricrete (laterite). In terms of the profiles shown in Figure 11.12 the CWI values correspond as in Table 11.8.

Volumetric indices

Where there is fabric evidence for isovolumetric weathering, the bulk density of the weathered rock is a direct indicator of the degree of chemical weathering, and the ratio of the weathered rock density to the parent rock density provides a measure of the extent of weathering.

Eggleton *et al.* (1987) examined basalt weathering from fresh rock to saprolite where the fabric of the parent rock was conserved, and showed that the assumption of constant volume implied immobility of Ti, Zr, Nb. Ollier & Pain (1996) describe the principles of isovolumetric weathering, and Nahon (1991) gives some detail of the processes involved.

Comparisons of several weathering indices for the data by Eggleton *et al.* (1987) for Baynton basalt corestones and for unpublished data from the Weipa bauxite deposit show that all chemical indices give very comparable estimates of the degree of weathering.

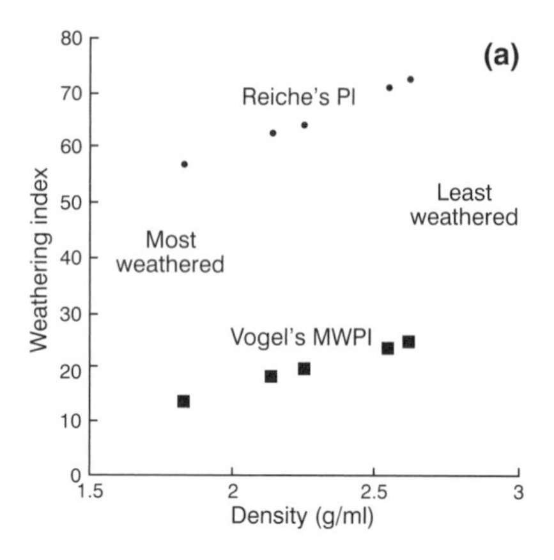

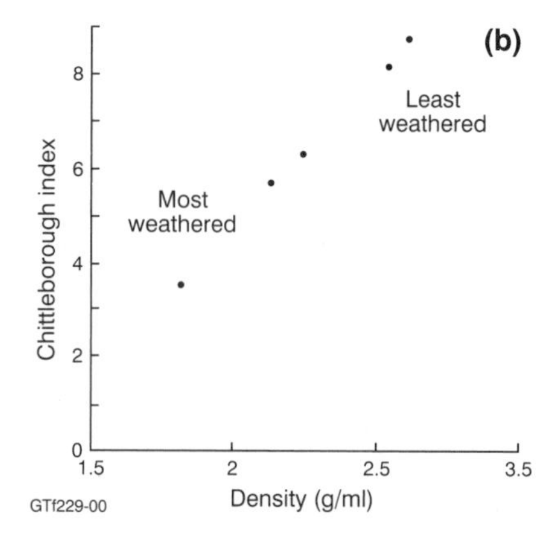

Figure 11.14 *(a) Vogel's MWPI and Reiche's PI plotted against density for 5 Baynton basalt samples. (b) Chittleborough's index of weathering plotted against density for 5 Baynton basalt samples*

Density provides a direct measure of dissolution and consequent mass loss, provided there is no collapse of the weathered rock, as does the immobile element method. The chemical indices of weathering of the basalt correlate closely with either density or with the ratio of TiO_2 in the weathered rock to TiO_2 in the fresh rock. Though Chittleborough's index was designed for a particular size fraction in soils, its emphasis on the three most soluble oxides (which decrease with weathering), and its use of a divisor (Zr) which increases with weathering, makes it a very sensitive index for the early stages of weathering (Figures 11.14a, b).

Comparison of methods

For deeply weathered regolith, such as bauxite, where density cannot be measured, both the MWPI and

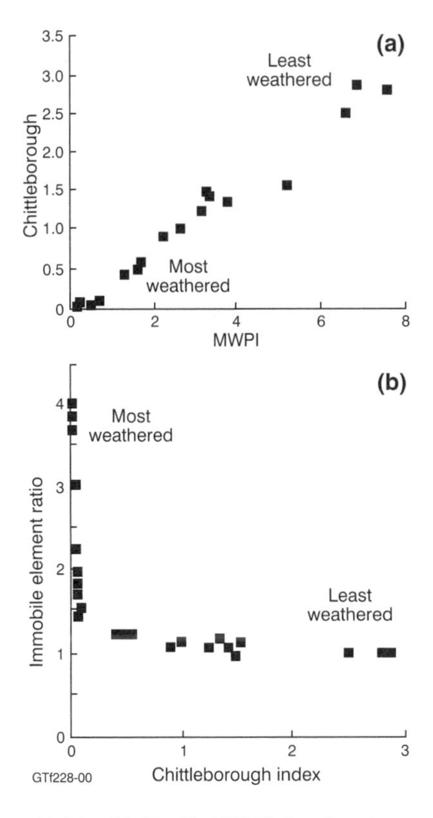

Figure 11.15 *(a) Vogel's MWPI plotted against Chittleborough's index of weathering for 31 Andoom bauxite samples. (b) Chittleborough's index of weathering plotted against an average immobile element ratio for 31 Andoom bauxite samples*

Chittleborough's index become insensitive, as Figures 11.16a, b show where these indices are plotted against an average immobile element ratio (see below). Figure 11.15a shows that both the MWPI and the Chittleborough index successfully separate the less weathered samples over this 30 m profile, but the most weathered samples group near zero. Again, Figure 11.15b shows that the Chittleborough index gives good discrimination in the early stages of weathering, but is insensitive later, and the MWPI would show the same features.

The immobile element method provides better discrimination over the upper part of the bauxite profile (Figure 11.16a).

Sueoka's CWI index gives a very good correlation with density for samples from the Baynton basalt (Figure 11.16b). By ratioing all the elements that increase (Al, Fe, Ti by relative enrichment, H by absolute enrichment) to those that are lost, Sueoka's CWI index, like the Chittleborough index, is sensitive

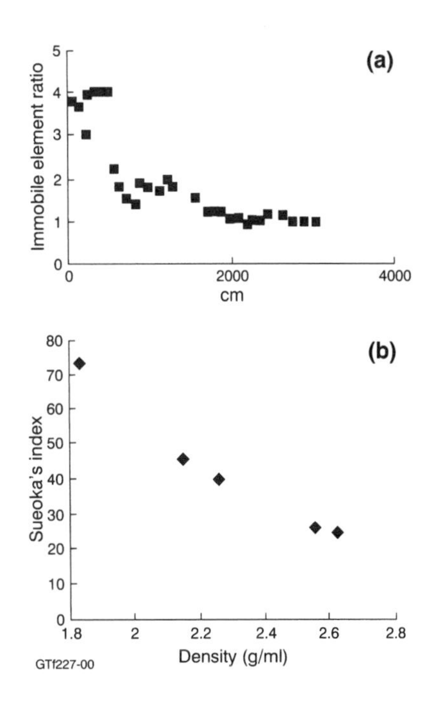

Figure 11.16 *(a) Average immobile element ratio versus depth for 31 Andoom bauxite samples. Samples at 3000 cm are fresh Rolling Downs Group. The cluster of samples from the surface to 600 cm are bauxite. (b) Sueoka's CWI index plotted against density for 5 Baynton basalt samples*

in the earlier stages of weathering. When weathering reaches the stage where there is little left of the more soluble elements that provide the divisor, the CWI index must lose accuracy, but it will still be informative (see Figure 11.17a, b showing the application of the CWI index to Andoom bauxite samples).

It is clear from the foregoing lengthy discussion of weathering indices that no one index serves to estimate the degree of weathering undergone by a rock. As an example, Sueoka's CWI index has a value of 25 for unweathered Baynton basalt and 12 for unweathered Bemboka granodiorite, thus chemical weathering indices are unable to define fresh rock. If degree of chemical weathering is defined as percentage mass loss, when Baynton basalt has lost 30% of its original mass, its CWI is 74, whereas when Bemboka granodiorite has lost 32% of its original mass its CWI is only 40; thus again, the chemical weathering index yields very different values for the same degree of weathering.

Despite the apparent scientific precision of all the weathering indices discussed, in many instances a direct assessment of relative degree of weathering in a

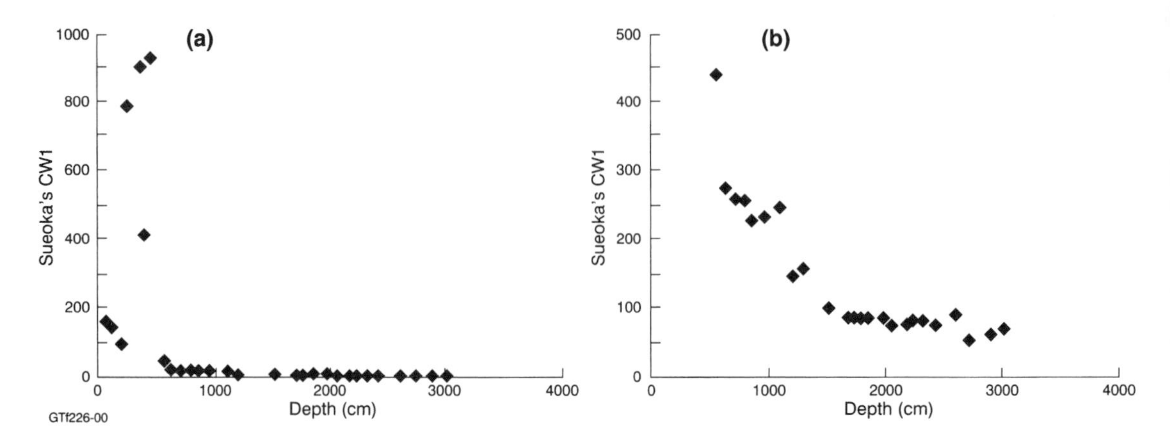

Figure 11.17 *(a) Sueoka's CWI plotted against depth for Andoom bauxite samples. Very low content of the more soluble elements leads to huge values for CWI in the bauxite (shallow depths) and there is little other discrimination on this scale. (b) Sueoka's CWI plotted against depth for Andoom bauxite samples, rescaled to show the changes in index lower in the profile*

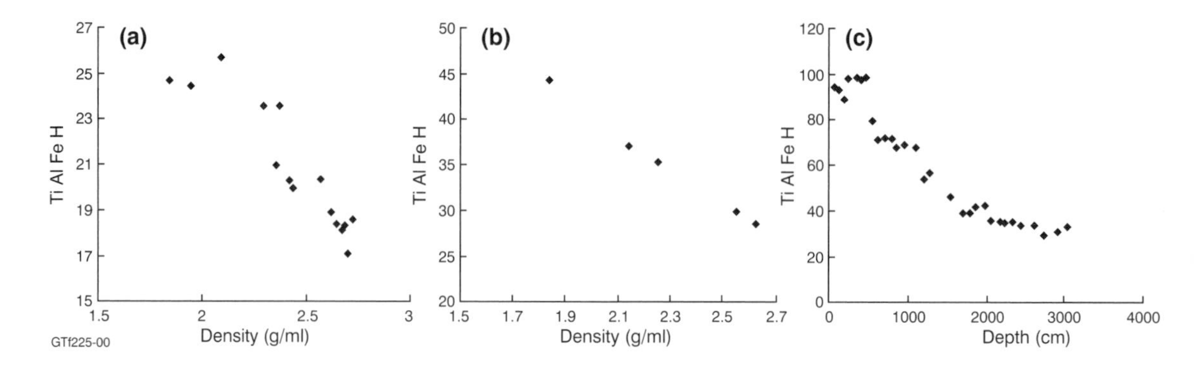

Figure 11.18 *(a) Change in ($TiO_2 + Al_2O_3 + Fe_2O_3 + H_2O$) = TiAlFeH with density for Bemboka granodiorite. (b) Change in ($TiO_2 + Al_2O_3 + Fe_2O_3 + H_2O$) = TiAlFeH with density for Baynton basalt. (c) Change in ($TiO_2 + Al_2O_3 + Fe_2O_3 + H_2O$) = TiAlFeH with depth for Andoom bauxite on sandstone*

profile can be made by remembering the simple principles of chemical weathering: that weathering is the replacement of soluble ions by hydrogen. If that fact is combined with the recognition that Ti, Al and Fe are essentially immobile, a figure can be constructed, similar to that of Sueoka, simply from the sum (weight %): ($TiO_2 + Al_2O_3 + Fe_2O_3 + H_2O$). Figure 11.18a, b, c shows this figure plotted against density for Bemboka granodiorite and Baynton basalt, and against depth for the Andoom bauxite. It is instructive to note that this figure for the most weathered granite (25) is the same as that for the least weathered Rolling Downs Group under the Andoom bauxite. It is a reminder that sedimentary rocks are the products of weathering; their compositions begin where the weathered igneous rocks leave off.

Depth of weathering

Deep weathering is the name coined for weathering profiles of exceptional depth. Many authors would argue that deep weathering is a tropical phenomenon, but there are many examples of deep weathering recorded outside tropical terrains. In many currently arid parts of inland Australia weathering profile depths of >100 m on arkosic sandstones are common (Senior 1978). Ollier (1965) quotes weathering depths of >300 m in Queensland and many authors report depths >100 m from various parts of Africa. Thomas (1994) reports weathering to depths of >150 m in Zaire with some minor weathering to >600 m, where current rainfalls are 1200 mm/year. On the Brazilian Plateau where present rainfalls are

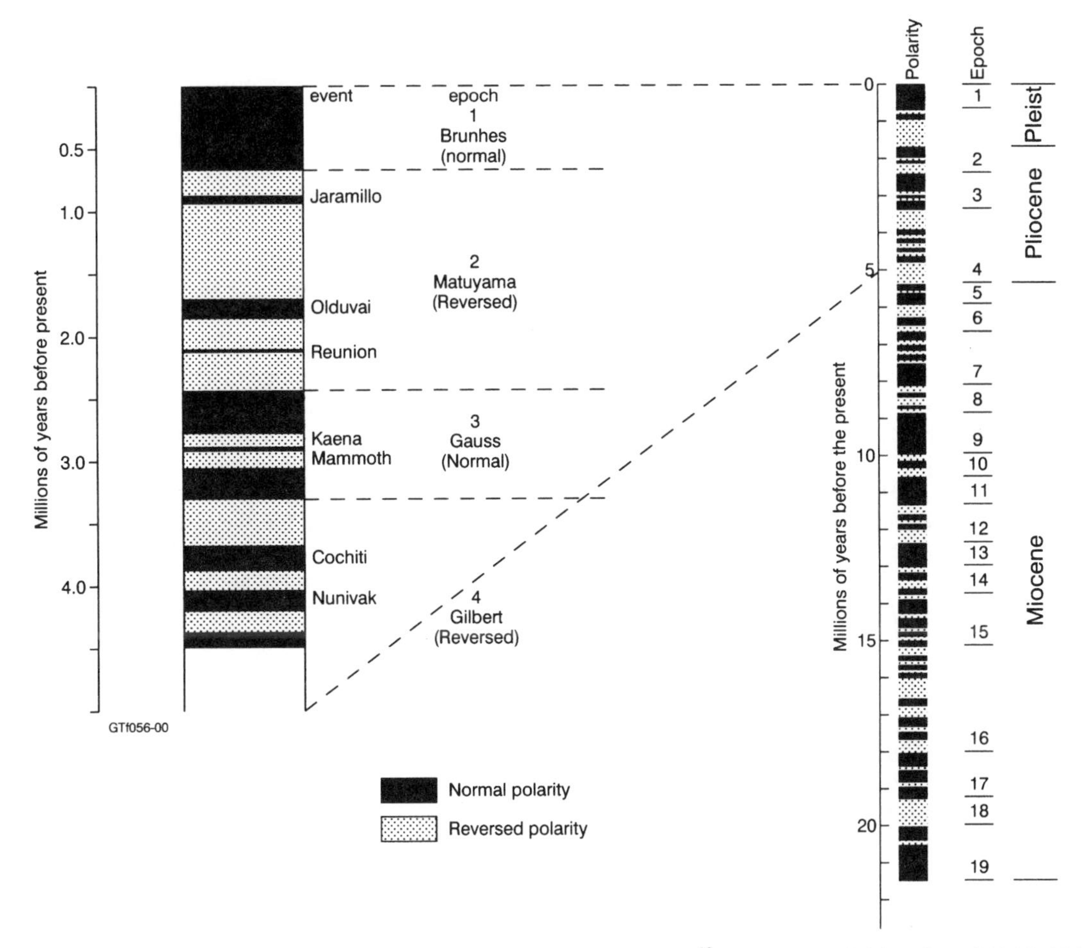

Figure 11.19 *Ages of kaolin samples from various Australian sites determined by $\delta^{18}O$ values (compiled from data of Bird & Chivas 1989)*

between 1800 and 3500 mm/year, profiles on migmitites attain depths of 50–120 m. Similar deep weathering profiles also occur in Sweden (Lidmar-Bergstrom 1989), Finland (Lidmar-Bergstrom 1989, 1995) and Tasmania (Beavis 1985). Across much of southern Australia deep weathering profiles are preserved; at Wilson's Promontory where they are about 50 m thick now, but could have been up to 100 m (Hill 1996). On the Yilgarn in Western Australia profiles attain a thickness of about 60 m on granitic bedrocks and deeper on ultramafic rocks (80+ m) (Butt & Zeegers 1992a) and on the Monaro Plateau where profiles >30 m thick are common on granitic rocks and deformed Palaeozoic sediments (Taylor *et al.* 1990b). All these examples come from climates anything but tropical at present; in fact many are dry and cool most of the time.

Dating weathering profiles

Dating weathering events is difficult as weathering can occur over millions of years, particularly when stable tectonic conditions and well-vegetated landscapes are involved. There are many techniques to date weathering products, but this does not give the age of weathering, only the age at which the minerals used were formed.

Stable isotopic techniques have proved useful in dating weathering products where the rates of plate motion and the altitude are known. During weathering the isotopic composition of δD and $\delta^{18}O$ in the water in equilibrium with the new-formed minerals is preserved. By correcting for the effects of altitude and latitude the age of formations of these minerals can be broadly determined. Bird & Chivas (1989) have shown that Permian, Cretaceous, Palaeogene and Neogene dates can be determined for kaolin formed by weathering (Figure 11.19). They specifically dated the time at which the kaolin preserved in the present profile formed and as we have described above, kaolin may form from parent minerals and later be re-formed. As the kaolin retains the $\delta^{18}O$ signature of the water with which it was in equilibrium at the time it formed, the date will either reflect the last time kaolin was formed, or more likely, some average between old and new kaolin. This does not mean all the minerals in the profile formed at that time, or that only one time phase of kaolin is present in the profile. Although the several phases of kaolin have not been isotopically detected in weathering profiles, petrographic evidence (Robertson *et al.* 1998; Nahon 1991) shows they exist.

Giral *et al.* (1993) have studied $\delta^{18}O$ values of kaolinite from a lateritic profile in Brazil. Their work, in contrast to Bird & Chivas (1989), shows no age trend in the kaolinite through the profile because, they suggest, of continual reaction of the clay minerals with the soil solutions. While this may be the case, the lack of systematic variation of $\delta^{18}O$ may be due to the erratic values of $\delta^{18}O$ in tropical climates.

Perhaps the most common regolith-dating methods involve palaeomagnetism simply because magnetic Fe-oxides are very commonly associated with weathering. There are a number of palaeomagnetic methods which utilize the magnetic properties of minerals. Magnetostratigraphy records the polarity of the Earth's magnetic field at the time the minerals were formed or deposited. The magnetic field has changed polarity throughout the Earth's history and the pattern of these changes is well understood, at least back to 100+ Ma (Pillans 1998). By comparing the reversal pattern in a relatively continuous sedimentary sequence with a global reversal stratigraphy it is possible to date layers in the sediments. A good example of these techniques has been demonstrated from Lake George in south-eastern New South Wales.

Work by Singh *et al.* (1981) suggested the sedimentary fill in the fault-dammed Lake George near Canberra was mid-Miocene based on a limited magnetostratigraphic investigation. Later McEwan-Mason (1991) again using magnetostratigraphy with palynological investigations, dated the sequence as 3.05–3.15 Ma during the Mammoth Chron, or Late Pliocene.

In cases where a single (and isolated) weathering sample is available, all that can be said, if the polarity is reversed, is that the sample is older than the Brunhes–Matuyama boundary (Figure 11.20). If the polarity is normal, very little can be inferred from the result.

This technique has also proved very useful to date periods of ferruginization in much younger Quaternary materials. Parna (aeolian clay deposits) at Junee in central New South Wales have been dated by this technique (Pillans, personal communication 1996) as >780 ka, but magnetic noise in the sample may indicate remnant palaeomagnetism has been disrupted, perhaps by lightning strikes (Figure 2.18).

The continents' drift over time has changed their position with respect to the poles and this allows the construction of time-calibrated apparent polar wander paths (APWP). This means that magnetic samples from the continents can be positioned and the age of the magnetism determined by comparison with the apparent polar wander path for that continent. This is particularly important for the dating of the age of magnetization in regolith samples. It has been widely used from all continents to age weathering and other events. The age of precipitation of Fe-oxides in the weathering profiles developed on the Deccan Traps shows they began to weather almost as soon as they were erupted (Schmidt *et al.* 1983).

These techniques are particularly applicable in Australia where weathering has continued over very long periods. The oldest profile dated is Carboniferous (Pillans 1998) in the Parkes area of central New South Wales where deep weathering has altered Ordovician volcanic rocks. This study also shows the weathered surface was incised by palaeochannels that were filled and also weathered during the Oligocene or Miocene. Figure 11.21 illustrates this situation in one of the mine pits at Northparkes mine.

A classic study of the ages of weathering in an area where multiple weathering profiles developed on Cretaceous and inset Tertiary sediments of the southern Eromanga Basin, southern Queensland, was described by Idnurm & Senior (1978). They identified two major weathering *events* (our italics) as Late Cretaceous and Miocene. It was this work which demonstrated that palaeomagnetic APWP techniques were applicable to dating weathering. The results of this work, however, led to extrapolation of their dates for similar weathering surfaces over wide areas of the continent on lithostratigraphic principles without any confirmatory dating. The readiness of researchers to do this has led to entrenchment of quite erroneous interpretations of regolith evolution over large areas of Australia.

Théveniaut & Freyssinet (1999) have used palaeomagnetism to document the ages and rates of forma-

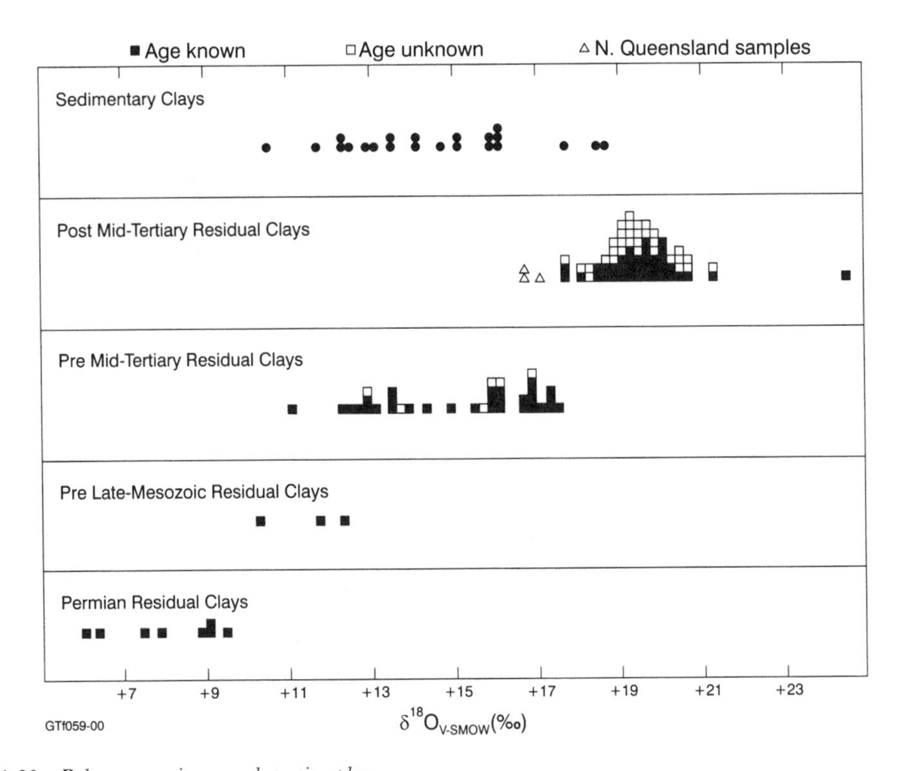

Figure 11.20 *Palaeomagnetic reversal stratigraphy*

tion of a 'lateritic' profile in French Guiana. The profile is some 54 m thick consisting of saprolite under a duricrust comprising a mottled facies overlain by a massive facies and an uppermost pisolitic facies. The base of the saprolite contains a natural remnant magnetism (NRM) consisting of an inherited Proterozoic NRM from the magnetite. Within the saprolite profile five magnetic reversals are recorded, the youngest (Bruhnes/Matuyama) being at a depth of 8.8 m. This gives an average rate of saprolitization of 11.3 ± 0.5 m/Ma. The massive base of the duricrust shows palaeomagnetic directions that fit on the apparent polar wander path for South America with the present-day pole position, but the upper posolitic facies have a position on the apparent polar wander path corresponding to one older than 10 Ma. This Théveniaut & Freyssinet (1999) interpret as the duricrust being formed in two stages of lateritization. It is interesting that the upper duricrust facies is older than the lower massive part. They do not offer an explanation for this, but confirm their findings by noting the two facies have significantly different haematite crystal morphologies.

Radiogenic isotopic techniques also provide a number of opportunities to date regolithic materials.

Figure 11.21 *A photo of a mottled palaeochannel fill over deeply weathered Ordovician volcanics at Northparkes Au/Cu mine near Parkes, central New South Wales. The weathering in the volcanics dates from the Carboniferous and the channel fill from the mid-Tertiary (Pillans 1998) (photo Graham Taylor)*

The most established technique is K/Ar used initially to date basalt flows covering palaeolandscapes to give a minimum age of the landscapes and their contained regolith (e.g. Taylor *et al.* 1990b; Young & McDougall 1982). Many other cases of K/Ar dating of regolith using lavas and ashes are well documented.

Examples of the use of K/Ar and $^{40}Ar/^{39}Ar$ for dating regolith materials are also now relatively common. Vasconcelos *et al.* (1994) use these techniques to date K-bearing Mn-oxides formed during the lateritization of Precambrian rocks in Brazil. They showed ages clustered at 65–69, 51–56, 40–43, 33–35, 20, 24, 12–17 Ma and at zero age. Despite their defining 'clusters' of ages this age spread could almost be read as continuous weathering between 72 and 20 Ma. Similarly, Dammer *et al.* (1996) date the Mn-ores of Groote Eylandt (see Chapter 15) as having formed during several episodes at 6–18 Ma, 30 Ma and before 43.7 Ma. Their data also clearly demonstrate that the Mn-minerals they dated are of a weathering origin rather than the diagenetic origin previously considered most likely. Another similar study using $^{40}Ar/^{39}Ar$ and $^{87}Rb/^{87}Sr$ in Mn-oxides in Brazil has been equally successful in dating regolith-formed materials.

Bird *et al.* (1990) demonstrate the usefulness of K/Ar dating in alunite from regolith across Australia. They show it works successfully for alunite ranging in age from <1 to 62 Ma.

In some circumstances it is possible to date carbonate and ferruginous nodules and coatings using radiometric techniques. Short *et al.* (1989) have dated Fe-nodules from within the Alligator River U province in northern Australia using Th–U disequilibrium series. They show that the U-rich Fe-nodules were useful to date Fe/Mn indurated soil layers <350 ka old. Mathieu *et al.* (1995) used short-lived U and Th-isotopes to assess weathering rates of a granite high in U and Th. Their work concluded that a 15 m thick saprolite profile containing nodules and pisoliths has developed over about 300 ka. While it is possible also to date carbonates using U-series dating, regolithic carbonates are generally considered to be unreliable because of the incorporation of detrital Th. Shell and coral materials are, however, ideal (Stirling *et al.* 1995) as are speleothems (Ayliffe & Veeh 1988). One example of the use of U-series to dating lake and dune deposits was conducted by Herczeg & Chapman (1991).

12 Physically transported regolith

Introduction

Most geologists when thinking of transported regolithic materials immediately turn to sediment deposits in the landscape. They are of course correct; such deposits do form part of the regolith. In fact one of the major problems in regolith geology is distinguishing transported from *in situ* regolith. Surface terrestrial sedimentary deposits are, however, not the only physically transported regolith of importance. Material is physically, as well as chemically, moved within the regolith as fluids pass through it.

In this chapter we discuss some of the sedimentary deposits important in regolith studies including colluvium, alluvium, glacial deposits, aeolian deposits, terrestrial coastal and volcanic deposits (and lavas, although they are not deposits). We will not discuss these in detail, as there are many good sedimentology books on many of these subjects. Reading (1996), Walker (1984), Pye (1994) and Selley (1988) provide excellent overviews of sedimentary facies. Selby (1993) summarizes colluvial and other hill slope deposits in more detail than we can here, and similarly there are reviews of alluvial sediments (Miall 1996), aeolian sediments (Pye 1993), glacials (Ashley *et al.* 1985; Gurnell & Clark 1987) and volcanic ash deposits (Cas & Wright 1987).

The criteria for distinguishing between regolith formed in place and that deposited over it are not summarized well in any readily available source, and we will treat that in some depth.

Physical transport of materials in the regolith is well treated from the perspective of soils (e.g. Brewer 1964) and the regolith as a whole by Nahon (1991), but we will draw some threads of this aspect of transported materials together here.

Sedimentary deposits of the regolith

COLLUVIUM

Colluvium is a mass deposit from transport down a slope under the influence of gravity. It lacks the internal bedding structures and is generally of more variable grain size than alluvium. It is largely derived locally and is composed of materials derived from upslope. Colluvial deposits are almost ubiquitous on hill slopes, sometimes culminating in colluvial fans or lobes (landslide lobes) at the foot of slopes.

Colluvium and the soils developed on them can move continuously downslope by hill creep, but this is more obvious on steeper slopes. On steep slopes colluvium often moves by landslides at times of increased rain or when slopes are denuded of vegetation. Such processes may also occur because of climate change. This continuous and episodic movement causes the colluvial lobes to build at the foot of the slope. Thomas & Thorp (1995) record such changes from Sierra Leone as a result of climate changing from drier during the Pleistocene to a wetter Holocene (Figure 12.1). A feature of the formation of these foot slope lobes is that many episodes of slippage can produce a sequence of colluvial deposits separated from each other by palaeosols. If palaeosols are not recognizable then it may still be possible to separate units by their varying degrees of weathering, those lower in the sequence being more weathered than recent additions. In Hong Kong three classes of colluvium of different ages can be recognized (Lai & Taylor 1984) on the basis of their composition and degree of weathering.

In many regions of Australia colluvial mantles are common, occurring from the tropical north to the southern regions of Tasmania. At Cobar in the west

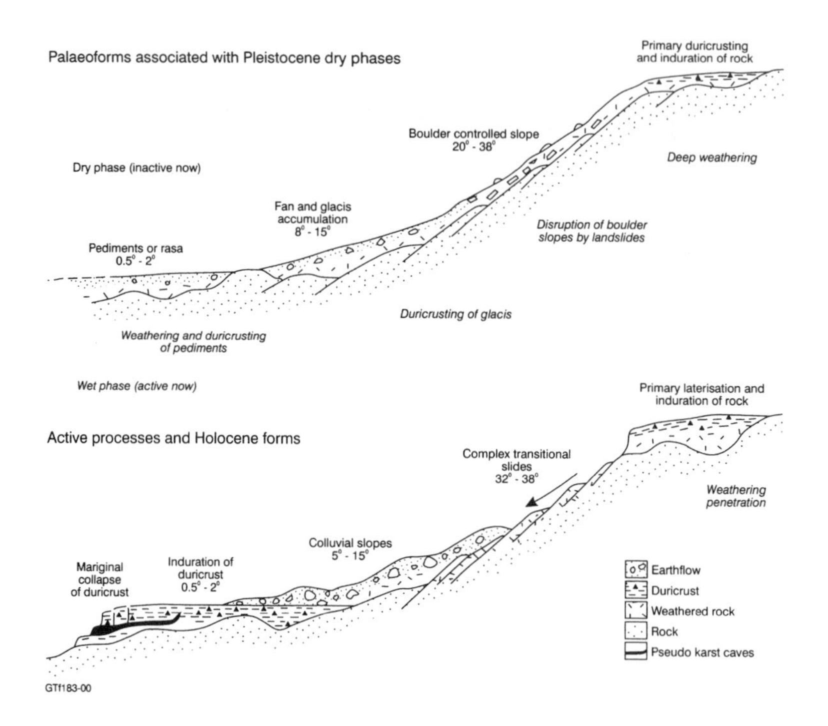

Figure 12.1 *Hillslope and footslope deposits in the Pendembu Valley near Freetown, Sierra Leone, showing recent landslide debris overlying ferruginized materials (from Thomas 1994)*

of New South Wales colluvium mantles most slopes, often with several episodes of colluviation being preserved. Individual colluvium layers are separated by well-developed soil profiles. In some cases the colluvium is stabilized by widespread and massive carbonate cementation (Figure 12.2). These deposits are interpreted as being Late Quaternary, probably deposited during a glacial maximum arid climate when the slopes from which the debris was derived were substantially devegetated. Joshi & Kale (1997) from the north-western Deccan in India describe a similar situation where basaltic colluvial deposits are attributed to drier glacial maximum climates. Interestingly Whitney & Harrington (1993) attribute colluvial deposits separated by well-developed palaeosols in southern Nevada as forming during interglacial humid phases of the Early to mid-Quaternary and the soils forming during drier glacial maximum climates.

Colluvial processes progressively move weathered debris downslope until it enters the zones of alluvial activity. Interbedded colluvial and alluvial deposits often mantle lower slopes. This is most commonly seen in alluvial fans where mass flow and alluvial deposits make up the body of the fans (Figure 12.3). These deposits grade out from the slope into purely alluvial deposits (Figure 12.4a). There are several major points to note in this simple model:

- colluvial sediments derive from the slope on which they occur; and,
- colluvium is incorporated into alluvial fan and alluvial deposits.

Surface and subsurface water flow may rework colluvial materials. Deposits formed by surface reworking or sheet flow are often bedded and are composed of the finer fractions of the colluvial materials. Subsurface flow causes erosion generally resulting in piping and deposition of the finer fractions of the colluvium in voids and pockets in the colluvial mass.

Figure 12.2 *Carbonate (calcrete) cemented colluvium west of Cobar, New South Wales. Sheetwash sands and minor gravels overlie the colluvium (photo Melissa Spry)*

Figure 12.3 *Colluvial and sheetflow/alluvial facies interfingering at the junction of a fan and valley alluvium from near Cooma, New South Wales (photo GT)*

ALLUVIUM

Alluvial sediments are materials deposited on the land surface from transport by flowing water confined to a channel or valley floor. As Figure 12.4a shows, there are two main types of alluvial sedimentary tracts, alluvial fans and valley alluvium. These are rather different in many ways.

Alluvial fans consist of debris derived from the valley slopes deposited as a wedge along the valley sides. They are composed of mixed sediments ranging in grain size from gravels to muds. The grain size of the deposits decreases away from the fan head and is finest at the fan toe. The fans grow as debris being transported from the lateral streams decrease gradient at the valley floor. As sediments accumulate the channel shifts across the fan, building a biconvex fan lobe (Figure 12.4a) consisting of coarse channel sediments (alluvial or debris flow) and finer overbank or sheet flow sediments (Figure 12.4b). In regions distal from the main axis of sedimentation soil formation modifies the sediments.

Alluvial fan sediments interfinger with colluvial deposits on their lateral margins and with valley alluvium along their toe. They may also contain beds or channel fills of debris- or mud-flow deposits. These are generally distinguishable by their lack of bedding and massive nature compared to the fan deposits. Through time, as a result of local base-level changes or climatic shifts, fans incise themselves, producing a set of 'inset fans' that are readily distinguished by the terracing which results (Figure 12.5).

Where a series of alluvial fans have coalesced along a valley margin to form an extensive fan wedge or bajada, have undergone a series of incisions and have developed a series of discontinuous palaeosols within them, the landscape developed and their internal stratigraphy can become very complex. Such is the case along the western margin of the Broken Hill Block in western New South Wales, where sediments from the highland Broken Hill region spill out on to the Mundi Mundi Plain (Figure 12.6).

Valley alluvium is equally complex but very different from fan alluvium in most cases. It comprises many facies ranging from very coarse-grained gravels to very fine-grained muds. Which facies occur where depends essentially on the energy levels operating in the system that are in turn controlled by stream gradient. Basically two facies are important to us in considering the regolithic aspects of valley alluvium:

- channel facies are relatively coarse-grained, composed of gravel and sands which occur as lenses or sheets. Channel facies represent the locus of sedimentation in the stream system and it is here that the most rapid deposition occurs. It is uncommon for soils to develop on channel facies as their accumulation rate exceeds that of pedo-

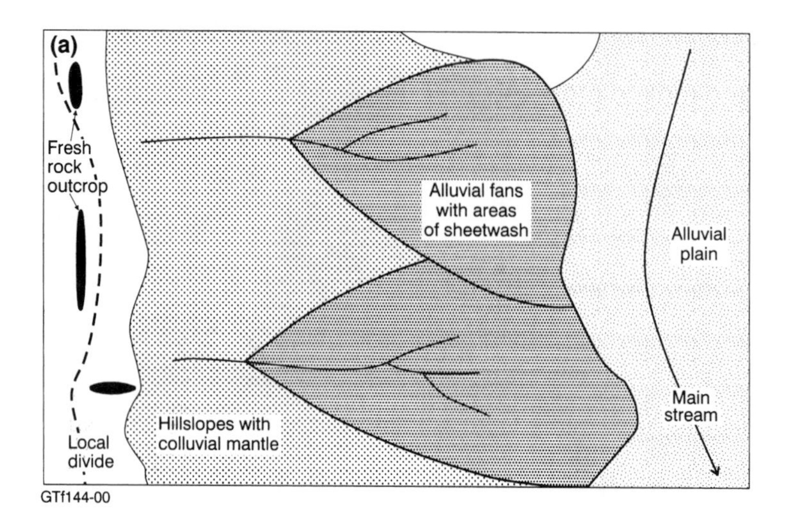

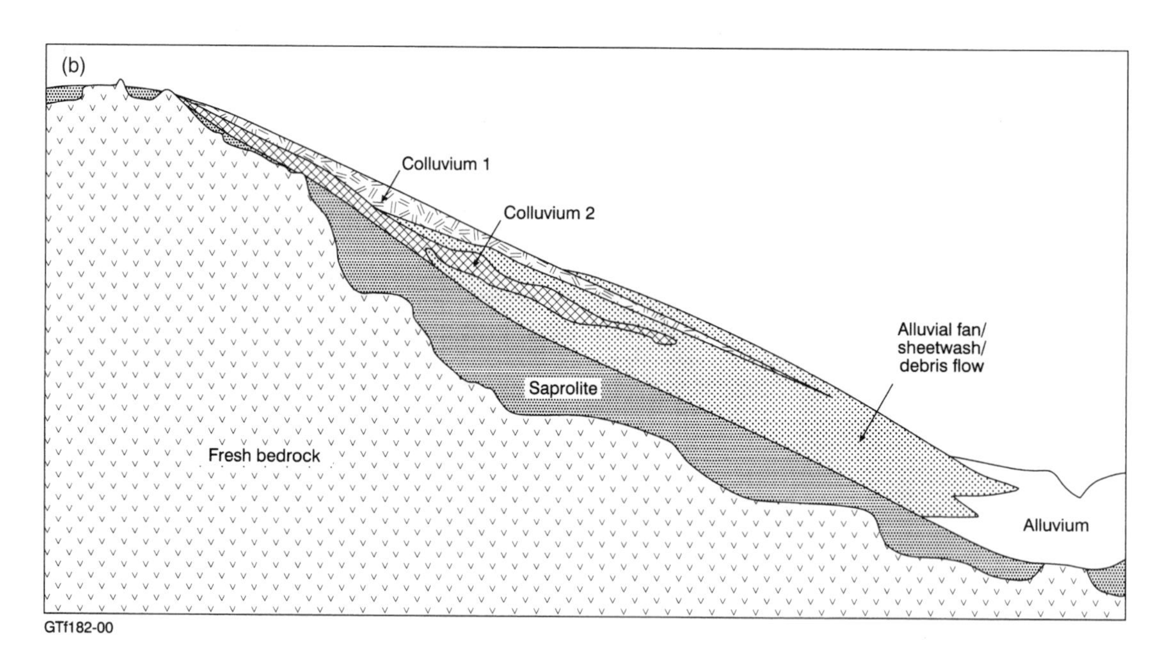

Figure 12.4 *(a) Diagrammatic plan and section of a hillslope and valley illustrating the spatial relationships between colluvium, alluvial fan and alluvium. (b) Sedimentary facies variations in alluvial fans (from Harvey 1997)*

genesis. It is in this facies that heavy minerals may accumulate (see 'Placer deposits', this chapter). These coarse-grained facies may fill the whole valley in higher gradient systems, but at lower gradients they are usually flanked by floodplain facies; and,

- floodplain facies are composed of relatively finer-grained sediments, often muds and in some very low gradient streams, clay-sized deposits. They deposit laterally to the channel facies, generally at significantly lesser rates than the channel sediments. Being away from the locus of sedimentation,

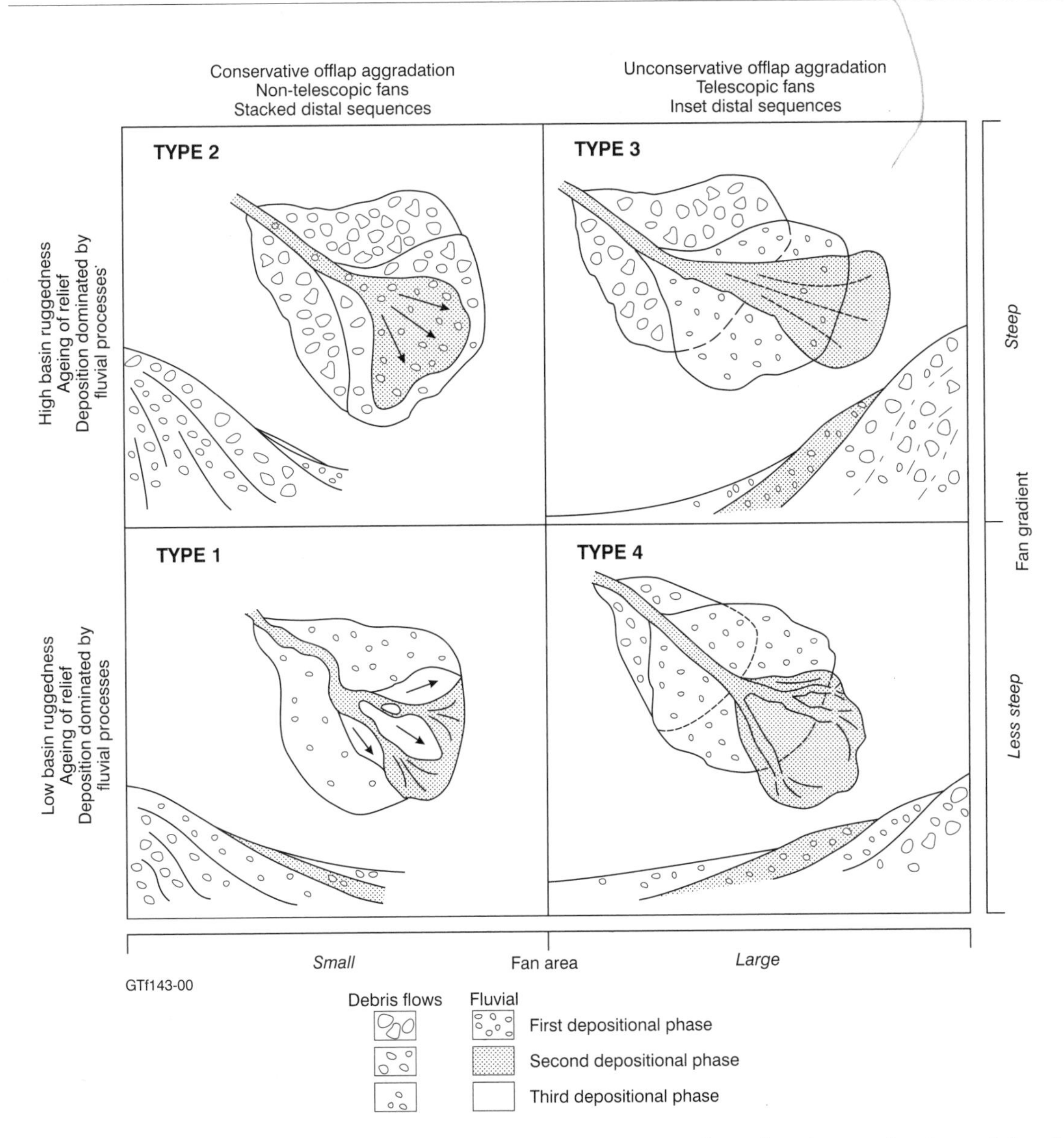

Figure 12.5 *Alluvial fan spatial geometry models reflecting varying sedimentary and tectonic environments based on work from the Murcia fan in Spain (after Silva* et al. *1992 from Harvey 1997)*

rates of pedogenesis may overtake deposition rates and well-developed soils are typical of the more distal parts of floodplain sediments.

Alluvial facies may deposit in all sizes of stream valleys from small, relatively steep gullies to very extensive continental scale basins (e.g. the Mississippi, the Ganges or the Murray–Darling). In small, steep valleys the deposits usually fill no more than the bottom metre or so, often discontinuously. In large systems such as those noted above the sediment pile can reach hundreds of metres thick and fill valleys hundreds of kilometres wide. In moderate- to large-scale alluvial accumulations channel and floodplain facies occur together with soils and palaeosols (Figure 12.7), rather like those in alluvial fans.

As an example of how this understanding can be brought to bear on a better understanding of regolith geology, we discuss the situation in the Drummond Basin, northern Queensland. The region is one in

Figure 12.6 *High angle colluvial/alluvial bajada at Euriowie near Broken Hill (photo Steve Hill)*

which extensive areas of Tertiary alluvial sediments overlie Palaeozoic basement rocks. These Tertiary sediments were divided into two fluvial formations, an older and rubified Southern Cross and a younger bleached Campaspie. One of very few localities where one is seen overlying the other is at Red Falls where the formations were defined (Wyatt & Webb 1970). The usual interpretation is that following the Southern Cross Formation deposition, which extended through most of the Tertiary, weathering reddened (rubified) the sediments and widespread ferruginous duricrusts were developed. Landscape rejuvenation led to incision of the Southern Cross and deposition of the Campaspie in the incised depressions (Grimes 1979, 1980; Henderson 1996) (Figure 12.8a). The alluvial blanket of the 'Southern Cross/Campaspie' has been widely incised, leaving a landscape dotted with flat-topped plateaux and mesas (Figure 12.8).

Aspandiar (1998) has shown there are in fact not two formations, but one alluvial blanket and that the internal variation is due to the evolution of the blanket, reddening or rubification resulting from prolonged pedogenesis during long periods of non-deposition in an area of the alluvial plain distal from the channels. The development of ferricretes on the mesa and plateau margins is due to post-incision mobilization of Fe (similar to the explanation provided by Rivers *et al.* 1995).

Australia is unique with respect to many of its alluvial systems. It is the driest continent, apart from Antarctica, with more than 70% of its land mass being arid to semi-arid. It is also a very flat continent and this enables occasional access to the centre for moisture-laden winds. This results in erratic but extreme rainfall in the continental interior. In the Lake Eyre Basin rainfalls vary from 500 mm around its northern and eastern margins to <100 mm in its centre (the Simpson Desert). The combination of all these factors leads to very erratic stream discharge. Major discharges occur on average every 20 years (Knighton & Nanson 1997).

The major stream in the Lake Eyre Basin is Cooper Creek, Australia's largest drainage system. The extreme variability of discharge when combined with

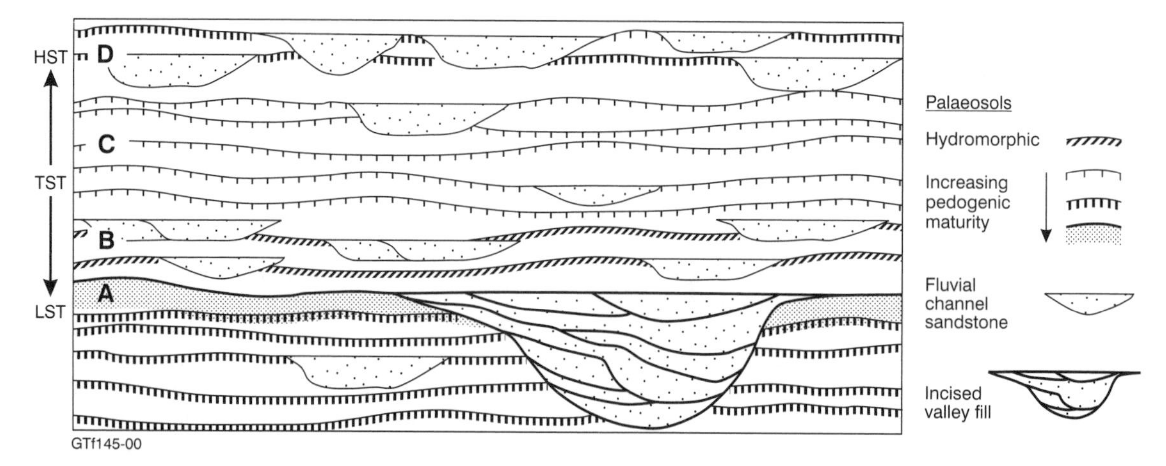

Figure 12.7 *Schematic diagram showing Wright & Marriott's (1993) model of pedogenetic development related to a sea-level cycle. (a) The lowstand systems tract (LST) is characterized by channel cross-incision and strongly developed well-drained palaeosols that form on terraces. (b) Early in the transgressive system tract (TST) hydromorphic palaeosols may form because base-level is rising; channel sands may overlap due to reduced accommodation space. (c) Increased accommodation space in the TST produces rapid sediment accumulation and weakly developed soils. (d) The high system tract (HST) is characterized by low accommodation space hence well-developed palaeosols (from Kraus 1999 modified after Wright & Marriott 1993)*

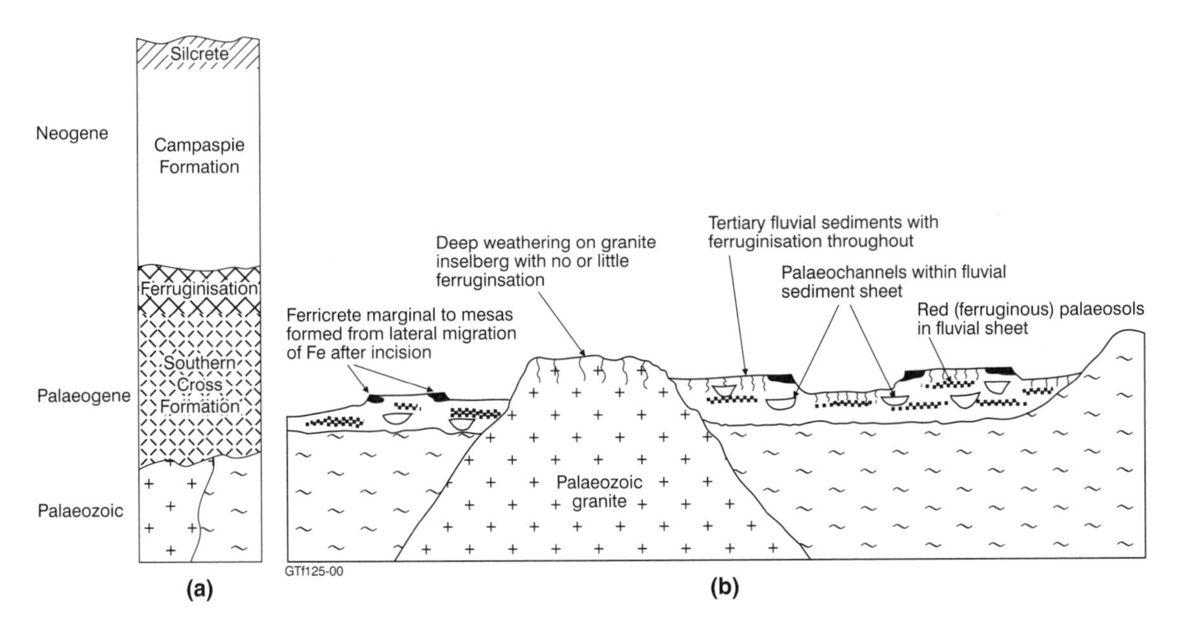

Figure 12.8 *(a) A schematic stratigraphic section of the Charters Towers region showing the traditional interpretation of the stratigraphy and regolith formation. (b) Shows the interpretation of the same sequence from the work of Rivers* et al. *(1995) and Aspandiar (1998)*

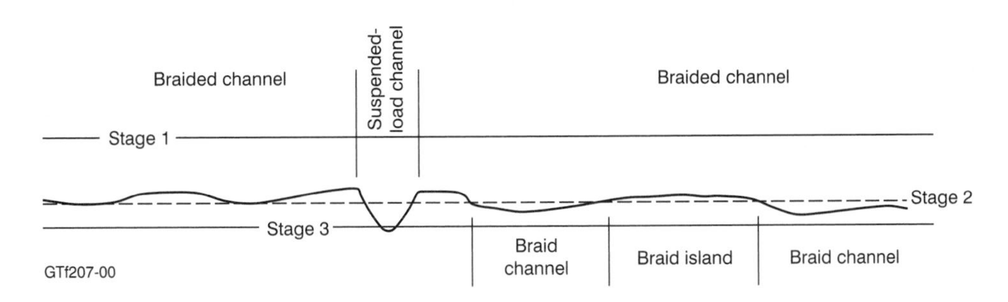

Figure 12.9 *Cross-section of the Cooper Creek (not to scale) illustrating the relationship between stage and channel morphology. At flow stage 1 sediment across the whole braid plain is in motion forming the braid islands. At flow stage 2 water is draining from the braid islands and erodes the back-flow channels (Figure 12.10) and moves along the braid channels, eventually entering the suspended-load channels. Flow stage 3 mostly represents flow in the suspended-load channels, or as it lowers it represents the regional water-table when flow in the suspended-load channels ceases and they are represented by a series of waterholes*

the unusual sediment load creates unique alluvial tracts. They are composed of clays with minor sands overlying sandy facies that relate to climates about 100 ka ago (Nanson *et al.* 1986). The contemporary sediments owe their unique character to the subplastic nature of the kaolinitic clays. This causes the clay to form aggregates about the size of fine to medium sand that are stable in water.

The Cooper Creek acts in a schizoid fashion. At high discharge the channel, about 20 km wide near Kurnaoma, is a braided system in which the clay aggregates behave as sand that forms braid islands. As the flow decreases the creek occupies sinuous channels and behaves as a suspended load stream with a sandy bed and muddy banks. The clay aggregates eventually disperse after long periods of wetting to form the suspended load. These channels form the waterholes typical of the creek (Figure 12.9).

Similar clay-rich plains with braided or anastomosing channels systems are typical of many other inland Australian stream valleys. In Australia they are referred to as 'channel country' (Figure 12.10).

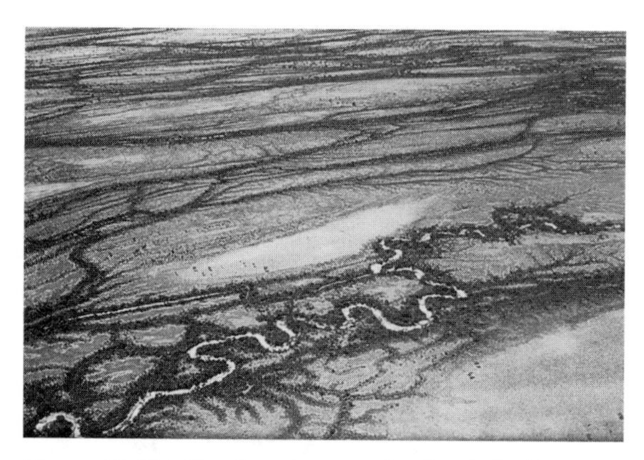

Figure 12.10 *The channel country showing the braid islands and channels and the sinuous low discharge suspended-load channels (photo GT)*

Other major inland drainage systems lie across very extensive clay plains. Most of these streams rise in the Eastern Highlands but flow westwards into semi-arid and arid zones, most losing discharge with increasing drainage basin area (Figures 5.15 and 5.18). These streams, after leaving the highland belt, carry and deposit predominantly clays derived from the weathered highland rocks, particularly the Cainozoic basalts that are distributed along them. The plains are composed of kaolin and smectite clays which form highly shrinking surfaces scattered with gilgai. The shrink–swell capacity of these deposits also leads to considerable mixing so the original alluvial nature of the sediments is completely destroyed. They simply form highly structured (pedal in the soil sense) grey, brown or black clays. They also commonly contain significant calcrete and soil carbonate, and gypsum. Also scattered across the area are large and small lake basins, most of which represent former outcrops of saline groundwater (Evans & Kellett 1989). Most of the lake basins contain lake clays, sands and dispersed evaporites and are flanked on their eastern margins by lunettes (aeolian dunes).

As regolith materials these clay plains have major engineering problems associated with them because of their shrink–swell capacity. Building roads, bridges and housing on these materials is a challenge. Additionally, their salt contents provide many challenges for agriculture and engineering.

GLACIAL DEPOSITS

Over the last 2 Ma or so glaciers have periodically covered varying areas of the Earth and before that many Gondwanan continents experienced glaciations during the Carboniferous and Permian. Because the Northern Hemisphere experienced the most widespread glaciation during the Quaternary, there is an extensive literature on this. In Australia, Permian glacial deposits remain at the surface at some localities such as central Victoria and form part of the contemporary regolith (Craig & Brown 1984; Macumber 1978). Quaternary glacial deposits are restricted to the highest parts of the mainland around Mt Koscuisko and to the higher parts of central Tasmania.

Glacial sediments may be derived from afar or near. In the case of continental ice sheets such as extended over much of Europe and North America during the Quaternary, debris was transported for up to several hundred kilometres, but the median dispersal distance of boulders from Finnish examples was 3.0 km before deposition (Salonen 1986). Their distance of transport is a function of grain size, lithology, width of original outcrop and topography over which the ice moved (Kauranne *et al.* 1992). Teale *et al.* (1995) report pebbles of cordierite-bearing volcanic rocks over 100 km from their source in Cretaceous sediments of south-eastern Australia. They conclude that their transport over this large distance was glacial or fluvioglacial. In valley glacial systems the deposits were mostly derived from the immediate glacial catchment. In Victoria, Australia, some of the Permian glacial deposits have been shown to have been transported over distances of up to 'hundreds of kilometres' (Craig & Brown 1984). Playford *et al.* (1976) record transport distances of erratics in the Permian rocks of the Perth Basin up to 250 km. These deposits have, however, experienced long periods of lithogenesis, weathering and pedogenesis and where they now occur at the surface are probably chemically partially equilibrated with underlying materials.

The Quaternary glacial deposits of the Northern Hemisphere have almost certainly not had time to achieve much of the chemical character of their underlying materials. Ross *et al.* (1983) point out there has not been sufficient or sufficiently intense post-Wisconsinian weathering at sites on Hudson Bay to form kaolinite or gibbsite, but they are present in soils in Canada formed on tills of pre-Wisconsinian tills. Where such minerals are found in Wisconsinian tills it implies they are composed in part from redistributed earlier tills. Such findings have important implications when attempting to source the till materials.

Glacial deposits comprise tills, subglacial glacioaquatic deposits and outwash deposits. As well as these regolith deposits directly related to glaciers there are many periglacial deposits, caused by repeated freezing and thawing. Till is primarily made up of sediment dumped by a glacier. It comprises poorly sorted particles ranging in size from boulders to silt- and clay-sized debris. They may be deposited beneath, beside or at the snout of a glacier as moraines. Glacio-aquatic deposits are sediments laid down by water flowing beneath or through a glacier. They are usually well bedded and, compared to tills, well sorted. They mainly occur interbedded between layers of till. Glacial outwash deposits include features similar to alluvium, but they are usually made up of relatively freshly eroded glacial debris washed from tills by melting ice. Again compared to tills they are well sorted and bedded sediments. A comprehensive review of glacial regolith is contained in Kauranne *et al.* (1992).

Periglacial deposits consist predominantly of angular boulders and pebbles with no signs of bedding. They form by ice fracturing of the bedrock and its movement down slope by solifluction, gelifluction or as block streams. In modern periglacial environments aeolian transport of sediments is very common and it is the fine-grained material from periglacial environments that form much of the loess blankets in North America and Asia.

AEOLIAN DEPOSITS

Aeolian deposits are also very common in the regolith, but often go unrecognized. Most are familiar with the common aeolian sands of desert terrains, but these often make up surprisingly little of the regolith. Although they form prominent landforms they constitute a relatively small component of wind-blown regolith. Dust, clays and silts are much greater component of regolith in both arid and humid landscapes.

Aeolian sediments can be transported over large distances. In New Zealand and the south-western Pacific significant aeolian accession occurs as a result of dust transport from Australia (Glasby 1971) (Figure 12.11). Records from the Tasman Sea (Hesse 1994) show dust accumulation forms a significant component of the marine sediments. Similarly, McTainsh *et al.* (1997) record dust being moved and depositing over >500 km in West Africa. Similar records of long-distance transport of dust (Middleton 1997) are common, and therefore it can form a significant part of regolith materials over wide areas a long way from the source of the material (Figure 12.12). In sandy deserts, sand is not transported over such huge distances, but does appear to travel hundreds of kilometres from its original source (Pell & Chivas 1995); Wasson (1983) also finds no evidence for long-distance transport of sand in the dune systems. Windom (1970) has shown that dust originating in Australia adds significantly to the trace element content of New Zealand snowfields. Of particular importance are Ca, Mg, Fe, Mn, Cu, Ni, Co, Cr and Sr. He also found similar elevated trace element values in sediments of the South Pacific Ocean. In a study of the North Atlantic Ocean Chester & Johnson (1971) found elevated trace element values in marine sediments including Fe_2O_3, Mn, Cu, Ni, Co, Cr, V, Ba and Sr. These they concluded were mainly derived from the deserts of North Africa with minor contributions from the UK and Europe.

These data from New Zealand and the Atlantic clearly illustrate that dust accumulating on landscapes has the potential to alter the regolith geochemically, and perhaps mineralogically, compared with that which may be derived locally. A good example of this is the high proportion of quartz silt that occurs in basalt-derived regolith on the Monaro of southern New South Wales and the 70% exotic quartz accumulating in soils of the south-eastern highlands of Australia (Chartres *et al.* 1988).

In Australia most sandy aeolian regolith occurs as longitudinal dunes which occur as an anticlockwise gyre around the continent (Figure 12.13). These dune-fields cover something like 40% of Australia's surface and represent 38% of the world's aeolian landscapes. Most of the dunes have a simple longitudinal form and can extend for up to 300 km with a spacing of about 200 m and heights up to 35 m. They occasionally bifurcate, particularly in the Simpson Desert. The dunes are separated by clay pans that originate from clays or clay pellets blown in with the sand. The dunes are complex, comprising dunes of various ages from Holocene to 274 ka (Nanson *et al.* 1995). They are generally red coloured, but where dunes move across watercourses they emerge white, reddening again downwind.

Pell *et al.* (1999), studying the Great Sandy Desert in the south of the continent, have shown sand can be subdivided into three groups based on their physical and chemical characteristics. The three are western, central and eastern groups with the changes between them corresponding to major changes in the underlying Tertiary geology. They consequently surmise this is because the sands are locally derived with very little aeolian transport. The sands of the western

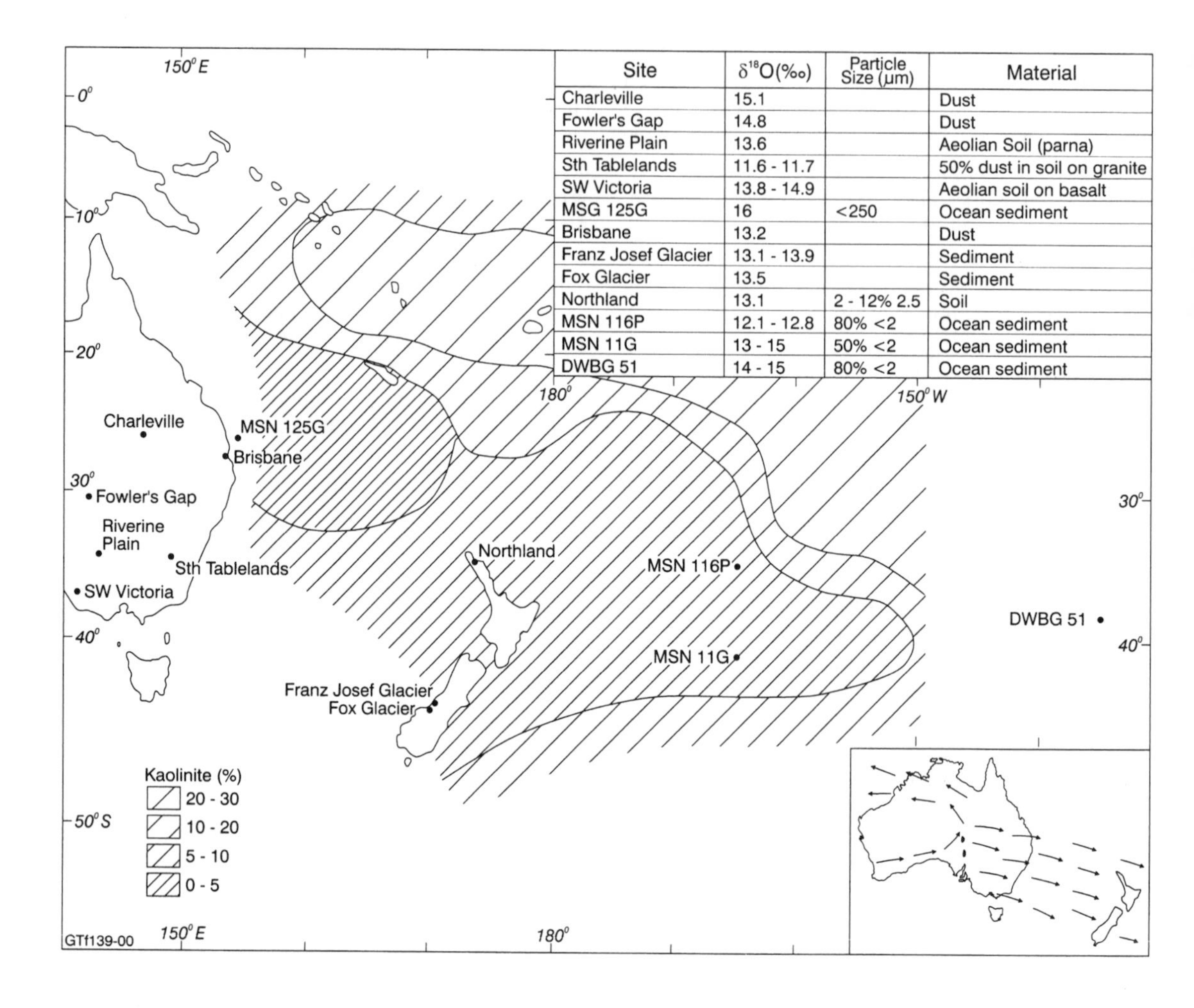

Site	$\delta^{18}O$(‰)	Particle Size (μm)	Material
Charleville	15.1		Dust
Fowler's Gap	14.8		Dust
Riverine Plain	13.6		Aeolian Soil (parna)
Sth Tablelands	11.6 - 11.7		50% dust in soil on granite
SW Victoria	13.8 - 14.9		Aeolian soil on basalt
MSG 125G	16	<250	Ocean sediment
Brisbane	13.2		Dust
Franz Josef Glacier	13.1 - 13.9		Sediment
Fox Glacier	13.5		Sediment
Northland	13.1	2 - 12% 2.5	Soil
MSN 116P	12.1 - 12.8	80% <2	Ocean sediment
MSN 11G	13 - 15	50% <2	Ocean sediment
DWBG 51	14 - 15	80% <2	Ocean sediment

Figure 12.11 *$\delta^{18}O$ characteristics of dust soils and ocean in the Australasian region sediments (from Kiefert & McTainsh 1996)*

group derive directly from the Officer Basin, which in turn derived its sediment mainly from the Yilgarn Craton, Albany–Fraser Orogen and Musgrave Complex. The central group sands are immediately also from the Officer Basin, but ultimately from the Musgrave Complex to the north. The eastern group derive from the Arckaringa Basin that derives originally from the Gawler Craton, Curnamona Block and the Adelaide Geosyncline. The sediments in the Tertiary basins were transported several hundred kilometres from their original sources, but post-Tertiary transport and reworking have been minimal.

The other major sandy deserts of the world are smaller than those of Australia. They may differ in having larger dunes and different dune forms, e.g. parabolic, transverse and irregular, but in other aspects they are similar. In many deserts a coating of haematite and clay minerals reddens the sand. Walden *et al.* (1996) attribute this, based on studies in the Namib, primarily to the nature of the source materials, but Australian evidence suggests it occurs in the dunes during the time they become rereddened after 'cleaning' in a watercourse or lake. Perhaps it is due to grain ferriargillans developing from fine clay-rich pellets being included with the sand.

Source-bordering dunes of various types are common aeolian regolith deposits in many places. There are two common types, those bordering marine and alluvial deposits where sand is predominantly deflated from the beaches or bars respectively and deposited downwind close to the source. Common examples of these occur on windward coasts (Figure

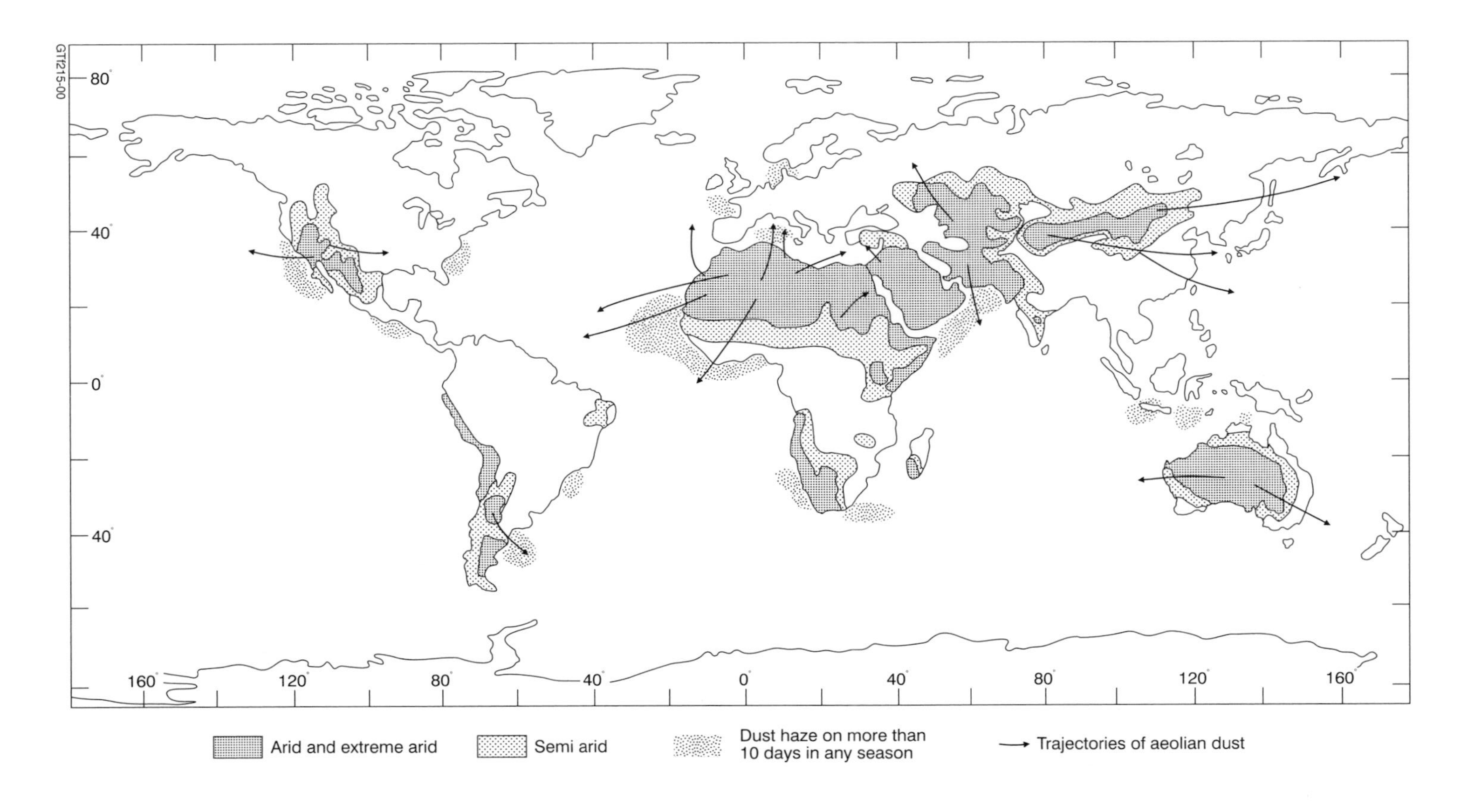

Figure 12.12 *Global distribution of major dust storm regions with the main seasonal dust trajectories (from Middleton 1997)*

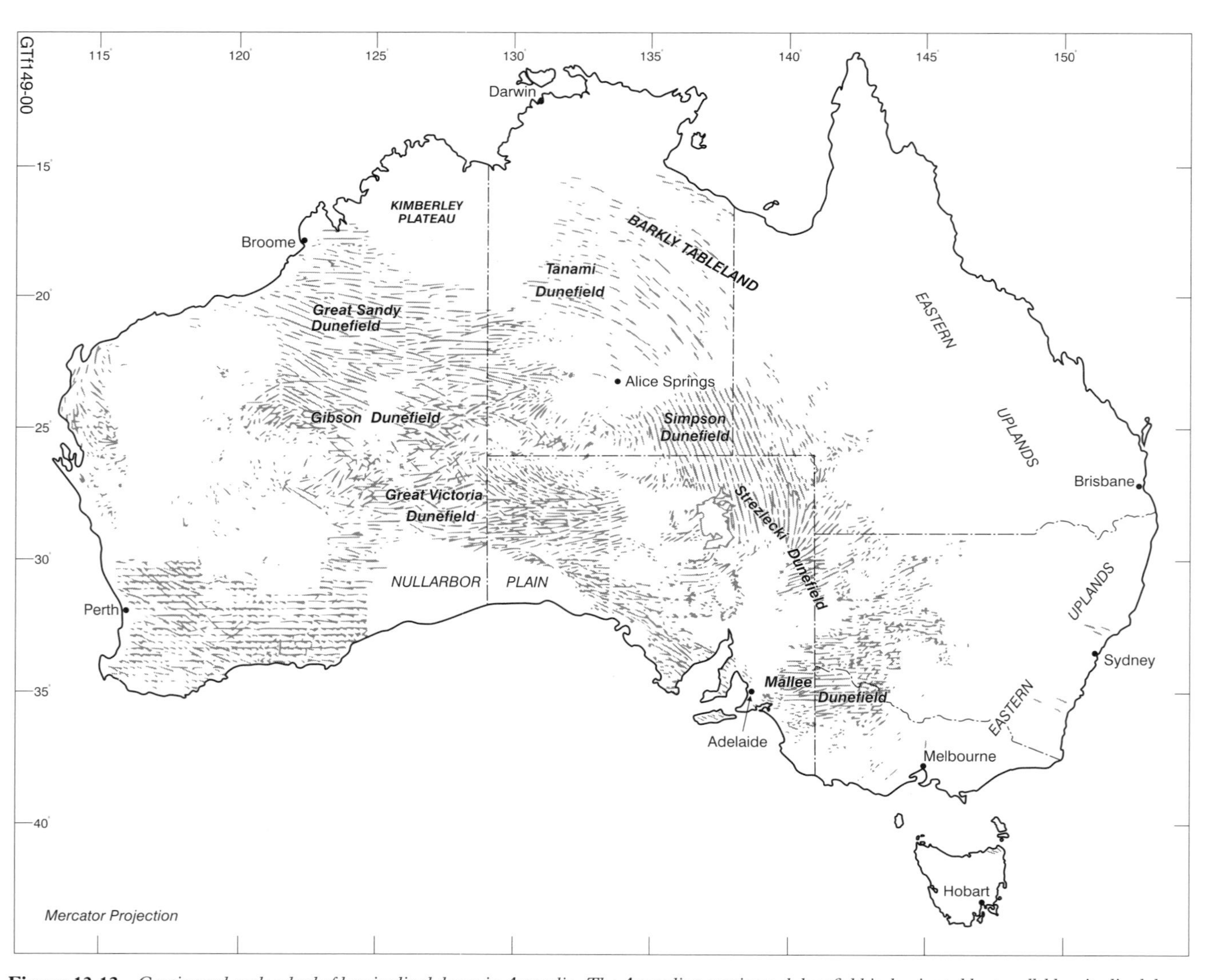

Figure 12.13 *Continental-scale whorl of longitudinal dunes in Australia. The Australian continental dunefield is dominated by parallel longitudinal dunes and consists of seven individual but interconnected deserts (from Cook 1997)*

Figure 12.14 *Looking west along the coast of the Great Australian Bight. Easterly drifting sand climbs cliffs about 50 m high and forms an inland migrating source bordering sand-sheet (photo Ken McQueen)*

12.14) and downwind of sandy rivers and lakes forming extensive sheet and dune-fields adjacent to the sand source.

Another major aeolian feature are lunettes that occur widely in Australia. They are crescent-shaped dunes on the downwind side of lakes or palaeolakes. They typically consist of a mixture of materials, each of which is derived from the lake margin or floor during different lake levels. Bowler (1990) summarized these changes and their effects on the nature of the lunettes (Figure 12.15). Many studies of lunettes in Australia and elsewhere have been useful in reconstructing palaeoenvironmental and palaeoclimatic histories (e.g. Chen *et al.* 1991; Magee *et al.* 1991; Croke *et al.* 1996; Pillans 1987).

Loess

Loess is an aeolian deposit consisting of mainly quartz, feldspar, mica and clay minerals with varying proportions of carbonates. Its average grain size is about 0.02–0.03 mm and it forms a significant part of the regolith over about 10% of the Earth's land area. The world's most extensive and thickest loess deposits are in China where it deposited between 22 and 18 ka at rates up to 260 mm/ka (Pye 1987). Loess is also widespread in central USA and in Europe. In the USA, like China, most loess is considered to be post-Wisconsinian (Birkeland 1984). Loess sequences commonly contain many palaeosols and develop thick surface soils.

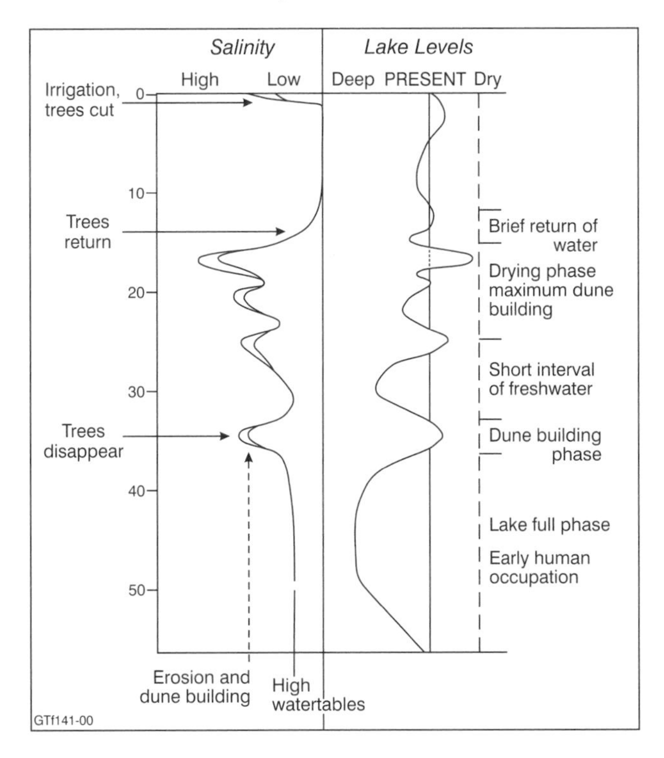

Figure 12.15 *Generalized record of lake level oscillations in south-eastern Australia over the last 50 ka (from Bowler 1990)*

Parna

In Australia there is no loess, or at least no materials known as loess, but aeolian materials are very widespread over much of the continent. Parna is a word coined by Butler & Hutton (1956) to describe a silty clay carbonate-rich mantle deposited uniformly across much of the Riverine Plain in south-eastern Australia. Since then the material has been described from a large number of other localities across the continent. In many ways it is similar to loess, but its source is not from glacial materials but rather dry lake beds and alluvial plains. From these, clay pellets and silt were blown to accumulate downwind as sheets across large areas or, where deposition rates were lower, incorporated into existing regolith (e.g. Walker *et al.* 1988). At several localities three layers of parna separated by palaeosols have been described (Churchward 1961; Beattie 1970, 1972) and more recently Chen (1997) has mapped large areas of parna and documented its accession in other regolith over wide areas around Wagga Wagga in central New South Wales.

Parna is red, commonly contains carbonates and is kaolin and quartz dominated. It has or imparts a characteristic red coloration, probably due to haematite, in well-drained positions, but is yellow in less well-drained circumstances due to goethite providing the common pigment. The source of these minerals is probably the weathering of clay minerals deflated from arid zone lakes. The lake sediments may include kaolinite, smectite, sepiolite and palygorskite, as well as gypsum and calcite or dolomite. These minerals, deposited as parna in areas with a very different regolith environment from an ephemeral saline lake, will undergo weathering and contribute new materials to their depositional site (see calcrete section, Chapter 13).

TERRESTRIAL COASTAL DEPOSITS

Marine-derived sediments are frequently blown inland or depositional relics of former marine processes occur well inland from contemporary marine influences. Perhaps one of the best-known examples occurs on the north coast of Papua New Guinea where fringing reefs front marine terraces now occurring up to several hundred metres above sea level (Chappell 1974) (Figure 2.40). Chappell (1974) describes a series of terraces, each containing a fringing reef, preserved on a hill slope. The oldest of the terraces is about 320 ka. Similar features occur on many tropical coasts in tectonically active areas.

Probably more importantly from a regolith perspective are the dune sequences built up during the Quaternary around many coastal fringes. Along much of the Australian coast sand dunes have blown inland from coastal regions, covering earlier regolith materials. Typical examples occur along sections of the Western Australian coast, the southern coast along the Great Australian Bight (Figure 12.14) and along parts of the northern Australian coast. Lees *et al.* (1992, 1993, 1995) showed transgressive dunes across northern Australia date from 171 ka, including at Cape York (11 ka), the east Kimberley coast (3 ka) and Cape Arnhem (19 ka).

In south-western Africa the Namib sand sea derives much of its sand from sands delivered to the coast by rivers, the most important of which is the Orange River (Rogers 1977).

Coastal barrier systems are important along many coasts, including the east coast of the USA, around the 'low countries' of Europe and along various parts of the Australian coast, particularly in Queensland and Western Australia where these sequences have built up during Quaternary eustatic changes to form extensive sand chenier plains. Many host extensive heavy mineral deposits (see 'Placer deposits' later in this chapter).

Sircombe (1999) has shown that the heavy minerals concentrated in coastal deposits of eastern Australia are primarily derived from a multiplicity of sources. These include older distal Precambrian rocks, an exotic Pacific Gondwanan source and locally from the Palaeozoic rocks of the adjacent highlands. This means these minerals have been recycled through many weathering and erosion cycles yet they have persisted to accumulate as significant economic deposits on the contemporary continental margin.

In regions where substantial marine regressions have occurred over the last few million years beach ridge complexes may be preserved, similar to those preserved around shrinking lakes (Abell 1985). Such a situation has occurred in the Murray Basin of south-eastern Australia where beach ridges extend inland some 400 km. The oldest (furthest from the sea) date back to the Miocene and become progressively young towards the coast (Figure 2.43). The inland ridges are predominantly quartzose sands and fine gravels which contain significant heavy mineral deposits while those closer to the coast are more calcareous. This decrease in carbonate with increase in age is probably due to leaching.

Some marginal marine sediments, particularly those associated with estuaries and mangrove areas, although only just qualifying as regolith are worthy of mention here as they do occur well inland, buried by later alluvial covers on coastal plains. The estuarine deposits are often very reduced and commonly contain significant amounts of pyrite. After draining, oxidizing solutions or air enters them causing the pyrite to oxidize forming goethite, jarosite $\{KFe_3(SO_4)_2(OH)_6\}$ and sulfuric acid (Fanning *et al.* 1993). The acid aggressively attacks many minerals including clays. This releases significant Al^{3+} that is very toxic. In many ways the effects are very similar to acid mine drainage problems experienced in sulfide and coal mines.

Relict coastal margin deposits are also important features of the landscapes and regolith in many places. In the Murray and Eucla basins of southern Australia significant sequences of ancient beach ridges remain part of the extant landscapes. They date back to the Neogene and yet still form significant depositional landscape features (Figure 2.43).

Volcanic regolith

In many parts of the world volcanic debris and lavas make very important components of the regolith. In islands like Iceland or Hawaii the regolith is entirely derived one way or another from volcanic materials. Obviously as lavas weather they form regolith, its nature depending to a large degree on the composition of the lavas. As it is possible to date lavas and other volcanic debris, either from historical records or isotopically, these landscapes provide excellent places to study weathering and to determine the rate at which it occurs (e.g. Ruxton 1968).

During explosive eruptions tephra composed of ash lapilli, bombs and other volcanic debris are spread over large areas. In Iceland most landscapes are overall accretionary because of the continual addition of tephra and lava to the surface. This is not to say erosion does not remove much of it, but tephra blankets much of the island from frequent eruptions of their many volcanoes. In addition, because of the Icelanders' remarkable historical records dating well back into the first millennium, they have a well-documented record of eruptions and distribution of fallout from them. This is well illustrated in Figure 12.16 where 14 tephra blankets dating back to 1104 are documented. Similar features are common in New Zealand, but the long historical records are lacking there. Again the sequence of tephra blankets in a landscape provides the potential to date events and landscape evolution from the tephra sequence, using what is today called tephrochronology.

Placer deposits

Placer deposits are concentrations of minerals more dense (>3.0 g/cm^3) and more resistant to chemical and physical destruction than the sediment normally moved in a transport/depositional system. The minerals so concentrated are usually referred to as 'heavy minerals'. Heavy minerals commonly concentrated include ilmenite and rutile, monazite, zircon, Au, diamond, sapphire and ruby, cassiterite, magnetite, chromite and other spinels, garnets and many others.

Placers accumulate in positions where sedimentary processes remove the bulk lower-density minerals (quartz, feldspars, micas and rock fragments) leaving concentrations of the heavy minerals. Reworking is one way this may occur and is most commonly seen in littoral deposits such as along beaches. Eddy currents in zones of flow-separation in fluvial currents are also favourable sites. This includes behind rock bars in channels, and among gravel deposits where quartz sand is readily moved on, but higher-density sand- or silt-sized grains like Au are stable.

Several well-documented examples of fluvial placer accumulations include the Main Leader Reef in the Precambrian of the Wiswatersrand, South Africa (Figure 12.17) in which Au placers occur, and at Ballarat in Victoria (Figure 12.18), again where Au has accumulated as placers. Fluvial placers of Sn occur off the southern Malay Peninsula where, although many are now below sea level, they are fluvial deposits (Gupta 1987) formed during low stands of the sea (Figure 12.19). Diamond placers occur also in Sierra Leone in Quaternary alluvial sediments of the headwaters of the Bafi–Sewa river system (Thomas 1994). There are excellent examples of beach placers along the east and west coasts of Australia and along the south-western coast of Africa. In southern Australia as the sea regressed during the Cainozoic it left behind a series of beach ridges (Figure 2.43) along the inland margin of the Eucla Basin and across 400 km of the Murray Basin.

Transported versus *in situ* regolith materials

Differentiating *in situ* from transported regolith is most easily accomplished in the field using readily observable physical characteristics, but there are a number of other ways as some of the discussion above indicated. Below we discuss a variety of methods to differentiate *in situ* from transported materials.

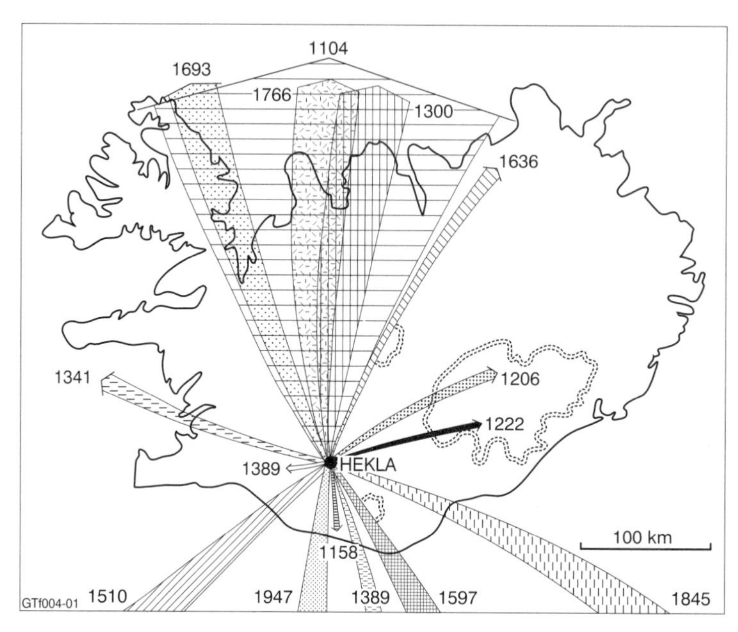

Figure 12.16 *Direction and extent of the initial phase of each of 14 historical eruptions of Hekla. The width of the arrows indicates the relative volume of tephra fall from each eruption (from Thorarinsson 1967)*

PHYSICAL PROPERTIES

The presence of rock fabrics (textures) in the saprolite and C-horizon clearly indicates the bulk of the material is *in situ*. Care needs to be exercised in identifying the extent of the rock fabric in the field, as gravel- to boulder-sized alluvial and colluvial pebbles may weather over extended periods producing local patches of saprolite.

The presence of bedding and other sedimentary features in any part of the soil profile clearly indicates a transported origin for the parent material, unless these features are preserved in saprolite developed on a sequence of sedimentary rocks.

Many soil profiles contain unconformities where an older soil material has been partially eroded, new parent material added to the surface and a new soil formed on the surface. The newer soil may merge with the older or it may be separated from the erosion surface by newly laid unpedogenized regolith. In the latter case it is easy to recognize the possibility of having transported regolith with a soil over the older regolith, but the former may be very hard to recognize in the field, core or cuttings. The presence of a stone line formed as a lag on the erosional surface is one way such an unconformity may be recognized. Figure 12.20 illustrates this.

Silt-sized products do not commonly form from *in situ* weathering of most rock types. The most common source for silt-sized materials is from aeolian materials. The origin of high levels of silt in soils is therefore due to the addition of wind-blown sediment to the profile. In Australia particularly, this is often also accompanied by the addition of clay-sized material, because of the prevalence of clay pellets in aeolian sediments derived from the arid and semi-arid continental interior. These clays will often be of a different composition from the locally produced ones, but they are generally indistinguishable in the field (core or cuttings).

Another easy way to physically differentiate *in situ* and transported parent materials is by the composition of gravels or coarse fractions of the soil materials. If the gravel corresponds in composition to the local bedrock then the regolith (and soil) are probably local, but if they are different then the regolith is probably, at least in part, transported.

The shape of the mineral grains in the soil compared to the parent material may also indicate whether the grains, despite weathering, are locally derived or foreign (transported).

MINERALOGICAL AND CHEMICAL PROPERTIES

Because during soil formation (including weathering, deposition and erosion, and biotic activity) the

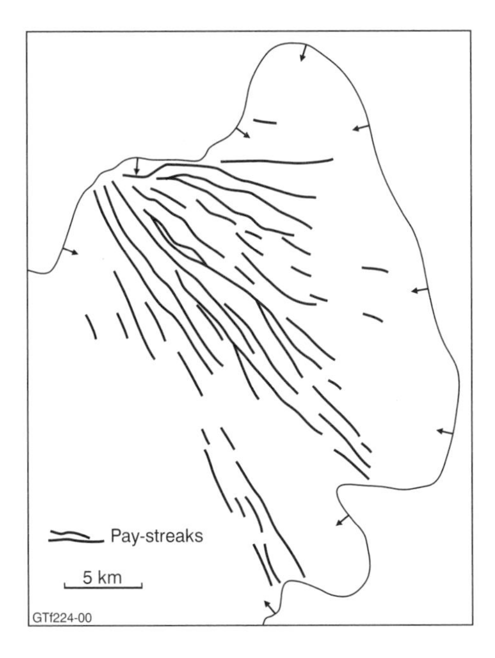

Figure 12.17 *Distribution of 'pay-streaks' (placer deposits or channel deposits) in the Main Leader Reef, Witwatersrand, South Africa. The arrows indicate the dip of the sub-crop or the depositional basin basement (after Du Toit 1954)*

original parent material is destroyed, a number of assumptions are commonly used to estimate the nature of the soil parent and whether it is *in situ* or transported or mixed.

Assumption 1: That the material identified as the parent has been unaffected by weathering.

In the case of soils formed on igneous or metamorphic rocks it is possible to be confident of this by petrographic examination, but for sedimentary parents, whether rocks or unconsolidated sediments, this assumption is obviously wrong. We need to be sure of the actual nature of these sedimentary parents to understand mineralogical and chemical changes associated with pedogenesis.

Typically, soils developed on granites are enriched in Fe_2O_3, Al_2O_3 and SiO_2 relative to their assumed parent material and depleted in alkalis (CaO, MgO, K_2O, Na_2O). If these data are normalized to a stable component under weathering (TiO_2 in some circumstances, ZrO_2 in others), the Fe, Al and Si-oxides generally remain near constant through the profile and are similar to the parent rock, but appear to be enriched because the alkalis are strongly leached. Even this approach depends on another assumption; that there has been no substantial volume change between the parent rock and the soil (see below).

Assumption 2: That the parent material is uniform in composition within and below the soil.

As we have seen above there are no such guarantees. The very processes of soil formation homogenize the regolith so any variations tend to become obscured. Checking will show if the mineral composition follows expected trends given normal weathering and soil-forming processes. In a granitic terrain one would normally expect clay minerals to follow a normal weathering trend from the fresh rock to the surface, i.e.

Surface	kaolinite > illite
	kaolinite = illite, smectite
Saprock	kaolinite < illite, smectite

or some similar trend. The presence of smectite or vermiculite in the upper horizons would suggest additions of clays from another source, perhaps eroding lower parts of weathering profiles further upslope or from aeolian accretion. This shows that although the soil is apparently formed in a granite regolith there is a chance that there has been addition and that the parent material for the soil was not uniform.

Similarly, in a region of basic or ultrabasic rocks the presence of quartz in the soil would suggest the additions of transported material to the profile, as quartz is uncommon in these rocks. Again, although the underlying rocks may be apparently uniform in composition, the parent material for the soil is clearly not compositionally uniform (see also Figure 9.19 and the associated text).

The composition of the heavy mineral suite in the soil as compared to the parent regolith/rock may also prove useful. Many of the heavy minerals tend to survive weathering, and thus are useful to demonstrate whether the soils were formed on a uniform or *in situ* parent as opposed to a mixed parent material. The shape of the zircon grains is one commonly used criterion, as is the trace element ratio of Zr to some other element contained in zircon (e.g. Y) (Figure 12.21).

The ratios of stable isotopes can also be used to assess whether the soils relate to local or transported sources.

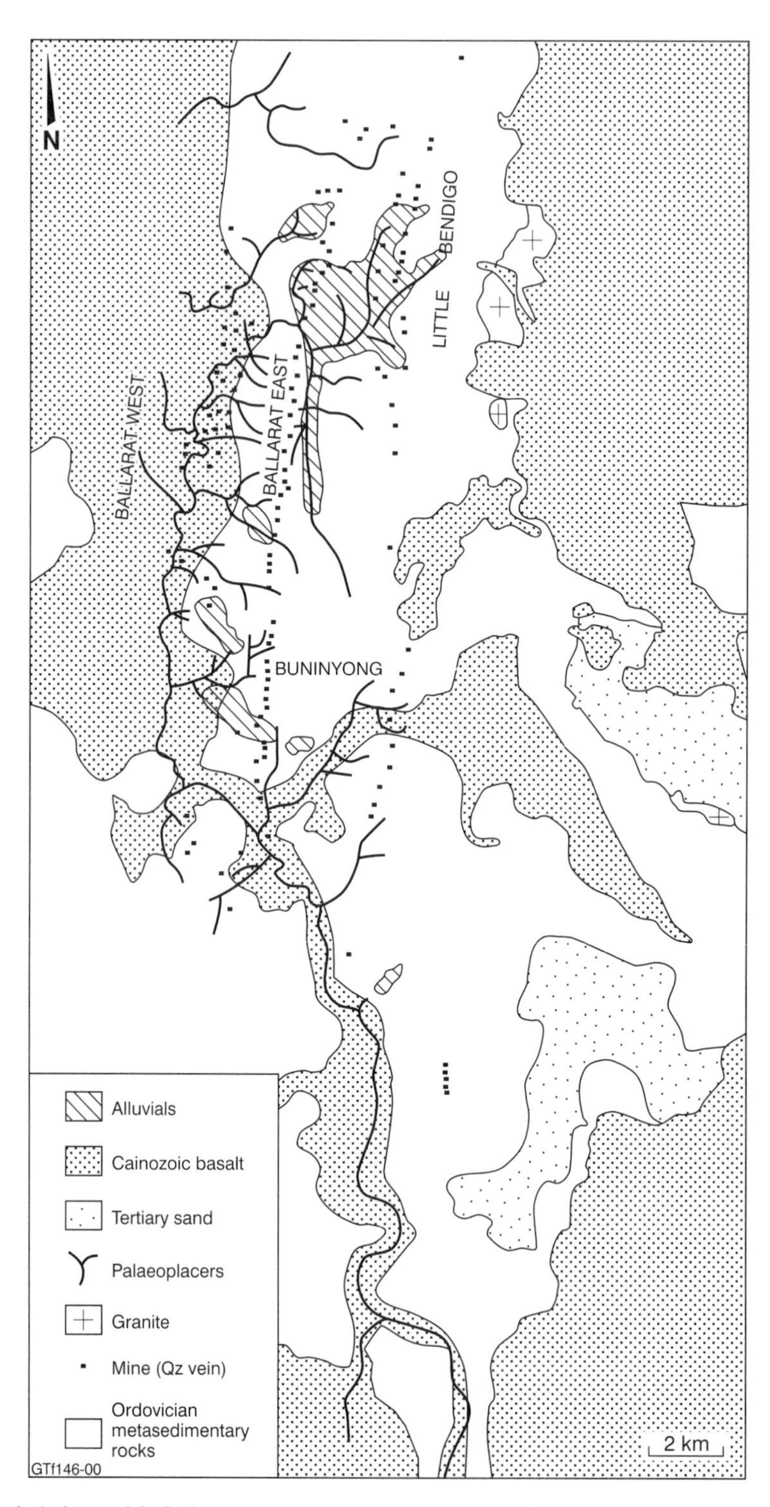

Figure 12.18 *Geological map of the Ballarat area showing the close spatial association between lode Au and placer deposits both below basalt and in modern deposits (from Phillips & Hughes 1996)*

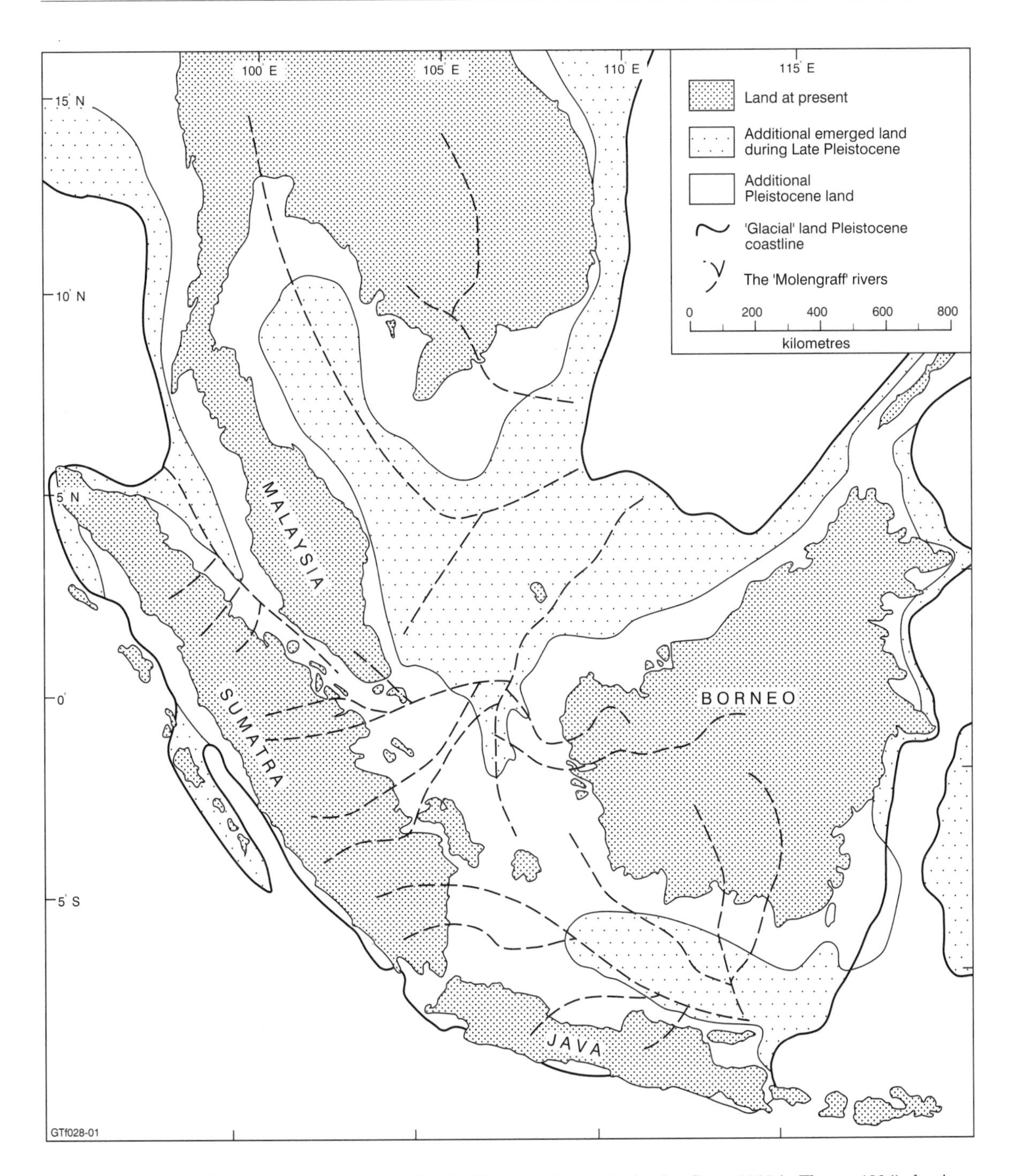

Figure 12.19 *Sundaland, south-eastern Asia, during the Pleistocene low sea level (after Gupta 1987 in Thomas 1994) showing extensions of the present drainage, which host the alluvial Sn placer deposits*

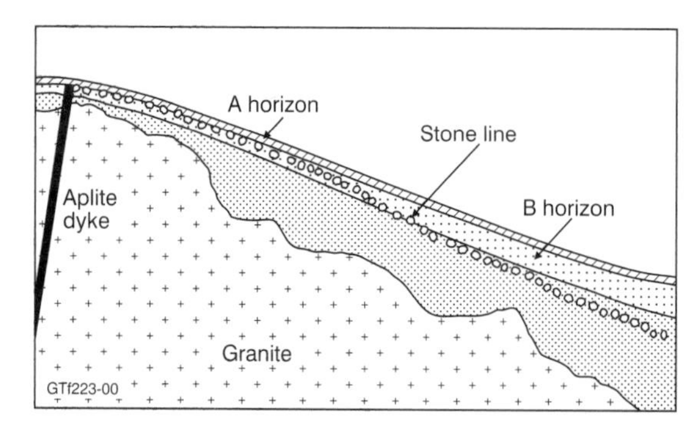

Figure 12.20 *Hillcrest situation on a granite substrate with an aplite dyke cropping out to produce a stone line marking an unconformity in an apparently single soil profile. Note how the stone line cuts across the B-horizon and passes into the C-horizon/saprolite downslope. This clearly indicates that this soil profile (a red podzolic/chromosol) is composed predominantly of* in situ *material at the top of the slope, a mixture of* in situ *and transported material in mid-slope areas and entirely of transported material at the base of the slope*

Assumption 3: That there are components of the parent material that remain unaffected by weathering and soil formation.

Much of what is discussed above makes this assumption. As we have seen, this is useful to assess whether soils have an absolute or relative accumulation or depletion of certain components, and whether the soil is formed on *in situ* or transported regolith. The question is – are there any components in soils which are unaffected by weathering? Many argue that zircon is unaffected; however, in very alkaline environments there is evidence that it is mobile, and evidence for zircon solution is reported from some silcretes. In neutral to alkaline weathering environments TiO_2 is thought to be relatively immobile, and in the weathering of basalts in dry cool climates this is so; however, in more intense leaching environments this may not follow. The intimate association of SiO_2 and TiO_2 in silcretes, and the significant absence of other minerals, is presently an enigma not well explained by our current understanding of their solubilities.

The use of ratios of components eliminates this assumption as far as determining whether a soil is composed of *in situ* or transported material goes, but it does not take into account the possibility of differential element mobility in different parts of the profile. Different chemical environments in different parts of profiles are common and are most easily demonstrated by highly acid A-horizons overlying alkaline B-horizons, perhaps over neutral or acid C-horizons.

Micro-scale physical transport

Physical transport of regolith components through the regolith is common. Material is physically moved by a number of processes related to water movement and bioturbation and physical churning and collapse. Many of the observations and processes related to the physical movement of materials in regolith come from the pedologists studying the micromorphology of soils. Brewer (1964) in his book on the petrology of soils provided a technique for the microscopic description of soils and his techniques can readily be extended to the petrology of regolith. Nahon (1991) draws heavily on Brewer and extends his work into the realm of regolith successfully. Brewer's techniques have also been widely used in the area of palaeopedology as is well illustrated by Retallack (1990, 1997).

WATER TRANSPORT

Water permeating through a regolith will almost inevitably carry fine particles with it. In most regolith the finest materials are clay minerals and Fe-oxides. This is clearly demonstrated from soil studies in the development of clay-rich B-horizons. Brewer (1968) showed the proportion of clay washed into a B-horizon (Figure 12.22) as 'illuvial clay'. Illuvial accumulations are common in regolith to great depths (50 m+) where clays are washed down through profiles eventually depositing on void and grain surfaces as oriented clay mineral layers called clay cutans or argillans. The Fe-oxides also wash downward through regolith to form similar coatings, ferruginous cutans or ferrans. Physical transport by water will only occur vertically or laterally downwards in a regolith profile.

TRANSPORT BY COLLAPSE

Collapse in regolith essentially results from three mechanisms: the collapse of highly weathered regolith where the remaining mineral grains are no longer able to support the fabric of the original material; by the development in cracks in the regolith as it dries and some surface material (or the walls of the cracks) collapse into the void; and by the collapse of material into bio-voids. By this process material from higher up in the regolith is moved down through it causing mixing in a vertically downward sense.

TRANSPORT BY PHYSICAL CHURNING

Churning is a process whereby material can be moved not only downwards in a profile but also up through it.

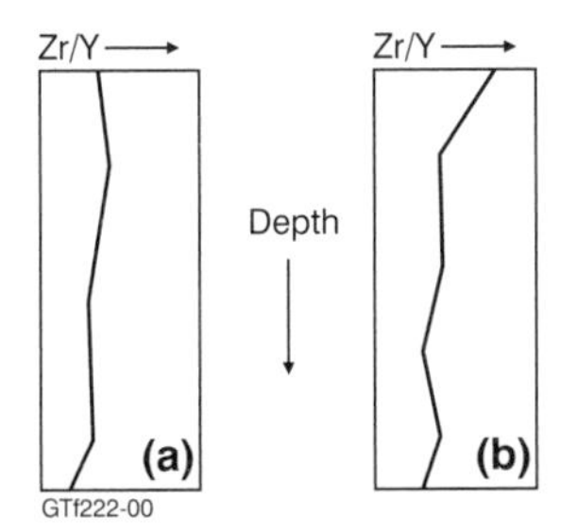

Figure 12.21 *Zr/Y ratios plotted against depth below show in the case of (a) that the parent materials are probably uniform and thus also* in situ, *but (b) shows a marked change in the ratio near the surface suggesting non-uniformity and perhaps a transported addition to the profile*

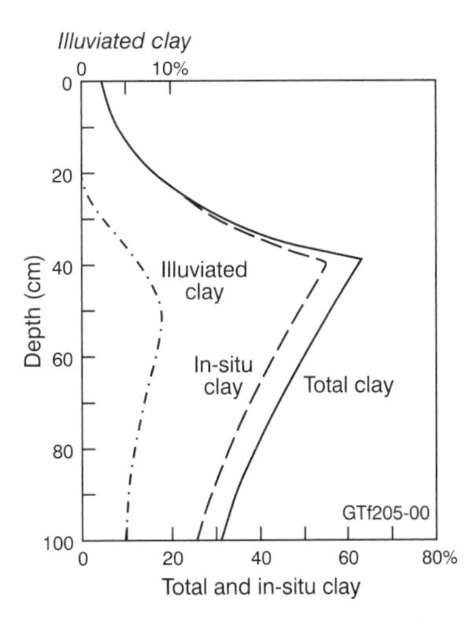

Figure 12.22 *Relationship among total clay, illuviated clay (recognized by oriented clay films or argillans), and by difference,* in situ *clay. Because only some clay distribution is accounted for by illuviation, the clay in place represents that formed by weathering and/or that originally in the soil parent material (from Brewer 1968)*

While this process is not common, except where organisms are involved, gilgai formation is one relatively common example as is the formation of such things as pingos in periglacial environments.

Gilgai (an Australian Aboriginal word for 'small waterhole') is common in Australia and occurs elsewhere but less commonly. It is manifest by a series of depressions and ridges at the surface in various patterns (small or large circular depressions and surrounding ridges or linear ridges and depressions) with a relief up to 3 m, but more usually of about 1 m. Gilgai is formed by the shrinking and swelling of a clayey subsoil and its movement to or near the surface. Figure 12.23 illustrates several examples.

Pingos form by the penetration of unfrozen but saturated sands to the surface through a layer of permafrozen ground. The effects are again to bring lower layers of regolith to the surface.

BIOTURBATION

Organisms that live in the regolith move large volumes of material around within it. In other parts of the book (Chapters 9 and 11) we have discussed their role in some detail. Here we just give a quick summary of some of the biological processes that cause physical transport in the regolith.

The major biological agents physically transporting materials within the regolith are:

- *earthworms* that continually move through the soil creating burrows, ingesting and casting mineral and organic matter (at rates up to 260 kg/km^2 per year in soils of the Nile Valley, Edwards & Lofty 1977);
- *termites* physically move regolith materials mainly from the lower parts of the regolith to the surface, but subterranean species move it around below the surface. Rates of movement vary depending on the species, but rates of surface accumulation from termites have been estimated at 0.0125 mm/yr (Williams 1968), 0.10 mm/yr (Lee & Wood 1971) from the Northern Territory and 0.025 mm/yr (Nye 1955) and 0.20 mm/yr (de Ploey 1964) from Africa;
- *ants* move materials of selected sizes around within the regolith and many species move material from depth to the surface;
- *other invertebrates* also move regolith materials as they burrow during feeding and in building habitation;
- *trees* when they fall cause massive disturbance of the regolith surface to depths up to 1 m when their root bowls lift large masses of soil (Figure 9.11);
- *tree roots* move regolith laterally as they grow, and on death, decompose leaving often deep (20–30 m) open channels as the organic matter decomposes, and these allow for physical transport downward in the regolith; and,
- *vertebrate animals* (gophers, wombats) move very large volumes of soils and regolith in making their living burrows and in hunting for food.

During all of the bioturbation large volumes of organic matter are introduced to the regolith as it is physically moved about.

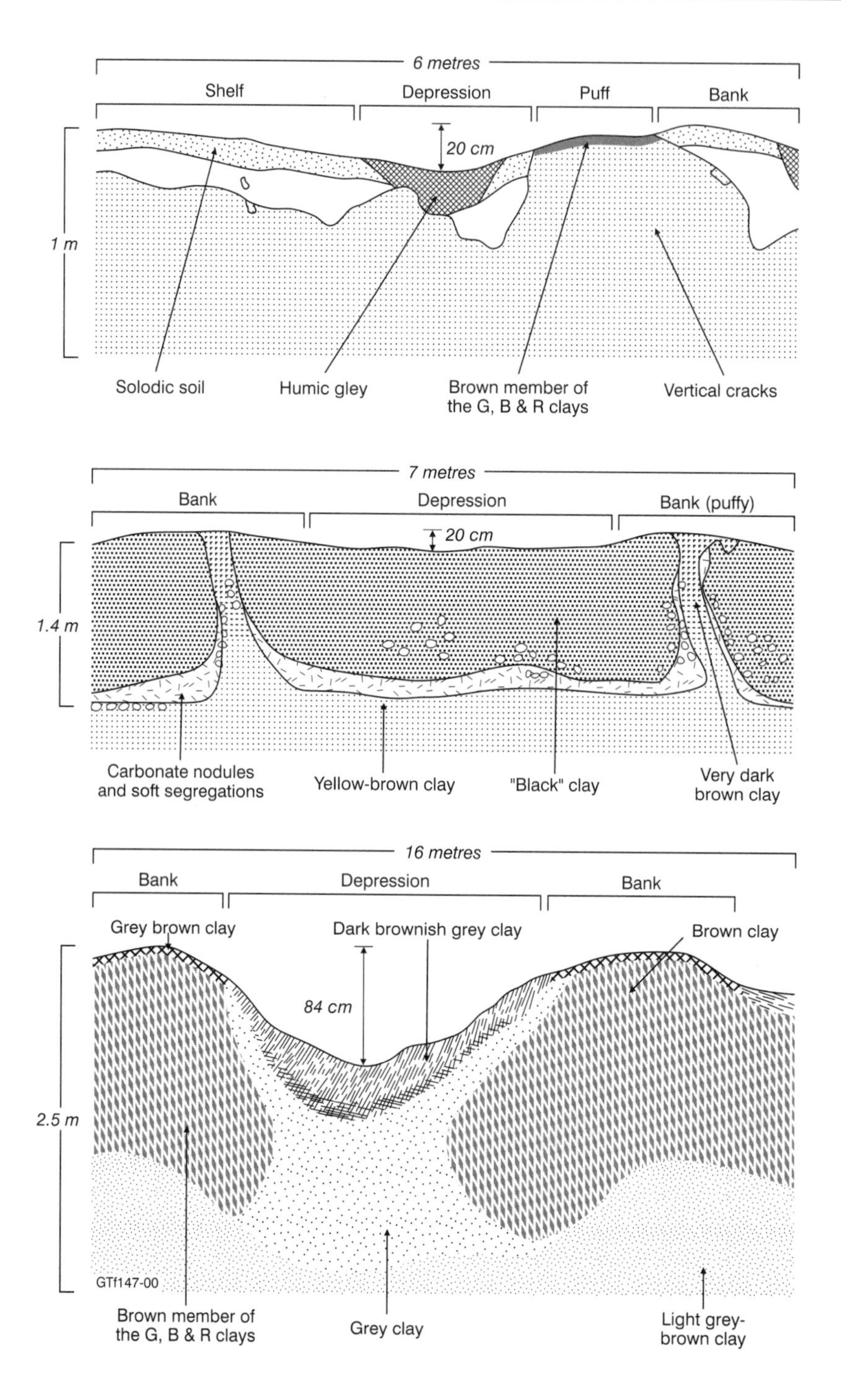

Figure 12.23 *Schematic sections through various types of gilgai. (a) crabhole gilgai from Launceston, Tasmania; (b) linear gilgai from the Darling Downs in southern Queensland; (c) melonhole gilgai from Chinchilla, southern Queensland (from Hubble* et al. *1983)*

13 Solute transport and precipitation in the regolith

Introduction

Water is obviously essential in moving the soluble weathering products around in the regolith and ultimately to the ocean or inland drainage termini. The gross movements are controlled essentially by the groundwater system discussed briefly in Chapter 5, but the system is much more complex than this.

From the chapters on weathering it is clear that water and its dissolved ions, polymers, organic ligands and perhaps even very fine grain-sized solids are responsible for the alteration of minerals and the liberation of soluble components. What happens to them from the point of liberation is critical to understanding the evolution of the regolith and its character.

To encompass all this in a short space it is necessary to think of the mobility of solutes in terms of scale. Movements can occur at various scales in the regolith. Very short distances of movement (nano-scale, nm), short distances (micro-scale, μm–mm), medium distances (meso-scale, cm–m × 10^2) and long distances (mega-scale, $\geq 10^3$ m).

Although solutes are created during weathering many do not leave the regolith, or at least not for a very long time. If we think of a river sediment transport system, sediment enters the channel, and is transported during floods. Some sediment forms in-channel bars, floodplain deposits, and other depositional entities, features that will remain in the river system until floods large enough to erode them occur. At this time they re-enter the transport part of the system. Some floodplain or channel deposits were never eroded, hence we have geological records of fluvial deposits. So too with regolith solutes, if solutes were all removed from the regolith we would have no record of chemical precipitates in present or fossil regolith materials.

Weathering-limited and transport-limited landscapes

Carson & Kirkby (1972) introduced the concept of weathering- and transport-limited denudation regimes. Weathering-limited regimes occur where transport processes remove solutes (and physical detritus) produced by weathering from slopes faster than new ones are produced. Weathering is selective: the more easily weathered minerals are selectively weathered, the less susceptible minerals are generally removed by physical processes. Solute concentrations coming from weathering-limited slopes are high because of extensive reactions between the water and rocks, the element ratios of the water being different from those in the rocks; Na and Ca are enriched relative to K and Mg. In such regimes the regolith is thin and bedrock is not isolated from incoming rain. Transport-limited regimes occur where weathered products are removed more slowly than, or at the same rate as, weathering occurs. In these situations the regolith tends to be thick and the weathering rate is controlled by the rate at which weathered materials are removed chemically or physically. In these regimes the water element ratios are about the same as in the rock. The ratio of SiO_2 to (Mg + Ca + Ma + K) is high because the regolith is dominated by Fe, Al and Si in the form of kaolinite, gibbsite and goethite or haematite. These concepts are well illustrated in the Amazon catchment (Drever 1988). Figure 13.1 shows the data of Gibbs (1967) for solute concentrations in the Amazon catchment. The waters can be divided into four groups:

- <200 μeq/l – these drain silicate rocks under transport-limited regimes. Cation ratios are similar to the rocks, pH is acid and relative concentrations of Fe, SiO_2 and Al are high

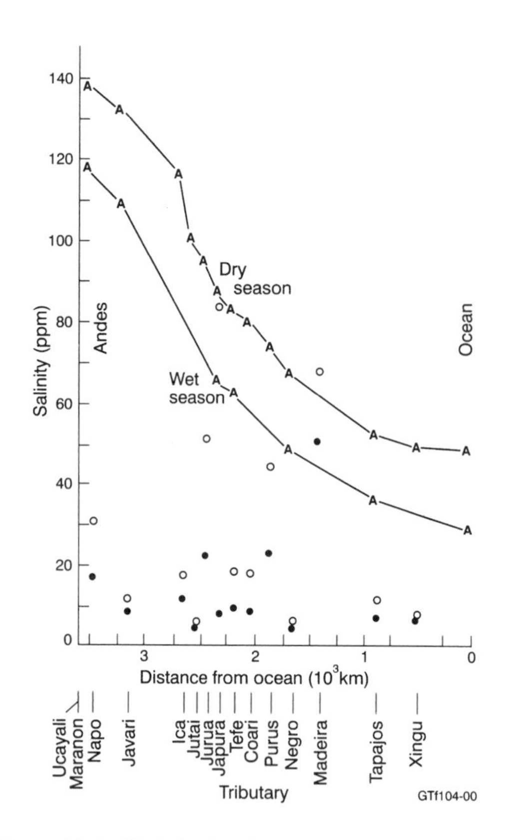

Figure 13.1 *Variation in salinity of the Amazon River and its tributaries with distance from the ocean. 'A' indicates concentration of the main stream, open circles indicate concentrations in the tributaries during the dry season, solid circles indicate concentrations in the tributaries during the wet (from Gibbs 1967)*

- 200–450 μeq/l – drain silicate rocks under weathering-limited regimes. Compared to the rocks, Na is concentrated compared to K, Ca and Mg
- 450–3000 μeq/l – drain marine sediments or red beds with carbonates and minor evaporites. These waters are alkaline, Ca and Mg are relatively high and SO_4^{2-} is high from weathering of pyrite in the marine sediments and gypsum beds
- 3000 μeq/l – occur in association with marine evaporites

The water chemistry of the Amazon reflects both bedrock type and erosion regimes within the catchment.

Solutes in regolith systems

The regolith water system carries solutes derived from the atmosphere, biosphere and the regolith. This

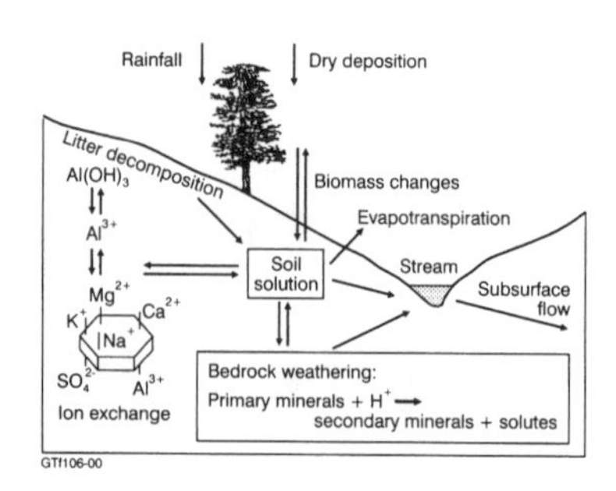

Figure 13.2 *Schematic sketch of some processes affecting solute fluxes in a catchment (from Drever & Clow 1995)*

water system is illustrated in Figure 13.2. Dry deposition consists of particles (e.g. dust, ash) and gases from the atmosphere (e.g. SO_2, NO_x).

In a recent study Drever & Clow (1995) attempt to separate the solute composition of water falling from vegetation and that falling directly to the ground. They quote figures from a study in the Vosges Mountains in France to illustrate the composition of rainwater from deforested areas and that falling through a spruce forest cover (Table 13.1). These data clearly show that there is a significant amount of solutes contributed from trees compared with those from rain falling on open country. They also show the types of fluxes of elements introduced to the regolith from precipitation, although a comprehensive list was not analysed. It should also be noted that occult deposition is very dependent on the total biomass and the types of trees. Conifers are more efficient in trapping solutes than deciduous trees, but less so than grasses.

Plants grow in the regolith and absorb nutrients and trace elements from it. Schnoor & Stumm (1985) show the process is approximately:

$$\left.\begin{array}{l} 800CO_2 \\ 6NH^{4+} \\ 4Ca^{2+} \\ 1Mg^{2+} \\ 2K^{+} \\ 1Al(OH)^{2+} \\ 1Fe^{2+} \\ 2NO^{3-} \\ 1H_2PO_4^{4-} \\ 1SO_4^{2-} \\ H_2O \end{array}\right\} \Rightarrow \text{photosynthesis} \Rightarrow \text{biomass} + 16H^{+} + 804O_2$$

Although this is part of the solute cycle in the regolith, in the long term it is neither a net sink nor source for

Table 13.1 *Mean concentration of rainfall and rain falling through the forest canopy from the Vosges Mountains, eastern France (from Probst* et al. *1990)*

	NH_4	Na	K	Mg	Ca	H^+	Cl	NO_3	SO_4
Concentration (μeq/L)									
Bulk precipitation	19.1	10.0	2.8	4.5	11.9	33.9	12.5	24.1	41.5
Sub-canopy	36.9	46.4	52.7	17.8	65.5	114.8	63.4	78.3	185.0
Flux (mol/ha per year)									
Bulk precipitation	270	142	39	32	84	480	177	340	290
Sub-canopy	385	484	550	93	342	1197	661	817	966
Difference	115	342	511	61	258	717	484	477	676
Occult deposition[a]	115	342	102	31	206	1282	484	477	676

[a] Occult deposition is the composition of the water that falls through the canopy corrected for elements translocated through the trees.

elements as plants grow and die, so their elements are continually recycled through the regolith. In the shorter term, however, imbalance can and does occur due to catastrophe, fire, human influence and so on. Also as many trees in particular are deep rooted, they extract elements deep within the regolith and shed their death products at the surface, so they act as an internal element cycling mechanism within the regolith. Additionally as part of their growth process trees consume CO_2, an important component in weathering solutions, but yield H^+ and O_2, also both important in this cycle.

These waters are added to the regolith with their solutes and then are able to react with the regolith materials and underlying rock. Some components of the solution are used in the reactions, others in precipitates and others move on through the regolith. The gross balance of all these reactions can be measured in modern catchments by measuring the outflow from a catchment. Such a balance only determines what is occurring at a macro-scale. There are a plethora of new-formed minerals in the regolith which have varying residence times and these will not be recorded in the geochemical balance calculations at a catchment scale.

Regolith water been recorded as containing most elements, implying that most elements, given the right environment, are soluble, at least to some extent.

SOLUTE TRANSPORT AT THE NANO-SCALE IN THE REGOLITH

As minerals weather they release elements to solution. This may occur at an atomic scale or as polymers. It is questionable whether the weathering occurs by solution or by simple movement of ions across a semi-coherent interface from the primary to the new-formed mineral (Hochella & Banfield 1995). It is only atoms which are not required in the new-formed minerals that leave the weathering site and it is only those which truly go into solution. A typical example is the formation of smectite following the weathering of amphiboles, where a poorly crystalline or amorphous intermediate phase forms at the interface (see Figure 8.6).

At this nano-scale the solution conditions may not be anything like those of the fluids in the larger weathering environment. Concentration gradients from the weathering front to the solutions in the larger system may be such that viscosities are so high at the front that diffusion of elements from the front is slowed. Some elements never make it more than a few nanometres away from the front before they are fixed as a new-formed mineral. Eggleton (1984) has shown that during the weathering of olivine, Fe-oxides form pillars in the narrow channels formed during weathering and little Fe^{2+} leaves the nano-environment. Solutions must, however, play some role as these channels are commonly filled with smectite as well, and the Al necessary for this to form must come externally as olivine contains none. Titanium becomes fixed as anatase after only a few nanometres of movement during the weathering of titanite (see Chapter 8).

MICRO-SCALE WATER IN THE REGOLITH

In the plasmic microsystem, the water with its solutes exists in pores and as sheaths around grains or voids, held in place by surface tension in unsaturated regolith and filling voids in saturated material. In the latter case solutions are mobile if the regolith is permeable, and it acts to transport solutes away from the weathering face of minerals either down gradient or to plant roots where plants, by exuding acids, are able to take up nutrient and trace elements.

Precipitates (new-formed minerals) that occur by these processes are commonly observed in SEM micrographs, for example as clusters of clay minerals on the surfaces of etched feldspars (or other weathering minerals) (see Figures 8.5–8.9). Other mineral clusters also precipitate close to the origin of their components – Fe-oxides, Al-oxyhydroxides, quartz and alunite are common examples (see Figures 8.5–8.9). These precipitates, called by some 'armouring precipitates', occur as three types according to Velbel (1996).

Porous coatings allow rapid diffusion of weathering products through fluids in the micropores and thus do not limit weathering rates. Clay coatings on feldspars, particularly from the vadose zone, are typically cracked and lifting from the parent grain and they do not form a tightly adhering surface coating.

Non-porous polycrystalline coatings consist of non-porous pseudomorphic clay-oxide materials. These types of coatings are rare in highly weathered regolith. Non-porous monocrystalline coatings occur as single-crystal pseudomorphs of the primary mineral (Velbel 1996) or the epitaxial overgrowth of primary minerals by new-formed minerals. The most common example of the latter is quartz. These coatings on grains of new-formed minerals are commonly thought to slow the reaction of primary minerals during weathering (Berner 1978, 1981), but Velbel (1996) has shown that this is not the case. Weathering products diffuse through these coatings more rapidly than weathering rates are observed to occur in the laboratory or in the field.

Another common phenomenon at this scale of solute mobility is the pseudomorphic replacement of minerals. Feldspars are commonly replaced in all their detail by kaolin. This is particularly well observed in basic dykes intruding the granites of the Yilgarn near Perth at Jarradale (Figure 13.3), where the labradorite in the dykes has been totally replaced by gibbsite and in so doing it has preserved all the structural features of the original feldspar. A similar example comes from the sediments of the Late Neogene Lake Bunyan (Taylor & Walker 1986a, b) where basaltic ash deposits have been perfectly pseudomorphed. Plagioclase has been replaced by kaolin, preserving all the twinning and cleavage, olivine crystals are replaced by Fe-oxides and clays, again preserving fractures typical of olivine and pyroxenes are similarly replaced with twinning and cleavage preserved.

Velbel (1993) reports most pseudomorphic replacements of aluminosilicates formed by intense weathering under oxidizing conditions are composed of 'box-works' of clay minerals. Similarly, he reports that ferruginous pseudomorphs are composed of rigid frameworks of secondary products. Despite this, these pseudomorphs preserve structural elements of the original mineral. This microporosity allows weathering to continue as the new-formed minerals occur as porous masses through which element diffusion is rapid.

Figure 13.3 *Replacement of feldspars by gibbsite in a dyke at Jarradale, south of Perth, Western Australia*

Many of the earliest stages of weathering are observed in the field at this scale of solute transport. In granites the earliest stage of weathering commonly observed is the oxidation of Fe^{2+} to goethite around biotite grains and within the rock flanking fractures. This is the result of Fe^{2+} being released from biotite and encountering water with a positive Eh soon after release (see Figure 9.2).

The development of clots or 'blasts' of kaolin in the upper parts of saprolite of deep weathering profiles (Robertson *et al.* 1998) is another example of solute mobility at this micro-scale. The mechanism suggested for this phenomenon is that kaolin begins to grow as small aggregates in the saprolite and as they grow they gradually overprint the saprolite fabric and exclude other minerals, such as quartz, that are not consumed. This process results in the progressive textural differentiation in saprolite as clays and quartz sand and silt are segregated (see Figure 9.4).

MESO-SCALE SOLUTE MOVEMENT IN THE REGOLITH

Meso-scale mobility (centimetres to some hundreds of metres) of solutes is common in regolith materials.

Features that fall within this category are the seasonal vertical movement of water-tables, the movement of solutes in the capillary fringe of the water-table, the movement of water through the vadose zone by unsaturated flow, and lateral flow of groundwater in small catchments. Many of the effects of these have been described in Chapter 5.

The vertical movement of the water-table most commonly leads to the formation of Fe-oxide accumulations either as bands or mottles. It also extends the zone of maximum weathering over the zone of oscillation of the water-table. Transfers of solutes then can move over time with the water-table, so any solutes are potentially precipitated within the capillary fringe. Examples of this, other than Fe-oxides, are carbonates, silica (opal or quartz), clay minerals, calcite and gypsum.

Vertical movement of water through the vadose zone transports many solutes gained from the atmosphere, pedosphere (soil layer) and from the upper parts of the weathering profile. These waters transport weathering solutes from soil minerals and organic compounds. These move vertically through the regolith to the water-table, changing on their way, dissolving more solutes, and leaving others as precipitates. Kaolinite, gibbsite, carbonates and sulfates are commonly precipitated in the regolith from solutes, at least in part from the soils, in the lower regolith. Organic compounds are efficient in transporting trace metals from the upper regolith to the lower zones where Fe-oxides and clay minerals may fix them.

At this scale solute transport through the regolith is not generally uniform. Most fluids move through macropores (joints, fractures, root tubules, insect galleries) and do not penetrate the whole regolith. As a consequence solutions may not be gathering solutes from all the regolith nor transferring materials into all of it. The majority of solutes are dissolved from around the macropores and precipitates or eluvial material is deposited in or near them. If precipitation or deposition occurs preferentially along these fluid pathways then large parts of the regolith may be excluded from interaction with fluids for long periods of time. This is one possible reason that solute exports from catchments are much lower than would be predicted by laboratory experiments.

In North Carolina Vepraskas *et al.* (1991) tested the hydraulic conductivity of deep saprolite developed on schist using undisturbed cores. Using dyed water, not only could conductivity be measured, but the water paths could be mapped. They found that the saturated hydraulic conductivity ranged from 0.01 to 0.71 cm/ha with a mean value of 0.74 cm/ha. The cores contained 1.9% by volume of channels varying in diameter from 0.1 to 0.5 mm and these pores contributed 93% of the saturated hydraulic conductivity. In unsaturated flow conditions at a soil suction of −10 cm these same pores contributed 50% of the flow. When suction was ≥30 cm the channels conducted no water and the unsaturated hydraulic conductivity was 0.02 cm/ha (Figure 13.4). The implications of this study clearly show that the bulk of the regolith has little water pass through it. Integrated over time, significant water may move through the bulk of it, but much more slowly than it does through the whole mass of regolith where it is routed preferentially through larger voids or macropores.

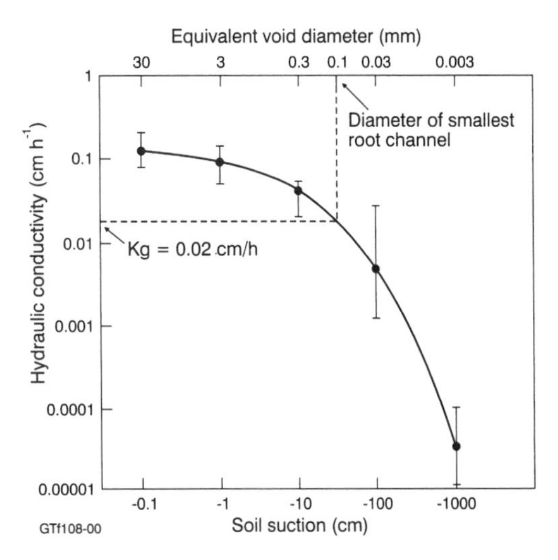

Figure 13.4 *Means and ranges of unsaturated hydraulic conductivity of saprolite formed on schist for different soil suctions. Equivalent pore diameters are shown. The maximum diameter of soil channels (from roots) is 0.1 mm.* K_g *is the limit of saturated hydraulic conductivity (from Vepraskas* et al. *1991)*

This water also transports solids, which provided flow is slow enough, can be dissolved en route from the zone of soil moisture to the water-table. It is also interesting to note that the transport of these solids (mainly fine-grained oxides and clay minerals) coat the main solution pathways from the surface to the water-table forming cutans. These may develop into quite thick coatings on the voids, perhaps up to 1 or 2 mm thick. In the case of clay particles they deposit with their *z*-axes normal to the void surface forming smooth and well-aligned clay coatings (argillans, Brewer 1964). The argillans may serve to confine the water to the conduits and slow or prevent diffusion of solutes from the mass of the regolith to the infiltrating water.

This is important as elements residual in the regolith through which the water passes will not be readily transported from their place of residence. This, in turn, has important implications for sampling regolith materials for geochemical purposes.

Vepraskas *et al.* (1991) found this to be the case in the deep saprolite of North Carolina. Here the voids created along foliation and shears in the saprolite developed on schists were plugged by Fe and Mn-oxides, and as a consequence their effect on hydraulic conductivity was negligible.

Many small catchments are of this scale and some have been monitored for solute exports. As discussed above, the elements that move at this scale are dependent on many factors, but in general there is an excess of silica, alkalis, alkaline earths and trace elements exiting catchments through the soil water, groundwater and stream systems. What is important from a regolith perspective is what happens to the solutes moving through it. Many elements pass out of the regolith system readily, except in arid and internal drainage situations. These include such elements as Cl^-, Na^+, much of the K^+, and Ca^{2+}. Stallard (1985) had shown this to be the case in the Orinoco Basin and Douglas (1968) in north Queensland and eastern New South Wales. Gunn (1985), on the other hand, working in eastern New South Wales, has suggested that in areas of deeply weathered regolith many of the solutes accumulated from rainfall and from weathering are retained, at least for extended periods, in these profiles and that this is reflected in groundwater compositions.

At the meso-scale Billett *et al.* (1996) have shown stream-water solute discharge is directly dependent on weathering. They showed in small catchments in Scotland and Norway that the Na : Ca : Mg ratios change in stream solutions in response to changes in the geochemistry of the underlying bedrock. This demonstrates that in areas of high rainfall solutes transported may change rapidly downflow in response to weathering release of elements. Whether the same can be demonstrated for drier environments is questionable. It can, however, be shown that groundwater in small catchments in other environments does change similarly.

Gunn (1985) showed that in the southern highlands of New South Wales groundwater chemistry remained the same regardless of lithology or seasonal rainfall variations – NaCl dominated groundwater in weathered granite, acid volcanic rocks and Ordovician turbidites and concentrations were highest on the granite and acid volcanic rocks. Domination by NaCl in Australian catchments and groundwater systems is common across much of the southern half of the continent. Most of the salt has accumulated from both aerosols and rock weathering and it is stored in the groundwater in highly weathered regolith. Douglas (1968) showed that in eastern Australia local geology was more influential in controlling river water chemistry than climatic difference between north and south, but precipitation does significantly influence chemistry of low-discharge river systems.

MEGA-SCALE (>10^3 m) SOLUTE TRANSPORT

Transport at this scale is mainly through groundwater and stream flow. Many examples of long-distance transport are documented, particularly those at continental scale rates of chemical erosion (Garrels & MacKenzie 1971; Tardy 1992). Thomas (1994) shows the solutes transported from tropical rivers to the ocean are highly variable, but in most cases the solute load is less than the solid loads carried (Table 13.2). This is a reflection of the higher chemical weathering rates producing deep and relatively easily eroded regolith in tropical climates compared with more temperate climes. Solid loads in the Zaire are low because the river has virtually no high mountain catchment and it is a large river basin capable of storing large volumes of sediment in its catchment.

Some of Thomas's (1994) data point to the importance of geomorphology in considering solute loads. Stallard (1985) shows that in the Amazon and Orinoco basins of South America the chemistry of the water can be clearly related to the geology of the catchments. In weathering-limited catchments denudation rates are lithologically dependent and significant removal of cations is observed. In transport-limited catchments erosion rates depend on tectonics (lowering rate) and lithological susceptibility to

Table 13.2 *Solute/solid load ratios for selected tropical rivers (from Thomas 1994)*

River	Load (t × 10^6/yr)		Ratio suspended/ solute
	Suspended	Solute	
Amazon	900	290	3.1 : 1
Zaire	43	47	0.9 : 1
Ganges/Brahmaputra	1670	151	11 : 1
Niger	40	10	4 : 1
Zambezi	48	15.4	3.1 : 1
Orinoco	210	50	4.2 : 1
Orange	17	12	1.4 : 1

weathering is a minor factor. The composition of the rivers' loads can be explained by these (weathering- and transport-limited) models. Thus neither biomass nor soils can be either aggrading or degrading or this would not be the case.

Although in absolute terms tropical rivers have significant solute and suspended sediment yields, many in cold climates also yield significant solute loads. In Iceland Stefansson *et al.* (2001) reports a suspended to solute load of 1.2 : 1 from the island's largest river. This shows a higher proportion of solute load than the tropical rivers in Table 13.2. Millot *et al.* (1999) report that the solute loads of the Mackenzie River in the North-West Territories of Canada are about half those of the Amazon River and that solute loads in its tributaries are five times lower than those of the Amazon, suggesting that temperature is an important determinant of solute load. Why then are the values from Iceland so extraordinarily high? Perhaps because they are draining basalt, and in particular young basaltic regions containing a lot of glass.

By contrast with tropical rivers the Murray–Darling River system in south-eastern Australia exports only some 3.7×10^6 t/yr of salts and virtually no solid detritus to the Southern Ocean. It has over one-third of its catchment as highlands and two-thirds as lowlands in which any sediment derived from the highland tract are stored, but salts accumulated by weathering and from cyclic salt accession are flushed from the basin.

Another example of long-distance solute transport was inferred by Stephens (1971) and earlier by Dury (1968) on the basis of the distribution of silcrete in central Australia and ferricrete in the moister continental margins. This they attributed to the solution of Si during the formation of the ferricretes, and its lateral transport, accumulation and desiccation in the central areas of the continent, to form silcrete. More recent studies of landscape evolution suggest that both silcretes and ferricretes have rather local sources, but these earlier hypotheses highlight the importance of solute mobility to duricrust formation.

Duricrusts

Duricrust is a term coined by Woolnough (1930) to define

> A product of terrestrial processes within the zone of weathering in which either iron, and aluminium sesquioxides (in the case of ferricretes or bauxites) or silica (in the case of silcrete) or calcium carbonate (in the case of calcrete) or other compounds in the case of magnesicrete or the like have dominantly accumulated in and/or replaced a pre-existing soil, rock, or weathered material, to give a substance which may ultimately develop into an indurated mass.

The formation of all duricrusts is the result of the transport of solutes, whether by the addition of precipitates or the removal of elements to leave well-cemented crusts atop or within deeply weathered profiles. In this context it is useful to distinguish duricrust formed by *absolute accumulation* (the addition of an element to the cemented regolith) from *relative accumulation* (the concentration of cementing elements by the leaching of others). The difference is critical in the interpretation of a duricrust in its landscape context. If a duricrust forms by absolute accumulation then it follows that the cementing elements must have moved to the site of precipitation in solution, which must be down a hydraulic gradient, except within the capillary fringe. It follows then that the site of precipitation must be lower in the regolith or landscape than the source of the solutes. On the other hand, if a duricrust forms by absolute concentration it is the non-cementing elements that are removed to lower in the regolith of landscape sites, or out of the system all together. It is thus critical in interpreting duricrusts to know whether they are cementing *in situ* or transported materials as this will determine which cementing model applies (see below).

The American Geological Institute glossary (Jackson 1997) defines duricrust as

> A general term for a hard crust on the surface of, or layer in the upper horizons of, a soil in a semi-arid climate. It formed by the accumulation of soluble mineral products deposited by mineral-bearing waters that move upward by capillary action and evaporate during the dry season.

This definition is very different from that in common usage among most, if not all, workers in the field. They are not necessarily formed in semi-arid climates, not always formed by capillary water evaporation and are not always associated with the surface or a soil horizon.

Many books on duricrusts have been published, including Goudie & Pye (1983) and Goudie (1973) on various duricrusts, Langford-Smith (1978) on silcrete, Wright & Tucker (1991) on calcrete, McFarlane (1976), Bardossy & Aleva (1990), Tardy (1993) and Valeton (1972) on ferricretes, laterites and/or bauxites.

The nomenclature of duricrusts is difficult and fraught with danger for the uninitiated. The term 'duricrust' was coined by Woolnough (1930). The word essentially means a mass of hard material formed within the regolith by natural cementation. Specific names are given to various types of duricrust according to their cementing mineralogy.

In a recent glossary of regolith terms (Eggleton 2001) duricrust is defined as

> Regolith material indurated by a cement, or the cement only, occurring at or near the surface, or as a layer in the upper part of the regolith. The cement may be, e.g. siliceous (silcrete), ferruginous (ferricrete, lateritic duricrust), aluminous (alcrete), gypseous (gypcrete), manganiferous (manganocrete), calcareous (calcrete), dolomitic (dolocrete), salty (salcrete) or a combination of these.

Common forms of duricrust include:

- Ferricrete where the dominant cementing cation is Fe^{3+} (either as goethite or haematite)
- Bauxite (alcrete), Al_2O_3 as gibbsite, boehmite or other Al-rich minerals
- Silcrete (grey billy), SiO_2 as opal A, opal CT, microquartz, epitaxial quartz or chalcedony and varieties
- Calcrete (caliche), $CaCO_3$ as calcite, high Mg-calcite
- Dolocrete, $CaMg(CO_3)_2$ as dolomite
- Gypcrete, $CaSO_4.2H_2O$ as gypsum
- Others, e.g. manganocrete ($MnO_2.nH_2O$) as various forms of Mn-oxyhydroxides

There is some confusion in the literature over the use of the words 'laterite' and 'ferricrete' as different authors use these words to mean different things. Laterite was originally used by Buchanan (1807) to describe a softish regolith material cut in Kerala, India, to make building bricks, which Ollier & Rajaguru (1989) report does not harden on exposure but is used directly. They describe this material as typical vesicular mottled saprolite, hard enough to maintain shape while being dug from the ground with an axe. The original 'laterite' of Kerala occurs in a deeply weathered profile, but does not seem to be the hardened, or Fe-oxide cemented, cap to the profile, but that is the sense in which the word is used by innumerable scientists. In our view, and that of Thomas (1994), the term '-crete' should be used to avoid this confusion. In Chapter 10 we described 'lateritic profiles' as weathering profiles similar to those from Kerala. Ollier (1984) suggest the term for these should be a 'Walther profile' after those described by Walther (1916) from Western Australia to avoid the confused and emotional reactions the word 'laterite' often provokes.

Eggleton & Taylor (1999), regarding the use of the term 'laterite', proposed the following.

- Laterite is not a good word to use for a rock. Ferricrete is preferred as it has none of the historical baggage of 'laterite'
- Laterite profile can be used to refer to a weathering profile consisting of some or all of soil, ferruginous crust, mottled zone, pallid zone and bedrock
- Laterite is a word that probably can never be rooted out of the geological lexicon. We suggest its use be always informal or broadly descriptive, never defining

FERRICRETE AND BAUXITE

Ferricrete is the term we use to describe surface or near-surface masses of regolith cemented by Fe-oxides and oxyhydroxides. Ferricrete is a hard material, not to be confused with ferralitic soils or other 'soft' ferruginous regolith materials. Some authors use the term 'ferruginous regolith', but this does not distinguish between cemented and hard material and ferruginized regolith that is soft.

Many authors refer to Fe-oxide cemented duricrust as 'laterite', which they are clearly not if the material from Kerala is indicative of what laterite is. This is not to say that 'laterite' profiles are not capped by ferricrete; Walther defined them as having such hard ferruginized caps developed in both *in situ* collapsed saprolite and in transported debris overlying saprolite (Figure 13.5). For a lengthier discussion of these matters look at Eggleton & Taylor (1999).

Almost as many researchers as have worked on ferricrete and bauxite have offered classification. To review all here would be a huge job and is not our aim, rather we present some models of ferricrete formation for consideration.

In profiles that have been described as having formed *in situ* in seasonally tropical climates, Nahon & Tardy (1992) describe the ferricrete (cuirasse) as the result of relative accumulation. They suggest it forms from the mottled zone by the removal of kaolin and quartz, leaving nodules of ferric and alumina oxyhydroxides. Such ferricrete thus has a nodular structure. In the upper part of the ferricrete a cortex of Al-goethite

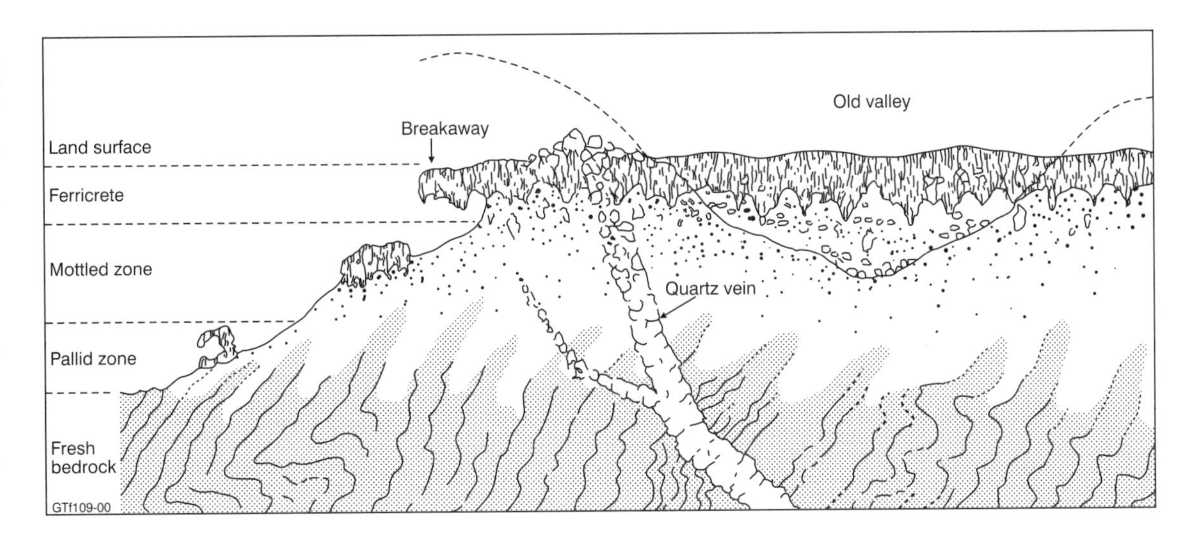

Figure 13.5 *A figure from Walther (1916) showing his concepts of deep weathering and 'laterites' from Western Australia*

develops on the nodules resulting in the formation of a pisolitic structure. This pisolitic zone is cemented, but with continued weathering the Al-goethite is leached of Al_2O_3, the pisolites consequently become smaller and separated to form a loose layer of free pisolites. Nahon & Tardy's experience is mainly in tropical West Africa. At Weipa on Cape York, Queensland, the situation is different, the ferruginized nodules forming a duricrust under loose aluminous pisoliths. Similar ferricrete has been described widely from the Yilgarn Craton in Western Australia by various authors, including Butt & Smith (1992), Anand (1995) and Smith (1989) who call these *in situ* ferricretes 'lateritic' duricrust. Anand (1995) also refers to other ferricrete types within the Yilgarn, almost all of which he attributes to reworking of the 'lateritic duricrust'. These are, when recemented genuine ferricrete, but formed by erosion of former ferricretes. Anand (1995) also includes ferricretes derived from lateral transport of Fe in his classification, but they are very minor components of the landscape, examples include 'bog iron ore' and ferruginized bedrock.

On the north-western shores of Cape York, Queensland, at Weipa sits one of the world's largest bauxite mines. Here mining concentrates on an upper 2–5 m of pisolitic bauxite which overlies an *in situ* deep weathering profile developed in the arkosic shallow marine Cretaceous Rolling Downs Group sediments and the more quartzose fluvial Palaeogene Bulimba Formation. The 'lateritic' bauxites at Weipa occur on a very extensive plateau extending up to 60 m above sea level. The plateau is flanked on the west by sea cliffs and to the east by an erosional escarpment produced as the plateau was warped and the Wenlock River incised it.

Mapping in the mine pits at Weipa clearly demonstrates the ferricrete below the pisolitic bauxite is continuous and it forms a wavy surface with a relief of up to $\approx$2 m. The ferricrete grades downwards into a sandy mottled zone that contains weakly ferruginized patches and mottles. In places where the weathering profile has been incised along the major rivers and the coast the overlying pisolitic material has been removed leaving the ferricrete as outcrop, sometimes atop small escarpments. The extent of the ferricrete is shown in Figure 13.6.

Anand (1998b) views 'laterites' and ferricretes as genetically different and classifies ferruginous duricrusts developed in greenstone terrains on the Yilgarn Craton as shown in Table 13.3. Tardy *et al.* (1995) published another and rather different classification of 'laterites' (Table 13.4). Not only did Tardy (1993) classify the 'laterite' family of regolith materials, but he went on to ascribe the climates under which they formed and the changes in climate necessary to form them.

For Tardy (1993) *ferricrete* is a nodular Fe-rich accumulation in which the Fe-nodules are indurated or cemented concentrations consisting mainly of haematite in the porosity between kaolin crystals. Tardy (1993) sees the nodules as residual accumulations from the washing of the pallid material of the

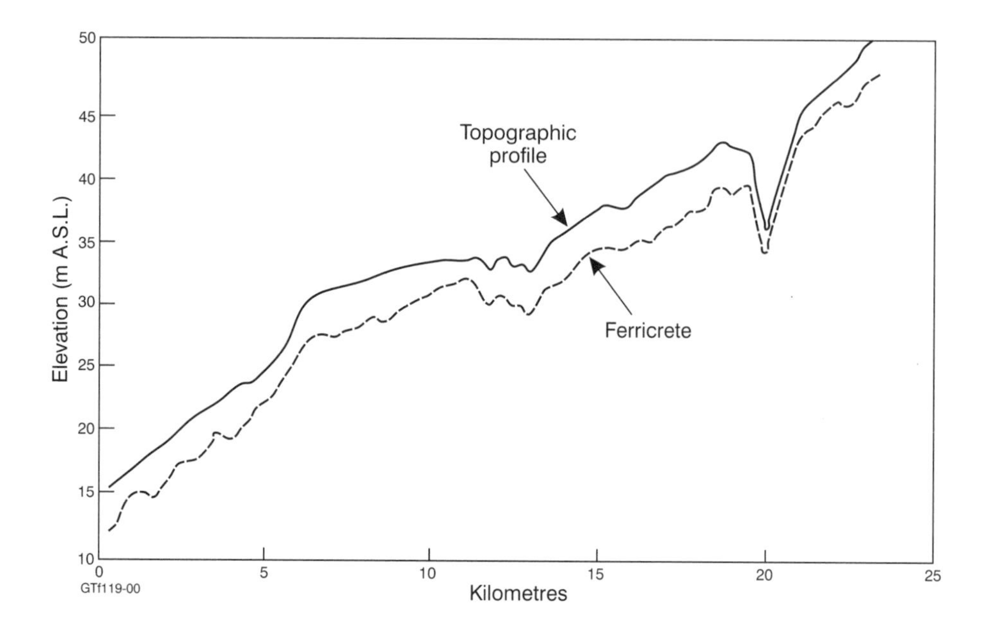

Figure 13.6 *Topography and thickness of bauxite (= depth to ferricrete) across the Weipa plateau from west (left) to east (from Eggleton & Taylor 1999). Data courtesy Comalco Ltd, processed by S. Laffan and A. McPherson*

mottled zone of a 'lateritic' profile. He refers to them forming in seasonal (contrasted) tropical climates.

Latosols form by the transformation of Fe-nodules into microglaebules by some of the kaolin being desilicified to gibbsite and some of the haematite hydrated to goethite under humid climates.

Conakrytes are massive Fe-accumulations that develop from non-aluminous rocks such as dunite. They form from any ultramafic rocks and are characterized by having a massive microcrystalline fabric with no nodules or pisolites. They are often referred to as nickel laterite.

Orthobauxite forms under rainfalls >1700 mm/yr and are typical of localities such as Surinam, Darling Ranges, Western Australia, and Latoka on the Ivory Coast. They are homogeneously red and have no nodules, pisolites or concretions and are composed of gibbsite, haematite and goethite.

Metabauxite is like the bauxite at Weipa, north Queensland. Metabauxites are typically less Fe-rich at the top but more so at the bottom. They are typically pisolitic and the pisoliths are more boehmite than gibbsite. Tardy (1993) suggests that these are formed by a change in climatic conditions from humid to arid, causing the change of orthobauxite into metabauxite.

Cryptobauxite is widespread in Amazonia. Tardy (1993) considers these bauxites to be formed from ancient ferricretes that have been dismantled under humid tropical climates. Biological agents such as trees and termites are thought to be important in their formation and in the mechanisms of SiO_2 redistribution.

The formation of ferricrete on deeply weathered profiles of this nature requires extraordinary amounts of chemical erosion or landscape lowering to achieve these concentrations of Fe-oxides. Tardy & Roquin (1992) suggest landscape lowering by chemical erosion to produce ferricrete of between 0.1 and 3.3 km/10^9 yr in Chad and Malagasy respectively. Trendall (1962) suggested that 4.25 m of granite was weathered to yield just 30 cm of ferricrete on the Yilgarn of Western Australia, meaning a landscape lowering of about 180 m to produce 10 m of 'laterite' (Thomas 1994). Nahon (1991) basically espouses this relative accumulation model for ferricrete formation and he suggests the life of 1 mm of fresh rocks be between 250 and 20 years (Table 13.5). Boulangé *et al.* (1997) quote landscape lowering rates from bauxite formation on granite in the Ivory Coast of 30 m to form 15 m of bauxite and up to a 70% volume reduction to form the pisolitic bauxite overlying the main bauxite. They report this taking from 3 to 5 Ma. From Brazil

Table 13.3 *Classification and description of ferruginous duricrusts related to greenstones on the Yilgarn Craton, Western Australia (from Anand 1998b)*

Duricrust	Description	Genesis
Lateritic residuum	The upper part of a Walther profile containing the following components in order from the top	Deep weathering and relative accumulation
Lag of nodules/nodular duricrust	Angular-platy ferruginous saprolite of haematite, goethite, kaolin, gibbsite and quartz. Syneresis cracks are common. Where cemented it is called nodular duricrust. Some pisolites occur	
Collapsed ferruginous saprolite	Large Fe-oxide cemented nodules of saprolite with irregularly distributed rock fabrics preserved in a plasma of yellow/brown kaolin, quartz and goethite. Nodules have cutans of goethite and kaolin. Some pisolites occur	Collapse of cemented (hard mottles) saprolite due to removal of silica and alumina with intense weathering. Cutans form by solution of haematite and precipitation of geothite. Cutans indicate the *in situ* formation of the nodules
Ferruginous saprolite	As above except rock fabrics are aligned with those in the parent rocks	Intense leaching and diffusion of Fe-oxides to form nodules
Fragmental duricrust	Rounded to subangular fragments of ferruginous saprolite cemented by gibbsite	Residual fragmented Fe-saprolite and removal of clays causing nodular lateritic residuum to collapse
Iron segregations	Fe-enriched bodies from 2 to 200 cm in saprolite breccias and joints composed of goethite, haematite and quartz. Pseudomorphs of sulfides occur. Solution cavities are common	
Ferricrete	Fe-oxide cemented reworked detritus	
Conglomeratic ferricrete	Ferruginous nodules and pisolites cemented by goethite	Reworking of lateritic residuum and recementation lower in the landscape than lateritic residuum
Pisolitic ferricrete	Reworked pisolites cemented by kaolin, gibbsite and goethite	Reworking and cementation of lateritic residuum
Vesicular ferricrete	Goethite, quartz and organic remains with vesicular mega-pores	Precipitation of Fe-oxides in poorly drained topographic lows
Ferruginized palaeochannel sediments	Palaeochannel sediments over lateritic residuum or saprolite containing mega-mottles of goethite and haematite and pisolites developed *in situ* around fine-grained quartz or fossil wood. Some detrital pisoliths	Mega-mottles develop around root channels in the palaeochannel sediments

they quote a 90 m lowering of Cretaceous sandstones and mudrocks to form 10 m of sandy residuum over bauxite. This they calculate took between 30 and 100 Ma to achieve.

McFarlane (1983) similarly uses extension of the weathering profile into fresh bedrock and concomitant landscape lowering to explain the concentration of Fe in the ferricrete on top of deep weathering profiles (Figure 13.7). These models all require huge downwasting of the landscape or mighty fluctuations in the water-table to obtain sufficient Fe-oxides at the surface to form ferricretes of any thickness.

Brimhall *et al.* (1991) using mass balance techniques in Western Australia demonstrated that relative accumulation has only a minor contribution in the enrichment of Fe_2O_3 and Al_2O_3 in 'laterites'. Rather they sought to explain the enrichment by the accession of local and foreign detritus being incorporated into the profile. They suggested this mechanism was so significant that it could contribute sufficient material to move the land surface upward. Only in the lower parts of the weathering profiles did they consider residual enrichment to be a significant process, and this zone is not associated with ferricretes ('laterites'). Similarly, Glassford & Semeniuk (1995) argue that the so-called *in situ* 'laterite' profiles of the Yilgarn Craton in Western Australia are the result of a succession of sedimentary accumulations, mainly aeolian and fluvial, over saprolite. Their main arguments are based on consideration of the mismatch of type and age of resistate minerals in the lower and upper parts of the 'lateritic' profiles, the presence of unconformities in the 'pallid' and 'mottled' zones of the profiles and the widespread presence of the profiles in a multitude of landscape positions and on essentially unweathered bedrock.

Table 13.4a *The classification of 'laterites' (after Tardy 1993)*

Name	Fabric	Contents		Haem.[a] (size)	Goe.	Gib.[a] (contents)	Boe.	Kao.
		Al	Fe					
Conakryte	Crystalliplasmic	p	✓✓	Large	✓	✓	x	x
Ferricrete	Nodular	✓	✓✓	Very small	✓	?	x	✓✓
Orthobauxite	Massive	✓✓	✓	Large	✓	✓✓	✓✓	x
Metabauxite	Pisolitic	✓✓✓	p	Very small	x	✓	✓✓	✓
Latosol	Microglaebular	✓	✓	Small	✓	✓✓	x	✓✓✓

Haem. = haematite, Goe. = goethite, Gib. = gibbsite, Boe. = boehmite, Kao. = kaolin.
✓✓✓ = very abundant, ✓✓ = abundant, ✓ = present or moderate, ? = possible, p = poor, x = absent.
[a] Haematite is always present, but in different sizes, gibbsite is always present but in different amounts.

Table 13.4b *The climatic conditions and climatic evolution necessary for the formation of 'laterites' (after Tardy 1993)*

Name	Tropical climate	Parameter		Palaeoclimatic evolution
		Humidity	Temperature	
Conakryte (ferroaluminous rocks)	Humid	Medium	High	Constantly humid tropical
Conakryte (ultramafic rocks)	Undifferentiated	–	–	Undifferentiated
Ferricrete	Tropical contrasted	High	Medium	Constantly contrasted
Latosol	Cool humid	High	Medium	From contrasted to humid
Orthobauxite	Humid	High	Medium	Constantly humid
Metabauxite	Arid	Low	Very high	From humid to arid
Cryptobauxite	Humid	High	Medium	From arid to humid

Table 13.5 *Rates of weathering and landscape lowering associated with weathering of various lithologies (from Nahon 1991)*

Rock	Climate	Rock life (yr/mm)[a]	Equivalent landscape lowering rate ($m/10^6 yr$)[b]
Acid rocks	Tropical semi-arid	65–200	15.4–5
	Tropical humid	20–70	50–14.3
	Temperate humid	41–250	42.4–4
	Cold humid	35	28.4
Metamorphic rocks	Temperate humid	33	30.3
Basic rocks	Temperate humid	68	14.7
	Tropical humid	40	25
Ultrabasic rocks	Tropical humid	21–35	47.4–28.4

[a] 'Life' of the rock means its conversion to kaolinitic saprolite.
[b] Calculated assuming all the kaolinitic saprolite is eroded.

McFarlane (1976) also describes 'groundwater laterite' formed in the vadose zone in the narrow band of water-table oscillation where Fe-oxides can segregate. Similar Fe-oxide accumulations (ferricretes) described by Smith (1989), Butt & Smith (1992) and many others play an important role in dispersing trace elements around ore-bodies and so creating enlarged exploration targets. Mann (1998) summarizes these types of ferricrete and their significance to the formation of oxidized Au deposits.

Many have questioned the idea that all ferricretes and bauxites atop deeply weathered 'lateritic' profiles are the result of relative accumulation. Maignien (1966) described the formation of ferricrete by the lateral transfer of Fe^{2+} in groundwater from upgradient (Figure 13.8). Although few have embraced this model, Pain & Ollier (1995a) use it to explain the distribution of ferricrete and the origin of landscapes in Cape York, northern Queensland. Ollier & Pain (1996) also produce a small cartoon to illustrate the

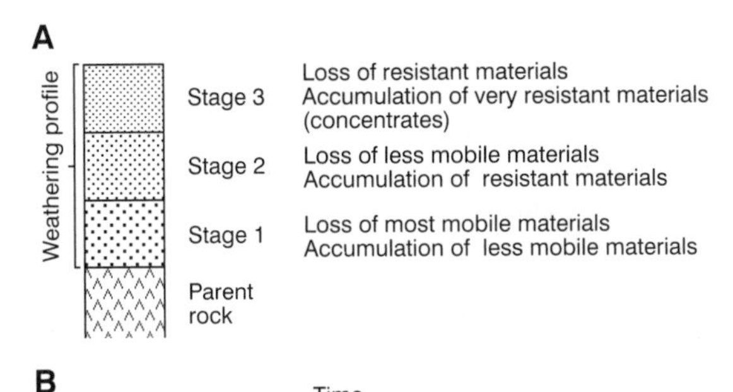

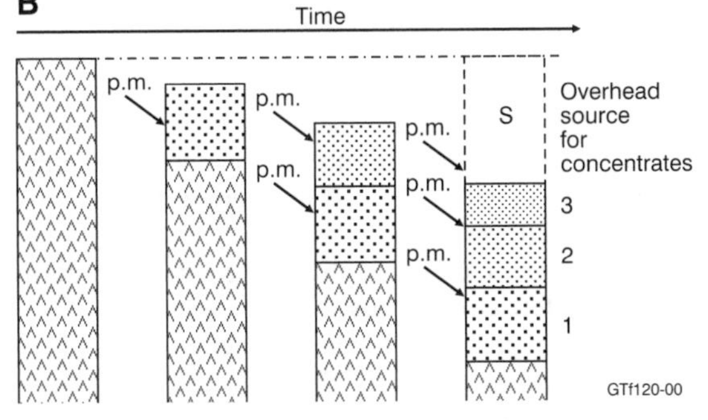

Figure 13.7 *The stages of weathering and the process of landscape lowering associated with intense weathering to form a lateritic crust (from McFarlane 1983)*

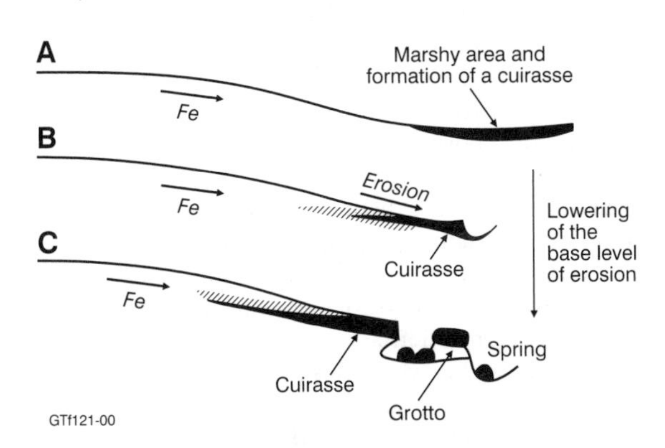

Figure 13.8 *Lateral migration of sesquioxides and the formation of a ferricrete crust. (a) Migration of Fe in groundwater by organic acids and accumulation in marshy valley floor. (b) Initial stages of incision of the valley floor causing water-table to fall. (c) Later erosional stage and the ferricrete becomes dissected and fragments are available for transport (after Maignien 1966)*

difference between the relative (vertical accession) and absolute (lateral migration) models of ferricrete formation (Figure 13.9). Similar lateral transport of Fe and its thicker accumulation in topographically lower parts of the landscape are observed at Weipa,

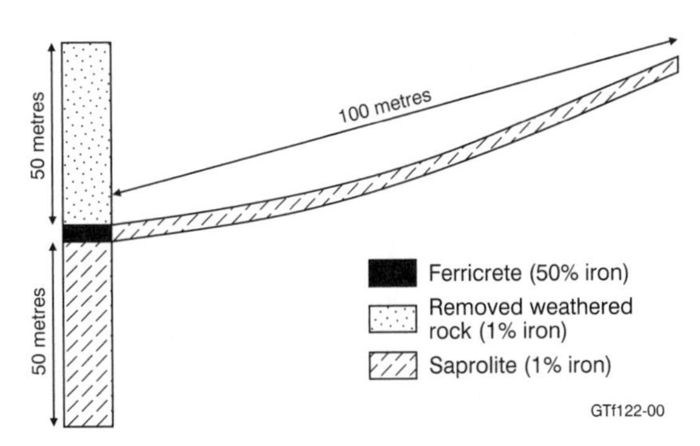

Figure 13.9 *Illustration of the difference between the relative (vertical accession) and absolute (lateral migration) models of ferricrete formation (Ollier & Pain 1996)*

where the Fe is fixed in ferricrete as goethite/haematite at the average wet season high water-table across most of the landscape. The bedrocks are homogeneous, so thicker ferricretes in lower situations demonstrate that Fe must be moving laterally from higher landscape/water-table situations. Bourman (1993) also suggests the Fe is provided by absolute accumulation by lateral migration in solution from upslope. Bourman (1995), in relation to the 'laterites' of South Australia, makes the point that there is no necessary genetic connection between deeply weathered profiles and ferricrete and that most are polygenetic, having been subjected to long and continuous periods of weathering interspersed with phases of erosion and sedimentation. Bourman's (1989) classification of ferricrete is given in Table 13.6.

Ollier & Galloway (1990) went so far as to suggest, from their field observations and literature research, that if not all cases, most ferricretes overlying saprolite ('lateritic' profiles) were separated from it by an unconformity. Their view is very much that nearly all ferricretes are developed in transported debris deposited on top of the saprolite and therefore the ferricrete can post- or pre-date deep weathering significantly.

Ferricreted bedrock retains bedrock fabrics and is enriched in Fe-oxides and usually overlies mottled zones. Bourman (1993) considers the Fe originated from absolute accumulation as a result of lateral transfer of Fe^{2+} in solution from upslope, as well as erosion exposing mottles and their being transformed into ferricrete.

Ferricreted sediments are colluvium, alluvium, marine and lacustrine sediments ferruginized by lateral

Table 13.6 *Ferricrete classification (Bourman 1989)*

I Simple ferricrete
1. Ferricreted bedrock
2. Ferricreted sediment
   (a) Ferruginized clastic quartzose sediment including blocks of reworked ferricrete (detrital ferricrete)
   (b) Ferruginized organic sediment displaying a massive to vermiform fabric (bog iron ores)
   (c) Ferricrete formed by *in situ* weathering of Fe-rich sediments containing siderite, glauconite and pyrite. Some of this contains voidal concretions

II Complex and composite ferricrete
1. Pisolitic ferricrete in which the pisoliths are important constituents
2. Nodular ferricrete
3. Slabby ferricrete
4. Vermiform ferricrete

migration of Fe-rich solutions precipitating Fe-oxides in the sediments. Lacustrine or bog Fe-ores typically have vesicular fabrics.

Pisolitic ferricretes are cemented accumulations of pisolites. Pisoliths are formed as suggested by Nahon (1991) in Chapter 11 or by Tilley (1998), or as clastic components from other materials (Bourman 1993), where they formed by deposition of lamina of various compositions (goethite, kaolin, gibbsite, haematite and maghemite with quartz grains, usually in preferred laminae). Their cementing materials are usually very different from the pisoliths, being composed predominantly of goethite and kaolin.

Nodular ferricrete contains nodules of earthy appearance that are smaller than pisoliths. They occur contiguously with pisolitic and vermicular ferricretes. Unlike pisolitic ferricrete these do not have significantly different nodule and cement composition, both being of high-Al goethite, kaolin and gibbsite.

Slabby ferricrete occurs in mottled regolith, on plateau margins where laterally moving groundwater nears the surface (Figure 13.10).

Vermiform ferricrete consists of vermicular tubules filled or lined with kaolin and gibbsite set in a surface lag of pisoliths, bedrock, ferruginous clasts and hardened mottles cemented by high-Al goethite. They typically occur in low relief areas where lateral drainage is postulated to have been poor (Bourman 1993).

Another possible method of ferricrete formation comes from the observation of the weathering of Cretaceous marine sediments around Darwin in the Northern Territory. Darwin has a strongly seasonal tropical climate and water-tables fluctuate over many metres. Exposed around the sea cliffs there are weathered sandstones overlain by transported alluvial fan debris which contain ferricretes. The underlying 'mottled' zone is sandstone saprolite from which Fe^{3+} is actively being leached along joints, fractures and bedding planes, leaving Fe-oxides in the centres of saprolite blocks bounded by these features (Figure 13.11). The mottling here is a process of deferruginization rather than one of growth. As the marine sediments contain significant Fe^{2+}, groundwater easily mobilizes it forming ferruginous coatings on the cliffs (Figure 13.11). Ferrihydrite is common in springs along the cliff lines and the Fe-oxide cementation of various bits of rubbish (car parts, mining equipment, 44 gallon drums) into ferricrete along the wave-cut platform (Figure 13.12) is testament to the rapid movement of Fe in this environment to form substantial ferricrete in less than 50 years.

Because of its intense colour, haematite is a very obvious pigmenting agent. This can lead to an overemphasis on the distribution of Fe induration and even the misidentification of profiles as lateritic when in fact they are cemented by another agent (silica, clay) and coloured by haematite. Many of the mesa edges in the Charters Towers region of north Queensland are very red at their upper 2–3 m (Figure 13.13). Aspandiar (1998) has shown that these lateritic-appearing profiles are slightly indurated red earths.

From the above summary of some of the types and interpretations of ferricrete it is clear that there is little consensus about whether ferricretes form from absolute or relative accumulation. It is our opinion that both models apply (Eggleton & Taylor 1999); Fe/Al-oxides and oxyhydroxides can accumulate by relative concentration across a variety of parent rock types, as at Weipa where this process is active now. It is equally clear that lateral transfer of Fe in solution and its precipitation in springs is an observable process that leads to the formation of ferricrete and to the deferruginization of other rocks.

There is no answer yet to these questions, and there are many other questions still surrounding the formation of ferricretes. We now address some of these.

Use of ferricrete and bauxite for litho-correlation

In Australia ferricrete has been widely used as a stratigraphic marker (e.g. Twidale 1983; McGowran 1978; Firman 1981). The discussion above shows that

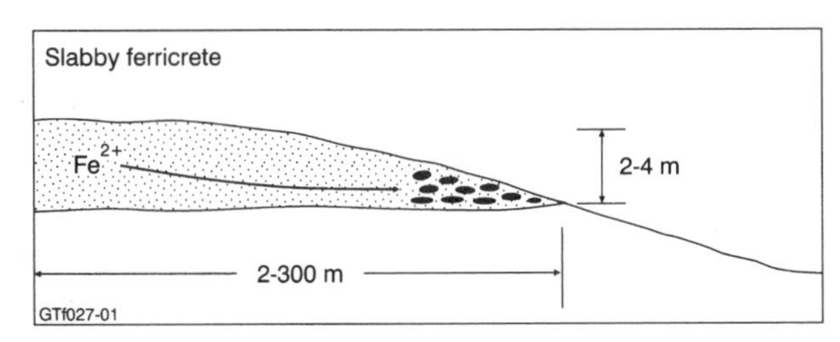

Hand specimen	*Mineralogy and Chemistry*
Slabs separated by clays	Hematite Goethite Gibbsite (minor) Fe_2O_3 45% 6% Al-substitution

(a)

(b)

Figure 13.10 *(a) The formation of slabby ferricrete (from Bourman 1993). (b) Photo of slabby ferricrete from the Mt Lofty Ranges, South Australia (photo Robert Bourman)*

ferricrete formation occurs in many ways and in many landscape positions. Taylor (1994) and Bourman (1995) show weathering has not occurred in periods but continuously through the Late Mesozoic and Cainozoic in Australia, so the likelihood of their being useful stratigraphic markers is minimal. It also seems to us that the use of weathering or pedogenetic overprints on sediments is a very poor basis for lithostratigraphic correlation. It is even unlikely that in one landscape all the ferricretes form synchronously, rather they will be of different ages depending on the evolution of the landscape and the groundwater history of the region. This latter point is ultimately related to the nature of parent lithologies and to palaeoclimates, mainly from the perspective of water availability.

Figure 13.11 *Deferruginization of the rocks at Darwin. Note the ferrihydrite crystallizing on the rock surface as Fe^{2+} in solution oxidizes where a spring issues from a cliff face (photo GT)*

Figure 13.13 *Strongly haematite coloured upper regolith at an exposed face of a deep red earth in the Charters Towers region, north Queensland*

Figure 13.12 *Photo of car parts in ferricrete at Darwin (photo GT)*

Climatic conditions for ferricrete and bauxite formation

The majority of researchers equate 'laterite' and bauxite formation with tropical and seasonal climates. Nahon (1991), Valeton (1972), Bardossy & Aleva (1990), Thomas (1994), McFarlane (1976), Butt & Zeegers (1992a) and Boulangé *et al.* (1997), for example, all subscribe to the tropical climate hypothesis, and only Boulangé *et al.* document that bauxite can continue to develop under varying climatic conditions. In fact, so ingrained is the idea that 'laterite' is related to tropical climates that many circular arguments exist in the literature relating to this. The most recent, and perhaps most blatant, is in a modelling study by Price *et al.* (1997) where they argue, based on modern climatic distribution of bauxites, that bauxites are found in those climates. Dissent from this view, although not uniquely Australian, has been mainly expressed in Australia. Taylor *et al.* (1992) and Bird & Chivas (1993) have demonstrated unequivocally that 'lateritic' weathering and the formation of ferricrete and/or bauxite can occur in cool climates. Others including Bauer (1959), Stephens (1971), Paton & Williams (1972), Geidans (1973), Glassford & Semeniuk (1995) and Young *et al.* (1987), have expressed doubt about the need for tropical seasonal climates to form these duricrusts.

There is no reason to have high temperatures to produce deep weathering, nor mobility of Fe, Al and Si to produce ferricrete or bauxite. All that is required is prolonged weathering with ample water over an extended period. Acton & Kettles (1996) report it taking between 10 000 and several million years to form a bauxite-capped profile 30 m thick on basalt at Inverell, eastern Australia, during the Late Oligocene to Early Miocene. This clearly shows thick profiles can be preserved in tectonically stable environments, but Boulangé *et al.* (1997) suggest bauxites can only be preserved in areas of positive tectonism.

Records of ferricrete, bauxite and weathering profiles in the regolith (or geological) record are, in our opinion, more a matter of preservation than a record of past climatic conditions. This is not to dispute that 'lateritic' weathering is more common in the tropics, but it is very common also to see ferricrete forming in temperate and semi-arid climates where springs deliver ferrihydrite to the surface. Thus it is impossible to equate ferricrete (or 'laterite') and bauxite exclusively to tropical climates or palaeoclimates.

Topographic requirements for ferricrete and bauxite formation

The influence of Davis's (1889) ideas on the evolution of landscapes still haunts us. Many early researchers were of the view that 'laterites' or ferricetes formed on peneplains, particularly in Australia where many authors refer to peneplains as being a prerequisite for 'laterite' development. Jutson (1914, 1934, 1950) and Woolnough (1927) were perhaps the most influential in this belief, which at the time was probably a reasonable view to hold, but there were critics including Craft (1932a, b, 1933) who showed that 'lateritic' bauxites in the eastern highlands did not necessarily relate to peneplains. Despite this early criticism Woolnough and Jutson's ideas persist in Australian geology. Another researcher to criticize Jutson's ideas was Geidans (1973) who incontrovertibly demonstrated that bauxite in the Darling Ranges of Western Australia formed across significant relief far in excess of what might be expected on a 'peneplain'.

Boulangé *et al.* (1997) suggest that chemical erosion with associated landscape lowering will lead to the formation of a plain; they call the process 'geochemical plaination'. There is little doubt this process of plain formation could operate in areas of Australia, Africa and South America, but definitive evidence for it having occurred is missing.

While we are of the opinion that ferricrete and bauxite can form in almost any topographic position, we recognize the need to separate residual accumulation (*in situ*) ferricrete from absolute accumulation ferricrete because of the implications for palaeolandscape interpretations and mineral exploration (see Chapter 15). This differentiation can only be achieved by examination of the framework materials within the ferricrete and their landscape or palaeolandscape position at the time of their formation.

Age of ferricrete, 'laterites' and bauxite

The age of laterites has been important in Australia where they have been widely used as stratigraphic markers. 'Laterite' in India, South America and Africa is widespread and provides useful clues to landscape evolution. Boulangé *et al.* (1997) record the age of bauxite at Porto Trombetas in Brazil as dating from between 30 and 100 Ma. Ollier & Pain (1996) note that bauxites in India range in age from Late Cretaceous to Early Tertiary. Bauxite associated with basalts in Antrim, Ireland, date from 64 to 62 Ma and on the Monaro, New South Wales, from 57.5 to ~35 Ma. Bird & Chivas (1989) show, using $\delta^{18}O$ techniques, that kaolin and gibbsite from weathering profiles associated with ferricrete and bauxites from across Australia range in age from Late Mesozoic to Neogene. Palaeomagnetic ages from ferricrete and bauxites around Australia date Fe-oxide precipitation from Mid-Palaeozoic to Holocene. There is, in fact, no 'time' of lateritization that emerges from these studies. The cumulative distribution of determined ages shows an exponential increase towards the present, clearly indicating that what has been sampled is preserved evidence from continuous weathering. It is well known that preservation potential decreases exponentially with time, so it is not surprising that the numbers of established dates do likewise.

As we note above, ferricrete is still forming in a host of environments around the world. It is an active process, probably most so in the tropics, but also in semi-arid and temperate climates. Ferruginization in deep podzolic soils is also common in climates ranging from tropical to cold, wet climates like those of the south-western coast of Tasmania.

In situ *versus transported regolith*

The distinction between *in situ* and transported regolith is critical in the interpretation of the evolution of duricrusts and particularly of ferricrete and bauxite. There are a number of methods to determine whether or not duricrusts are *in situ*. In Chapter 11 we mentioned stone lines in the regolith as indicating the movement of the regolith above the stone line. This is the simplest method to use, particularly in the field.

As an example of a stone line and its relation to the regolith let us look at a red podzolic soil developed over granite saprolite. At the Yass River just north of Canberra a stone line emanates from an unweathered aplite dyke that comes to the surface at a hill crest. Two hundred metres downslope fragments are buried by >1 m of soil/regolith (Figure 13.14).

The soil B-horizon is continuous, but thickening downslope. The stone line cuts across the B-horizon, beginning at the crest in the upper A-horizon, and ends up well within the C-horizon at the base of the slope. Clearly, if the stone line represents a break between *in situ* regolith below and transported regolith above then the soil developed on the slope is diachronous. It is older at the crest and younger at the slope foot, even though it appears with continuous soil horizons downslope and like most soils the profile thickens downslope. In reality the soil at the lower end of the slope has had regolith added to it through time,

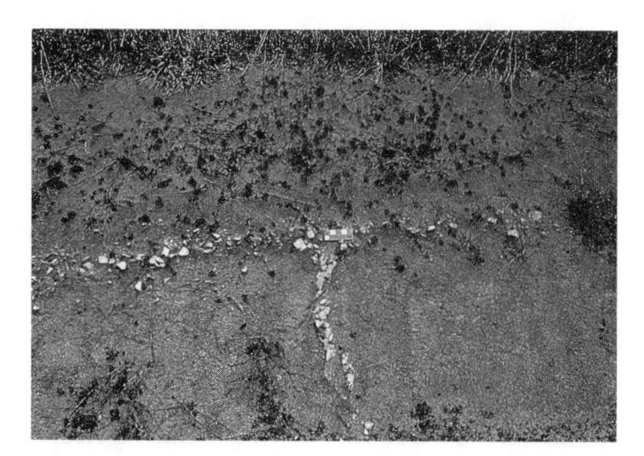

Figure 13.14 *Stone line between* in situ *regolith and transported cover over the Bega Batholith near Bega, New South Wales (photo GT). Scale 15 cm long*

perhaps from slope wash or from aeolian accretion, or both. The importance of recognizing this is that the upper part of the regolith at the crest is of a different age and origin from that on the lower parts of the slopes.

The presence of scour and fill structures in the regolith is another field technique for distinguishing between *in situ* and transported regolith. Obviously the material within a scour must be a younger transported part of the regolith than the material into which the scour-channel is cut. A superb example of this occurs at Northparkes Pit in central New South Wales (see Figure 11.21). A channel is cut into highly weathered Ordovician volcanic rocks, and its fill has been predominantly derived from the erosion of the same weathered rocks. Thus the differences in material are minor. The channel is marked mainly by a change in the mottling style; mega-mottles occur in the palaeochannel fill while the *in situ* weathering is marked by smaller and more uniform mottles. This is probably because trees preferred growing in the channel fill as it acts as a local aquifer and the mega-mottles are controlled by root channels as opposed to the *in situ* regolith where tree roots were less common. It may also be due to the huge age difference of the *in situ* weathering (Carboniferous, Pillans *et al.* 1999) and the weathering of the fill which is Miocene. Other indicators of the scour include minor lag gravels in the thalweg of the channel, different geochemistry (Table 13.7) and a slightly coarser grain size of the fill.

Mega-mottles are common in channel fills within weathering profiles and are noted across much of the Yilgarn Craton (Figure 13.15).

There are other more sophisticated methods to distinguish transported from *in situ* regolith. The $\delta^{18}O$ signatures of quartz have been used by Walker *et al.* (1988) to distinguish locally derived quartz from that derived from aeolian accession in the south-eastern highlands of New South Wales.

Table 13.7 *Comparison between the composition of average saprolite and average channel fill sediment at the Northparkes Pit, NSW*

wt%	Residual regolith	Transported regolith
SiO_2	47.6	45.0
TiO_2	1.1	0.8
Al_2O_3	22.3	21.3
Fe_2O_3	8.5	14.3
MnO	0.2	0.3
MgO	2.1	0.8
CaO	1.7	1.2
K_2O	3.1	0.4
Na_2O	1.1	0.2
ppm		
Au	80	40
Cu	1640	380
Zn	65	35
Pb	25	30
As	8	12
Zr	100	200

Source: Tonui EK (1998) Regolith mineralogy and geochemistry at Goonumbla, Parkes, NSW. PhD Thesis Australian National University, 260pp.

Figure 13.15 *Photo of mega-mottled channel fill over saprolite in a open pit mine on the Yilgarn Craton, Western Australia (photo GT)*

More recently Pell *et al.* (1997, 2000) have used U/Th dates from zircon grains to source the quartz sand in Australia's Quaternary dune-fields. This work showed that although the ages and morphologies of zircons did not match the underlying *in situ* material they had not been transported far. The sand was derived from locally occurring fluvial sediments and then locally reworked, but the original protoliths were the Palaeozoic rocks of the eastern highlands and older cratonic rocks of central and southern Australia with minor contributions from the Antarctic.

The matching, or otherwise, of the weathering-resistant minerals above and below suspected transported components of the regolith is important. For example at Weipa it is clear that between the Cretaceous Rolling Downs Formation and the Tertiary Bulimba Formation there is a marked change in the population of resistate minerals, particularly zircon. It is also noted that along the unconformity between the two there is a concentration of resistate heavy minerals that do not occur elsewhere in the sequence. In this case it is especially interesting as the resistates are used to differentiate two deeply weathered transported units in which the regolith is formed.

Similar techniques can be applied to ferricrete and bauxite to determine the nature of their parent material. In the Charters Towers region Aspandiar *et al.* (1997) have used palaeosols to differentiate one transported unit from another within the regolith or the Southern Cross/Campaspie Formations. Soils represent a significant break in deposition within a sequence and show that either deposition stopped altogether or that the locus of deposition moved elsewhere. At Weipa a palaeosol divides the Rolling Downs regolith from that formed on the Bulimba Formation.

SILCRETE

Introduction

Silica cementation of regolith (silcrete) is one of the more common forms of duricrust in Australia and it also occurs widely in Southern Africa and parts of Europe, less commonly on the other continents. In many parts of Australia they are one of the most commonly outcropping rocks and were widely used by Aborigines for producing stone tools, in Europe as palaeolithic monuments and large parts of Windsor Castle in England appear to be built of it. They occur across the Australian continent from contemporary humid to arid climates from Cape York (Pain *et al.* 1994) to Tasmania (Yim 1985), and from sea level on the south coast of New South Wales (Young 1978) to Western Australia (Singh *et al.* 1992).

Several good summaries of silcrete research exist, including Goudie (1973), Langford-Smith (1978), Goudie & Pye (1983), Martini & Chesworth (1992), and Thiry (1997) and Thiry & Simon-Coinçon (1999). Early work on silcrete in Australia and Africa defined a relationship between silcrete and land surfaces. In Australia the work of Woolnough (1927) related silcrete to a widespread peneplain and commented that they were all recent and perhaps even forming today. He also linked their formation to deep weathering on this vast peneplain. Kalkowsky (1901) and Stortz (1928) worked on silcrete in the Kalahari Desert and Lamplugh (1907) in Rhodesia. All these early researchers related their formation to the arid climates in which they now occur.

Browne (1914, 1972) considered 'grey billy' (silcrete) occurring beneath basalts to be genetically related to the extrusion of the basalt in some way (hydrothermal fluids, metamorphism). In Australia this led to the widely held belief that the presence of silcrete meant there had formerly been basalt over it. This idea persisted well into the second part of the century (e.g. Gunn & Galloway 1978; Browne 1972). It began to be broken down by Stephens (1971) who suggested that many arid zone silcretes were the result of long-distance transport of silica from deep weathering and its precipitation in the arid climates of central Australia. Earlier Woolnough (1927) suggested similar origins for the silcrete, but despite his suggestion many followed the Browne model.

More recent work, particularly by Thiry, summarized in his 1997 paper, Taylor & Smith (1975), Taylor & Ruxton (1987) and Thiry & Milnes (1991), has examined the geochemical and petrographic character of silcretes from various areas and shown that many of the earlier ideas, while having some merit, needed significant refining.

Definition

Lamplugh (1902) coined the term 'silcrete' for 'silicified sandstones and conglomerates like the sarsen stones of southern England' and in 1907, he described as silcrete 'hard sandstone knit together by a chalcedonic cement'. The term came into general use in Southern Africa and was imported to Australia by Williamson (1957).

Many definitions of silcrete followed, the most recent being in *The Regolith Glossary* (Eggleton 2001):

> Strongly silicified, indurated regolith, generally of low permeability, commonly having a conchoidal fracture with a vitreous lustre. Silcrete appears to represent the complete or near-complete silicification of precursor regolith by the infilling of available voids, including fractures. Most are dense and massive, but some may be cellular, with boxwork fabrics. The fabric, mineralogy and composition of silcretes may reflect those of the parent (regolith) material and hence, if residual, the underlying lithology. Thus, silcretes over granites and sandstones have a floating or terrazzo fabric and tend to be enriched in Ti and Zr in the forms of anatase and zircon respectively; silcretes with lithic fabrics (e.g. on dunites) are silicified saprolites with initial constituents diluted or replaced by silica.

Fresh silcrete gives a very particular ring when hit with a hammer. It may occur as sheets, lenticular or irregular pods with mamillated surfaces, or as completely or partially silicified beds, or as columnar or 'dog-tooth' outcrops. Internal structures are highly variable from preserved sedimentary structures or igneous fabrics to completely massive or stalactitic or flowstone-like surface textures. Anatase 'caps' on large particles are common in coarser-grained silcrete.

Classification

There are a number of classifications of silcrete. Some relate to the silcrete, others to a 'weathering profile'. Smale (1973) provided an early classification based on field experience in Southern Africa and Australia (Table 13.8). This classification is relatively simple and has not been used widely. Summerfield (1983) developed another classification based on South African experience using micro-fabrics and geochemistry as the essential factors for dividing up silcretes (Table 13.9).

The silcrete classification schemes of Smale (1973, Table 13.8) and Summerfield (1983, Table 13.9) are essentially descriptive and will allow classification without understanding genesis. Wopfner (1978) recognized some of the shortcomings of this type of classification and thus devised his from field experience in South Australia, north-western New South Wales and southern Queensland (Table 13.10). His classification is based on the mineralogical composition of the 'matrix' (plasma), macroscopic texture, retention of host rock fabrics and textures, and thickness and type of associated weathering profiles.

Thiry (1997) developed another silcrete classification, distinguishing pedogenic from groundwater silcretes and using fabric characters to further subdivide silcretes of each group (Table 13.11). In the light of this more recent work it would appear the classifications of Wopfner and of Thiry are the most useful. For a completely objective classification, Wopfner's is better, but Thiry's is the one most used now where most silcrete is simply classified as 'pedogenic' or 'groundwater', but this of course requires making genetic interpretations.

Table 13.8 *A classification of silcrete proposed by Smale (1973)*

Silcrete type	Description
Terrazzo	Framework of quartz grains form about 60% of the rock. Grains have solution cavities indicating the sediment underwent a phase of SiO_2 solution. Cement may be chalcedony, cryptocrystalline or opaline SiO_2. TiO_2 is common as a cream cloudy material. Complex fabrics occur in the TiO_2 material
Conglomeratic	Abundant siliceous pebbles in a siliceous matrix. Detrital grains have an oolith-like coating of Fe_2O_3 giving the rock a red-brown colour. The irregular shape of pebbles suggests they have not moved far, but may be later solution rounded
Albertina	Almost entirely matrix of the terrazzo type. The matrix is of crystalline quartz and chalcedony or opal
Opaline – fine grained massive	Layers of common opal, chalcedony or cryptocrystalline silica. They are homogeneous of chemical origin with no detrital component
Quartzitic	Identical to sedimentary orthoquartzite with cementation by authigenic overgrowth on the detrital quartz grains. Quartz grains are coated by Fe_2O_3 dustings

Silcrete description

There are essentially four types of silcrete based on their gross field features:

1. Massive
2. Structured
3. Skin
4. Pans

Massive silcretes

These are typically composed of siliceous gravel, sand and/or silt framework grains set in a plasma of

Table 13.9 *Micromorphological silcrete classification based on South African silcretes (adapted from Summerfield 1983)*

Type	Sub-type (matrix fabric elements)	Description
Grain supported fabric		Skeletal (framework) grains are self-supporting
	Optically continuous overgrowth	
	Chalcedonic overgrowth	
	Microquartz matrix	Includes microquartz, cryptocrystalline and opaline silica
Floating fabric		Framework grains 'float' in the matrix and do not support themselves. >5% framework
	Massive (glaebules absent)	(glaebules are used to describe aggregates)
	Glaebular (glaebules present)	of SiO_2 + anatase ± Fe-oxides
Matrix fabric		Framework grains <5%
	Massive (glaebules absent)	
	Glaebular (glaebules present)	
Conglomeratic fabric		Detrital component contains pebbles (>4 mm)

Note: Framework grains are >30 μm.

Table 13.10 *A classification of silcretes. A particular silcrete is identified by an appropriate combination of numbers and letters (summarized from Wopfner 1978)*

Group	Matrix	Habit	Bedrock features	Profile
I	Quartz			Overall kaolinitization of underlying regolith
	1. Irregular	(a) Blocky	Yes	
	2. Syntaxial	(b) Bulbous – pillowy	Yes	
II	Cryptocrystalline quartz			Intense kaolinitization of underlying regolith, usually with a zone of brecciation between kaolin and silcrete zones
	1. Massive	(a) Columnar, polygonal, prismatic	No	
	2. Pisolitic (pseudopebbly)	(b) Platy	No	
	3. Laminated	(c) Botryodal	No	
		(d) Pillowy	No	
III	Amorphous (opaline) cristobalitic opal C-T			No specific profile, but associated with gypsum and alunite
		(a) Breccious	Yes	
		(b) Conglomeratic	Yes	
		(c) Replacements and fillings	Yes	

siliceous cement. They form as sheets, lenses, blobs or are confined by sedimentary bedding (Figure 13.16) which crops out as massive features in the landscape.

Massive silcrete may retain the internal structure, fabric and texture of the regolith being silicified. Silicification may be confined to a particular stratigraphic unit in a weathered sequence of rocks (Figure 13.17), or massive silicification of granite may preserve the original granitic fabric (Senior 1978; Figure 13.18).

Many silcretes contain plant fossil impressions (Figure 13.19). One of the most spectacular examples is from Stuart Creek near Ferguson Hill in the Lake Eyre Basin of central South Australia where an extensive flora has been well preserved in bedded massive silcrete. Palm-like impressions, broad-leaved

Table 13.11 *Silcrete types based on field, micromorphological and silica mineralogy, summarized (adapted from Thiry 1997)*

Type	Silica accumulation	Profile	Description	Significance	Palaeogeography
Pedogenic					
1. Quartzose	Relative	Columnar	Quartz grains with irregular voids between syntaxial overgrowths grading to anatase-enriched microquartz – opal cutans coat column boundaries and bottoms of platy horizontal joints as illuvial Si reworking	Si migrates from top to bottom by illuviation and progressive Si reworking. Each step involves the crystallization series – opal → microquartz → quartz – microfabrics confirm this downward movement of Si and the landscape lowering associated with this type of silcrete	Residual outcrop (Paris Basin), indurates edges of plateaux (Australia) represent landscape lowering
	Relative	Granular pseudo-breccia	Granules of microquartz, opal and corroded kaolin crystals welded by Si-gel – matrix kaolin		
2. Hardpan	Absolute	No particular	Red-brown colour, opal cemented, no quartz or Fe-oxide dissolution, burrows, parent nodules capped with cutans, kaolin common	Si source unknown, form in semi-arid to arid seasonal climates	Closely related to present landscapes, form in subdued topography
Groundwater					
1. Sandstone	Absolute	Massive lenticular forms	Quartz sand with subhedral epitaxial overgrowth, >99.5% SiO_2, may contain etched pebbles and sand, often have mamillated surface features	Form by precipitation from groundwater at springs, Si concentration in water = 10–15 ppm (double quartz saturation) precipitation induced by physico-chemical change and evaporation	Around scarp (hill) margins and mesa tops where springs issue
2. Claystone		Upper quartzite	Microquartz replaces kaolin and kaolin-free quartz has overgrowths. Solution features common at field- and micro-scale, associated with gypsum, alunite, jarosite	Associated with deep weathering profiles, mainly in the Eromanga Basin of central and northern Australia. Climatic implications of deep weathering are that the profiles were leached by vegetation-induced acid groundwaters. Presence of Ca^{2+}, K and SO_4 suggest relatively acid and arid climates	Associated with groundwater movement under low relief landscapes

continued

Table 13.11 *(continued)*

Type	Silica accumulation	Profile	Description	Significance	Palaeogeography
	Local relocation	Lower opal-cemented sandstone	10–20% opal, SiO_2 replaces kaolin and fills pores, smectite replaced by SiO_2, associated with gypsum, alunite, jarosite		
	Local relocation	Precious opal	Occurs in deeply weathered clay and sand regolith below the silcretes, 50–70% opal-CT		
3. Limestone	Absolute or local relocation	Irregular massive or replacing limestone features	Microquartz, chalcedony, euhedral quartz crystals replace limestone, particularly in porous zones		
Evaporite	Absolute	Layers	Opal, Si-gel, Mg-calcite, dolomite, magnesite	High pH solutions precipitate gel and carbonates	Coorong, South Australia

and other leaves and fruiting bodies typical of rain-forest vegetation occur. At this site they are thought to be Eocene.

Another feature typical of many silcretes is a cap of anatase accumulating on the top surface of framework grains. Figure 13.20 illustrates a megascopic example from massive silcretes from Lake George in eastern New South Wales, but such features are observable at all scales (macro- to microscopic) in silcrete from almost all localities even though the content of TiO_2 may be highly variable. It is one of the few ways to identify a silcrete as distinct from other highly quartzose rocks.

Massive silcrete also occurs in palaeochannels ('deep leads') preserved beneath and between basalt flows (Figure 13.21) extensively distributed along the eastern highlands of Australia (Taylor & Smith 1975; Browne 1972; Young 1978; Francis & Walker 1978; Taylor & Ruxton 1986). The silcretes typically form as lenses, sheets and lumps in the upper parts of the palaeochannel sediments. The silica cements quartzose and other siliceous sediments, but despite occurring in basalt terrain, the silcretes never preserve any basaltic debris. These silcretes are also frequently ferruginous, particularly towards the margins of the palaeochannels. Plant fossils are also preserved in these silcretes.

Massive silcretes occur in many landscape positions from hill or plateau crests to valley bottoms. Whether this is where they formed is a different question, and their occurrence on topographic highs is probably a result of later topographic inversion, although some, which form a 'skirt' around plateaux, are doubtless formed there. Silcrete 'drapes' down hillsides are reported by Ruxton & Taylor (1982), illustrating the range of topographic positions in which they can occur. Massive silcrete along the Mirackina Palaeo-channel demonstrates the formation of silcrete in valley bottoms and later landscape inversion (Figure 13.22) (Barnes & Pitt 1976; McNally & Wilson 1995). Such silcrete is widely described from Europe (Wopfner 1983), Africa (Summerfield 1983) and Australia. The famous sarsen stones of England and France are massive silcrete.

Another form of massive silcrete is *porcellanite*. It is a very fine-grained silcrete, usually white, but often coloured by Fe-oxides. It is entirely made up of micro- or cryptocrystalline quartz. It has very evident conchoidal fracture and often occurs in a bedded form: fossils may be well preserved as impressions. It is always associated with weathered claystones or mudstones and is formed by the replacement of clay minerals by quartz. Typical examples occur in the weathered Cretaceous Bulldog Shale in South Australia. The Bulldog Shale is composed of quartz, illite and smectite and contains erratic boulders. On weathering the clays are transformed to kaolin, which is in turn replaced by quartz to form porcellanite.

Figure 13.16 *(a) Sheet silcretes on mesa top (formed in the Bulldog Shale) over a lower silcrete forming the prominent bench in the foreground. Central South Australia (photo GT). (b) Coarse conglomeratic silcrete occurring as a linear cemented zone in a pre-Late Miocene palaeochannel, Lake George, New South Wales (photo GT). (c) A lenticular body of massive silcrete with a sandy framework from Bendalong, southern coast of New South Wales. Note the unusual 'cockade' or 'glerp' structures on the surface of the silcrete. These features are only observed on the outside surface of massive silcrete bodies where they contact (contacted) unsilicified regolith (photo GT)*

Figure 13.17 *Silicified mudrock bed in the weathered Cretaceous at Darwin, Northern Territory, Australia. It forms the smooth-faced bed two-thirds the way up the cliff (photo GT)*

Figure 13.18 *Silicified granite with an* in situ *quartz vein from near Broken Hill at Boulder Tank (photo Steve Hill)*

Similar porcellanitic silcretes occur, replacing Cretaceous marine sediments at Darwin (Figure 13.17) south of Charters Towers in Queensland, and at Lake Cowan near Kalgoorlie.

Precious and common opal are also forms of massive silcrete. They occur in both weathered and fresh sedimentary rock (mainly Cretaceous and in some Tertiary alluvial sediments) as veins, nodules and replacing fossils. Opal occurs throughout the southern Eromanga Basin of Queensland, South

Figure 13.19 *Plant fossils in massive Miocene silcrete from Ferguson Hill, central South Australia (photo GT)*

Figure 13.20 *Photo of anatase caps on pebbles in massive silcretes from Lake George in eastern New South Wales (photo GT)*

Australia and Queensland at such famous localities as Coober Pedy, Lightning Ridge and White Cliffs.

Structured silcrete

This is a silcrete that has a distinct structure in outcrop, of which the most common is *columnar silcrete*, or 'dog-tooth' silcrete (cf. Wopfner type IIa) (Figure 13.23). These silcretes have an upper layer of stalagmite-like columns or prisms with rounded tops and internal convex-up layering (Figure 13.24). These

Figure 13.21 *Photo of intra-basaltic deep leads sandwiched between basalt flows. Pioneer Tin Mine, north-eastern Tasmania. These leads are commonly silica-cemented in various parts of eastern Australia where Tertiary basalts enclose or bury quartzose sediments (photo GT)*

have been widely described from Australia (Senior & Senior 1972; Milnes & Twidale 1983; Thiry & Milnes 1991), from the Paris Basin (Thiry 1981), and from the southern UK (Ullyott *et al.* 1998).

Another common type of structured silcrete is *pisolitic silcrete* (Figure 13.25a, b). These are composed of a layer of pisoliths (cf. Wopfner 'pseudopebbles') of varying size either loosely distributed at or near the surface or silica cemented pisolites. The pisoliths vary in size from 0.5 to 15 cm in diameter and are well rounded and subspherical to spherical. Internally the pisoliths are layered with alternate bands of microquartz and microquartz plus anatase (Figure 13.25a, b). Wopfner (1978), Watts (1978) and others describe pisolitic silcrete. They characteristically occur on low convex hilltops and plateau surfaces, but also as sheets of loose material on slopes and pediments. They are occasionally associated with columnar silcrete, and like them almost invariably overlie deeply weathered profiles.

So-called 'candle wax' silcrete commonly occurs in similar locations to columnar and pisolitic silcretes (Wopfner 1978) but there are few other references to its occurrence. It consists of thin rod-shaped irregular columns of nearly circular cross-section, with or without pisoliths, and with a wax drip-like coating of

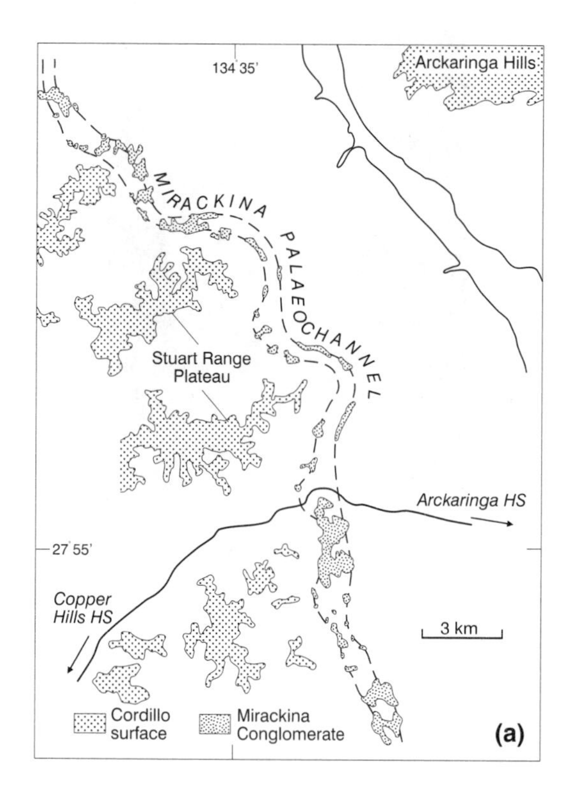

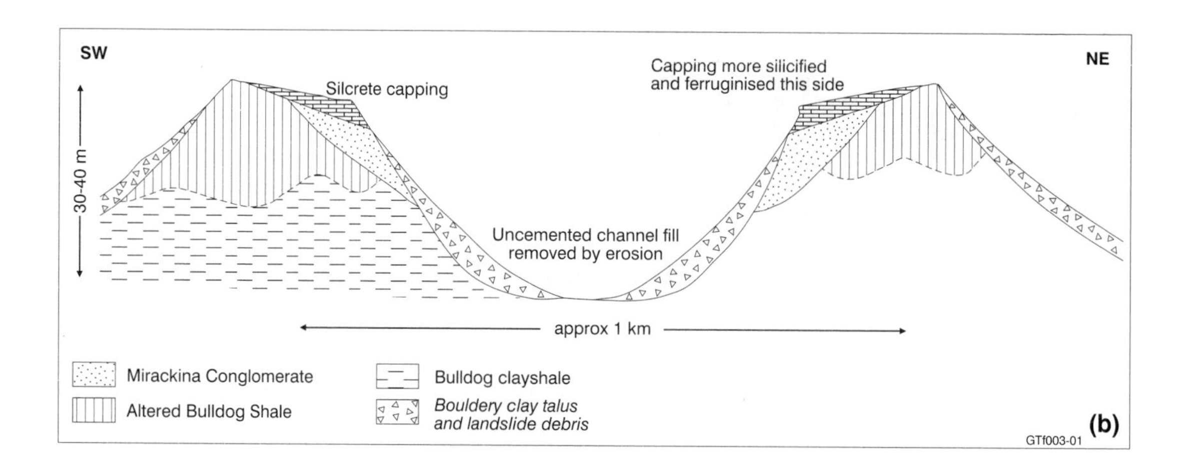

Figure 13.22 *Mirackina Palaeochannel landscape evolution from McNally & Wilson (1995). (a) Shows the course of the palaeovalley about 50km and (b) shows the silcretes to have been formed along the flanks of a now topographically inverted valley*

(a)

(b)

Figure 13.23 *Structured 'dogtooth' or columnar silcrete (a) from the Paris Basin near Fontainebleau, and (b) from near Garford Palaeovalley near Woldia in central South Australia (photo GT)*

microquartz which imparts their characteristic form. They generally overlie deeply weathered regolith.

Structured silcretes are generally thought to be formed at or near the surface under the influence of what can be grouped as 'pedogenetic' processes (cf. 'pedogenic silcrete' of Thiry 1997 and Thiry & Milnes 1991).

Skin silcretes

These occur on the surface of weathered rocks. They form by the diffusion of silica to the surface and

Figure 13.24 *Section through the centre of a structured (dog-toothed) silcrete column illustrating the internal growth structure outlined by variations in the anatase content (photo GT)*

(a)

(b)

Figure 13.25 *(a) Pisolitic silcrete lag over desert loam and cemented into structured (columnar) silcrete from Tibooburra, New South Wales (photo GT)*

precipitation as the water evaporates. Examples are documented from the Beda Pediment in South Australia (Milnes & Hutton 1974), and in the Bungle Bungle Ranges of Western Australia (Young 1986) where essentially hills of loose sand are held in place by silica skins cementing the outer margins of the rock. At Marlborough in central eastern Queensland chalcedonic quartz and chrysoprase drape hills of saprolite formed from ultramafic rocks.

Pans

These are widely reported from the Massif Central in France (Thiry & Turland 1985), from much of Australia (Bettenay & Churchward 1974; Hubble *et al.* 1983; Chartres 1985) and North America (Flach *et al.* 1974; Chadwick *et al.* 1989). They are generally red, grey or white coloured, hard sheet-like bodies (soil B-horizons) of quartz sand grains floating in a porous plasma with an overall rough columnar structure. The column boundaries are coated in Fe-oxides. They may be up to 30 m thick, but more normally are 1 m or so. In Australia they typically have a plasma of randomly interstratified clays, haematite and goethite. In France they are kaolinitic with Fe-oxides. The whole body is cemented by opal.

Pans are soil horizons cemented by clay minerals and Fe-oxides and opal. The origin of the silica is uncertain as neither the framework grains nor clay minerals show signs of solution. Phytolith reprecipitation is one possibility, but phytolith remnants are rare in these pans.

Processes of silicification

Millot (1970) led the way in studying silicification from a solution chemistry and crystallographic perspective, but still the processes of silica solution and precipitation are poorly understood. Thiry (1997) summarizes many aspects of silcrete chemistry and formation, but neglects many others.

Silica solubility is little affected by changing pH throughout the pH range 4–8 (Chapter 6). However, if alkaline silica-saturated waters are neutralized, silica precipitates. This in itself does not explain most regolith silcrete occurrences, because external evidence for pH fall is lacking, and such a change (from pH 9 to pH 7) would not separate silica from the more insoluble ubiquitous elements Al and Fe. Although highly acid water (pH 4 or lower) theoretically can dissolve Al and Fe but not Si, such waters are uncommon but silcrete is common.

Other ions in solution have been shown to increase silica solubility, particularly Na, and most Australian groundwaters are saline. The role of organic matter and microbes in the solution of silica is an open question, but there is evidence from bioremediation of petroleum that quartz is etched during that process (Bennett & Siegel 1987).

The rate of silica dissolution and precipitation is also important in coming to understand silcretes. Experimental data indicate substantial interaction of organic acids with silica and their presence in solution increases the dissolution rate of quartz. On the other hand, silica cannot precipitate from solution as quartz without nucleation. It can, however, nucleate on 'clean' quartz surfaces, but where quartz surfaces are 'poisoned' quartz does not precipitate. This may explain why most groundwater is oversaturated with respect to quartz. It may also explain why massive silcretes form as lenses and patches rather than regularly distributed masses, as its formation might be controlled by the distribution of 'clean' quartz, which in turn may be related to preferred conduits for groundwater flow or particular sedimentary facies.

Once amorphous silica is precipitated it can dehydrate and crystallize as microquartz, which may in turn eventually form quartz. Recrystallization into quartz is favoured when equilibrium is established between opal dissolution, quartz nucleation and quartz crystal growth rates (Williams *et al.* 1985).

Overall, the detailed mechanisms for the dissolution and precipitation of siliceous cements in silcretes remain obscure. The fact is it happens and it is somewhat ironic that we know so little about the geochemical behaviour of the Earth's second most common element.

Anatase is almost ubiquitous in silcretes. Milnes (1983) indicates that Ti (with Zr and Al) is introduced to the silcrete concomitantly with the SiO_2. Thiry (1997) and Thiry & Milnes (1991) suggest by implication that the anatase in massive and structured silcretes from the Stuart Creek opal field is residual. Again we know very little about the origin of the Ti in silcrete, even though it is generally the second most abundant cation.

From field evidence and geochemistry it is apparent that prior to, or during, silicification Al_2O_3 is removed from the parent materials of some silcretes. No basalt inter-flow silcretes have been recorded to contain any recognizable remnants of basalt-derived framework. Silcrete cropping out in lateral continuity with kaolin/quartz saprolite contains no kaolin, and only trivial amounts of Al_2O_3. It is clear there remains a problem relating to the dealuminification of regolith associated with silicification, a problem not yet addressed in the literature.

Geochemistry

Analysis of silcretes clearly shows the high SiO_2 content, the low Al_2O_3 and Fe_2O_3 and high TiO_2 values. Other cations are virtually absent. This contrasts with the few red-brown hardpan analyses given where these values are reversed. Lintern & Sheard (1998) report significant Au values in silcreted *in situ* regolith from the Gawler Craton in South Australia. They comment that although the Au is not mobilized during silcrete formation the silcrete fixes the Au in place.

Approximately 190 analyses of materials called silcrete from 26 authors[1] have been collated and analysed and some geochemical variable noted. An average analysis of these samples shows the following anhydrous chemistry:

SiO_2 92.15%, Al_2O_3 2.28%, total Fe as Fe_2O_3 2.88%, MgO 0.25%, CaO 0.18%, K_2O 0.30%, Na_2O 0.07%, TiO_2 1.60%, ZrO_2 0.00, MnO 0.01%, P_2O_5 0.05%

As the 'silcretes' are not all accurately described there are some materials that have higher than expected contents of Al_2O_3 and Fe_2O_3 and less SiO_2 than we expect for a silcrete, but the averages are representative of a large number of analyses of materials authors have called silcrete. Silica values ranged from 4% to nearly 100% with a modal value of 97.5%, Al_2O_3 from 39% to <1% with a mode of <1%, Fe_2O_3 from 69% to <1% with a mode of <1% and TiO_2 from 48% to <1% with a similar modal value.

In silcretes that have >96% SiO_2, TiO_2 tends to be more abundant than either Al_2O_3 or Fe_2O_3, and TiO_2 becomes most significant as SiO_2 is highest. In all the 190 samples SiO_2, Al_2O_3 and Fe_2O_3 constitute just about everything in the materials. There is no correlation between TiO_2 content, Al_2O_3 and Fe_2O_3. Only about a dozen analyses contained Zr, but from those there appears to be a positive correlation between Zr and TiO_2. This tends to suggest that TiO_2 accumulation in silcrete is a residual accumulation rather than absolute, however the evidence of Hutton *et al.* (1978) suggests otherwise.

The other interesting point is that as SiO_2 increases in silcrete the proportion of Al_2O_3 decreases, suggesting that dealuminification occurs during the formation of high silica silcretes. Whether this is a solution phenomenon or simply the physical removal of clay minerals is unknown, but where silcretes form on granite saprolite it is clear that fabrics are preserved, suggesting that chemical removal and replacement occurs.

By using Zr as an insoluble element in silcrete Webb & Golding (1998) have shown that significant silica is added to silcretes during their formation. In silcretes associated spatially with basalt they conclude it came from their weathering. In these silcretes the volume of SiO_2 added as cement has caused a significant volume increase causing, they say, a floating framework fabric. In silcretes not associated with basalt they surmise the SiO_2 was added from saline waters.

Age of silcretes

In Australia this has been a matter of some controversy for a long time. Few if any silcrete minerals are suitable for dating so dates are indirectly inferred. In the Paris Basin where silcrete is buried by other formations it is easy to date their formation (Thiry & Simon-Coinçon 1996). In Australia, although many silcretes are sandwiched between basalt flows, it is not possible to date the silicification, only the depositions of the sediments they silicify. These range through from the Late Cretaceous to the later Neogene. Taylor & Smith (1975) argue silicification occurred after the overlying basalts erupted, with the weathering basalts providing a silica source (Taylor *et al.* 1990b). Others have argued the reverse. Vasconcelos (1996) has dated Mn-minerals by $^{40}Ar/^{39}Ar$ techniques to show the silcretes, from central Queensland, were formed before 19 Ma. Palaeomagnetic dating has been used to date ferruginization in weathering profiles with which silcrete is associated in south-western Queensland as Late Cretaceous–Eocene and Miocene (Idnurm & Senior 1978). These two dates have been unjustifiably extended almost continent-wide by some researchers. There is little evidence to show silicifications were contemporaneous with ferruginization and hence these correlations may well be spurious. Despite this Taylor (1994) has shown that deep weathering

1. Alexander & Cady (1962), Butt (1983), Callender (1978), Dury & Habermann (1978), Fox (1936), Goudie (1973), Gunn & Galloway (1978), Hutton *et al.* (1978), Kerr (1955), Maignien (1966, 1968), Nash *et al.* (1994), Netterberg (1985), Sivarajasingham *et al.* (1962), Radke & Bruckner (1991), Senior (1979), Summerfield (1983), Taylor & Smith (1975), Thiry & Simon-Coinçon (1996), Webb & Golding (1998), van Dijk & Beckmann (1978), Wopfner (1978).

and silcrete dates reported for Australia in the literature spread through the whole of the Late Cretaceous and Tertiary, thus it would seem reasonable that silcretes formed throughout this period.

Palaeoclimatic and topographic implications

Many have written about the palaeoclimatic significance of silcrete. Summerfield (1983) suggested the TiO_2 content of silcrete was a palaeoclimatic indicator: high in wetter areas and low in drier. This certainly does not hold in Australia (Young 1985). Webb & Golding (1998), studying silcrete in Victoria and South Australia, found that Ti content has no climatic significance, but they do conclude that the silicifying solutions were at or very near ground temperatures. Webb & Golding (1998) conclude that the formation of silcrete depends on the complex interplay of climate and silica supply and it is impossible to generalize that the presence of a silcrete is indicative of any particular climate.

Massive silcretes, because they form by evaporation of groundwater, form most commonly near the watertable and where groundwater is able to evaporate. This implies that they form in topographically intermediate to low positions by absolute accumulation of silica. They are thus a feature of dissected landscapes. Structured silcrete, on the other hand, can form almost anywhere where pedogenetic processes occur faster than erosional stripping of the landscape. This is generally in relatively flat terrains.

CALCRETE AND OTHER CARBONATE DURICRUSTS

> Calcrete is a near surface, terrestrial, accumulation of predominantly calcium carbonate, which occurs in a variety of forms from powdery to nodular to highly indurated. It results from cementation and displacive and replacive introduction of calcium carbonate into soil profiles, bedrock and sediments in areas where vadose and shallow phreatic groundwater become saturated with respect to calcium carbonate (Wright & Tucker 1991).

This general definition is preferred because the term has been used loosely in the past (e.g. Anand *et al.* 1997) and it is not helpful to produce a terminological quagmire. Lamplugh (1902) introduced the term for calcium carbonate cemented gravels. He later used it to refer to calcium carbonate materials deposited as a result of evaporation of lime-charged waters in the valley of the Zambezi River. Milnes & Hutton (1983), like Netterberg (1978), used the term to describe terrestrial materials cemented by carbonates in general. We do not accept this generalization, preferring calcrete to be confined to accumulations of calcium carbonate. Consider a material hardened by siderite ($FeCO_3$): would you call it a ferricrete or calcrete? In keeping with Lamplugh's (1902) definitions we prefer the term '-crete' to reflect the dominant cations in the cementing material.

The most important calcretes are those in soils where they form a calcium carbonate pan. Yaalon (1988) estimates about 13% of the Earth's landscapes are covered by them. They are very common in landscapes with seasonal climates where the dry season allows calcium carbonate to accumulate. They are typically not part of thick weathering profiles as many other duricrusts may be, but they may be superimposed over earlier deep weathering profiles as climates shift through time. They tend to not be as hard as ferricrete or silcrete and as a consequence are more erodible. Consequently they do not provide the threshold to erosion that ferricrete and silcrete do.

Other carbonate duricrusts include dolocrete, magnecrete and siderocrete, but the latter two are relatively uncommon.

Classification

There are many types of calcretes and classifications schemes. Wright & Tucker (1991) provide one of the most recent, based essentially on classifications of Netterberg (1980) and Goudie & Pye (1983) (Table 13.12). This is a morphological classification, and although various forms have genetic implications they are not defined in the classification.

Calcrete profiles have for a long time been seen to have a developmental sequence or form a chronosequence where form changes as profiles become more developed or mature. This of course implies a given sequence of evolution which is certainly not always the case, but it is observed sufficiently for the concept to have some veracity.

Machette (1985) provides a classification based on evolutionary models of calcrete formation (Table 13.13). One problem with this classification is that lacustrine limestones can undergo weathering and transformation to mimic the fabrics described in stages 5 and 6 in Table 13.1 above. Discrimination is very difficult.

Calcretes can also be classified by the hydrological setting of their formation (Carlisle 1983 in Wright & Tucker 1991) (Figure 13.26). While this may be

Table 13.12 *Morphological classification of calcrete (from Wright & Tucker 1991)*

Type	Description
Calcareous soil	Very weakly cemented or uncemented soil with small accumulations as grain coatings, patches of powdery carbonate including needle-fibre calcite (pseudomycedia), carbonate-filled fractures and small nodules
Calcified soil	A firmly cemented soil, just friable; few nodules. 10–50% carbonate
Powder calcrete	A fine, usually loose powder of calcium carbonate as a continuous body with little or no nodule development
Pseudotubule calcrete	All, or nearly all, the secondary carbonate forms encrustations around roots or fills root or other tubes
Nodular calcrete	Discrete soft to very hard concretions of carbonate cemented and/or replaced soil. Concretions may occur with laminated coatings to form pisoids
Honeycomb calcrete	Partly coalesced nodules with interstitial areas of less indurated material between
Hardpan calcrete	An indurated horizon, sheet-like. Typically with a complex internal fabric, with sharp upper surface, gradational lower surface
Laminar calcrete	Indurated sheet of carbonate, typically undulose. Usually, but not always, over hardpans or indurated substrate
Boulder/cobble calcrete	Disrupted hardpans due to fracturing, dissolution and rhizobrecciation (including tree heave). Not always boulder grade (clasts are rounded due to dissolution)

useful in modern environments it is difficult to apply to calcretes no longer in equilibrium with their formative environment as the calcrete types which form in pedogenetic and phreatic zones may be very similar (Wright & Tucker 1991). It is, however, useful in bringing to our attention the fact that calcrete can form at considerable depths in the regolith and not necessarily in the vadose zone.

It is well known that the composition of calcrete cements is not 100% calcite, but also includes varying proportions of Mg. Netterberg (1980) thus provided a classification of carbonate-rich duricrusts to incorporate the idea of mineralogy in classification. We have modified his classification to include magnecrete as one possible end-member (Table 13.14).

Wright (1990) proposed yet another classification based on the micromorphological features of calcretes. This classification essentially differentiates between those with framework (parent material) grains floating in a micrite plasma with some sparry calcite crystals and those showing obvious organically formed microfabrics of micrite such as rhizotubules, pellets, microbial coatings and needle-fibre calcite crystals. Mixtures of these two types occur and can be described in terms of the two end-member types.

The various classifications demonstrate the range of thinking about calcrete and the various features various researchers consider important to describe in order that calcretes be understood and their origins interpreted.

Precipitation of carbonates

It is commonly considered that calcite (and other carbonates) precipitate by evaporation/evapotranspiration or by CO_2 degassing. There are other mechanisms, but whatever the mechanism the solubility of $CaCO_3$ is decreased by the removal of CO_2 or H_2O or by the addition of Ca^{2+} (or Mg^{2+}). Water is easily removed by evaporation or evapotranspiration and is the common cause of calcite precipitation in semi-arid to arid environments. It may also be the reason for the formation of micritic rhizotubules. Carbon dioxide degassing occurs because its partial pressure in soils is significantly higher than in the atmosphere. In the soils P_{CO2} ranges from an average of 0.9% in temperate soils to <0.1% in arid soils, but these values are high compared to the atmosphere with a P_{CO2} of 0.03%.

Other mechanisms of calcite precipitation include common ion effects and organic processes. The presence of high Ca^{2+} or Mg^{2+} in groundwater will enhance the possibility of calcite, high-Mg calcite or dolomite precipitation. Such situations occur near arid zone playas, or perhaps in areas around weathering sulfides where gypsum may occur as a secondary product in semi-arid and arid climates. Plant roots cause the precipitation of calcite by removing H_2O and CO_2 and, as a consequence, increasing pH. It is also possible that bacteria and fungi cause calcite precipitation by the removal of CO_2 and thus affecting pH, or by dumping Ca^{2+}.

Some light has been shed on these processes by stable-isotope studies. Wright & Tucker (1991) conclude that negative $\delta^{13}C$ values (−10) indicate the

Table 13.13 *Classification of pedogenetic calcretes based on stages of development (from Machette 1985). The high gravel content refers to >50% gravel, low is <20%. Percentage $CaCO_3$ refers to the <2 mm fraction.* K *refers to a carbonate soil horizon, and* $_m$ *to its being indurated*

	Stage	Gravel content	Diagnostic Features	$CaCO_3$ distribution	Max. % $CaCO_3$
Immature	1	High	Thin discontinuous coatings on pebbles, usually on underside	Coating sparse to common	Tr.–2%
		Low	Few filaments in soil or faint coatings on ped surfaces	Filaments sparse to common	Tr.–2%
	2	High	Continuous, thin to thick coatings on tops and undersides of pebbles	Coatings common, some carbonate in matrix	2–10%
		Low	Nodules, soft 5–40 mm in diameter	Nodules common, generally non-calcareous to slightly calcareous	4–20%
	3	High	Massive accumulations between clasts, fully cemented in advanced forms	Continuous in matrix to form K-fabric	10–25%
		Low	Many coalesced nodules, matrix is firmly to moderately cemented	Continuous in matrix to form K-fabric	20–60%
Mature	4	Any	Thin (<2 mm) to thick (>10 mm) laminae capping hardpan (K_m). Calcrete is impervious at this stage	Cemented platy to tabular structure. K_m horizon is 0.5–1.0 m thick	>25% in high gravel >50% in low gravel
	5	Any	Thick laminae (>10 mm); small to large pisoids above. Laminated carbonate may coat fracture surfaces	Indurated, dense, strong, platy to tabular structure. K_m horizon is 1–2 m thick	>50% in high gravel >75% in low gravel
	6	Any	Complex fabric of multiple generations of laminae, brecciated and recemented, pisolitic. Typically with abundant peloids and pisoids	Indurated, dense, thick, strong tubular structure. K_m horizon is commonly >2 m thick	>75%

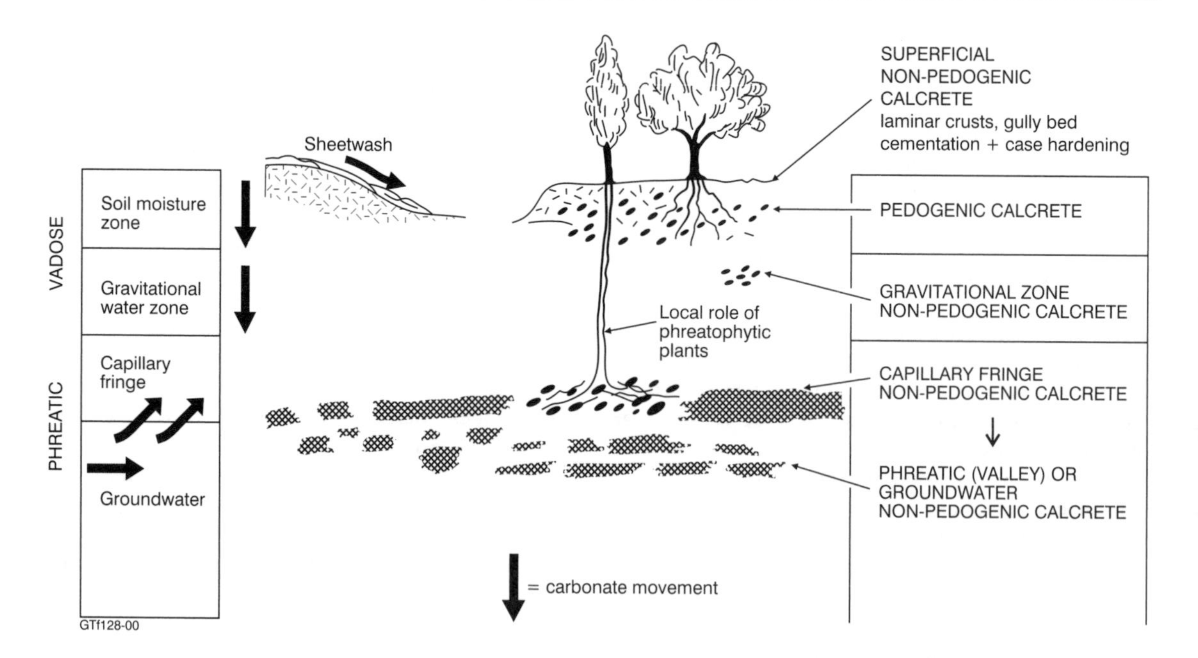

Figure 13.26 *Classification of calcretes by hydrological setting (from Wright & Tucker 1991, based on Carlisle 1980, 1983)*

Table 13.14 *Classification of carbonate-rich duricrusts (after Netterberg 1980)*

Name	% dolomite by mass of soil carbonates	≈ Equivalent % of $MgCO_3$[a]
Calcrete	<5	<2
Magnesian calcrete	5–10	2–5
Dolomitic calcrete	10–50	5–25
Calcitic dolocrete	50–90	25–40
Dolocrete	>90	>40
Magnecrete (magnesite)	0	100

[a] $\%MgCO_3 = \frac{MgCO_3}{MgCO_3 + CaCO_3} \times 100$

CO_2 has come from the respiration of excess ^{12}C by certain types of plants (C3). Higher $\delta^{13}C$ values are possibly caused by contamination from existing calcite in the regolith or, more probably, a lower respiration rate or plants (C4 or CAM) which have less negative $\delta^{13}C$ values. Values of $\delta^{18}O$ have been used to determine palaeotemperature (Suchecki *et al.* 1988; Naylor *et al.* 1989) and evaporation as the cause of precipitation (Drever *et al.* 1987).

Source of calcium (magnesium)

Although Ca^{2+} is relatively common in groundwater, HCO_3^- is not, except in areas where calcretes are forming or rocks like basalt or limestone are weathering. Many calcretes are formed in the pedosphere as well as in deeper parts of the regolith in the vadose, capillary and phreatic zones. The source of Ca for the phreatic and capillary zone calcretes may well be the groundwater. But those in the vadose zone, and particularly in the pedosphere, are not likely to form from groundwater-derived Ca as groundwater may be tens of metres below them and capillary rise will be only be up to 1 m at most.

Other potential sources of Ca include rainwater, marine aerosols, terrestrial biota (molluscs), vegetation, dust and rocks, and in lower landscape positions, groundwater. In arid and semi-arid climates dust is probably the most significant source of Ca and Mg added to the regolith surface. Many playas contain significant Ca- and Mg-rich clay minerals and carbonates which, during dry lake phases, are free to be blown from the lake surfaces and deposited downwind. Callen (1976) has demonstrated that a wide range of Mg/Fe/Ca-rich clay minerals occur in the Lake Frome depression in eastern South Australia. Similarly, Kiefert & McTainsh (1996) record smectite as a significant component of dusts sweeping across eastern Australia, across the Tasman Sea to New Zealand. These dusts contribute significant Ca/Mg/Fe to the regolith on which they deposit as the clay minerals weather.

These additions of Ca-containing materials to the regolith surface may then be moved through the vadose zone of the regolith depositing in various forms. Deposition is most probably due to evaporation and evapotranspiration or through solutions coming into contact with localized zones of increased concentrations of Ca^{2+}.

Groundwater calcrete

Calcretes precipitated from groundwater may be extensive (10 km wide by 100 km long and 10 m thick) (Arakel 1986). They form lenses, lumps and sheets of micrite cemented and replaced framework and plasma grains. They are most commonly formed where groundwater is forced to near the surface by bedrock highs or at springs. Evaporation, degassing or the common ion effect causes precipitation when Ca/Mg bicarbonate waters mix with Ca/Mg sulfate- or chloride-rich groundwaters often associated with playas. Arakel (1986) describes such a situation from central Australia at Lake Napperby (Figure 13.27). An interesting point about these calcretes is that they increase in Mg content downflow and downprofile. A similar trend in pedogenetic calcretes is also noted in many Australian profiles.

Groundwater calcrete is difficult to distinguish from other carbonate accumulations, but a few pointers can be given. They tend to be nodular, brecciated or massive and do not have the 'mature' profile of pedogenetic calcretes (Wright & Tucker 1991). As most groundwater calcrete forms below the zone of biotic activity it is rare to find biological features in them and similarly desiccation features will also be uncommon.

Pedogenetic calcrete

These profiles develop in stages as listed in Table 13.13; $CaCO_3$ accumulates in the upper regolith or soil by precipitation and replacement, often causing a separation of framework grains. A mature profile would thus consist of a layer of coalesced nodules of micritic calcite with displaced fabrics overlain by a laminated calcite layer which may itself be overlain by calcite pisoliths. Continued development may result in the chemical reworking of the calcite to form

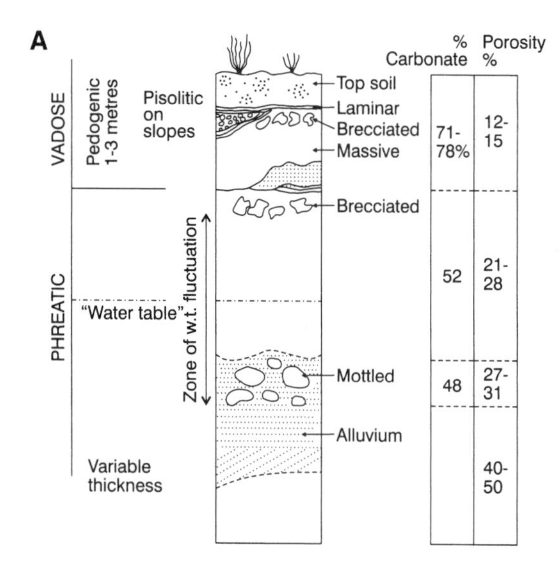

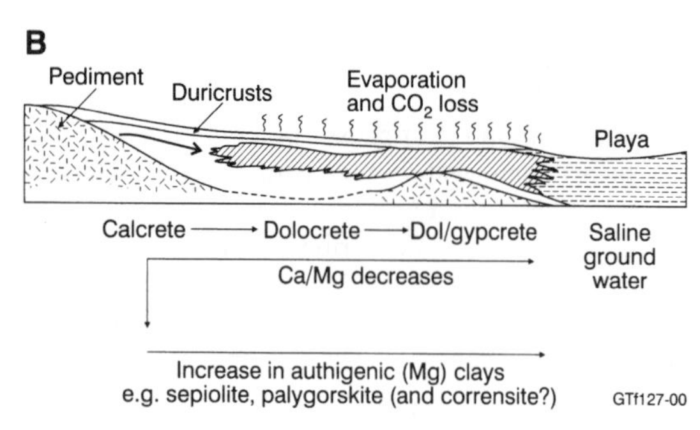

Figure 13.27 *Evolution of (a) groundwater and (b) its precipitates across an arid alluvial basin (from Wright & Tucker 1991)*

microkarstic features possibly filled with pellets and pisoliths as well as brecciated fragments of calcrete.

Calcrete geochemistry

The majority of geochemical data collected from calcrete, and other carbonate-rich duricrusts, is limited to analyses of Ca and Mg and in some cases to SiO_2, Al_2O_3 and Fe_2O_3. Other minor element analyses are rare. Goudie (1973) provides a summary of data from some 220 analyses from Africa, India, Cyprus, Libya and Australia. Milnes (1983) has compiled 330+ analyses of calcrete, calcrete components and associated regolith from southern Australia, and particularly the South Australia region. From these summaries it is most obvious that calcite is the dominant mineral in carbonate duricrusts, but high-Mg calcite is relatively common. Dolomite is rare, and none report magnesite as one of the potential end-members of the calcrete–magnecrete continuum. Where profiles have been analysed it is common for the Ca : Mg ratio to decrease downprofile. This trend has also been noted in western New South Wales (Hill *et al.* 1999; Smith 1996), and the southern Yilgarn Craton of Western Australia (Anand *et al.* 1997). Few calcretes grade downward into dolocretes. Anand *et al.* (1997) report dolomite in several profiles from Western Australia at depths of about 50 cm in calcrete profiles and Watts reports it from depths ranging from the surface to about 2 m in Kalahari calcretes.

Goudie (1973) compiled a table of world average calcrete compositions (Table 13.15). Although these data are now quite dated, similar values are still being reported from individual studies (e.g. Hay & Reeder 1991 from Olduvai Gorge in Tanzania, CaO $\approx$ 44%). Others, however, are reporting calcrete in which the CaO values are significantly less than the world average (e.g. Anand *et al.* 1997 from calcretes of the Yilgarn Craton in south-western Western Australia, CaO $<$20%).

This brings into question how the term 'calcrete' is being used by different authors. Anand *et al.* (1997) use the term to describe carbonate accumulations in soils where CaO concentration factors compared to bedrock range from maxima of +0.8% to +153.0%.

Trace element data on calcretes are rare and not widely reported. Some authors report relatively immobile element (Zr, Ti) data to indicate concentration/depletion ratios with respect to bedrock values, some report calciophile elements such as Sr, again to determine relative element mobility in the profile. Manganese values are reported occasionally, but the reasons for this are obscure.

Since the discovery by Butt & Zeegers (1992a) and Lintern & Butt (1993) that soil carbonates (calcretes) contain Au, calcrete trace element analyses have been more routinely collected in Australia where they are now used as a prospecting tool for Au. Few if any of these analyses have been published for commercial reasons.

Summary

Calcrete is a term referring to an end-member of the calcrete/magnecrete spectrum. It is composed of predominantly calcite concentrations in the regolith. It may form pedogenetically near the surface as a result of additions of Ca to the soil surface, or from groundwater precipitation in the capillary fringe or at springs.

Table 13.15 *World average calcrete compositions (from Goudie 1973)*

	Mean %	Ratio of CaO	Ratio of SiO_2
$CaCO_3$	79.28	–	–
SiO_2	12.30	3.47	–
Al_2O_3	2.12	20.07	5.8
Fe_2O_3	2.03	21.02	6.06
MgO	3.05	13.96	–
CaO	42.62	–	–

It is chemically relatively Ca-rich (≈40%) with varying amounts of Mg. Few recorded calcretes contain dolomite, but as research continues more dolomite is being identified at depth in profiles.

MAGNECRETE

Magnecrete is a rare duricrust composed of magnesite ($MgCO_3$); however, concentrations of magnesite in the regolith certainly occur. At Kunwarara 60 km north of Rockhampton, Queensland, significant magnesite deposits occur in Tertiary fluvial deposits overlying granitic bedrock. Deeply weathered ultramafic rocks surround the Tertiary basin that release Mg^{2+} in significant amounts. The Mg^{2+} precipitates as displacive magnesite nodules of high purity along drainage pathways. The nodules, although not cemented, vary from densely packed to forming pseudocolumnar features (Figure 13.28) in the sedimentary matrix. In places where erosion has exposed the magnesite nodules they form a highly erosion-resistant surface lag, similar in many respects to a duricrust (Figure 13.28) (see Chapter 7). Similar magnesite deposits occur elsewhere along the coastal ranges north of Rockhampton and associated with Ni-laterites in the Yilgarn Craton, Western Australia (Brad Pillans personal communication 1998).

(a)

(b)

Figure 13.28 *(a) Magnesite pseudocolumns in an alluvial matrix from Kunwarara, near Rockhamption, Queensland and (b) magnesite accumulations in weathered bedrock fractures from Marlborough, north of Rockhampton, Queensland (photo GT)*

MANGANOCRETE

Manganocrete is a duricrust in which the framework has been cemented or replaced by Mn-oxide minerals. Weber (1997) suggests there were two main periods of manganogenesis, Middle Proterozoic and from the Cretaceous to Oligocene. She also points out that weathering processes akin to 'lateritization' have enriched these Mn-rich rocks. An excellent example of this is at Groote Eylandt in the Gulf of Carpentaria where mid-Cretaceous marine sands and clays have been weathered to form significant Mn deposits. These Mn-ores are very similar in many respects to the bauxites at Weipa except, due to the relative abundance of Mn-source minerals at Groote Eylandt, Mn has formed the duricrust rather than Al.

The geological setting of the Groote Eylandt is summarized by Pracejus *et al.* (1988) in a cross-section of the deposit and a log of the deposits (Figure 13.29). The Cretaceous Mullaman beds (95 Ma from fossils and K/Ar dates) consist of quartzose sandstones conformably overlain by shallow marine glauconitic claystones. The deeper parts of the sequence also

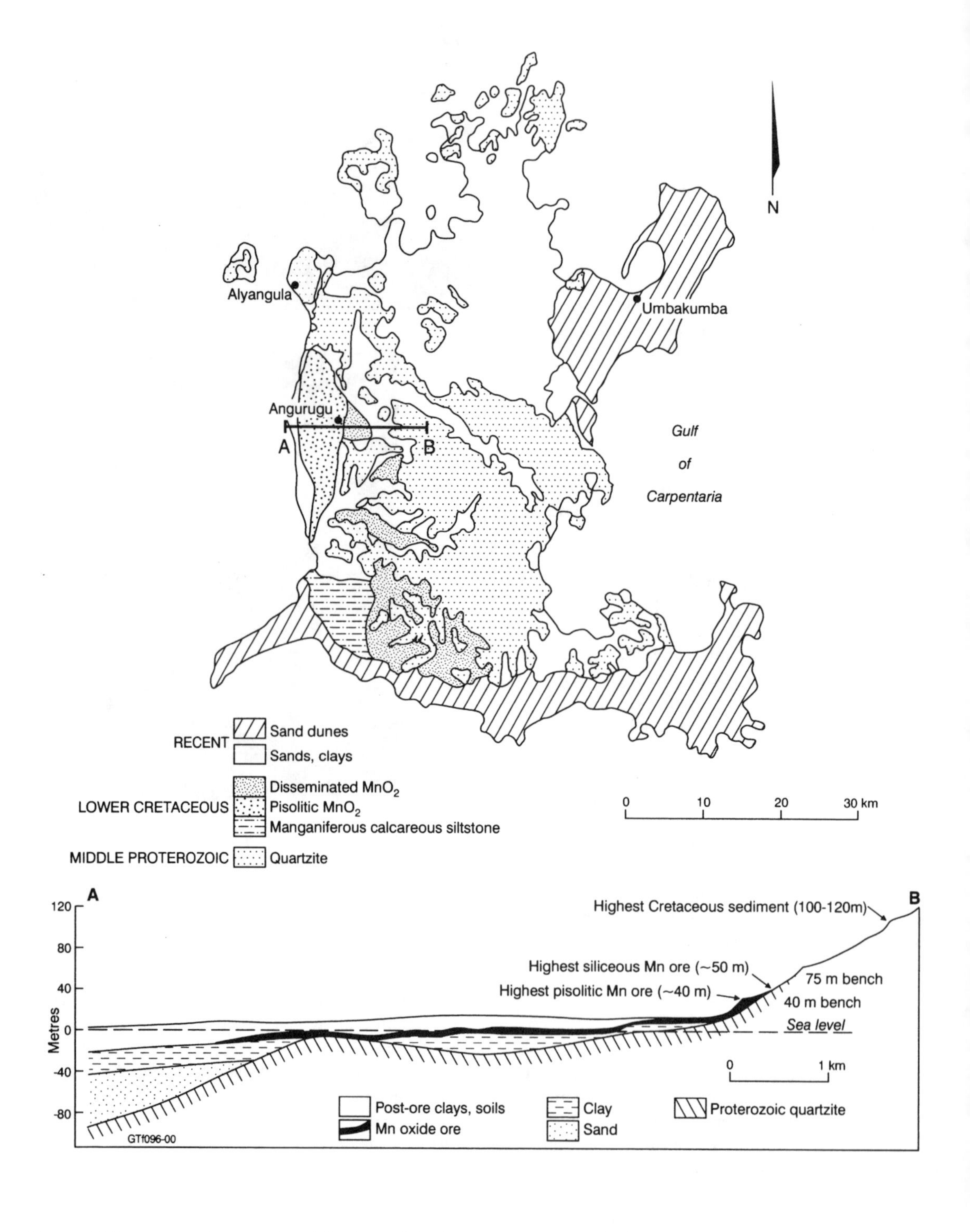

Figure 13.29 *Map and cross-section of the Groote Eylandt manganocretes (modified after Pracejus* et al. *1988 and Nott 1995)*

contain Mn-oolites. The Mullaman beds range from 0 to 100 m thick. The upper mangan–pisolite zone is inversely graded (2–3 mm at the base to 25 mm at the top) and contains well-bedded oolitic beds, suggesting a sedimentary nature to the Mn deposits. Bolton *et al.* (1990) suggest the deposits are shallow marine, forming in anoxic Mn- and organic-rich waters. As sea levels fell, oxygenated water caused the Mn to be fixed as pyrolusite (and many other Mn minerals). The deposit also contains kaolin and detrital quartz, with illite and smectite only being recorded in trace amounts in the surrounding sediments. The inverse grading and the growth of layered pisoliths and ooliths are said by them to reflect increasing marine current energy levels as the Cenomanian seas fell. The Mn deposits are overlain by 'laterite' consisting of 'ferruginous, soft and indurated, red, brown and yellow mottled clays and sands, goethitic and manganiferous pisolites, and pebbles and clasts of manganese oxide and orthoquartzite' (Bolton *et al.* 1990). They also describe secondary alteration of the Mn to form massive layers of Mn-oxides, Mn-spherulites, Mn concentrations and dendrites in areas where the 'laterite' is missing.

Frakes & Bolton (1984) and Ollier & Pain (1996) as well as others have suggested the Mn was derived from weathering and lateral transport to its site of precipitation. Frakes & Bolton would presumably suggest that the Mn was moved into the marine basin, stored in a reduced state in the marine sediments and later oxidized as seas retreated. Ollier & Pain (1996), however, imply a terrestrial process by suggesting the Mn deposits are 'reminiscent of foot slope ferricrete'. They do comment that Mn precipitation derives from chemical and biological sedimentary precipitation and weathering processes are probably restricted to minor natural benefaction.

We suggest that the remarkable similarity between the Groote Eylandt Mn deposits and the Weipa bauxites indicates that, like Weipa, Groote Eylandt could well be a weathering profile. Nott (1995) shows that Cretaceous sediments on Groote Eylandt occurred up to heights of about 120 m above sea level (a.s.l.) and Bolton *et al.* (Figure 13.29) show pisolitic Mn-ores only as high as 40 m a.s.l. and the highest Si–Mn-ore at 50 m. This means that at least 50 m of sediment have been removed from the western side of Groote Eylandt since 95 Ma. If that sediment had an average Mn content of 8% then it is possible if under weathering Mn was conserved locally ore grades could have reached 80% over 5 m (Figure 13.34). The overwhelming presence of kaolin as the major clay associated with the manganiferous facies of the deposit and the lack of illite (muscovite) suggest terrestrial environments for the clay mineral formation. On the basis of this we suggest that the Groote Eylandt Mn deposits may well be terrestrial, formed by weathering of the Cretaceous sediments under a variety of climatic conditions on a tectonically stable (Bolton *et al.* 1990) platform.

Ruxton & Taylor (1982) and Taylor & Ruxton (1987) describe manganocretes forming part of a duricrust catena from 80 km north-east of Canberra. Here MnO_2 (pyrolusite) cements colluvial deposits consisting of weathered Ordovician metasediments to form a slope mantle below bauxites that are all that remain of formerly extensive Tertiary basalts. The manganocrete has obviously formed by Mn^{2+} being leached from the basalt, along with trace elements (Co, Ni) and been precipitated among the colluvium lower on the slopes presumably due to its oxidation.

GYPCRETE

Gypcretes are duricrusts formed by the cementation or replacement of framework grains by gypsum or by the wholesale solution and reprecipitation of gypsum. They are relatively common in contemporary arid climate areas. Gypsum crystals grow in clastic sediments, either by enclosing or displacing the clastic particles. As precipitation continues the crystals coalesce to form a solid pan or layer of microcrystalline gypcrete. This most commonly occurs through the leaching and recrystallization of gypsum under subaerial evaporative conditions (Chen *et al.* 1993, Chen 1994) (Figure 13.30). At Lake Amadeus in central Australia Chen *et al.* (1993) describe gypcretes intercalated within lacustrine clays which represent periods of exposure of the clays and their cementation by gypsum.

In other localities around lakes across southern Australia gypsum (kopi) dunes form typically on the leeward side of saline lakes. The solution and reprecipitation of gypsum too may cement these.

Interpreting the regolith environment

It should now be clear that by observation, mapping and analysis, the environment of formation of a regolith profile can be, at least to some extent, interpreted. There are always hazards in interpreting from observation, for the interpretation depends very

Figure 13.30 *Gypcrete forming in aeolian gypsum dunes from Lake Cowan near Kalgoorlie, Western Australia (photo Jonathan Clarke)*

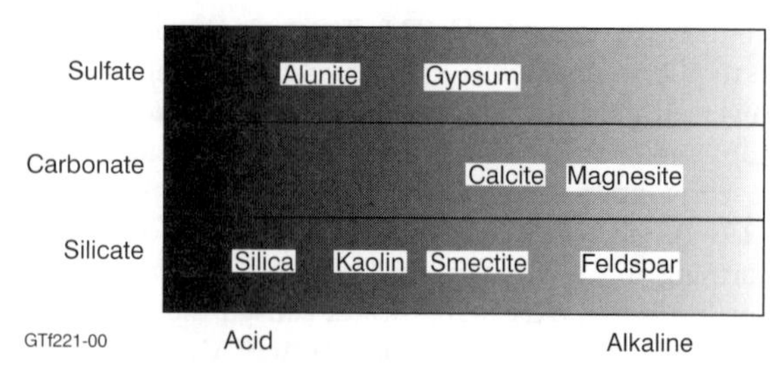

Figure 13.31 *Minerals indicative of pH*

strongly on the model, or paradigm, used to make that interpretation.

For example, a weathering profile having fresh rock at the base, then saprock, then saprolite and a thin interval of smectitic regolith before the topsoil might be interpreted as a rather young profile. A safer conclusion might be that it is an immature profile. Its character could result from a long period of very slow weathering, from a shorter period of mild weathering, or from an even shorter period of aggressive weathering. Interpreting the profile from within is always a risky business. Similar problems have been alluded to elsewhere in this book. Lateritic weathering is popularly thought to equate with tropical climates. By now you know that, while there are indeed many lateritic profiles that evolved in tropical climates, they also evolve where and whenever weathering goes on uninterrupted for long enough to create such a profile – hot or cold, wet or dry.

Nonetheless, there are many aspects of the regolith environment that observations can pin down; many ways to establish some parameters that will give us evidence of past environments, or that may show equilibrium with the present.

ACID OR ALKALINE

The pH of regolith waters has a very profound effect on the minerals that are stable. For example, in acid environments alunite may replace kaolinite, in alkaline waters magnesite or sepiolite precipitate replacing smectite and probably dissolving quartz. While quartz is found over a wide pH range, its greater solubility in alkaline waters means that it will precipitate if the water becomes acid. Since Al is more soluble in acid than neutral water, silcretes may indicate precipitation of silica in an environment when Al was mobilized, for if Al and silica are present together, kaolinite or smectites are the minerals most likely to form (Figure 13.31).

REDOX CONDITION AND pH

There are some minerals in the regolith that are indicative of redox state, minerals that can only be present in a restricted range of oxidation potential. Commonly their stability is also pH dependent, and examples are given in Chapters 6 and 7. Figure 13.32 summarizes this for the most common minerals.

DRAINAGE

Where groundwater flow is slow or impeded, the water has a longer residence time in the profile, and so has longer to react with the minerals and to achieve equilibrium, that is, to become saturated with some, if not all of the components being released. Thus where a mafic rock is weathering, Ca, Mg, Si, Fe and Al are all being released. If there is slow flow of the groundwater, all might reach saturation with respect to the regolith minerals, and with carbonate introduced from the atmosphere and decaying biotic material, a typical assemblage of calcite–saponite–goethite may develop. The same profile under free flow conditions may only reach saturation in Al, Si and Fe, leading to a kaolinite–goethite association. The expressions 'slow flow' and 'free flow' need to be treated with some caution. If there is very little water entering the system because of low rainfall, saturation can occur, leading to the same assemblages as under impeded drainage in a wetter climate. Thus smectites are indicative of either aridity or impeded drainage (swamps).

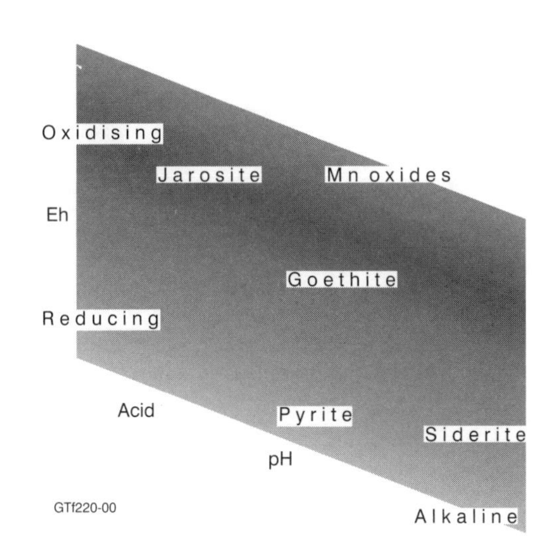

Figure 13.32 *Eh/pH mineral indicators*

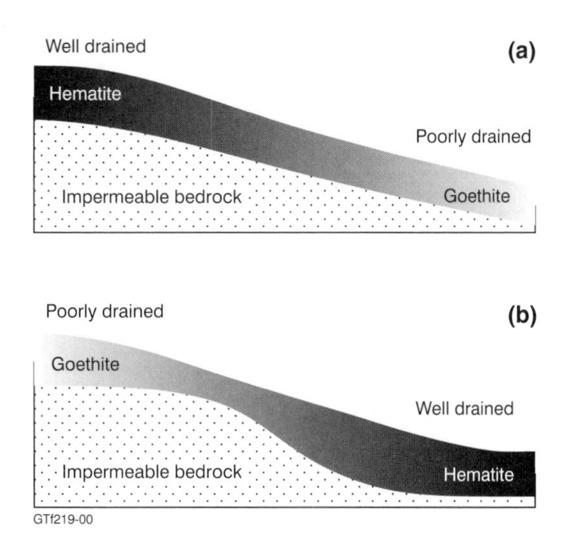

Figure 13.33 *(a) A common situation: well-drained crest and poorly drained valley. (b) Situation west of Charters Towers, poorly drained crest with relatively shallow depth to bedrock and better drained valley*

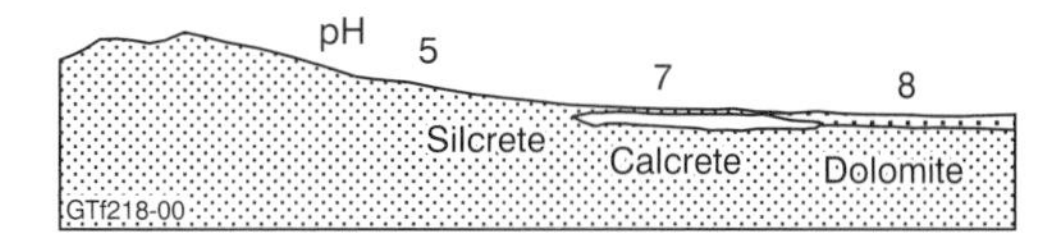

Figure 13.34 *Variation in regolith precipitates with landscape position and increasing pH. Example from Western Australian saline lakes after Carlisle (1983)*

CATENARY POSITION

Landscape position is a major factor in setting the local regolith environment. Generally, hill crests drain more freely than slopes and valley bottoms, and this affects the regolith mineralogy. Typically, the drier crests will have haematite > goethite, boehmite > gibbsite, whereas the wetter valleys will show the reverse (Figure 13.33a). Similarly, progressive desilification from smectite to kaolinite to gibbsite is a common trend both vertically up a lateritic profile, and up the catena. But again, caution in interpretation is needed. The real relationship revealed by the mineralogy is drier versus wetter, not necessarily higher versus lower. If the underlying bedrock impedes drainage more at hill crests than in valleys (Figure 13.33a, b), the crests can become goethite-rich compared to the valleys.

Another group of minerals, silica, calcite, dolomite and gypsum, precipitate in arid environments in response to, or in association with, pH change. Waters emerging from weathered silicate rocks tend to be mildly acidic, especially if the rocks contain any sulfide. In the arid climate of Western Australia, Carlisle (1983) shows that as the groundwater moves down towards salt lakes its salinity and pH rise and a sequence of precipitated minerals ensues (Figure 13.34).

The Al content of goethite is another pH indicator in the regolith. As was explained in Chapter 7, free-draining regolith is more acid than more hydromorphic regolith, and low pH water dissolves more Al. Goethite precipitated from Al-rich water will incorporate that element in its structure, so high Al-goethite is a sign of acid conditions during precipitation.

Indicators of the temperature of regolith formation are few and unreliable. At one extreme is the presence of maghemite, the spinel variety of Fe_2O_3, that can form near the surface from bush (forest) fires, and that requires a temperature of several hundred degrees. At the other is the evidence for the effects of ice on the regolith. Schwertmann (1985) found that the goethite : haematite ratio in soils increases in the Northern Hemisphere from equatorial regions to polar regions, the 50 : 50 latitude being at about 40°N, and he attributes this change to temperature change.

14
Regolith mapping

Introduction

Let us briefly return to what regolith is. It is all the material between fresh rock and the Earth's surface including weathered rock material, both *in situ* and transported, eroded and transported fresh rock materials, sediments, lavas and ash deposits and a wide variety of terrestrial sediments. These materials occur in landscapes. They may be moved around continually, being further modified as they are. They are also modified after they are deposited as terrestrial sediments. The idea of studying the regolith as a whole is a relatively new discipline, but aspects of regolith have been studied for a long time. Pedology, sedimentology and geomorphology are a few examples. Traditionally geologists have not turned their hand to regolith studies. Mapping of the regolith is different from mapping soils, landscapes or solid geology.

Regolith mapping is different from mapping solid geology for a number of reasons. Firstly, its characteristics vary over comparatively short distances in response to many of the factors discussed in the foregoing chapters. Secondly, complete sections of the regolith are rare, and although there are opportunities to examine it in road cuts, quarries and pits, and from drill data, it is very time-consuming to examine each exposure. The only way to examine regolith properly would be to dig it up, and this is impractical. Another major problem in mapping it is that genetic models for regolith formation are still open to argument as we have discussed. An example would be whether ferricrete caps on lateritic profiles were very extensively developed or whether they were restricted originally to topographic lows. After a landscape is eroded all we are left to work with are scattered elevated outcrops of ferricrete. As we map the regolith we could ask: 'Are these the remnants of a former extensive sheet or do they delineate former topographic lows and are, from their now elevated position, evidence for topographic inversion?' Such controversy affects the way one goes about mapping and interpreting maps. By contrast, in geological mapping we have robust models for granite intrusion and sediment folding that enable us to interpolate between outcrops with confidence and as a result produce maps with confidence. At the present and from this perspective, making a regolith map can be like making a geological map in the eighteenth century when granites were considered to have been deposited in the oceans. We exaggerate, but many of our models for regolith and landscape evolution are not well understood and not sufficiently researched for us to be confident they are robust.

As a result of the nature of regolith and our understanding of it, regolith mapping uses surrogates widely in mapping. A surrogate in this context is a feature of the land surface that is readily mappable and can be shown to have a close relationship with regolith properties. Examples include landforms, soils, magnetic properties, vegetation or radiometric response.

Ollier (1995) makes the point that early maps that showed regolith were devised to portray other things. Soils maps are of great antiquity and were initially devised by landlords or bureaucrats to enable taxation as for example in China about 2000 BC, Egypt or Mesopotamia. Ollier (1995) reports the earliest regolith map he has seen as being a soil texture map of Suffolk from 1797. Many maps representing aspects of regolith have been produced in the twentieth century. Thomas (1966) produced a map of regolith thickness in Nigeria. Woodall (1981) produced a very simple regolith map of Australia in the hope that his work would entice CSIRO to undertake research aimed at assisting mineral exploration in Australia (Figure 14.1). Maps of surface deposits have been produced for various purposes. Hunt (1967) published a small-scale map of the USA with three categories of surface deposits and three of

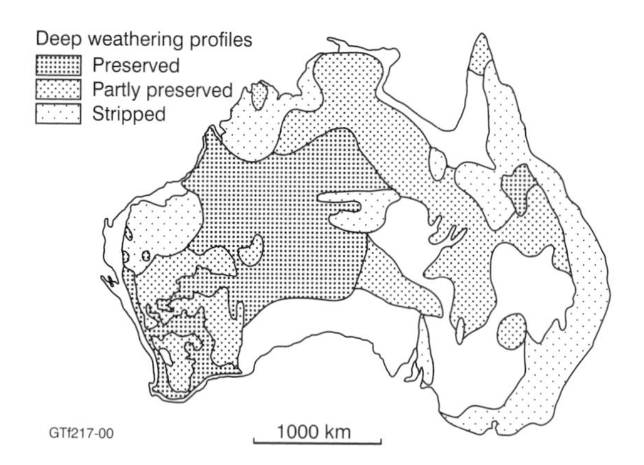

Figure 14.1 *The distribution of deep weathering profiles in Australia (from Ollier 1995 after Woodall 1981)*

weathering (Figure 14.2). Ollier (1995) regards this as 'the closest thing to a regolith map of the United States'. Butler *et al.* (1973) produced a map of the surface sediments of the Riverine Plain of south-eastern Australia. Hays (1967) and Mulcahy (1967) produced maps of the Northern Territory and parts of Western Australia respectively, which are both very close to what we think of today as regolith maps. They both related surface materials and landscape elements. Various maps of duricrust distribution in Australia have been produced, for example Prescot & Pendleton (1952) of laterite, Litchfield & Mabbutt (1962) of silcrete, Mabbutt (1984), Dury (1968), Stephens (1971), Twidale (1983) and Twidale & Campbell (1995) of silcrete and laterite. Few of these, despite their very small scale, are accurate, and all seem to be the result of the application of a particular model favoured by the author/s. Nonetheless they provide some picture of the Australian regolith. Unfortunately such maps have found their way into the international literature and are still being used as more or less indicative of the duricrust distribution in Australia.

Why map regolith?

Regolith maps can provide different information to different users, but it is equally important to be clear about the purpose for which a map has been compiled. Data that can be mapped are vast and just what is collected during the mapping process depends on the end use of the map. Ideally a map and the data on which it is founded should serve a multiple set of end users or the data collected should be adequate to provide special-purpose derivative maps on request.

What are the main uses of regolith maps?

1. To *record information* about the nature and distribution of regolith materials. Commonly recorded data include regolith type and its variation with depth, landform type and depth of weathering. These data are generally presented as a map of various regolith-landscape polygons and a legend to explain them. With the evolution of digital mapping techniques and spatial data storage and recovery systems (geographic information systems, GIS) links between the map polygons and the data they contain are becoming more sophisticated and easier to manipulate for extracting specific information.
2. To *provide derivative maps* of specific application such as engineering geological parameters of the regolith. Other useful maps may include depth of weathering, regolith geochemistry, thickness of transported regolith cover, distribution of duricrusts, regolith groundwater parameters and land-use parameters such as salinization or regolith erodibility.
3. To aid in the *visualization of regolith data*. The relationships between regolith types are difficult to elucidate in the field without producing a map, particularly when working over large areas (small-scale maps). Maps enable improved visualization of relationships. Chan *et al.* (1992) discussed the distribution of ferricretes in the Kalgoorlie region after mapping the area at 1 : 500 000 scale. They suggest that the distribution maps out former drainage channels, a hypothesis not previously entertained. Anand *et al.* (1998) have subsequently shown that parts of this model suggested by Chan *et al.* (1992) may apply on the Yilgarn.
4. *As the basis of much regolith research.* Although all regolith research is not based on mapping and maps, most is. The early work on the Yilgarn Craton by CSIRO scientists (Anand *et al.* 1991) was based on understanding the distribution of regolith materials. To achieve this they produced a regolith mapping scheme which they applied to further their research into the distribution of prospective regolith sampling media for the exploration of minerals.

Regolith mapping and scale

The view of regolith changes at different scales. When considering its variations at a continental scale, detail

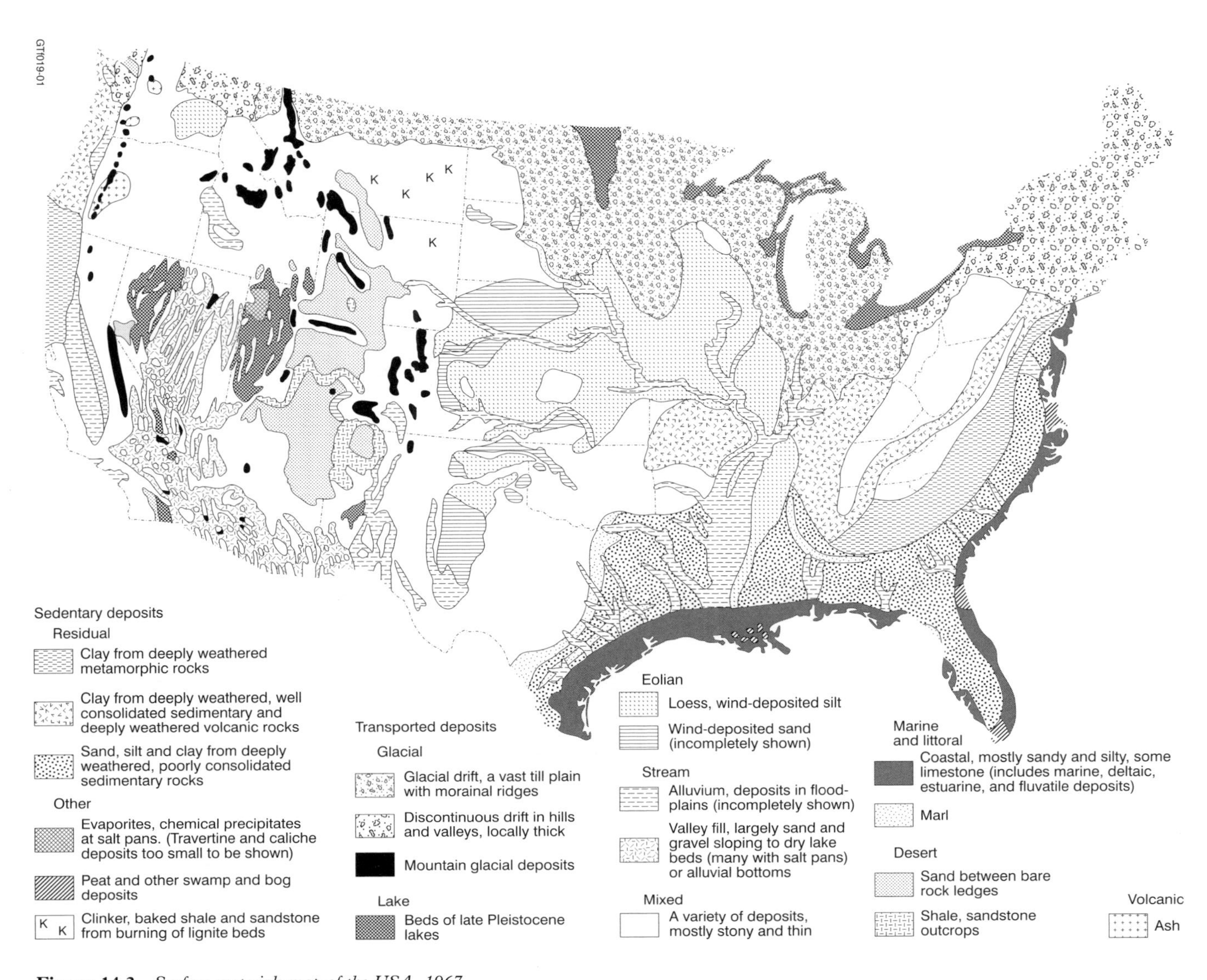

Figure 14.2 *Surface materials map of the USA, 1967*

becomes irrelevant to the 'big picture' (e.g. Taylor 1994; Taylor & Butt 1998; Chan *et al.* 1986; Figure 1.9). At this scale the polygons are very large so detail just does not show. Comparison of regolith maps at Broken Hill (Figure 14.3a, b) shows this clearly. Gibson & Wilford (1995) mapped the Broken Hill region at a very small scale of 1 : 500 000 and the total number of different regolith polygons on the map is 56. Foster *et al.* (2000) mapped parts of the region at 1 : 25 000 and the detail recorded is reflected in the number of different polygons on their map – 25. The average polygon size of the Gibson map is of the order of 100 km^2, while those on Foster *et al.* are about 1 km^2 or even less in areas of complex and spatially variable regolith.

Regional maps are generally produced at 1 : 250 000 to 1 : 100 000 scale, but there are many examples of maps at much larger scales produced for specific purposes. Many mineral exploration leases are mapped at large scale and in Hong Kong much of the urban and suburban areas are mapped at 1 : 2500 for planning and land stability purposes.

The scale of mapping, as well as depending on the purpose of the mapping, depends on the nature of the regolith and landforms in the region being mapped. For example, in many areas of Australia covered by extensive alluvial plains of grey clays there is little point at mapping at large scale as neither landforms nor regolith change significantly over short distances.

Regolith character and landscapes

One of the most useful surrogates for mapping regolith is landform because the development of regolith is generally intimately related to landscape evolution. Many regolith types follow particular landforms or landscape positions, but to map regolith/landform it is important to understand the underlying principles or models on which the mapping is based.

Some of the advantages of using landform as a regolith mapping surrogate include:

1. Greater speed and efficiency in mapping compared to checking the regolith physically at many more points.
2. Continuous coverage of the landscape by landforms.
3. Good suitability for representing spatial relationships between regolith units.
4. The possibility of transferring mapping models to other areas.

Potential problems using this approach that need consideration are:

1. The success of using landform depends on how well surface morphology correlates with regolith profiles (the relationships need to be established before and during mapping).
2. Landform is good for surface features (e.g. widespread thin alluvial cover), but does not map subsurface variations well.
3. Mapping on landform surrogates suffers from inbuilt interpretative problems relating to genesis or age of the regolith associated with the landform (e.g. alluvium occurring on hilltops).

Perhaps the most simple example is that rocks that are very resistant to weathering will invariably have very thin or no regolith, while those which are more weatherable have thicker regolith profiles developed on them (Figure 14.4). In a landscape sense this also is logical, as where softer rocks occur landscape lows will occur, so regolith tends to be thicker in valleys than on hills in many terrains (Figure 14.4). This situation also leads to the corollary that the regolith units mapped cannot naturally be placed in a time stratigraphic framework but only be used as lithostratigraphic units, and then only for transported regolith. Very detailed study is generally required to place these regolith units in a time stratigraphic context. To attempt placing *in situ* regolith in any stratigraphic order is fatuous and illogical as weathering creates an imprint on earlier-formed earth materials and none of the concepts of stratigraphy can apply. This also means that traditional geological mapping techniques are inappropriate to mapping regolith.

In most landscapes sedimentary regolith forms discontinuous bodies which, with several exceptions, form in the lower landscape positions. Fluvial, alluvial fan, lake and talus deposits (Chapter 12) form in the lower landscape positions and are relatively easily identified and mapped. The major exceptions are aeolian deposits, including dune sands and sand sheets and parna or loess and volcanic ash blankets or tephra. There are deposits more awkward to deal with which can mantle landscape from high to low positions and one such is colluvium. In most landscapes it only forms thin veneers, but in steep landscapes, particularly in tropical climates, colluvium can form thick slope mantles extending from near hilltops to valley bottoms, as it does in Hong Kong, for example, where for engineering reasons the three-dimensional

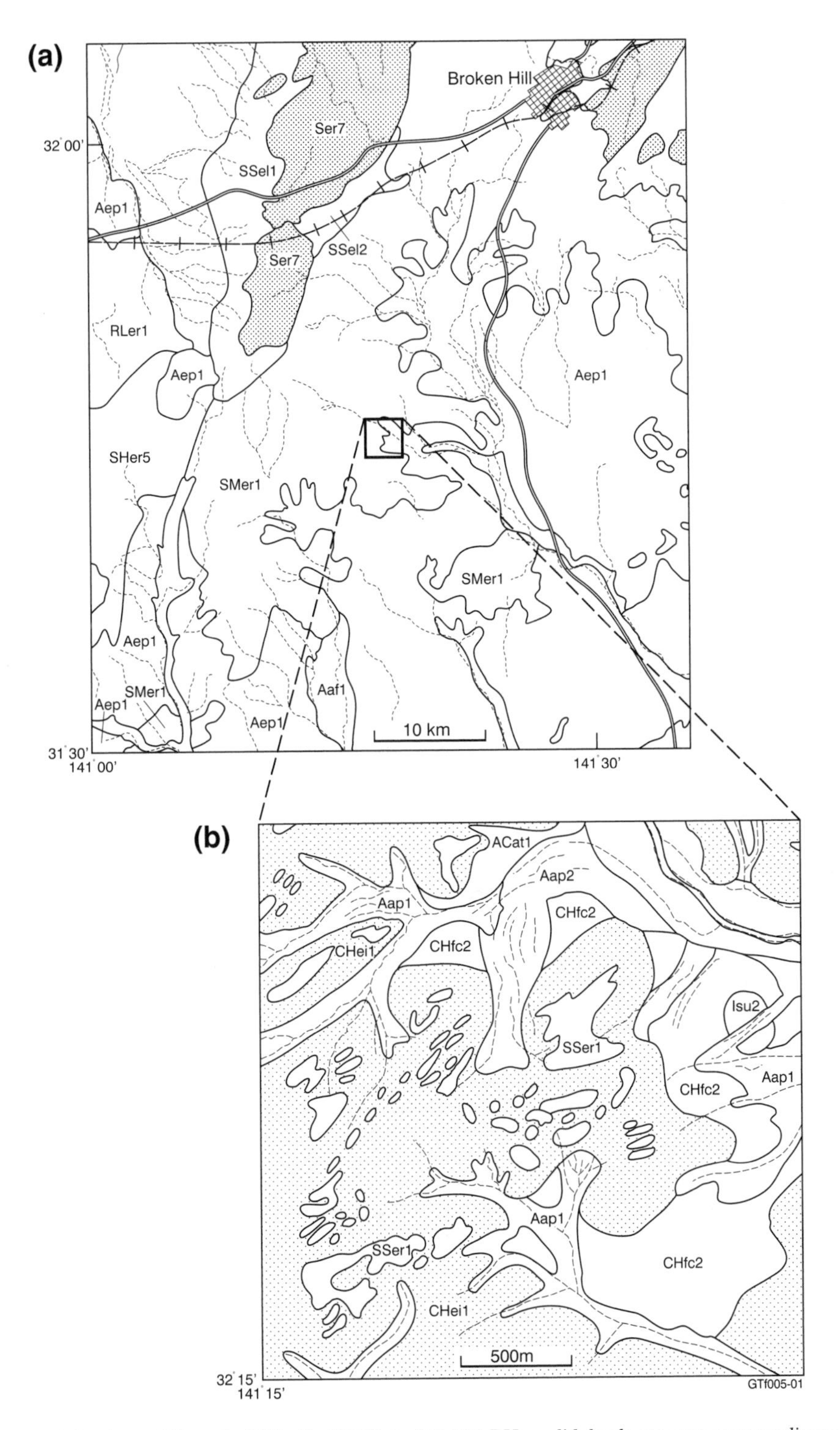

Figure 14.3 *(a) Section of Gibson & Wilford's (1995) 1 : 500 000 BH regolith landscape map corresponding to (b) Foster* et al. *(2000) 1 : 25 000 regolith landscape map. Note the huge difference in detail at these two scales*

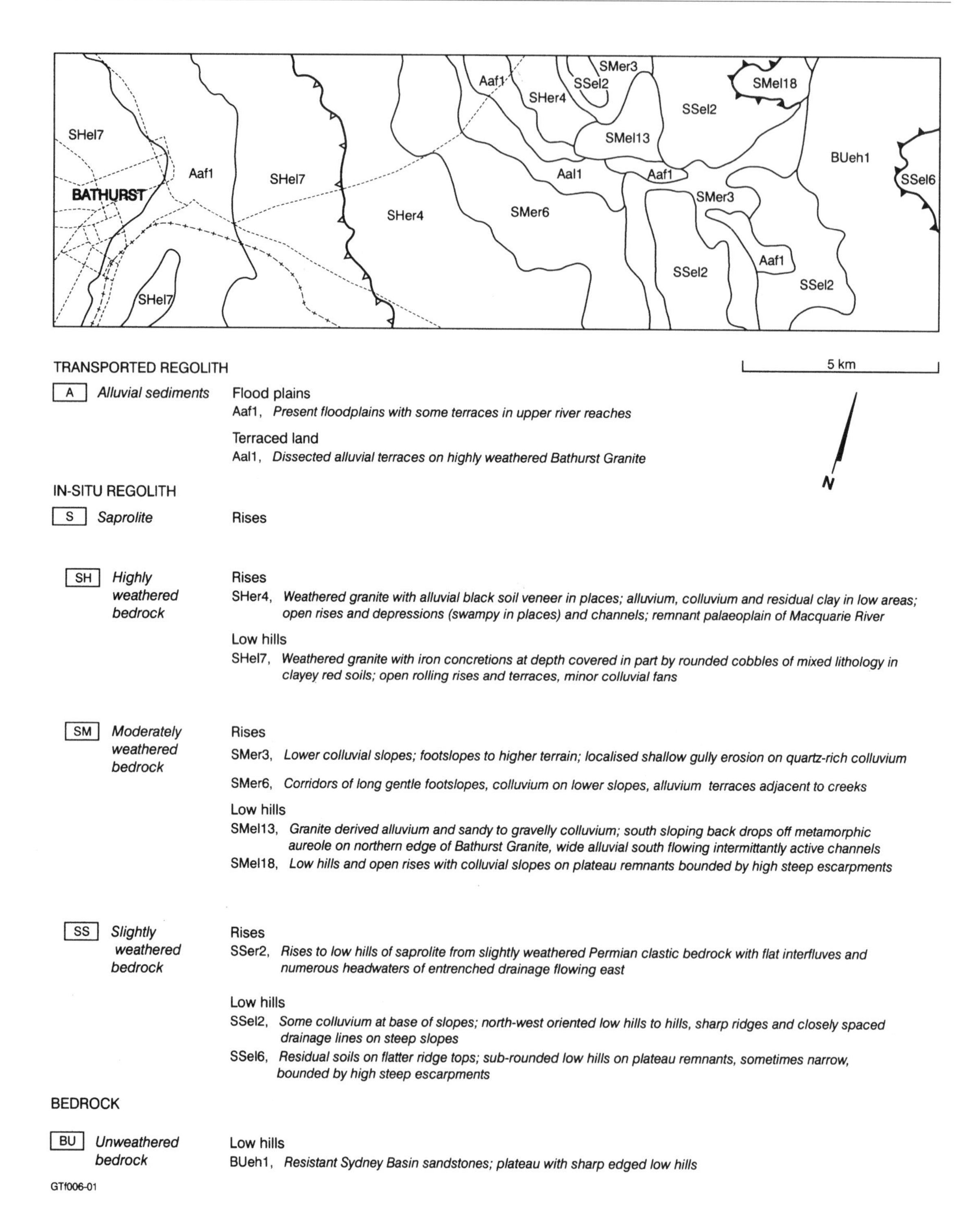

Figure 14.4 *A cross-section across the Bathurst 1 : 250 000 regolith terrain map (Chan & Kamprad 1995) illustrating the lack of substantial regolith over hard rocks and deep regolith over those readily weatherable which also form landscape lows*

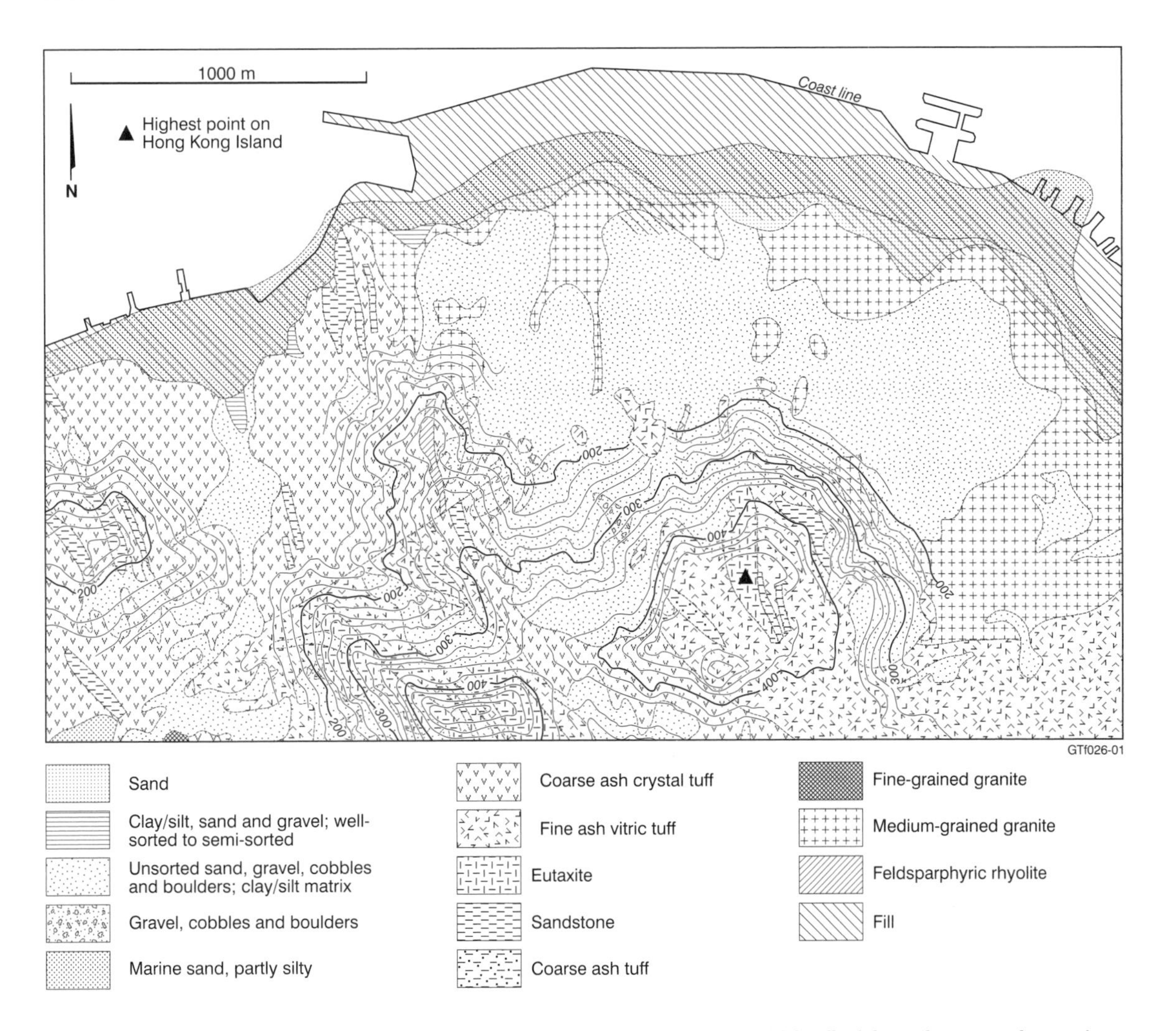

Figure 14.5 *Portion of the Hong Kong–Kowloon 1 : 20 000 geological map showing thick colluvial mantles to near the top of hillslopes (modified from Hong Kong Geological Survey)*

geometry of colluvial mantles has been closely studied (Figure 14.5).

When it comes to using landscape positions to map features formed by *in situ* weathering and duricrusts we strike the genetic model problem. They can easily be mapped in a particular region based on their characteristic on various remote tools (air photos, various satellite images, geophysical tools). But a full understanding only comes from an understanding of a robust model for the evolution of the regolith and the landscape in which it sits. As pointed out in various places in the foregoing text there is still a multitude of complex arguments relating to this problem. Until they are solved, and that will result from mapping as much as anything, mapping of *in situ* weathering and duricrust distribution will depend to a large degree on empirical observations. Take for example the origin of ferricrete on the Yilgarn Craton and its implications for making regolith maps. In its simplest form, one model holds that the region was once a peneplain covered by a lateritic profile (see Figure 13.5) capped by ferricrete. Figure 14.6a illustrates the basic model of landscape evolution of a 'laterite' covered peneplain incised by more recent valleys in which deposits, in part derived from the 'laterite', are deposited. Figure 14.6b illustrates a regolith cross-section interpreted from this model from Butt & Zeegers (1992a) and Anand & Smith (1993). The other model (Ollier *et al.*

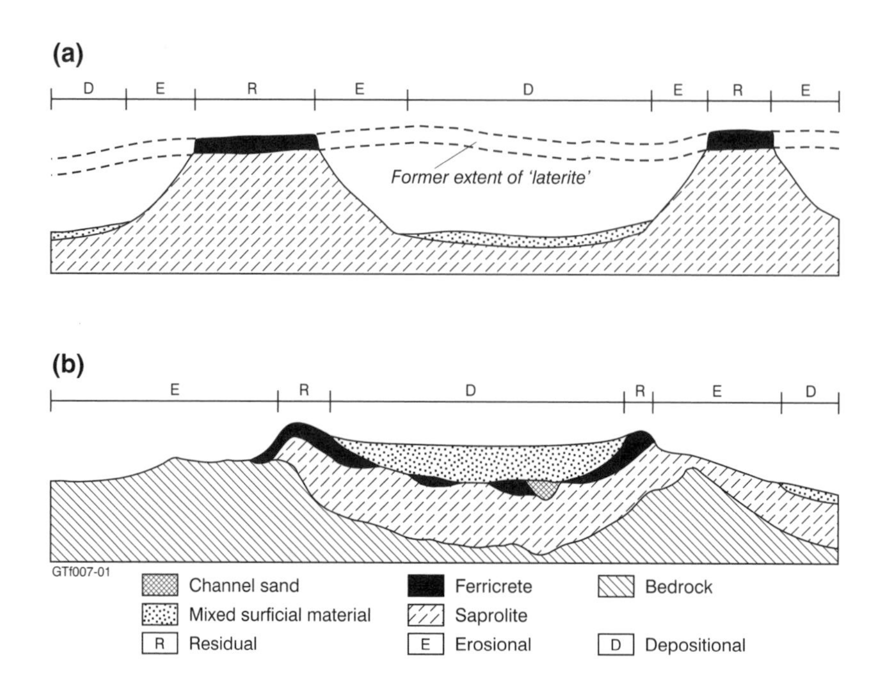

Figure 14.6 *Two sections to show the application of the residual–erosional–depositional (RED) scheme to a landscape of plateaux. (a) The generalized Jutson landscape and (b) the cross-section at Lawlers, Western Australia*

1988) suggests ferricretes formed in valley bottoms and landscape inversion has led to their now occupying topographically high positions (Figure 14.6). The consequences of adopting either of these end-member views for regolith mapping are:

1. Peneplain model – the valleys are enlarged at the expense of the ferricreted higher surfaces which the model assumes are relics of a formerly extensive surface of ferricrete. The flanks of the surfaces are erosional and the valley bottoms depositional regimes and can be mapped on these bases.
2. Inversion model – predicates there is more than one generation of deposition, one in the former valley now forming the upper surface and another later one in the adjacent valleys. In mapping on this model the erosional terrains remain as for the peneplain model, but depositional regimes of two ages occur.

Examples of where landform surrogacy does not work for mapping regolith are where landforms do not reflect the underlying regolith profile. This is particularly common in regions where shallow alluvium masks earlier regolith units or hides buried landforms. At Northparkes in central New South Wales mining of a porphyry copper deposit exposed sections through the regolith on an otherwise relatively featureless alluvial plain. Figure 11.21 illustrates the section exposed. The palaeochannel and deep weathering in this pit were not detectable at the surface and surrounding areas of rock outcrop had lost all indications of the deep weathering. The deeply weathered Ordovician volcanics that host the copper were deeply weathered during the Carboniferous (Pillans *et al.* 1998) and then weathering continued after the palaeochannel was filled during the Neogene. Thin Quaternary alluvium with deep red-brown earth soils mask the entire regolith and the former landforms associated with the palaeochannel.

Using landforms as a surrogate for mapping does not account for the law of superposition in mapping. Weathering profiles develop from the surface downwards so the layers, although similar to lithostratigraphic units, are formed in a completely different manner, and thus their interpretation in mapping regolith cannot be similar to that used in lithostrati-

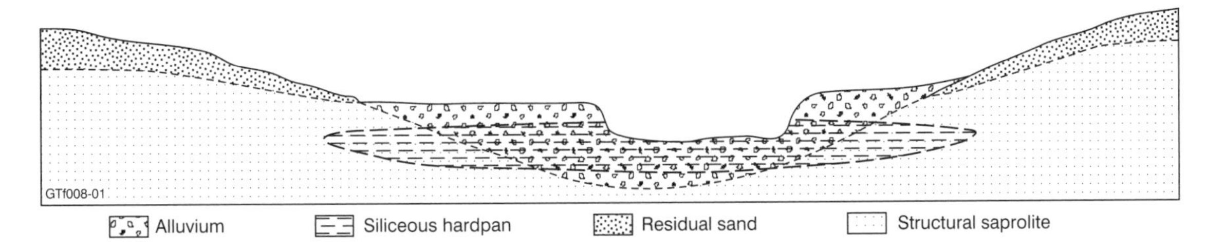

Figure 14.7 *Siliceous induration of alluvium and adjacent weathered bedrock on Cape York Peninsula (from Ollier & Pain 1996)*

graphic mapping. Glassford & Semeniuk (1995) attempt to use a classical lithostratigraphic technique in mapping weathering units from the Yilgarn Craton in Western Australia. If their hypothesis that many of the lateritic profile zones are due to deposition over a saprolite is not correct, then their whole approach is brought into question. Similarly Ollier & Pain (1996) make the same point with regard to weathering layers imposed on stratigraphic layers as occurs at Northparkes (Figure 11.21). They also refer to the imposition of a duricrust across different regolith materials as requiring a regolith mapper to separate the effects of other regolith processes from those of induration. They use an example from Cape York where silcrete has formed across alluvium and saprolite (Figure 14.7).

A number of landform maps or maps that use landform as a surrogate have been compiled over the years. Some we are all familiar with, maps of oceans and continents and topographical maps. Others we are less familiar with. These include the following map types.

GEOMORPHOLOGICAL MAPS

Geomorphological maps are more than landform maps in that they are usually based on a combination of the landform and the genetic processes which formed it. There are many examples of geomorphological maps of this type (Figure 14.8).

The RED scheme for regolith mapping developed by Anand & Smith (1993) is essentially a process geomorphological map (Figure 14.9). They define landscape elements on the basis of whether they are residual or remnant (R), erosional (E) or depositional (D). Their decision, however, on what is residual and what is not is based on the model that the areas being mapped were once covered by a peneplain across which lateritic profiles were widely developed. Thus not only is the geomorphological part of the map based on interpretation of the process, the definition of the process itself is based on another assumption about the former nature of the landscapes and the regolith. The RED scheme of regolith landform mapping is based on a series of assumptions. This in itself is not bad so long as the assumptions can be substantiated.

Other maps to use landform as the main surrogate for mapping are *soil maps*. Pedologists recognized early in the development of their science that soil types in a region bore a close relationship to topographic position. This realization led to the concept of a soil catena (Milne 1935). A catena is the systematic variation of soils with topographic position on hill slopes. There are many examples of this and it is a widely applied concept in soil/landscape mapping (Figure 14.10). An early example of catenary relationships comes from tills in Iowa (Figure 14.11). Similarly, Tardy *et al.* (1973) show that soil clay minerals vary systematically with topography in a variety of climatic settings (Figure 14.12).

Modern soil maps while using landforms as a surrogate also map soil materials rather than profiles. Each horizon or layer is considered a separate layer and mapped separately so that the distribution of individual material in the landscape can be outlined (Figure 14.10). Although the depth considered in soil mapping is generally only the top metre or two of the regolith, it is an attempt to map the three-dimensionality of the upper part of the regolith. Soil maps produced by New South Wales Land and Water in the 1990s are good examples of this approach. They used mapping technologies (GIS) which allowed the production of derivative maps from the base data. Maps of soil erodibility or suitability for plantation forests could, for example, be produced.

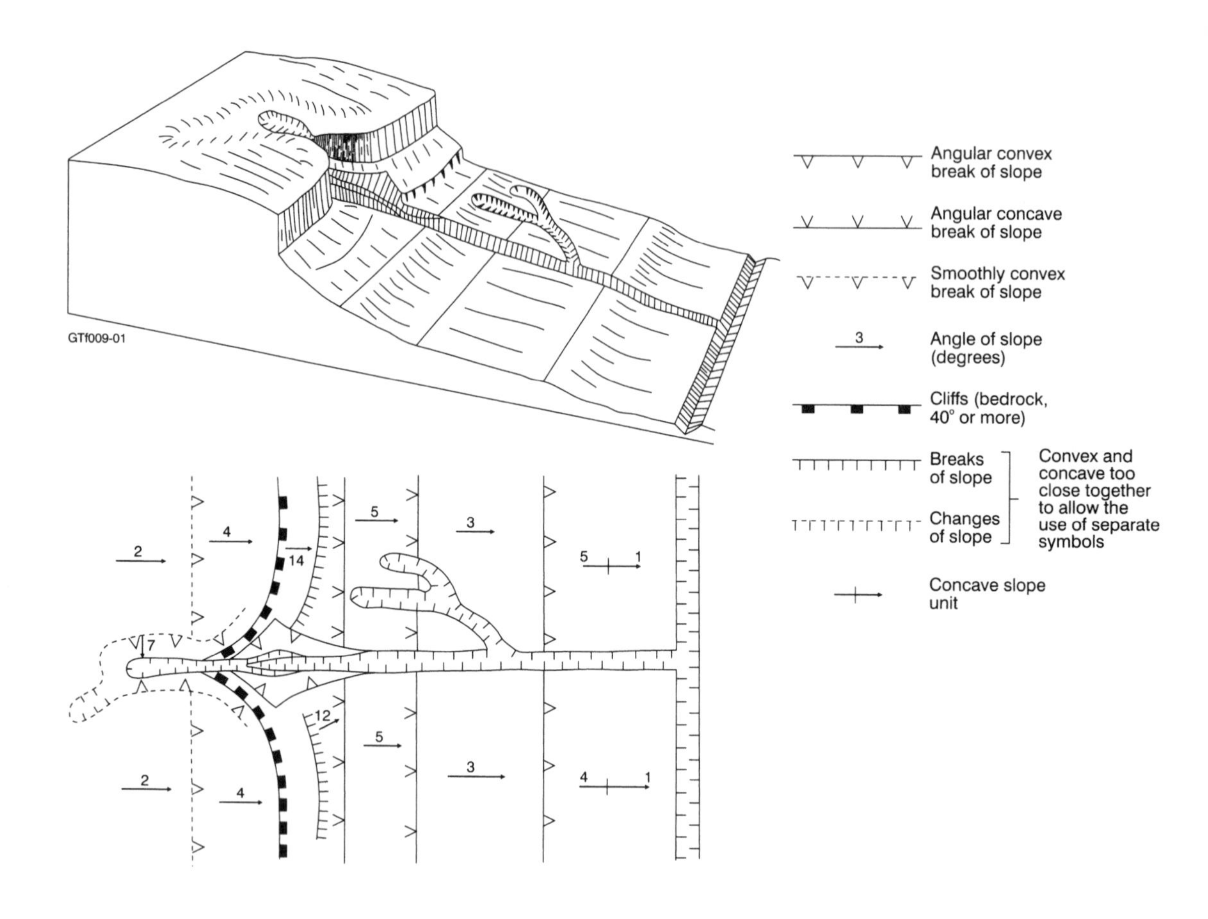

Figure 14.8 *A system of geomorphological mapping that is objective in the sense it uses no interpretive terms (from Cooke & Doornkamp 1974)*

Soil maps of much of Australia (Figure 14.13) and other continents provide a mass of data over large areas on the upper part of the regolith. These data sources are rarely used by regolith researchers. They generally provide a good starting point for regolith work and in some cases the data collected at the time of mapping (soil physical, mineralogical and chemical) provide insights to regolith materials not generally accessed by geologists.

Land system maps produced widely in Australia by CSIRO during the period from the 1950s to the 1970s outlined areas in which a recurring pattern of topography, soils and vegetation could be recognized. These different pattern units were called land system units that once established could be used to predict land properties in unknown areas. Figure 14.14 illustrates the land system approach to mapping landscapes in a regolith of south-western Queensland.

Land system maps contain data on underlying geology, landforms, soils, hydrology and vegetation. Regolith information is not specifically referred to, but soil data and landforms are useful in regolith mapping. It is also probable that the polygon distribution on land system maps would not be very different from those of regolith landscape maps of the same area. They are thus a potential source of data to be used in regolith studies.

Regolith character and vegetation

Vegetation can be a useful surrogate in mapping regolith in some instances. On the Yilgarn Craton

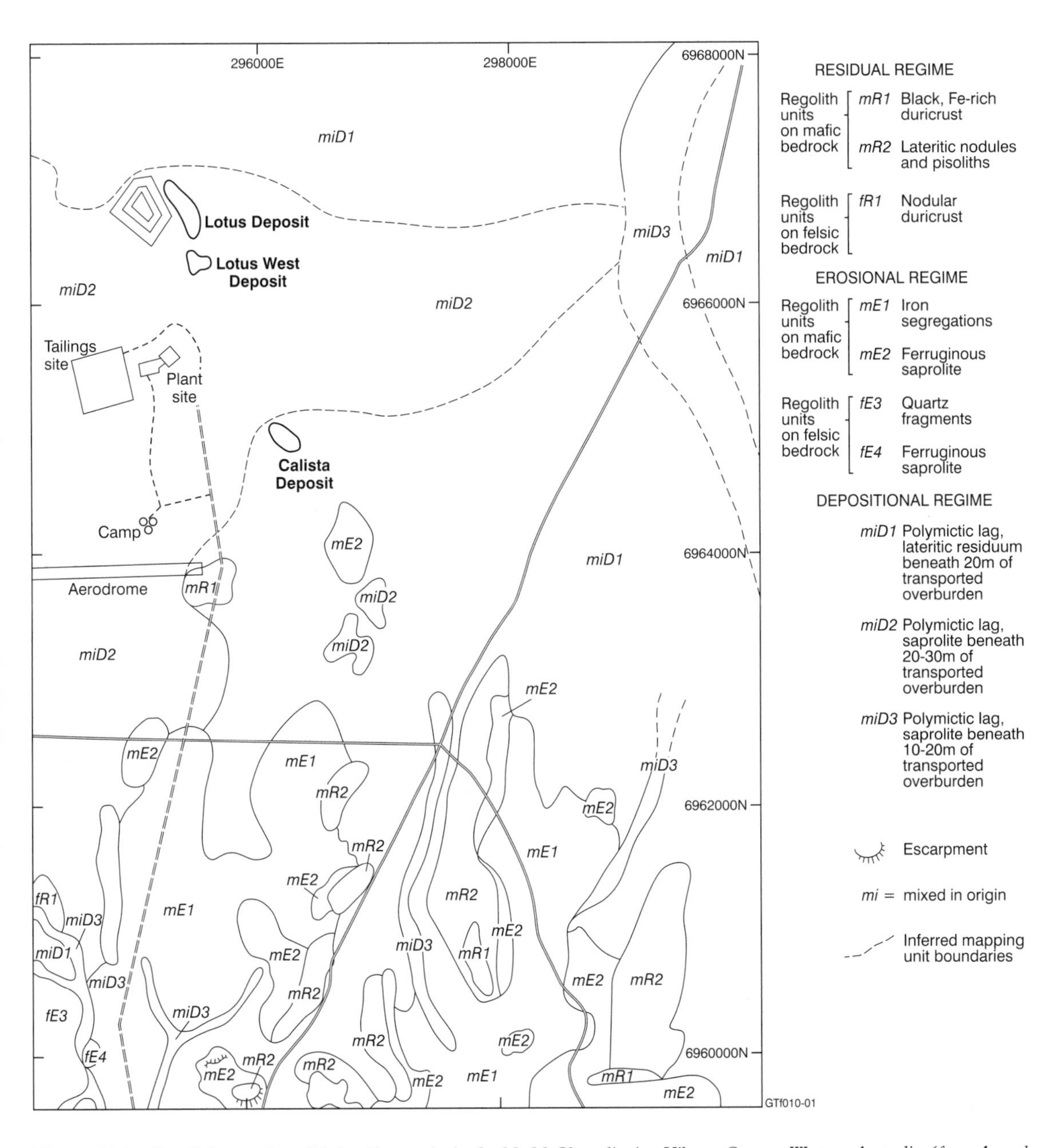

Figure 14.9 *Detailed map of regolith-landform units in the Mt McClure district, Yilgarn Craton, Western Australia (from Anand 1996)*

nickel bush is confined to weathering profiles developed on ultramafic rocks with relatively high Ni values. More generally, Brooks (1987) provides a review of vegetation associations with serpentinite rocks and regolith. There are many other examples of vegetation associations with regolith materials. Huggett (1995) summarizes much of the data available in geoecology in a wide-ranging and informative book. Bell (1982) summarizes the relationships between vegetation and soil (regolith) nutrient status for Africa. Similar data are available for much of the world.

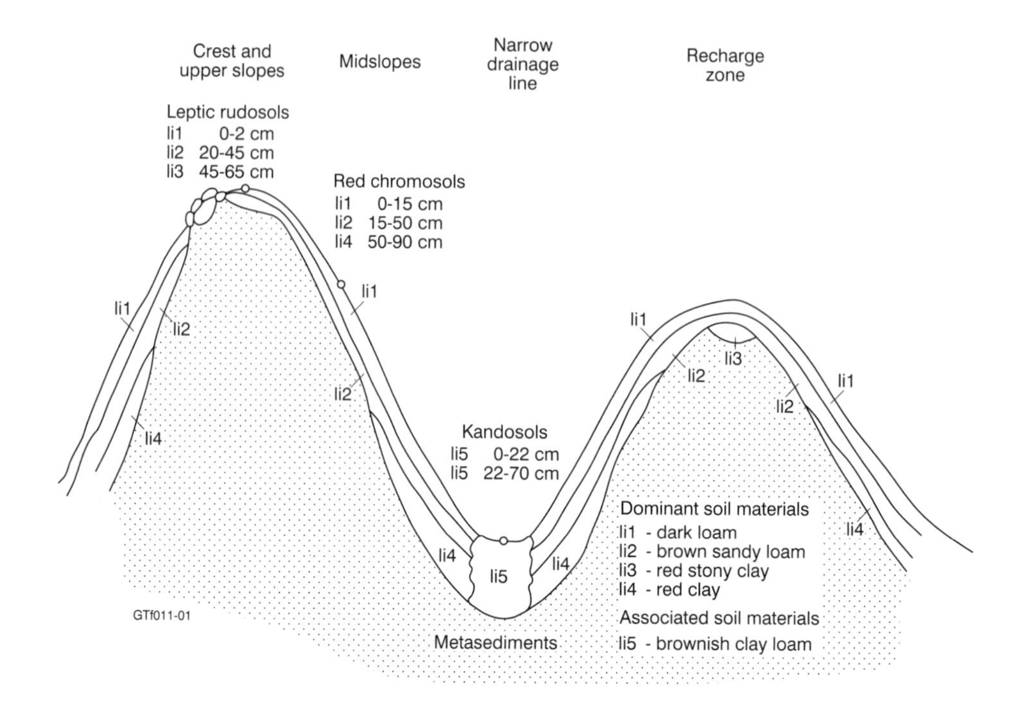

Figure 14.10 *A cross-section showing soil material–landform relationships from the Wagga Wagga 1 : 100 000 soil map (Chen 1997). This cross-section illustrates one soil-landscape polygon on the map*

At a more local scale in the Broken Hill region pearl bluebush (*Maureana sedifolia*) only grows where pedogenetic calcrete occurs within 1.5 m of the surface. Elsewhere the shrubby vegetation is dominated by black bluebush and saltbush (*M. pyramidata* and *Atriplex vesicaria*). Palaeochannels in the regolith, unrecognizable at the surface, are outlined at the surface by stands of she-oak (*Casuarina* sp.) that depend on groundwater stored in the channels for growth. A similar phenomenon occurs on the grey clay soil plains where native cypress (*Callitris* sp.) grows almost exclusively on sandy palaeochannels criss-crossing the plains. These channels are nutrient-poor compared with the grey soils and hold water more easily accessible to the trees. It is also interesting to note that many of these palaeochannels are also the plague locust's (*Gastrimargus musicus*) prime hatching areas in this part of Australia, because the sandy soils are easier for the locust to lay its eggs in than the harder grey clay soils.

Such relationships at both local and regional scales are very useful, once recognized, in assisting the mapping of regolith materials, even those not exposed at the surface.

Production of regolith maps

Modern regolith maps are produced by a combination of remote sensing of various types, topographic maps and fieldwork. The Australian Geological Survey Organization (AGSO) has developed an efficient, effective and accurate method for producing regolith terrain maps summarized in Pain *et al.* (1991). They map regolith terrain units (RTU) that are defined as

> . . . one, or more usually, several recurring landscape elements and their associated underlying regolith packages that together form a distinct regolith terrain entity. It is a landform area characterised by similar landform and regolith attributes; it refers to an area of land of any size that can be isolated at the scale of mapping.

These RTUs are different from normal rock stratigraphic units, and it is inappropriate to give regolith units stratigraphic names because sedimentary units in the regolith are usually thin and discontinuous, duricrusts are chemical overprints of previously formed materials, weathering is another overprinting of earlier materials and regolith is related closely to

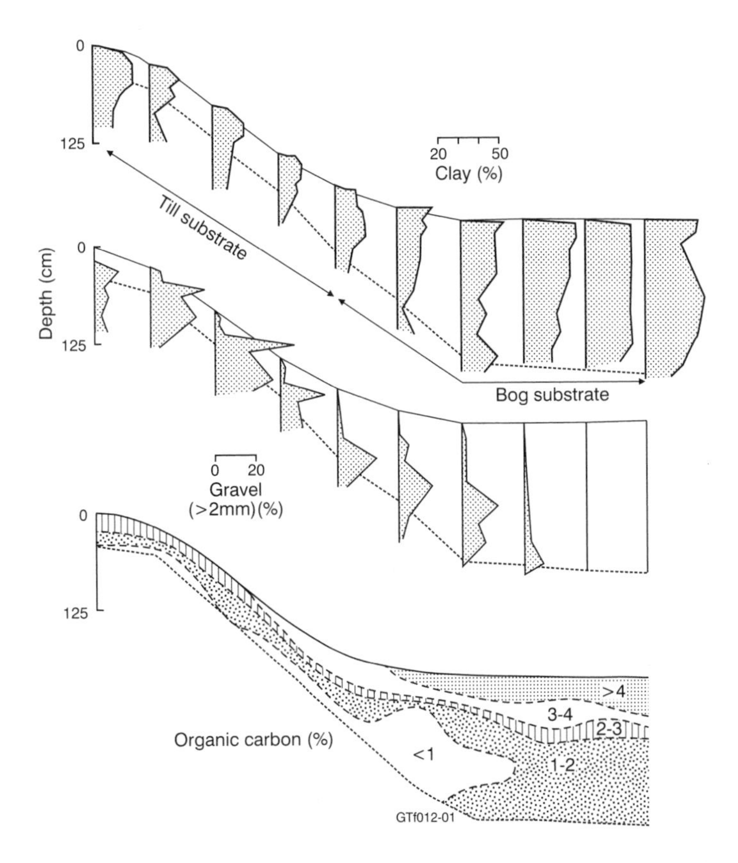

Figure 14.11 *Relationship of particle size and organic C with topographic position in a closed depression and on surrounding hill slopes, Iowa. The dotted line separates hill slope sediments from till on the hill slopes, and younger from older deposits in the low-lying area (Walker 1966)*

landform and thus its character varies with change in landform.

Because of the RTU being based on the concept of regolith and landform being closely associated, as with soils, mapping begins with the identification and separation of landforms into polygons in the area to be mapped. This is usually initially done using aerial photography and more recently using digital topographic data where available with adequate detail. Digital topographic data have the advantage of being capable of mathematical manipulation. Slope map polygons can be created and mapped automatically. In flat country the vertical scale of images can be exaggerated to emphasize small, but possibly significant, variations in elevation or landforms. The data can be readily viewed in three dimensions from any angle, providing the mapper with insights to the landscape previously unavailable. The availability of digital topographic data also allows them to be readily merged with other data sets used in mapping RTUs.

Other remotely sensed data that can be used to compile data for identifying RTUs include those given under the headings below.

MULTISPECTRAL DATA

Multispectral data of the Earth are regularly collected from satellites. Much of the data can be processed to produce false colour images emphasizing various regolith characteristics. While many researchers would argue that it is difficult to use these data because of vegetation and litter masking the regolith, it is possible in many areas to use these data, for example in arid to semi-arid regions (Figure 14.15).

Multispectral data may also be used in the more traditional photogrammetric way to provide stereo-

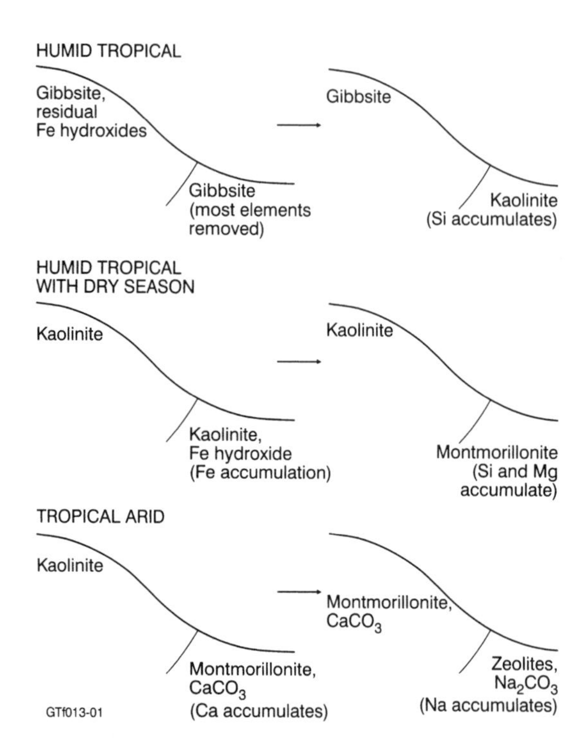

Figure 14.12 *Variations in clay mineral type in catenas under varying climates. The arrow shows trend expected as climates dry (from Tardy* et al. *1973)*

graphic coverage of large regions. In this application data are used in the same way as aerial photographs to determine areas of similar patterns, shapes and textures to initially define map polygons. The nature of these is of lesser importance than their spatial relationships at this stage. The nature of the regolith can be determined only rarely, or at least, an educated guess made, but it cannot be determined from this approach.

This technique is being used increasingly by AGSO and other survey organizations in Australia to map geology, regolith, soil salinization and soil contamination. High-resolution radiometrics were first used routinely in producing the Ebagoola 1 : 250 000 *Regolith Terrain Map* in humid north Queensland.

Where landform boundaries were indistinct, radiometrics allowed the firming-up of boundary positions. Its use also allowed an increased confidence in the extrapolation of field results to areas in the region not visited. Radiometrics also differentiates rock types so that in regions of uniform rock type they can be used to map regolith types in more detail. Clear distinctions between *in situ* and transported regolith and differences in the degree of weathering are easily distinguished. Because it is possible to map these features it is then very easy to differentiate erosional and depositional terrains. Figure 14.16 illustrates the radiometric image over part of the Ebagoola sheet.

The multispectral tool, HyMap™ measures 128 bands in the reflectance spectra from 400 to 2500 nm with a resolution of 15 nm and has proved to be very useful in mineralogical mapping of the landscape. Work by Willis (1999) has shown this aeroplane-borne system is capable of distinguishing clay minerals (kaolinite, illite, smectite), oxides (haematite, goethite), silicates (hornblende, quartz), calcite and gypsum. It is also effective in mapping vegetation abundance and different vegetation communities. Figure 14.16 illustrates the distribution of hornblende and calcite in an area near Broken Hill. It is interesting to note the coincidence of the high values where they occur over an amphibolite retrogressively metamorphosed from a greenstone. The use of these rocks as a local road base is also evident, as is the wash of material from the road in both images.

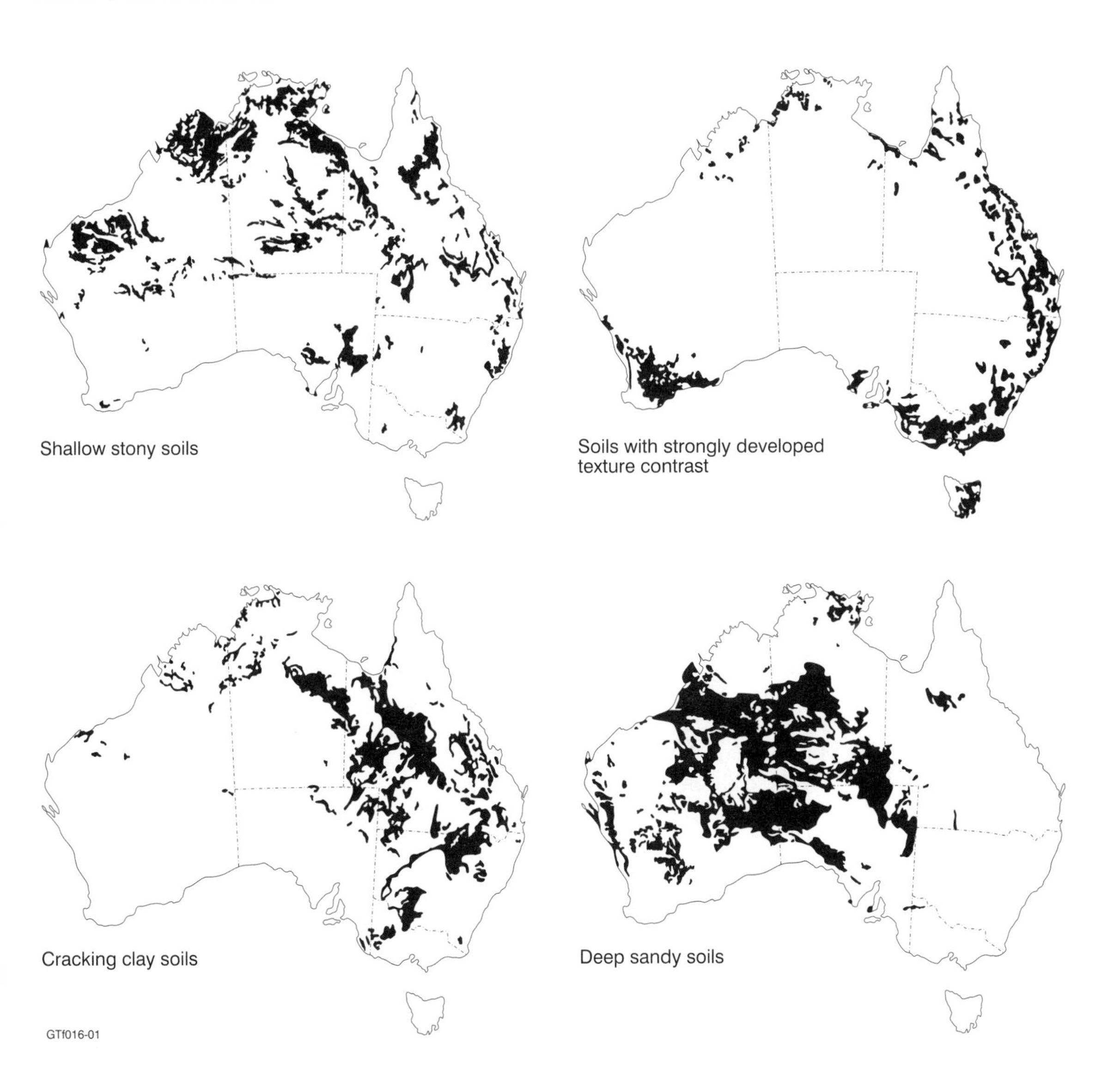

Figure 14.13 *Examples of the distribution of four prominent soil groups in Australia*

Other data are gradually becoming available for routine use in regolith mapping. There is increasing use of *radar data* to see through dry regolith materials. Tapley (1998a, b) has shown over a number of years that it is possible to pick up buried palaeochannels (Figure 14.17) and hidden bedrock structure, differentiate different phases of depositional regolith and delineate pre-cover landscapes.

A major benefit of imaging radar is the ability of long wavelength signals ($\lambda > 60$ cm) to penetrate loose overburden and map subsurface structures, provided that the volumetric soil moisture is <1% and the soils have fine to medium textures. This is a major advantage over reflectance remote sensing. In the Great Sandy Desert of Western Australia for example, where sandplains, dune-field and shrub–steppe vegetation obscure much of the subtle topography and underlying Proterozoic sedimentary rocks, enhancements of NASA-JPL AIRSAR data have demonstrated the benefits of polarimetric radar for revealing more infor-

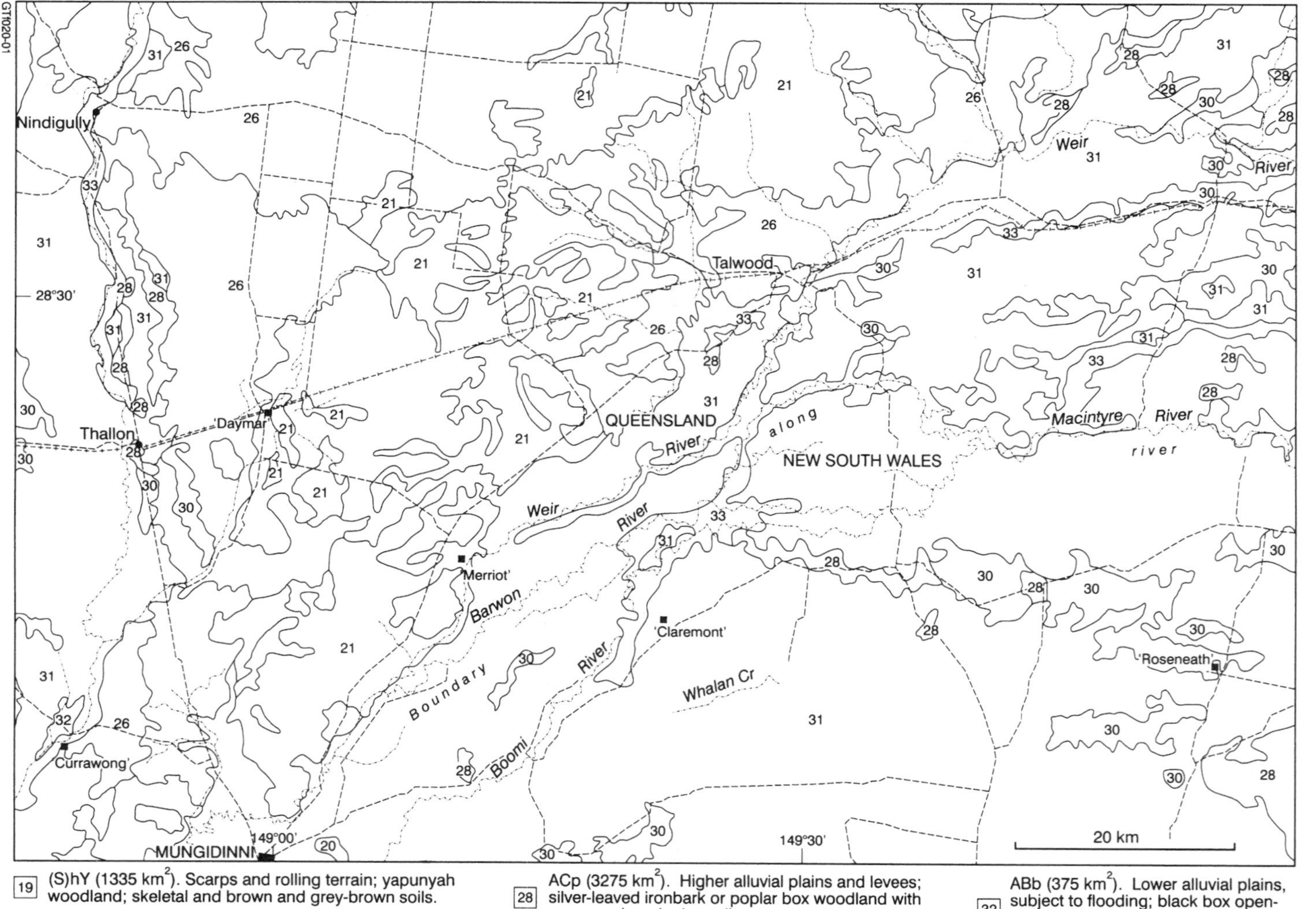

19	(S)hY (1335 km^2). Scarps and rolling terrain; yapunyah woodland; skeletal and brown and grey-brown soils.
21	(S)uBl (17,800 km^2). Lowlands; belah or brigalow open-forest; duplex soils and cracking clays with some gilgai.
26	CpBl (7230 km^2). Plains; belah or brigalow open-forest; cracking clay soils with frequent gilgai.
28	ACp (3275 km^2). Higher alluvial plains and levees; silver-leaved ironbark or poplar box woodland with cypress pine; duplex soils.
30	AX (5975 km^2). Alluvial plains; poplar box woodland with some belah; duplex soils.
31	AC (11,265 km^2). Lower alluvial plains, subject to flooding; coolibah open-woodland and grassland; cracking clay soils without gilgai.
32	ABb (375 km^2). Lower alluvial plains, subject to flooding; black box open-woodland and grassland; cracking clay soils without gilgai.
33	Af (1510 km^2). Lower alluvial plains and back swamps, frequently flooded; coolibah open-woodland with flood-tolerant shrubs; cracking clay soils.

Figure 14.14 *A land system description from the Balonne–Maranoa region of south-western Queensland (Gunn 1974)*

mation about the composition of the terrain than enhancements of either SPOT-PAN or Landsat TM data. Figure 14.17 compares a Landsat TM image of bands 2 : 4 : 7 (left) with a composite image of the vertical polarization response for AIRSAR bands C : L : P (right). The TM scene highlights the ubiquitous nature of the dune-fields and a composite history of fireburns, whereas the radar image 'sees through' much of this cover and confusion to reveal significant geological detail. Some of the more significant observations in the image include:

- The identification of the closure of a previously unmapped syncline (Location 1)
- Regional extensions of alignments of truncated sedimentary sequences, much of which is subsurface, within the sandplains and pronounced expression of the geometric alignment of individual stratigraphic units (Location 2)
- Morphological evidence of a complex fluvial history under a former climatic regime that permits the construction of a sequence of palaeoenvironmental events that produced the landforms associated with the palaeodrainage networks (Locations 3 and 4)

GEOPHYSICAL DATA

Aeromagnetic data are primarily collected and analysed for geological use, but they do provide information of use to regolith mapping. Aeromagnetic data reflect the magnetic properties of the rocks and their regolithic cover. Not many magnetic minerals occur in the regolith, but some, like maghemite, commonly occur in surface lags and are redistributed by erosion. These maghemite-rich deposits are easy to pick out and map on aeromagnetic images. At Cobar in central western New South Wales maghemite gravels are common and have been reworked a number of times by different drainage systems that are also different from those of the region at present. To map these palaeochannels by techniques other than by using aeromagnetic data would be very difficult if not impossible. Where detailed aeromagnetic data are available it may be possible to process the data to emphasize the regolith component and suppress bedrock effects, thus giving more detail on the internal regolith stratigraphy. This is well illustrated by recent studies on the Gilmore Palaeochannel in central western New South Wales (Lawrie *et al.* 1999) (see Figure 13.12).

Gamma-ray spectrometric (radiometric) data are useful in all types of country, even where vegetation masks other spectral responses. Radiometrics provide information on the U, Th and K content of the upper few tens of centimetres that make up the regolith, unhindered by vegetation cover (Figure 14.18).

Airborne electromagnetic (EM) data are being used to obtain the nature of the conductivity variations with depth in the regolith. Anderson *et al.* (1993) used the technique to map in two and three dimension variations in the conductance of the regolith. The three-dimensional results portray the geometry of regolith units of different resistivities. While the technique is in its infancy it is possible that in the not too distant future it will become a routine tool for gaining additional data on regolith distribution in the landscape.

Combining all these data sources provides a powerful set of tools that allow significant improvements over the more traditional mapping using air photos and fieldwork. Because all of it is digital it opens the possibility of producing relatively accurate regolith terrain maps to be produced before embarking on fieldwork in an area, although ideally both should proceed together. Because the data are digital it also means they can be rapidly and easily combined in various forms, which enables field relationships of materials and topography to be partially defined before fieldwork. For example, it is possible to drape aeromagnetic data over three-dimensional topographic models allowing the position and flow direction of maghemitic palaeochannel fills to be determined. This will also allow the beginnings of landscape evolution models to be worked out. In areas of flat country the draping of radiometric data over a vertically exaggerated topographic model allows relationships of materials to very slight topographic variations to be readily determined. An example of this is using U responses to map calcrete and determine its landscape positions with much less effort than would be required to do the same job in the field.

Fieldwork

Despite all the digital data available for compilation of regolith terrain maps they must be field checked and detail attributed to the polygons derived prior to fieldwork. Field data for regolith materials are often difficult to collect, as discussed at the beginning of the chapter.

Regolith is most readily observed at the surface. This will, however, only give a two-dimensional picture of how regolith varies with landscape. Road cuts, creek banks, breakaways or jump-ups (escarpments), quarries and opencast mines are often available to provide the third dimension to the regolith at selected sites. In many more remote areas, however, sectional views of the regolith are rare. In such situations borehole data are the only resort of the regolith mapper. Borehole data are available from mineral exploration drilling, water bores and, more rarely, seismic shot holes. While borehole data are not ideal they provide a third dimension missing in surface-only studies. Using bore data can also be tricky. Often no samples are available, only drillers' logs. These are difficult to interpret accurately, but experience in the region is invaluable, particularly when combined with some sectional data from road cuts and the like. Mineral exploration data are better and samples from the hole are frequently left beside the holes for examination. Since most exploration drilling is done by rotary air blast (RAB) or air core (AC) only chips and powder are available for examination. Again, skill is required to use these data to interpret regolith profiles correctly and regional experience is critical.

Data recorded in the field should be collected systematically using a system such as AGSO's RTMAP (Pain *et al.* 1991). Use of such a system enables data to be directly computerized and then automatically combined with the other data discussed above to correct map polygons derived from remote data and finally produce a map with its polygon attributes. The use of portable computers to collect field data, linked to a global positioning system (GPS), provides computer-ready data to compile maps. Figure 14.19 illustrates a data entry page from the AGSO RTMAP system to give an idea of the types of data collected and the way they are collected.

The map

In Australia, due mainly to demand from the mineral exploration industry, a number of regolith terrain maps (RTM) have been completed and are being used to aid in mineral exploration. Other land users and managers in such areas as soil salinity management and erosion management are also increasingly using them. An example of a typical hard-copy map and its legend is shown in Figure 14.4.

It is now common that regolith maps are electronically published with the various sets of data that make them up including:

- digital topographic models (DTM);
- bedrock geology;
- regolith polygons;
- multispectral data;
- radiometric data;
- aeromagnetic data; and,
- located field descriptions.

All of these are registered to the same map grid.

Using one of the many computer software packages available (e.g. ARCINFO, MAPINFO, ARCVIEW, ER MAPPER) it is possible then to view and interrogate the data sets in any combination. Thus instead of having a paper map and legend, which must owe much to the geologist who compiled it, map readers are now able to combine and assess the data from their own perspective and use their own prejudice to interpret the data.

Intelligent systems

Given the modern method of making regolith maps we have briefly outlined, it is now possible to think of producing maps of 80–90% accuracy without fieldwork provided the geologist is familiar with the landform/regolith relationships of the region.

Figure 14.20 illustrates the idea of using intelligent systems to produce maps using the digital data and a series of paradigms to guide the computer in making decisions. This process has been used in reverse to map the Ebagoola 1 : 250 000 sheet (Figure 14.21a) (Pain & Wilford 1994) after the original map was produced (Figure 14.21b). The similarity between these two maps is extraordinary and it may even be that the intelligent system map is picking up more detail than the original one. Intelligent systems approaches to RTM will permit the rapid production of reliable maps perhaps less subjectively than in the past. It all depends on the accuracy of the knowledge base used to guide the computer.

Summary

While models of regolith evolution are not as robust as we may like, it is now possible to produce regolith terrain maps that enable the generation of hypotheses to test. It is also possible to make maps remotely once

ORIG(______) SITE ID(__________) DATE (_-_-_) STATE(__)
REGION (_______________) LOC DESCR (________________)
1:100K (______) AMGEAST (________) AMGNORTH(________)
LOC METHOD(____) ABS ACC(______) AIRPHOTO(_________)

- -

EXPOSUREC(_____) SLOPE(________) ASPECT(________)
LANDFORM (__________________) GEOMORPH1(_________)
GEOMORPH2(________________) STRATUNIT(___________)
ROCKTYPE(_____) QUAL_1(________) LITHNAME(__________)
SOIL (____)
(____)
VEG(____)
HAZARDS (____)
PHOTO(____)
ABSTRACT (____)
(____)

- -

ZONE(_) THICKNESS(__) DEPTH TO LOWER BOUNDARY(___)
FRESH BEDROCK BELOW(__) REGOLITH(________________)
DEGREE OF WEATHERING(__) DESCR(___________________)
(____)
SKETCH

ATTRIBUTE	VALUE	DESCRIPTION
(_______)	(_______)	(___________________________)
(_______)	(_______)	(___________________________)
(_______)	(_______)	(___________________________)
(_______)	(_______)	(___________________________)
(_______)	(_______)	(___________________________)
(_______)	(_______)	(___________________________)
(_______)	(_______)	(___________________________)
(_______)	(_______)	(___________________________)
(_______)	(_______)	(___________________________)
(_______)	(_______)	(___________________________)
(_______)	(_______)	(___________________________)

SAMPLE ID	DESCRIPTION
(__________)	(________________________________)
(__________)	(________________________________)
(__________)	(________________________________)
(__________)	(________________________________)
(__________)	(________________________________)
(__________)	(________________________________)

GTf015-01

Figure 14.19 *Examples of RT MAP data collection systems*

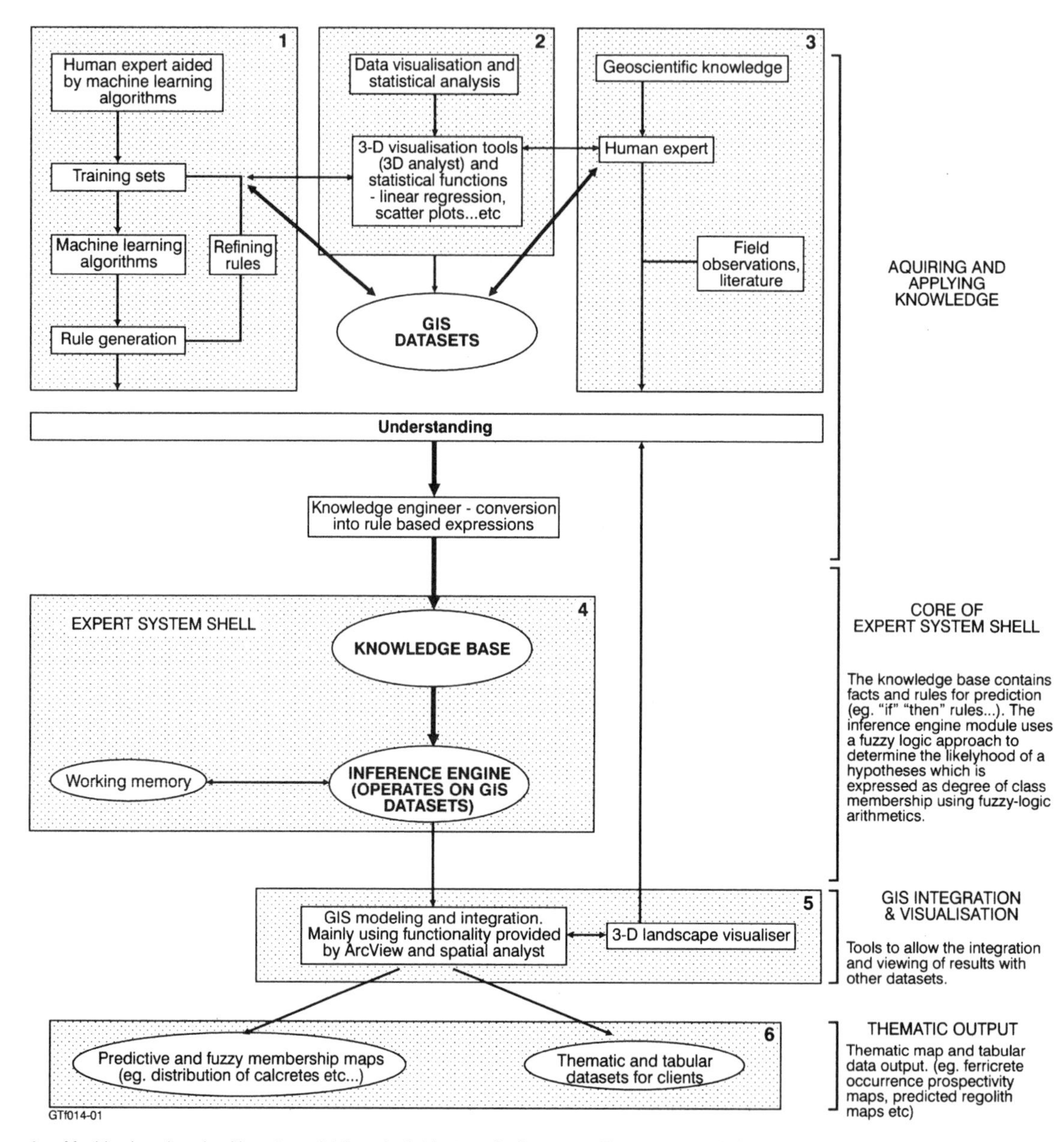

Figure 14.20 *Expert system flow chart (from John Wilford)*

some experience in an area is gained. The fact that data used to compile regolith terrain maps are increasingly available in digital form allows researchers to 'play' with the data to enhance particular aspects of the region to generate and test hypotheses. Having looked at the potential for mapping in this way let us also state our view that none of this replaces good fieldwork, it simply complements it.

15

Importance and applications of regolith geology and landscapes

Introduction

We finish this book where we began: regolith is the terrestrial interface between life on Earth and the solid planet. For humanity it is one of the essentials of life. It feeds us, it provides the fundamental resources for our homes, it provides many of the metals for our technology, it forms the foundations for our engineering infrastructure, it nurtures our biological environment, buffers our atmosphere against all its components, and finally we are buried in it. An understanding of the regolith is imperative for disaster management of such things as landslides or mudflows, for remediating soil erosion and land degradation, or for understanding the effects of earthquakes. The study of the regolith is essential in the exploration for mineral resources, and indeed many of the resources we use daily are derived from the regolith (sand, gravel, Al, gypsum, Fe) and many are hidden by it (Cu, Pb, Au) so we need to understand it to access these resources.

Figures for resources extracted from the Australian regolith are provided in Table 15.1.

The following discussion of various aspects of regolith, landscapes and our use of them is not comprehensive, but rather is intended to give some insights to their importance in our lives and future.

Ecology

Ecology is the study of how organisms fit in their home (*οικοσ* = house, Greek). It is related to how energy and materials move between organisms and the natural environment, and most terrestrial organisms depend to a large degree on the regolith, as well as the atmosphere and hydrosphere.

Regolith provides many of the nutrients essential for plants to grow including phosphate, S, K, Ca and most importantly trace elements (Zn, Mo, Co, Cu) essential for healthy growth. Regolith provides the substrate in which plants root. It also stores much of the water required for plants to maintain growth over dry periods, and availability of this water depends on the type of regolith and its position in the landscape. The nutrient status and physical characteristics of the regolith are largely controlled by mineralogy. Additionally many animals have particular requirements to maintain themselves apart from food. Many insects require particular types of regolith in which to lay eggs. Other animals (e.g. wombats) require particular regolith characteristics to build their homes. Some examples are given below.

Cracking clay soils developed on basalts of the Monaro in south-eastern New South Wales control the vegetation ecology of the region. Figure 15.1 shows the dominant species growing on the basalt-derived soils, as opposed to those formed on Palaeozoic rocks cropping out beneath. *Eucalyptus pauciflora* grows on south-facing valley sides where soils dry out less than on the north-facing slopes. The trees only survive where soils do not dry and crack, because here their taproots are not broken during early growth stages. Gilgaied soils in much of semi-arid Australia support chenapod shrubs for the same reason. This is the reason so much of Australia's black and grey soil plains tend to be treeless and steppe-like.

Examples of the relationship between regolith and vegetation are given in Chapter 14 and there are many

Table 15.1 *Economic figures for extractable resources from the regolith in Australia for 1996 (from Bureau of Resource Sciences 1997 and personal communication)*

Commodity	Reserves	Production (t)	
Metals			*Export Value ($ × 10^6)*
Bauxite	3024 t × 10^6	43.1 t × 10^6	88 (5185)[e]
Gold[a]	4454 t	289	5607
Iron ore[b]	17.8 t × 10^9	0.147 t × 10^9	2863.2 (4373.2)[f]
Magnesite	179.9 t × 10^6 ($MgCO_3$)	0.31 t × 10^6 ($MgCO_3$)	n/a
Manganese ore	118.0 t × 10^6	2.1 t × 10^6	220
Ilmenite[c]	135 t × 10^6	2.0 t × 10^6	110 (895)[f] } 1142
Rutile[c]	14.9 t × 10^6	0.18 t × 10^6	137 } 1142
Zircon[c]	21.4 t × 10^6	0.46 t × 10^6	223
Industrial minerals			*Value ($ × 10^6)*
Diamond[d]	175 c × 10^6	42 c × 10^6	519
Kaolin	n/a	285 000	61
Bentonite	n/a	239 200	n/a
Attapulgite	n/a	18 795	n/a
Gem minerals			*Value*
Opal	n/a	n/a	84.9 (1993)
Alluvial sapphires	n/a	n/a	14.1 (1993)
Chrysoprase	n/a	n/a	2.7 (1993)

[a] Gold is included here as much of the exploration for Au is regolith based and significant production is from Au-enriched supergene caps.
[b] Iron ore is included as significant amounts of the ore mined is naturally beneficiated by weathering and sedimentary concentration.
[c] Mined as heavy mineral sands from regolith materials.
[d] Only minor amounts of placer diamonds are produced from the regolith (includes industrial + gem).
[e] Figure in brackets includes bauxite alumina and Al export value.
[f] Includes processed ore products.

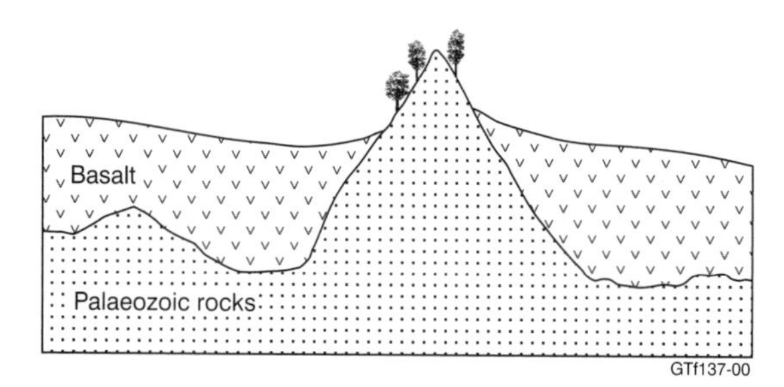

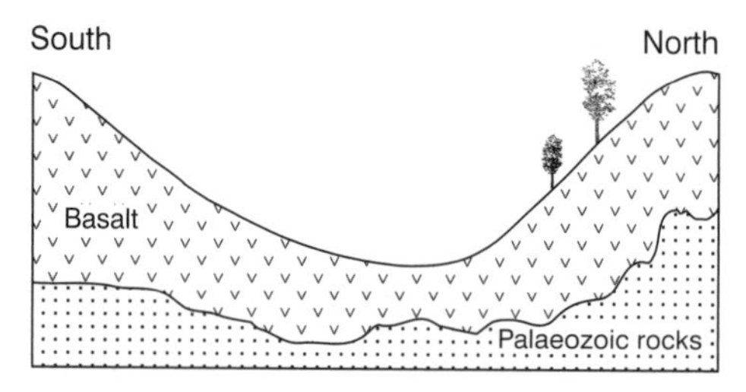

Figure 15.1 *Figure of* E. pauciflora *on the Monaro (a) is a section showing a bedrock island supporting* E. pauciflora *in a plain of basalt-derived regolith with no trees, only tussock grasses, and (b) shows the situation in a north–south section*

more. The sedge *Gahnia microstachya* and its dependent butterfly the silvered skipper only grows on sandy regolith in the New South Wales coastal ranges.

An interesting example of how the regolith affects fauna is the distribution of pink-tailed legless lizards around Canberra. These lizards live under small-sized rocks that only occur on mid-slopes of hills on acid volcanic bedrock (Hill 1995). They do not live under larger rocks higher up slopes or smaller ones lower in the landscape. This distribution is the result of the thermal properties of the mid-sized slabs of volcanic rock suiting the requirements of the lizards. The surface rock size in the landscape is related to colluvial processes and physical and chemical weathering, which break up the originally larger outcrops.

Another example of the relationship between regolith and an animal is the case of *Buramus*. It lives only in the Snowy Mountains in areas where the regolith is dominated by angular blocky boulder-fields formed by periglacial activity and as névé deposits. The close relationship between *Buramus* and these deposits has enabled a good approximation of the extent of such conditions during the last glacial maximum based on the distribution of its fossils (Ride *et al.* 1989).

Many specialized ecological niches exist because of a landscape or regolith peculiarity. There are many examples in Australia and elsewhere. One of the more interesting examples is the discovery of the wollemia pine (*Wollemia nobilis*) in eastern New South Wales. Here these auriacurian pines have been sustained as a small pocket in a region now dominated by dry sclerophyll forest. They are thought to be remnants of a formerly extensive rainforest extending along the eastern coast, but they remain because of a deep gorge that has maintained the wet conditions they require and protected them from fire, which is foreign to their ecological regime. Another example are the mound springs of eastern central Australia and their remarkable ecology which depends entirely on the presence of the springs and their unique regolith mineralogy and chemistry (as well as water) for their existence.

Landscapes, regolith, national parks and monuments

Most national parks depend on unique scenery or regolith-controlled ecology for their existence. The world's oldest national park, Yosemite, depends on glacial landscapes in granite for its remarkable status and its unique environment. Others depend on spectacular events that form the regolith such as Pompeii where volcanic ash buried and preserved a Roman city. Other famous landscapes now protected by parks include the Grand Canyon, Hawaii's Volcano National Park, Uluru (Ayers Rock) in central Australia and the exceptional landscapes of Kakadu in northern Australia.

Many parks exist because of their unique flora and fauna, which in turn depend on the regolith and landscapes. The Coonge Lakes/Innaminka Reserve of north-eastern South Australia is a region of seemingly endless longitudinal dunes with deeply weathered Cretaceous marine sandstones capped by silcrete through which the Cooper Creek flows at Innaminka only to lose itself in the dune wastes of the Simpson Desert. Within this arid country lives a unique flora and fauna adapted to sandy dunes and clayey inter-dune pans. The Coonge Lakes system marks one of the termini of the Cooper Creek where groundwater from its buried channel maintains permanent lakes in the desert. Again this region supports a very different flora and fauna from the desert around it. While not spectacular in the same sense as the Grand Teton National Park or Uluru, the Simpson region owes its unique status to the landscapes and regolith geology of the region.

Many other national parks and monuments owe their significance to regolith materials used by humans to build structures of significance. The sarsen stones of eastern France and Britain are silcrete. They were used to build rings and lines of large upright stones for some purpose, presumably religious. The most famous are the structures at Stonehenge where upright sarsens are capped by lintels of single large silcrete slabs. Sarsen stones are also common in many castles and churches in Britain, including Windsor Castle. Regolith materials were easily collected and used by societies prior to the development of sophisticated engineering equipment.

Many ancient cultures, like that of the Australian Aborigines, used regolith materials widely as it was those which lay around at or near the surface and they were consequently readily available. The Aborigines used silcrete widely to flake tools from, unlike their Northern Hemisphere contemporaries who used flint from marine limestones. They used ochre of various colours from weathering profiles. Redder colours were haematite-rich, yellows and orange-reds goethite-rich. White pigments came from kaolin and black from pyrolusite (Mn-oxide), all commonly formed in weathered materials. They used pigments for body, rock, bark and sand painting. Slightly weathered sandstones were widely used as grindstones (mortars) with a harder grinder (pestle) of harder rock, often silcrete. Many Aboriginal artefacts can be sourced using the characteristics of pigments and silcrete showing they had quite extensive trade networks.

Most cultures prior to the industrial revolution used naturally occurring, regolith-derived pigments, tools and ores. For example, the Roman civilization used many regolith-derived pigments for art and make-up among other things. Many colourful oxides of toxic elements like Hg, As and Pb were used. Equally hazardous was the use of Pb pots to boil wine to sweeten it and in the process producing lead ethanoate ($(Pb(CH_3COO)_2)O$.

Agriculture and land management

One of the most significant threats to agriculture in Australia is soil salinization. Much of the salt causing the problems is stored in the regolith and in its groundwater. As water-tables rise, salts are brought to the surface; this results in soil degradation and erosion and takes good agricultural land out of production. Many soils contain exchangeable Na so that when the soils are exposed to water they disperse easily and may erode. That which remains hardens on drying, forming a barrier to new growth seeking emergence.

The antiquity of the regolith and the resultant high Fe-oxide contents in many places cause problems with the use of phosphate fertilizers. The Fe-oxides adsorb the P making it unavailable to plants. This problem is often addressed by overfertilizing, resulting in excess P_2O_5 being washed off land and contaminating waterways or remains in the soil causing acidification. In turn eutrophication leads to algal blooms and problems in water quality. Many other regolith minerals interact with elements in or added to the environment. Clay minerals adsorb both heavy metals and pesticides and hold them within the environment until chemical changes cause their release.

Expansive clays in the soils and regolith may lead to the formation of gilgai. This is a problem in areas where cultivation is pursued as the continual heaving of the regolith to form depressions and rises localizes water which in turn leads to variations in crop quality and height and difficulties in harvesting. Even when gilgai is planed flat by ploughing, the microrelief returns after a few years, requiring additional expenditure to maintain land for cropping.

Land degradation is a global problem resulting from misuse of our land and its consequent loss of productive capacity. Common forms of degradation include:

- soil and regolith erosion by water and wind;
- loss of soil structure due to overcultivation or compaction;
- salinization;
- increasing soil acidity; and,
- desertification.

There are many good books that summarize soil and land management problems in Australia and elsewhere, including McTainsh & Broughton (1993) and Charman & Murphy (1991). Table 15.2 gives an indication of some of these problems in Australia, and it is worth noting that many are common to other agricultural systems, whether in developed or underdeveloped nations.

These figures are 20 years old. The situation with respect to salinization has worsened since, and the effects of wind erosion are more clearly understood now, so it is likely that >50% of Australia's arable land requires some sort of treatment to restore its productivity.

Table 15.2 *The forms and extent of land degradation in Australia requiring treatment in June 1975 (from McTainsh & Broughton 1993)*

Form of degradation	Area affected ($\times 10^3$ km^2)
Area in use	1804
Area not needing treatment	987
Water erosion	577
Wind erosion	57
Vegetation degradation	92
Dryland salinity	10
Irrigation salinity	9
Other	14
Total degraded	815 (45%)

Regolith and the greenhouse effect

There is a significant relationship between silicate mineral weathering and atmospheric CO_2 levels. We know there are connections between CO_2 levels and global climate producing feedback between weathering and global climate.

If global climate warms as a result of CO_2 increases in the atmosphere then there will be an increase in the rate of chemical weathering, which consumes additional CO_2. A negative feedback loop is created. On the other hand, as global temperatures increase, sea levels rise causing a reduction in the area of continental land masses. This in turn means a decrease in weathering of silicates, providing a positive feedback loop. Vascular plants accelerate the rate of chemical weathering as they extract nutrients by secreting organic acid into the regolith to release elements they need from minerals. This compares to more primitive plants that show little evidence of increasing the weathering of silicate minerals.

As a test of the above models, a plot of time against CO_2 in the atmosphere shows values decreased as vascular plants evolved during the Devonian–Carboniferous (Figure 15.2) and the associated global cooling led to the Permo-Carboniferous glaciations. Thus there can be little doubt of the connection between climate, weathering, plants and atmospheric CO_2 levels. See Kump *et al.* (2000) for a more extensive discussion.

Engineering and infrastructure projects

It is on the regolith that we must construct our homes, roads, factories and power stations. We must transport our goods across it physically and much of our information goes across or through it electronically.

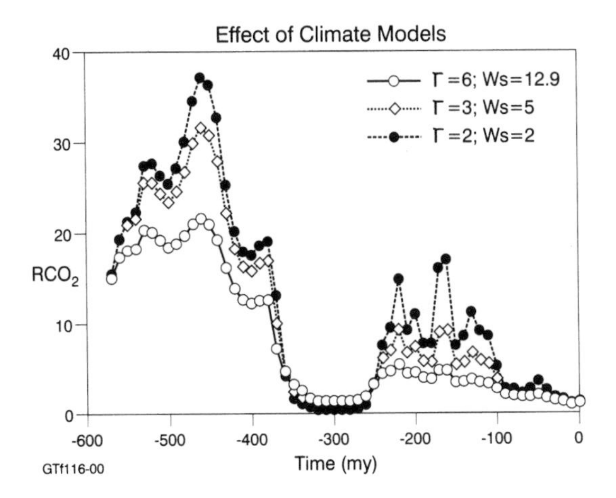

Figure 15.2 *Plot of atmospheric values of CO_2 against time showing the decrease in CO_2 associated with the evolution of terrestrial vascular plants (from Berner 1995).* Γ *and* W_s *represent the sensitivity of global mean temperature to atmospheric CO_2 and level of solar radiation respectively. The three values are different values used in the global circulation model used to calculate the curves*

We rely on the regolith to support just about all the infrastructure that supports our lives. The development of pipes to carry petroleum from the northern edges of North America to ports and facilities further south depended on research into permafrozen regolith and how that influenced the integrity of the pipe. The installation of buried fibre-optic cables between Melbourne and Adelaide required they be buried far across long stretches in expanding clay regolith. This required new methods of cable laying to avoid the possibility that glass fibres would be snapped by the heaving of the soil, more or less the same problem as in North American pipelines.

Examples of failure to take regolith and landscape parameters into account in engineering projects litter the literature. Vaiont Dam in Italy (Selby 1993), landslides in Hong Kong (Burnett *et al.* 1987), the Huascarán rock avalanche in Peru in 1970 (Plafker & Ericksen 1978) which killed some 18 000–20 000 people and many others are all disasters related to landscape and the regolith on it.

The following case study from engineering geology illustrates the way regolith geology, landscapes and materials are used in regional studies (Taylor & McNally 1992).

STUART RANGE, SOUTH AUSTRALIA

The Stuart Range is a 60 m high escarpment, or 'breakaway', which trends NW–SE through the town of Coober Pedy in the far north of South Australia (Figure 15.3). It is the major topographic feature in this region, and acts as a watershed between the Lake Eyre hydrological basin to the east and an upland plateau and sand plain to the west. The underlying Cretaceous sedimentary rocks are stratigraphically equivalent to those of the Eromanga Basin further to

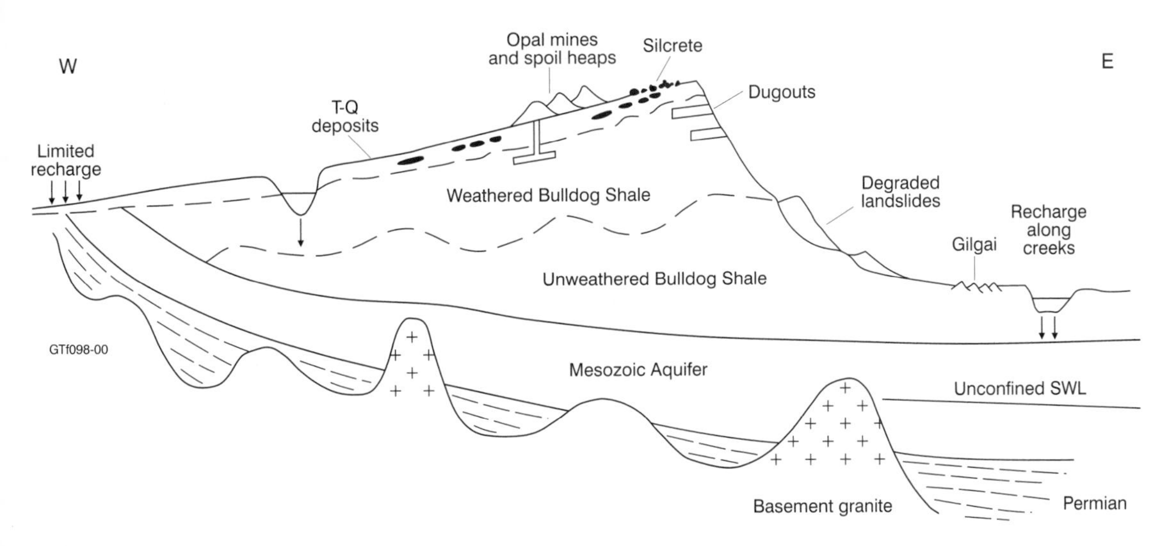

Figure 15.3 *Schematic cross-section of the regolith and Cretaceous sediments at Coober Pedy illustrating the engineering and groundwater related to the regolith and geology. T–Q refers to Tertiary–Quaternary (after Greg McNally 1977)*

the east, but due to their location at the rim of the basin are not fully saturated. The sand aquifer is therefore hydraulically unconfined, though generally capped with shale, and depths to standing water may be over 100 m.

West of the escarpment the Mesozoic sequence is about 150 m thick, overlying Permian coal measures and granitic basement. To the east this reduces greatly, where the upper weathered (and opal-bearing) portion of the Bulldog Shale has been stripped by erosion. Coober Pedy owes its location to the opal seams or 'levels' that were once exposed there in the breakaway face.

The 110 m thick Early Cretaceous Bulldog Shale is remarkable because intense weathering has rendered it harder and more permeable than the parent rock. In its unweathered state this formation is a dark-grey, very stiff clay-shale or weak mudstone, which is pyritic and carbonaceous in part and contains boulders towards its base. Some fossiliferous limestone bands and siderite nodules are also present.

The weathered Bulldog Shale is a white and mottled kaolinitic claystone, which resembles chalk but is referred to locally as 'sandstone'. The extent of leaching is apparent in its low bulk density. The weathering profile is about 50 m deep and its base is roughly coincident with that of the escarpment.

Throughout its 80-year history, Coober Pedy has been plagued by water supply problems – both of quantity and quality. In the late 1960s it became the first sizeable Australian community to be served by a desalinization plant. At first a solar still was installed, but this proved unsuccessful and was soon replaced by reverse osmosis equipment. The town's permanent population has increased twentyfold since the 1950s, to about 5000, and the tourist numbers have grown even more rapidly. At the same time living standards for both the residents and visitors have greatly improved, and one manifestation has been increased per capita water consumption. This demand has been met by a combination of saline bore water for toilet flushing, low-salinity (reverse osmosis) water for showers, and tank water for drinking. In addition, a small proportion of potable surface water is harvested from clay pans and ephemeral pools, and trucked up to a hundred kilometres by contractors.

Other environmental issues related to the regolith in the Coober Pedy area include:

- excavatability and stability of mine tunnels and residential caverns ('dugouts');
- availability and quality of roadbase and aggregate;

Figure 15.4 *The result of opal mining at Coober Pedy, South Australia (photo GT)*

- landslides and potential foundation problems on swelling clay; and,
- safety hazards caused by the presence of thousands of abandoned shafts, open cuts and mine waste dumps (Figure 15.4).

Groundwater supplies

Groundwater has been obtained around Coober Pedy from the Mesozoic sand, the weathered Bulldog Shale and alluvial sand aquifers, though only the first is a significant producer (McNally 1977). The Mesozoic aquifer is a composite 60–80 m thick, poorly cemented sand unit made up of the Late Jurassic Algebuckina Sandstone and the Early Cretaceous Cadna-Owie Formation. The overlying unweathered Bulldog Shale is nominally a confining bed, though the sand aquifer is in most places either unsaturated or saturated but with only a few metres of head. This hydraulic geometry is illustrated on Figure 15.3.

The salinity of the groundwater at Coober Pedy is quite variable, in the range 1000–17 000 mg/l total dissolved salts (up to half that of sea water). The low-salinity water that occurs to the north-west of the town (Mason 1975) is usable for domestic supply, either untreated for washing or desalinized for drinking.

The apparent paradox of an aquifer within the normally impermeable Bulldog Shale results from leaching and shrinkage during Palaeogene weathering, which has created clusters of open spaces and saturated joints. Although storage within this aquifer is small, its permeability is locally high. Leakage of surface waters through fractures and cavities is believed to be the main cause of lower salinity in the underlying sand aquifer. A number of low-yielding, low-salinity wells in the region are located adjacent to surface depressions in the weathered Bulldog Shale which become swamps in wet seasons.

Underground excavation

Another surprising aspect of the weathered Bulldog Shale is its capacity to sustain unsupported excavations 10 m or more wide beneath shallow depths of cover (only a few metres generally). In other weathered sedimentary rocks, artificial reinforcement provided by shotcrete, mesh and steel arches would be required to ensure public safety. Possible explanations for this cavern roof stability include the lack of bedding and of persistent jointing in the bleached and ferruginized kaolinitic 'sandstone', negligible *in situ* stress, and a tendency for the mottled zone material to self-cement on exposure.

Shaft sinking for opal mines is now generally carried out using truck-mounted Calweld-type bucket augers. These are drilled at 1 m diameter for exploration and reamed out to 2 m if completed as production shafts. This type of machine is primarily a soil auger used for large-diameter bored piles, so it can only operate in weak and non-abrasive rock such as the weathered Bulldog Shale. The main impediments to what is otherwise an economical form of drilling are silicified bands, which have to be blasted through. Mine tunnels were formerly blasted laterally from the shaft bottom, but are now largely driven by small tunnel-boring machines. The latter are also used in paired horizontal headings to excavate some dugouts, the upper and lower rock wedges between the circular tunnels being later removed to create an ovoid cross-section room.

Aggregate and road base

The Coober Pedy area is entirely devoid of aggregate quality rock, with the partial exception of silcrete boulders and cobbles. These one-stone-thick lags could, in theory, be 'broomed up' by bulldozer blading and then crushed and screened, but the environmental damage to the ground surface would be out of all proportion to the amount of product obtained. Generally the desert loams that occur below the gibber lags are highly dispersive and erode very rapidly after exposure. Concentrated silcrete deposits are more attractive, but these are uncommon and seldom cover more than 2–3 ha, and yield less than 100 000 t of road base; the proportion of screened and sized aggregate would be much less. Such deposits occur at sites of former groundwater discharge; many appear to be fossil mound springs or small silicified fans (McNally & Wilson 1995).

Furthermore, silcrete is only of marginal quality as an aggregate source. Its chips tend to be sharp-edged and brittle, causing tyre wear and being prone to fracture under wheel impacts. Particles are potentially reactive in concrete, forming an expansive gel with excess alkalis in the cement paste. They also are prone to strip (de-bond) from bituminous surfaces and to polish (become less skid-resistant) under traffic. Crushing and screening silcrete materials generate large quantities of siliceous dust that may constitute a health hazard to workers. Quarry sites are shallow, hence extensive, and difficult to rehabilitate because of the low rainfall and self-hardening pallid zone beneath the duricrust.

Road base for unsealed pavements is, nonetheless, available in large quantities from partly cemented Tertiary and Quaternary regolith deposits. This material, especially nodular calcrete, is also suitable for select fill or even sub-base in sealed roads, though base course for these higher-standard pavements (like aggregate) has to be imported from outside the area.

Expansive clays

The unweathered Bulldog Shale is composed largely of smectite. Self-mulching 'popcorn textured' soils and gilgai on treeless plains are characteristic surface features. The gilgai are defined by hummocky ('Bay of Biscay') microtopography and mounds of silcrete gibbers on the plains, and by ridge-and-furrow topography on lower slopes. The latter, informally referred to as 'hilgais', are caused by a combination of shrink–swell ground heaving and downslope creep due to gravity. Gypsum crystal clusters are abundant in the crumbly soil, sparkling in the sun like thousands of broken glass panes. Any buildings placed on these soils would be subject to large seasonal movements, salt damp and sulfate attack on concrete foundations. The main engineering developments to date in the swelling clay areas north-east of Coober Pedy have been unsealed dry-weather roads, which are relatively tolerant of subgrade heaving, but any future sealed pavements will require more protection.

Extensive landsliding has occurred further north in the Mount Barry–Arckaringa area, where unaltered Bulldog Shale is exposed in the lower slopes of the Stuart Range escarpment. These circular slips (Figure 15.5) at first give the impression that multiple duricrusts are present, but are in fact repetitions of

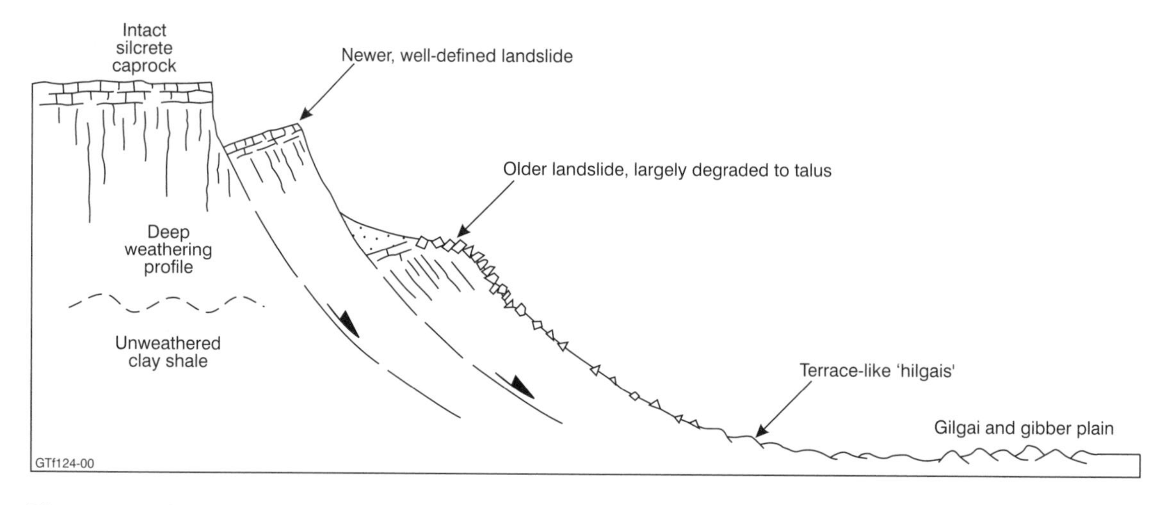

Figure 15.5 *Section showing the distribution of weathered and fresh Jurassic and Cretaceous rocks in the Coober Pedy region illustrating the effects of hydraulic variation*

the same surface. The shale is dispersive as well as expansive, gullying rapidly where the surface veneer of gibbers is disturbed.

Abandoned mining hazards

The Coober Pedy Precious Stones Field (PSF) covers 4700 km^2, though only a fraction of this area, mostly within 10km of the town, has been extensively mined. However, this activity has left literally thousands of derelict mine shafts and tunnels, open cuts, waste dumps and minor excavations. Almost none of these have been backfilled, and few have collapsed naturally. The extent of dereliction is such that no site rehabilitation appears to be economically feasible at this stage. The main safety measures are warning signs erected along main roads (Figure 15.4), plus light fencing of especially dangerous areas close to tourist routes.

Mine shafts are typically about 2 m in diameter and 10–20 m deep. They are nearly always dry and the walls appear to be stable for a decade or more. Their location is indicated by aureoles of white mullock, which contrast sharply with the surrounding grey-green scrub and provide a nominal safety barrier around the rim.

Open cuts range in size from shallow bulldozer scrapes a few metres deep to large costeans up to 30 m deep. These are usually ramped down at one or both ends and vertically sided. Like the shafts, these walls in the weathered Bulldog Shale are remarkably stable and will certainly stand for many years. The main risk is from loose debris and rocks falling from the surface on to visitors within the pits.

Mullock heaps are extremely variable in size and range in height up to about 25 m. The waste is a more or less free-draining mixture of silt and rock fragments, generally less than 0.3 m in diameter and lying at its angle of rest (around 40°). 'Noodlers' frequently rework dumps and this presumably causes some decrease in particle size and loss of compaction, but not to the extent that their stability is diminished. One environmental nuisance associated with waste dumps is the dust plume downwind from 'blowers', truck-mounted vacuum cleaners that draw up tunnel muck and separate coarse fragments (including opal lumps) from fines. Revegetation of mine dumps in this arid climate is practically non-existent.

URBAN DEVELOPMENT IN HONG KONG

Hong Kong is an exceptionally urbanized community. It is a city state of some 6 million people living on steep unstable slopes and on land reclaimed from the sea. This presents a unique set of development problems to city managers. The steep slopes are mantles in colluvium of varying ages and degrees of weathering. Much of the colluvium, particularly on granitic slopes, contains very large corestones in a matrix of finer-weathered material. On volcanic slopes the colluvium is finer, generally less weathered but no more stable (Figure 14.5). Construction on these slopes requires great caution as undermining may cause slippages at any time, but mainly during wet weather. The Po Shan

Road slip in the early 1970s (Figure 2.32) caused several apartment blocks to collapse and a large number of lives were lost. This slide was caused by construction work undermining a large section of slope. It is not uncommon for lives to be lost each year in the Special Economic Zone, so government has strict controls on development and any earthworks require assessment by a geotechnical engineer to avoid such problems occurring. Despite such precautions slides still occur. In the early 1990s the longest slide in the history of Hong Kong occurred on Castle Peak, bringing about 600 m of the hillslide down next to a housing estate completed only a year earlier. This slide occurred as a small rock slide high on Castle Peak developing as it descended into a debris flow in the zone of colluvial cover. No lives were lost, but as land close to the base of the slope was under consideration for further development it illustrates that even though geotechnical reports are done, the hazard may still exist.

Exploration in the regolith

The importance of regolith as a host for mineral deposits and its importance in mineral exploration have been recognized for some time. Exploration efforts over the last 10 years in regolith-dominated terrains such as Australia, Africa and South America as well as Canada, have led to an increased interest in regolith and landscape evolution (e.g. Butt & Zeegers 1992a; Kauranne *et al.* 1992). The weathering and accompanying geochemical processes which occur within the regolith, while masking many bedrock-hosted ore-bodies, also provide an increased chance of finding them using sound geochemical models in combination with landscape evolution models. In this book we have provided some insights to the processes and products of landscape and regolith formation and we now provide a few examples of how this knowledge may be applied to mineral exploration. Butt & Zeegers (1992a) and Kauranne *et al.* (1992) provide detailed examples of many exploration techniques and successes using regolith geology, and books like that by Evans (1995) provide excellent overviews of mineral exploration, which we will not do here.

GOSSANS

Gossans are discussed in some detail in Chapter 9. They are very important indicators of sulfide mineralization, and the dispersion halo of pathfinder elements around a gossan can provide an enlarged exploration target. Many of the world's sulfide ore-bodies were discovered by recognizing their presence from a gossan that cropped out.

FERRUGINOUS MATERIALS

Taylor *et al.* (1980) noted the difference between *in situ* gossans and what they called 'ironstones' formed by different processes from the weathering of ore-bodies, and that these were also capable of hosting ore or pathfinder trace elements. These 'ironstones' or ferricretes and hard mottles from the mottled zones of lateritic profiles may also act as sampling media for the detection of buried ore deposits. To make use of these other ferruginous materials in the regolith however, it is important to understand their origin in a landscape context so it is possible to trace anomalous ore elements back to their source. Much of this work was developed in studies on the Yilgarn Craton by CSIRO (summarized by Smith 1983 and later in Butt & Zeegers 1992a).

Their exploration strategy for the Yilgarn depends on a landscape model in which target elements were dispersed along the water-table, perhaps many times, as the water-table fell during the past-Neogene arid phase. This occurred below an essentially low-relief landscape underlain by lateritic profiles and capped by ferricrete. Where a complete lateritic profile is preserved the materials, including the ferricrete, are essentially the result of *in situ* weathering, and the ferruginous components (ferricrete, pisoliths, nodules and mottles) reflect the geochemistry of the underlying rocks including any mineralization (Figure 15.6). Here, because of lateral dispersion of the target elements by groundwater, the target anomaly is weak but much larger than the original gossan or ore-body (Figure 15.7).

As this old ferricrete landscape has been eroded the ferruginous materials have been removed and redistributed in a landscape now consisting of residual hills (breakaways) and ferruginous lags in colluvial and alluvial deposits in the lower portions of the landscape. Such redistributed materials may give false anomalies in the sense that they are not over an ore-body, but reflect materials transported from an area underlain by one. It is in this situation that an understanding of the evolution of the landscape is essential to interpreting anomalous element distributions. Equally, if the original model of landscape evolution is not correct anomalous element distributions may also be misinterpreted.

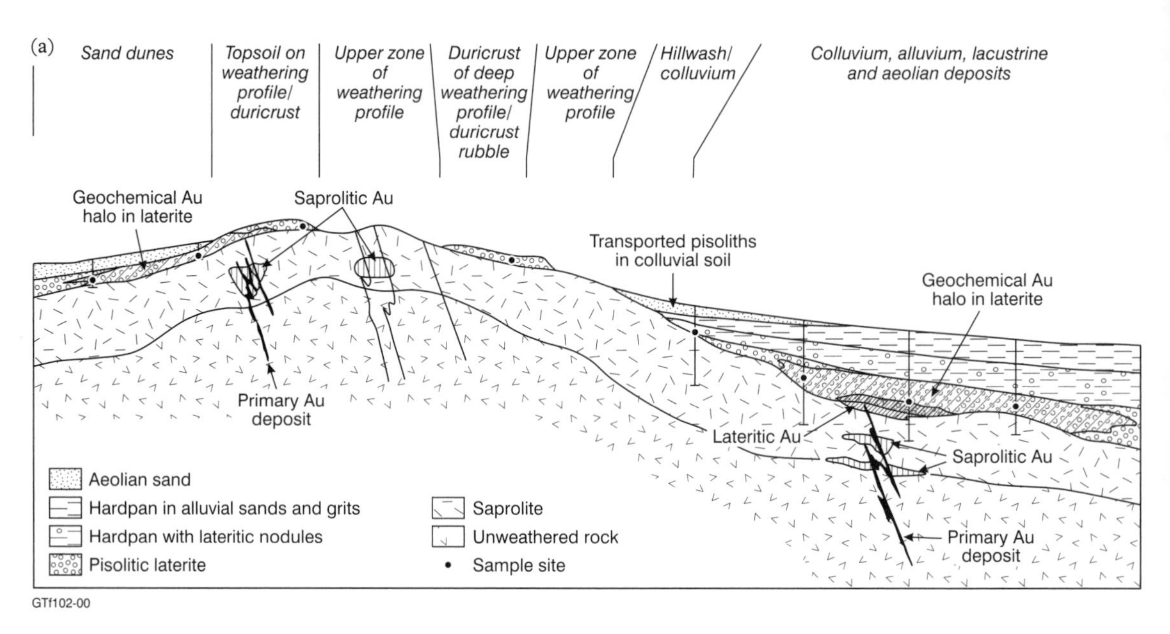

Figure 15.6 *(a) Landscape section illustrating the interrelationships of dispersion models in partly eroded, arid, 'lateritic' terrain and appropriate laterite exploration sampling locations. The dispersion and secondary accumulation of Au is used as an example (from Butt & Zeegers 1992a). (b) An expanded landscape model showing the increased Au anomaly caused by dispersion in mica and ferricrete residual and by the groundwater-controlled supergene halo in the saprolite (diagram from R. E. Smith, CSIRO)*

The CSIRO model has essentially been based on a more or less pervasive ferruginization of a landscape prior to its erosion (Anand & Smith 1998; Anand 1998a), but others (Ollier & Galloway 1990; Ollier & Pain 1996) have suggested (as we do in Chapter 11) that ferruginization may not have been extensive, but rather confined to topographically low positions. In this model the upstanding nature of the ferricrete-dominated hills represents a post-formational topographic inversion. The consequence of this model is that trace elements would have originated laterally to the ferricrete, and not below it.

In Burkino Faso at the Diouga prospect Zeegers & Lecomte (1992) show how Au dispersion has followed a palaeovalley. The Au is fixed by the Fe_2O_3 in the 'Cuirasse' or ferricrete mantling the old topographically inverted valley, but the ore-body is outside the remaining ferricrete duricrust (Figure 15.8). This example clearly demonstrates the necessity for an accurate and robust landscape evolution model in interpreting geochemical anomalies.

As we discussed in Chapters 5 and 13, ferruginization may also occur due to lateral leaching and cementation at escarpment margins as is the case at many sites in the Charters Towers region of Queensland (Rivers *et al.* 1995; Eggleton & Taylor 1999). Such accumulations of Fe reflect the chemistry of the uplands through which the water moves rather than the chemistry of the immediately underlying rocks. They may or may not contain anomalous element concentrations depending on the origins of the lateral seepages. Anand & Smith (1993) show such ferricrete occurring at Lawlers on the Yilgarn Craton and they occur elsewhere such as Mt Isa in Queensland.

Work on Au prospecting in West Africa, and Ghana (Bowell *et al.* 1996) describes four situations in an area of extensive 'lateritization'. Four profiles have been described and the geochemistry studied in detail. The weathering profile at Syama in Mali shows (Figure 15.9a) a laterite profile which is, in their terminology, complete. At Senoufo in northern Côte d'Ivoire (Figure 15.9b) a similar profile occurs, except there is complete to partial truncation of the upper horizons. Another profile in southern Côte d'Ivoire at Hire is a complete 'laterite' profile similar to that at Syama. The presence of complete and intact profiles at these two sites, but the lack of an anomaly in the upper parts (Figure 15.9c) is attributed to Recent intense weathering that has remobilized and dispersed previous anomalies. At another site in western Ghana, Kubi, the previous laterite sequence has been almost completely removed and very recent soils lie on saprolite (Figure 15.9d).

(b)

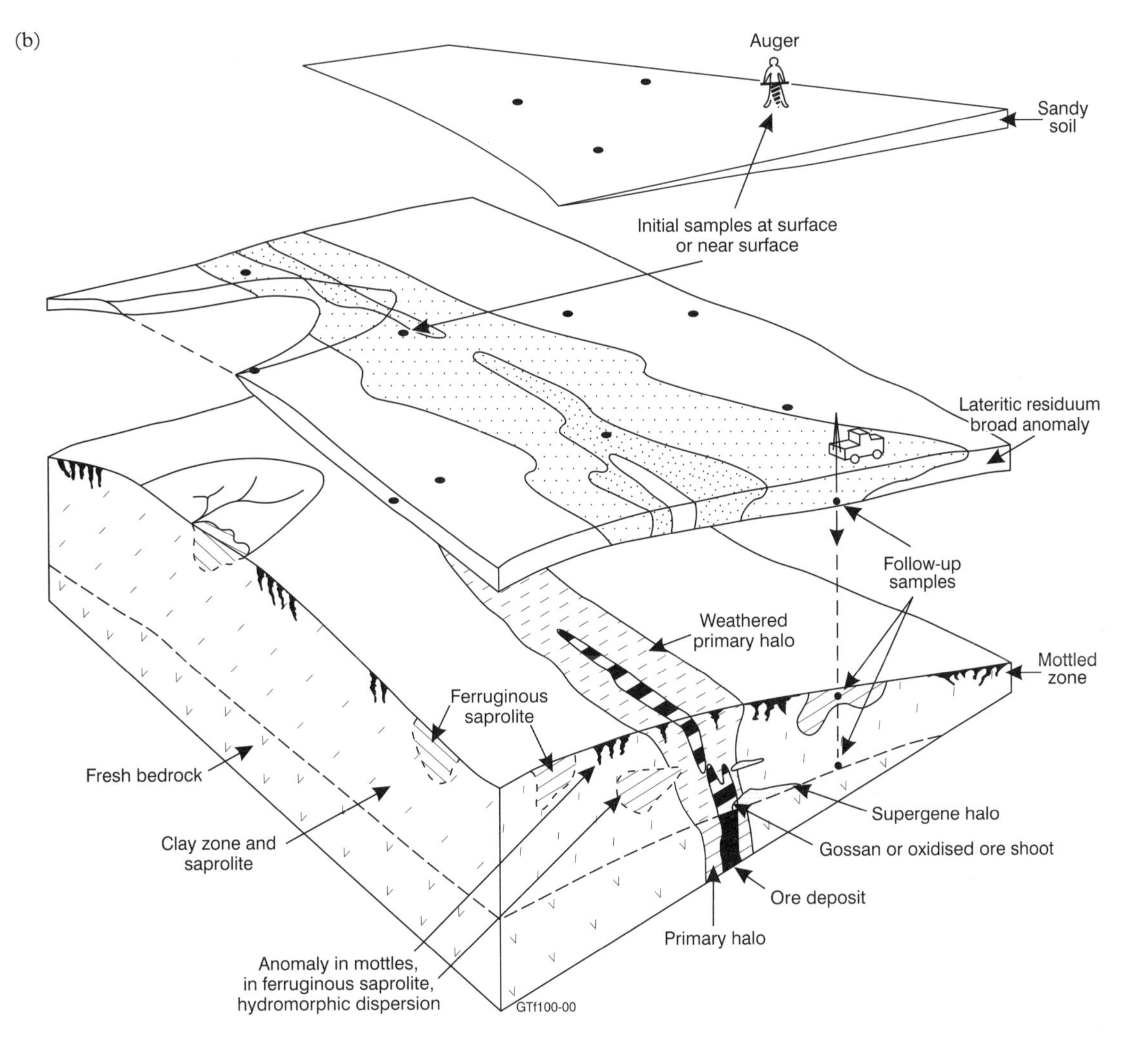

Figure 15.6 *(continued)*

Bowell *et al.* (1996) suggest the humid savannah Syama profile contains significant Au and base metal anomalies in the lower horizons that are fixed in smectite and in the ferruginous horizons (cuirasse and mottled horizon) by Fe-oxides. The Au secondary dispersion anomaly is small (~20 m wide) due to the contrast between the horizons. At Senoufo, where the profile is partially truncated and overlain by a thin transported cover, anomalies are preserved in remnants of residual cover and the dispersed anomalies are again small, as they are at Syama, but the transported cover dilutes and widens them in the surface horizon. In the rainforest profiles at Hire there is more intense leaching than in the savannah and this results in the formation of a stone line of ferruginous material. Where erosion is limited, anomalies are preserved in the stone line and Au lower in the profile is mobilized to the middle parts of the profiles as organic complexes. Secondary dispersion in the surface samples is still minimal (10–20 m). At Kubi where the profile has been stripped, weak Au anomalies occur along the present water-table where a poorly developed ferruginous stone line has formed. In the saprolite Au enrichment is insignificant.

The dispersion of trace elements in a groundwater system can vary depending on the tectonic and climatic history of a region. These changes are well illustrated by Gray *et al.* (1992) as shown in Figure 15.10. In these figures they show the situation of a single dispersion halo that results at times of rising

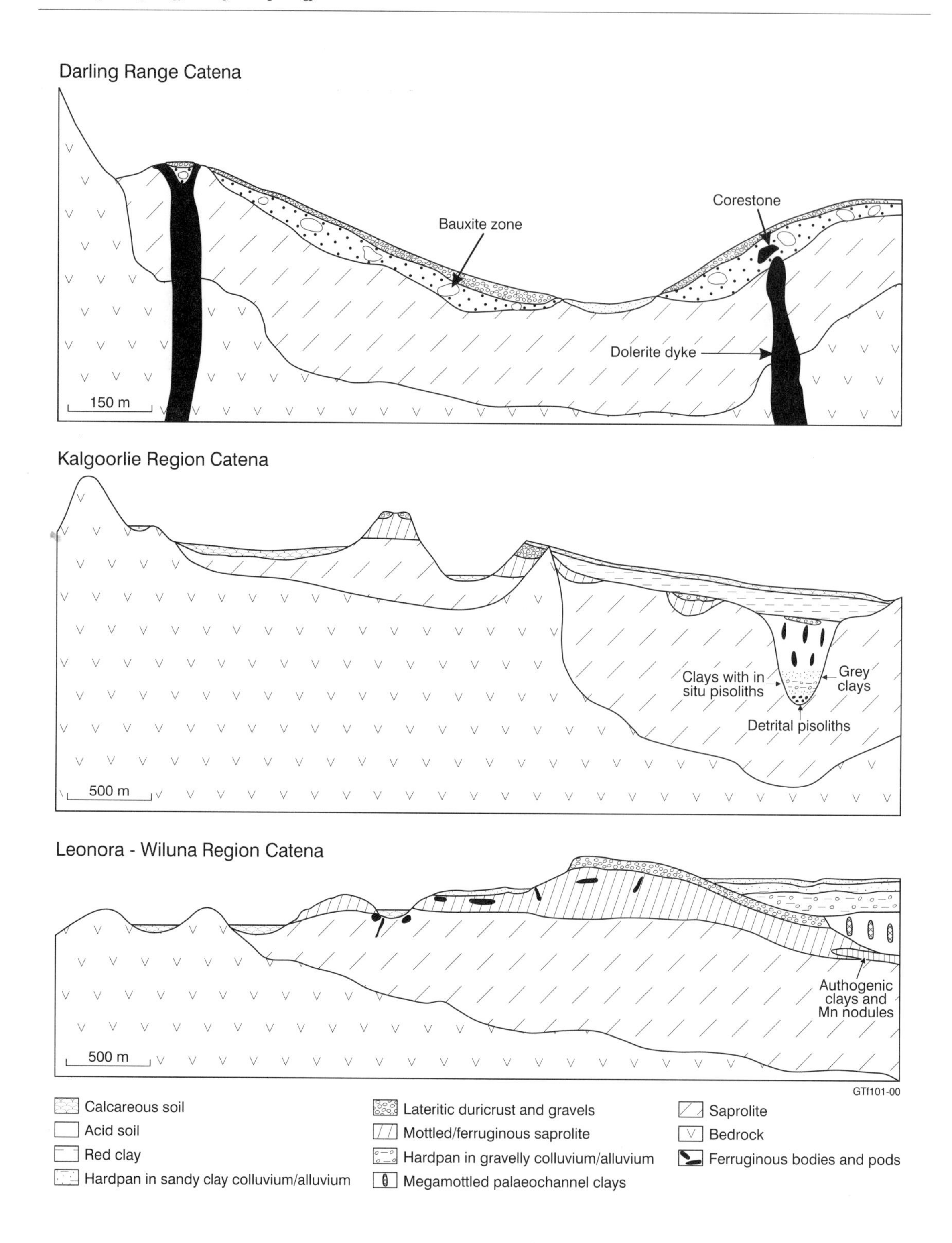

Figure 15.7 *Typical regolith catenas formed on the Yilgarn Craton (from Anand & Smith 1998)*

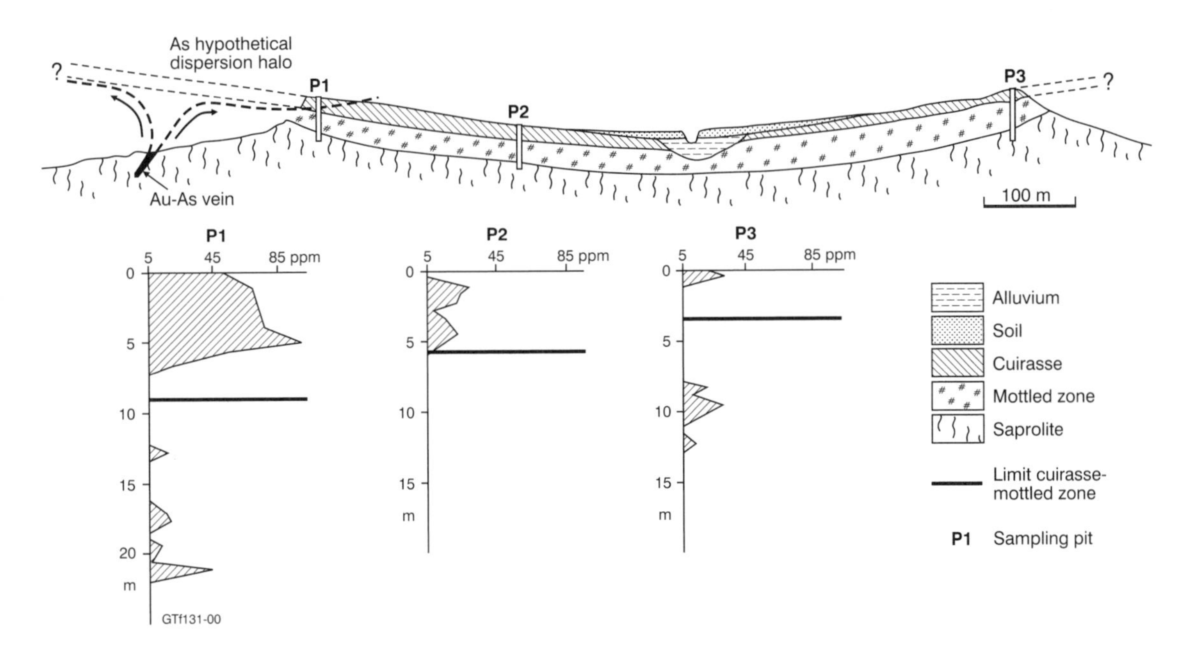

Figure 15.8 *Diouga Au prospect, Burkina Faso. Cross-section showing the dispersion of As from Au–As mineralization in different positions relative to the lateritic surface (from Butt & Zeegers 1992a)*

water-tables, whether climatically or tectonically induced (in some terrains this could also be caused by eustatic change) and the multiple halos formed from the opposite situation.

In all these examples it is Fe-oxides which are the preferred hosts for trace elements, and as they are reasonably long-lived in landscapes it is these which should be sampled to find anomalous metal concentrations that may lead to the discovery of an ore-body. Whether the discovery is made, however, depends on the correct interpretation of the regolith and the landscapes in which it occurs. The Fe-oxides provide larger targets for mineral exploration, but those targets are useless unless considered in the context of landscape evolution.

CALCRETE

Calcrete as a sampling medium for Au exploration has become important since the mid-1990s. At the Panglo Mine north-east of Kalgoorlie, supergene Au (1.5 Mt at 2.7 g/t) is mined from flat-lying bodies up to 16 m thick in a saprolite developed in shales and ultramafic volcanics. The primary mineralization is disseminated pyrite and arsenopyrite in strongly sheared shales and ultramafic volcanics. The presence of the Au mineralization is clearly indicated by the presence of 250 ppb Au in soil calcrete (Lintern & Scott 1990) (within the upper 1–2 m of the profile) above the deposit but not in the Fe-rich soil components. Gold does not appear in the transported or *in situ* regolith below the calcrete until the ore is encountered. Biogeochemistry shows that while *Eucalyptus* spp. do not reflect the Au concentration in the soils that of Emu Bush (*Eremophilla* spp.) does in a general way. What is important from this study is the presence of Au indicators in the calcrete at shallow depths whereas there are no values above background in the deeper regolith, which is often sampled for ferruginous materials where they are buried by transported regolith. Similar studies have revealed calcrete to be a useful sampling medium for Au exploration across southern Australia.

In South Africa at Jacomynspan and Putsberg Garnett *et al.* (1982) defined Cu and Ni mineralization from overlying calcretes in which they formed anomalies very little larger than the bedrock deposits. Nickel deposits are reflected in overlying calcareous soils at Pioneer, Western Australia where calcrete occurs below a thin residual soil and overlies a saprolite of lithorelics in a matrix of Ca-smectite. In

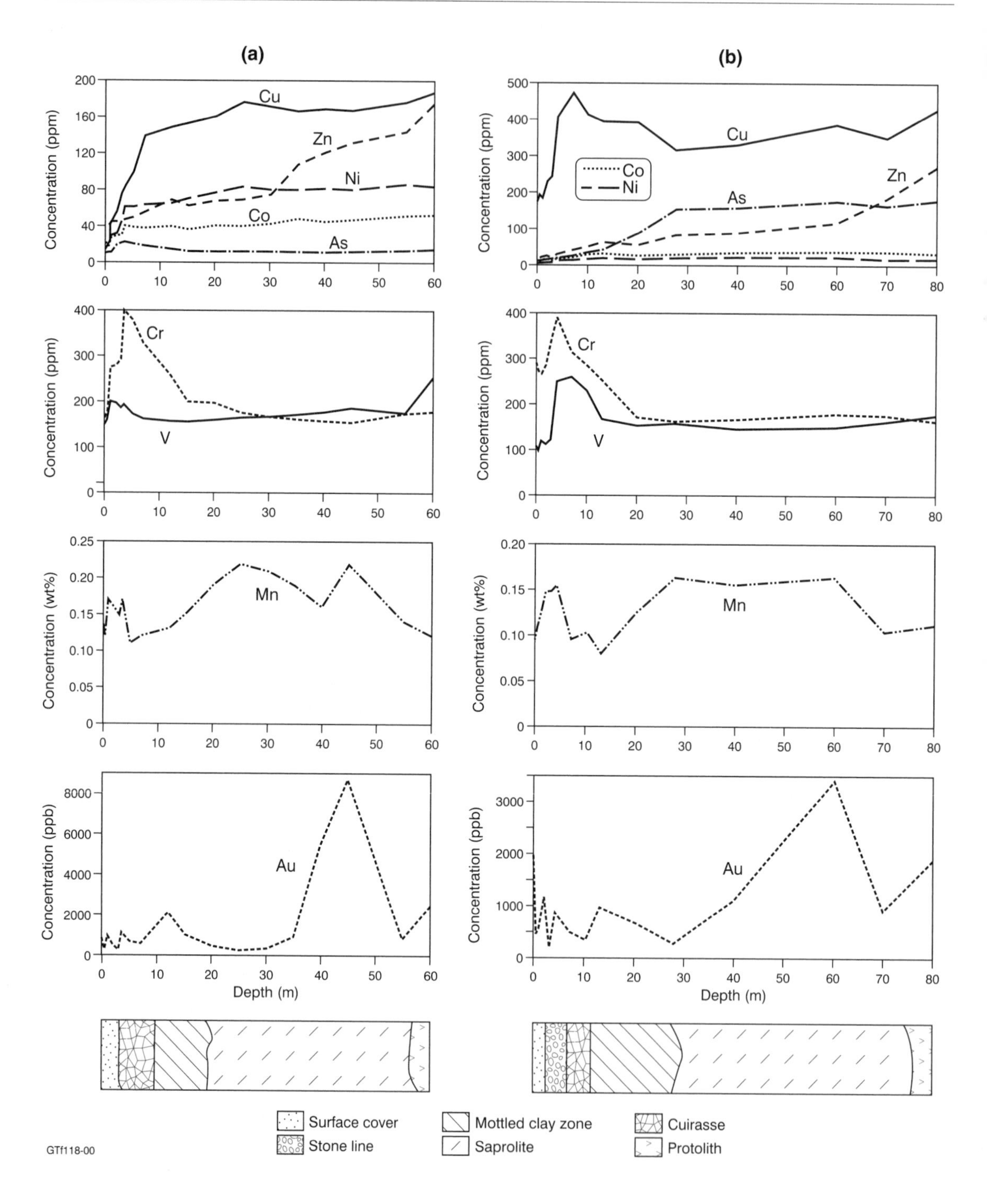

Figure 15.9 *Depth profiles in the weathered mantle at (a) Syama in Mali, (b) Senoufo in northern Cote d'Ivoire, (c) Hire in southern Cote d'Ivoire and (d) at Kubi in western Ghana. The profiles and geochemistry are shown. (e) A model showing the evolution of pre-existing duricrust profiles (Syama and Hire) and erosion of pre-existing profiles (Senoufo) and erosion of pre-existing profiles with present-day modifications (Kubi) developed by Bowell* et al. *based on Butt & Zeegers (1992a) (from Bowell* et al. *1996)*

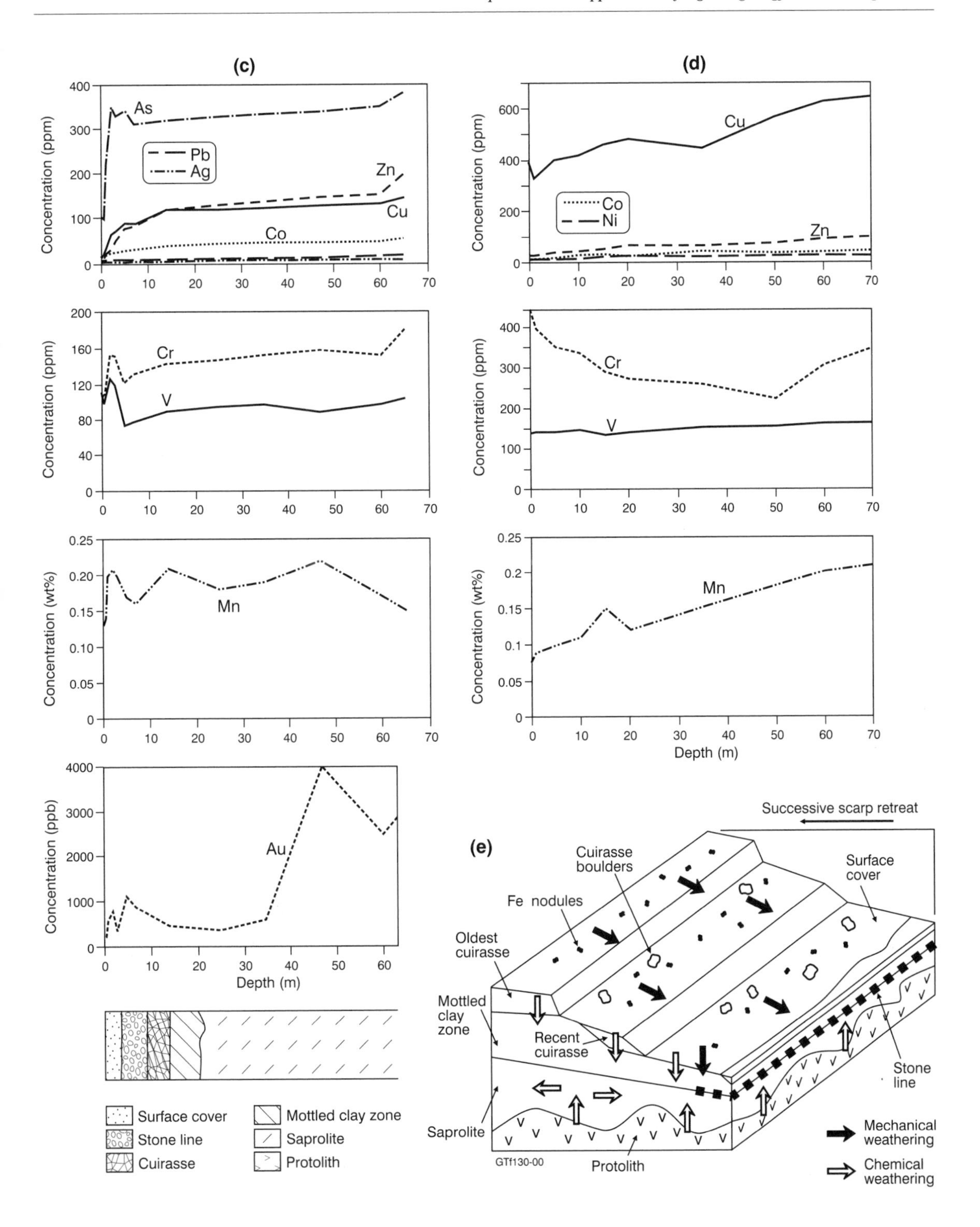

Figure 15.9 *(continued)*

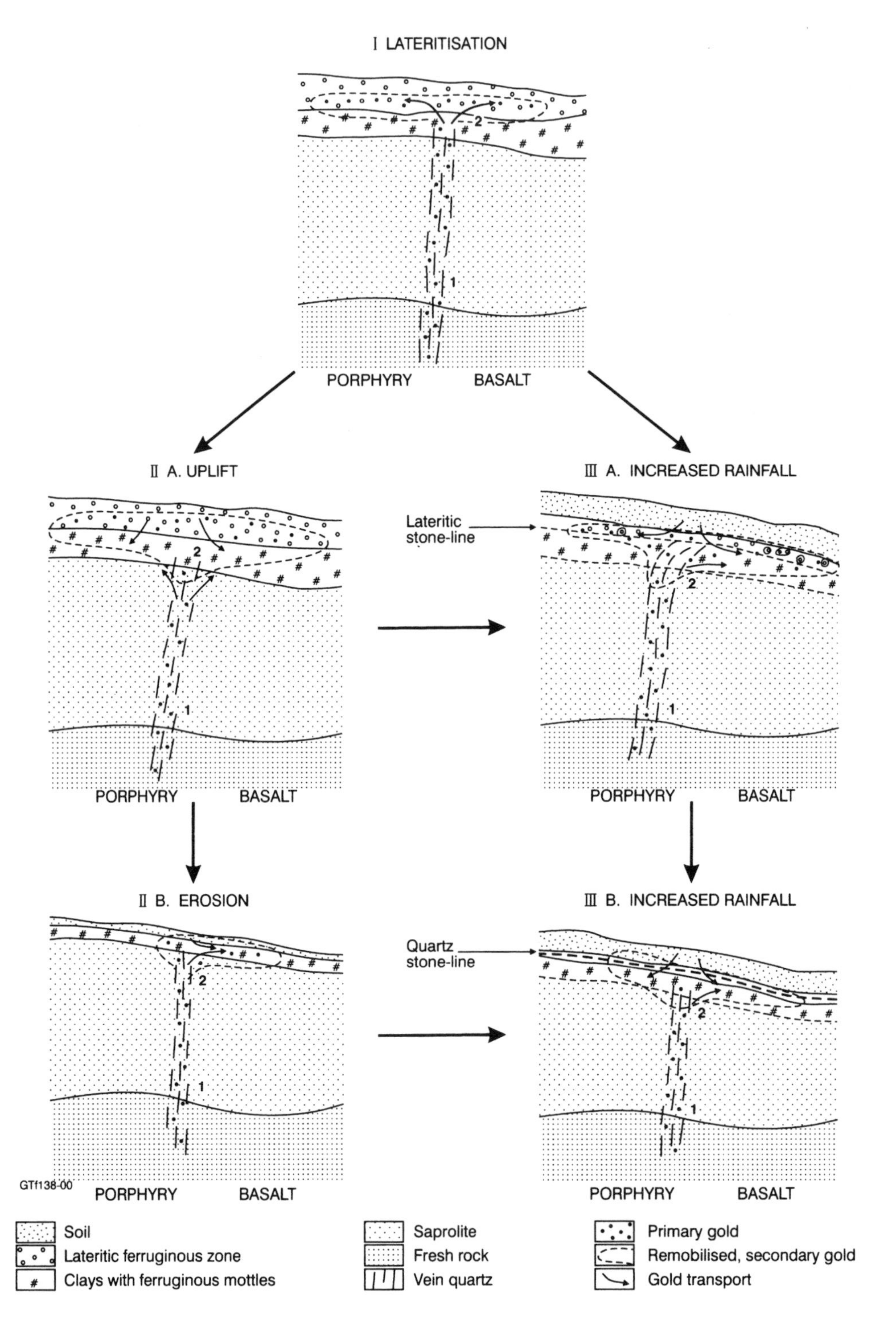

Figure 15.10 *Models illustrating Au groundwater dispersion during lateritization and modifications due to uplift and a change to more humid climate (from Butt & Zeegers 1992a)*

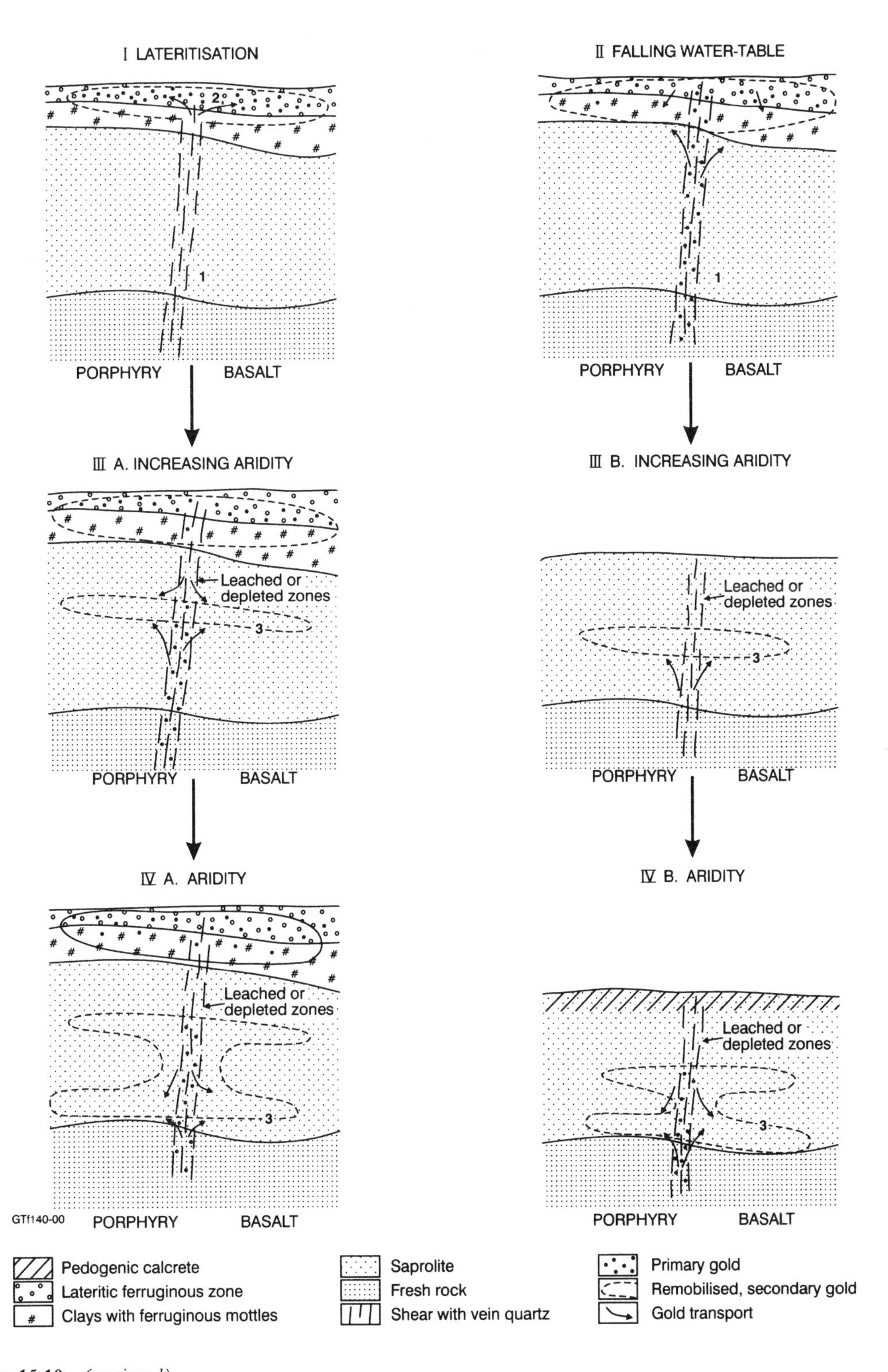

Figure 15.10 *(continued)*

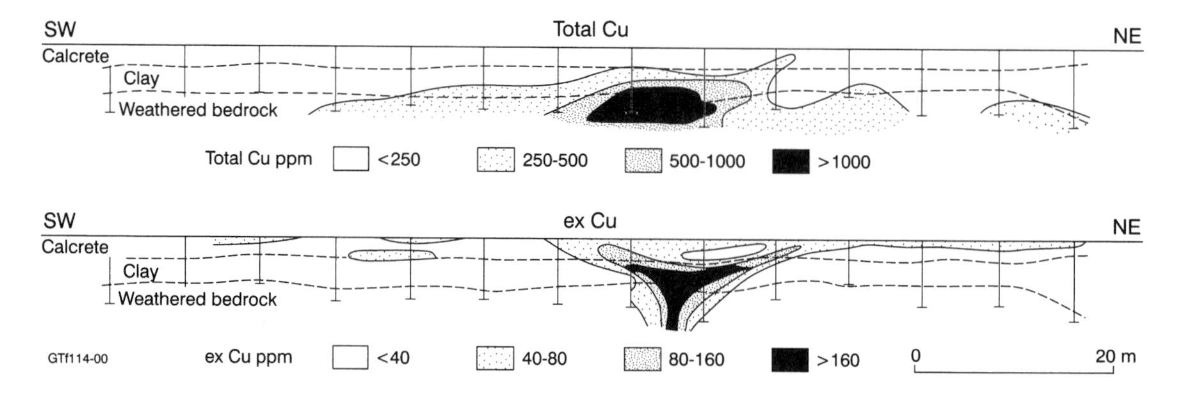

Figure 15.11 *Distribution of total and cold extractable Cu (exCu) in transported overburden and weathered bedrock, Kadina, South Australia (from Mazzucchelli* et al. *1980)*

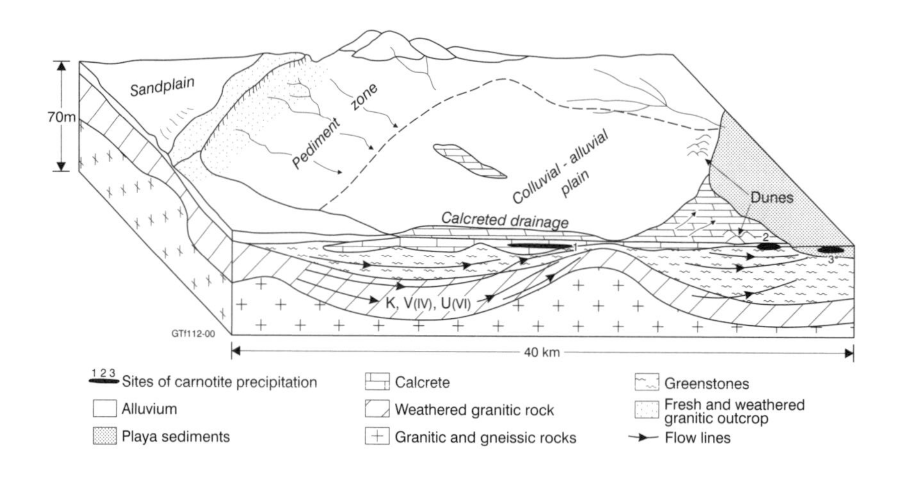

Figure 15.12 *Geomorphological settings of carnotite mineralization in drainage sediments on the Yilgarn Craton, Western Australia (from Butt & Zeegers 1992a)*

the Bushveld Complex calcrete anomalous As and Hg are indicative of the underlying Pt group element deposits (Frick 1985). Mazzucchelli *et al.* (1980) report cold-extractable Cu from the calcrete/clay saprolite boundary at depths of 0.6–2.0 m above 40 m of saprolite above a Cu-sulfide deposit at Kadina, South Australia. Here the Cu deposits in the acid saprolite below the more alkaline calcretes, giving a dispersion halo about 90 m across (Figure 15.11).

Calcrete is also often associated with U deposits in arid terrains such as Africa, the Americas and Australia. The U is transported from diffuse sources (usually granite), concentrated and deposited as carnotite ($K_2(UO_2)2(V_2O_8).3H_2O$) in fluvial palaeovalleys, lakes and playas, and in pedogenetic environments with carbonate or gypsum. Figure 15.12 illustrates the various environments in which U deposits with carbonates on the Yilgarn Craton (after Butt 1998).

CLAY MINERALS

Smectitic and kaolinitic clay minerals represent favourable conditions for the retention of trace elements leached from ore-bodies. Kaolin may fix large amounts of Cu in the saprolitic zones of lateritic profiles. The fixation of these elements in the weathering profile minerals may result in limited dispersion halos. According to Zeegers & Lecomte (1992) weathering conditions in these profiles do not favour wide hydromorphic dispersion.

At Petite Suisse Cu prospect in Burkina Faso, kaolin hosts the dispersed Cu in the upper parts of the weathering profile, while at depth smectite is the host. Kaolin reached Cu values of 2000 ppm and smectite 8300 ppm Cu.

GYPSUM

Although there are few records of gypsum hosting trace elements, its presence may indicate the proximity of sulfide ores below the regolith. In areas of semi-arid climates, sulfates from the oxidation of sulfides may occur within the regolith in the vicinity of the ore deposit. At Parkes in central New South Wales, gypsum is common in the soils over and around the underlying porphyry Cu ore-bodies. Other deposits in central western New South Wales have similar gypsum concentrations. Although it has been argued that the gypsum is aeolian and derived from the playa lakes downwind of these sites, it is only in the vicinity of the ore-bodies that the gypsum occurs.

No geochemical data from gypsum associated with underlying ore-bodies have been reported, and this could be a fruitful avenue for exploration research.

BIOGEOCHEMISTRY

Biological materials can make useful sampling materials. The majority of organisms take in trace amounts of elements useful in mineral exploration. Plants, because they are stationary, have been considered by many as exploration media. Plants can be used in two ways in mineral exploration. Firstly, in a geobotanical way. Particular plants have particular ecological requirements to grow successfully. Thus the presence of a particular plant or plant community may relate to particular regolith chemistry.

Several famous examples exist where geobotanical indicators are useful. In Zambia *Beccium homblei* (a small violet flowering plant called the copper plant) requires 50–1600 ppm Cu in the soil to thrive (Reedman 1979). Similarly in Western Australia the nickel bush is indicative of elevated soil Ni values. There are other indicator plants of many elements in the regolith, but they are local and generally not useful over wide areas. One exception may be pearl bluebush (*Maureana sedifolia*) that indicates the presence of calcium carbonate in the upper parts of the regolith across much of southern Australia.

With the increasing sophistication of remote sensing methods it is becoming possible to map variations in vegetation ecology. This may well lead to an increasing importance in geobotany in mineral exploration and in managing ecosystems over large areas.

Biogeochemistry describes the use of biological or biologically cycled materials as geochemical sampling media. Plants take up elements that are toxic to them in large amounts and consequently they excrete them by leaf-fall, in fruiting bodies and in bark-shed. These accumulate as biolitter on the regolith surface and as they break down, the elements are released back into the regolith from the surface. They may be fixed in the regolith by any of the mechanisms discussed in Chapters 6, 7 and 9. Many other organisms accumulate trace elements, some in soft tissue that can be recycled in the regolith, some in hard parts (bones and shells) that may also reflect higher concentrations in organisms that live in areas of high trace element concentrations.

In Chapter 12 we discussed the role of organisms like termites and ants in bioturbation of regolith materials. Several researchers and mineral explorationists have attempted to use termite mounds (e.g. Watson 1970; d'Orey 1975) and ant mounds have potential where they consist of selected size grains of a suitable sampling medium (e.g. haematite nodules) although we are not aware of any published studies of using this form of sample medium.

PARTIAL EXTRACTION

Mobile metal ions (MMI) may be released from metallic ore-bodies migrating to be fixed in the soil (Mann *et al.* 1993). Their measurement can be used to detect anomalous metal concentrations below regolithic cover. Mann *et al.* (1993) describe successful detection of ores below 70 m of weathered profiles. At Nepean west of Coolgardie, Western Australia, ferricretes west of the mineralized zone show the highest anomalies, but MMI responses were at their highest over the blind ore deposit. Gray (1998), on the other hand, has MMI results at Curara Well in Western Australia that show concentrations of Cu and Cd over the blind ore, but he comments that these elements show no association with the primary or secondary ores. He suggests that the partially and selectively extractable Cu and Cd concentrations over the Au ores at Curara are coincidental. MMI is a controversial technique; it has had successes.

Another partial extraction method developed in Russia to detect regolith-covered ore deposits geochemically is 'method of diffusion extraction' (MDI). This technique has been tested in Israel (Levitski *et al.* 1995) where they claim success in finding ores beneath transported regolith. They even note that the method may be useful for detection in surface regolith of petroleum concentrations at depth.

Regolith-based resources

METALLIC RESOURCES

Australia is currently the third largest producer of Au (289 t in 1997) after South Africa (490 t in 1997) and the USA (325 t), much of the Australian production coming from regolithic accumulations.

Gold is one of the major regolith resources mined in Australia and South America, and supergene (weathering concentrated) Au provides essential added value to gold and other metallic ore mines the world over. Butt & Zeegers (1992a) give a very good summary of the accumulation of Au where they consider examples from most continents.

Many Cu/Au deposits have significant supergene Au-rich caps. At Northparkes (New South Wales), Oktedi (Papua New Guinea), Olympic Dam (South Australia) and many others, Au from the supergene zone was a significant factor in developing the mines that exist there now. In other cases mining has been based entirely on the supergene enrichment of Au in Fe-rich parts of the weathering profiles. Boddington in south-western Western Australia is an example of a gold mine based on supergene Au. At Boddington, Au is enriched in saprolite over a Cu/Au porphyry system. The low-grade Au mineralization forms a more or less continuous blanket in bauxite and saprolite of a lateritic profile (Anand 1994). Of the Au, 30% is hosted in the duricrust and 70% in the saprolite. Gold is homogeneously distributed in the duricrust, but is more erratic in the saprolite. Figure 15.13 illustrates the profile and distribution of major and trace elements in the Boddington deposit.

Many similar deposits occur in deeply weathered terrains, and are well documented, from West Africa, Brazil and India (Freyssinet 1994; Olivera & Campos 1991; Santosh & Omana 1991, respectively). In most cases the Au is concentrated in the ferricrete as relatively large accumulations of very fine Au.

In other examples, such as at Dondo Mobi in Gabon (Lecomte & Zeegers 1992), Au anomalies occur in the saprolite despite intense weathering over a concealed ore between 30 and 120 m deep. The bedrock contains most Au as quartz lenses within amphibolite over a width of about 50 m. The saprolite (H3) shows no evidence for dispersion producing a larger anomalous halo, but anomalous Au values extend over 50 m with no significant leaching, dispersion or enrichment. A stone line (H2) occurs between the deeper saprolite and overlying sandy clay (H1). The dispersion halo in the sandy clay is much larger but of lower Au values. Gold grains show increasing dissolution from the base to the top of the profile. Grains in the saprolite are similar to those of the bedrock but they are obviously more dissolved near the top of the profile showing Au to be chemically mobile in the upper zones. Additionally the Au is also physically mobile; as Figure 15.14 shows, placer Au occurs downstream of the weathering profiles with considerable concentration factors.

The erosion, reworking and final deposition of Au grains from the adjacent hillsides at Dondo Mobi have resulted in a four-times enrichment of the Au in valley-bottom placers. This physical concentration of Au (and other commodities) makes placer deposits potentially of very high value. While not associated with the modern regolith, the Precambrian placer Au on the Witwatersrand makes them some of the richest Au mines in the world. The 'pay streaks' clearly show these deposits to be of fluvial origin.

Placer deposits occur where the heavy metal or metal-bearing grains physically transported are hydraulically concentrated and deposited. This is discussed further below. Placer Au deposits can be very rich, as illustrated by the deposits discussed above. At Kiandra, New South Wales, placer Au was mined from 1859 to the 1940s, but the field reached its peak in 1860 with some 2 t being shipped out officially, and significantly, more unofficially. These deposits are concentrated in alluvial Neogene channels now buried by Miocene basalts. Much of the Au has washed from the channels (or 'deep leads') occurring in contemporary alluvial systems. The largest nugget found there weighed almost 13 kg. It was one of Australia's richest ever Au mining districts for its peak year, 1860. In terms of total production, Ballarat in Victoria has produced 408 t of Au, most of it from deep leads as placer Au. Here Neogene alluvial channels concentrated the Au and were covered by basalts and as the landscape eroded Au also accumulated in contemporary alluvium.

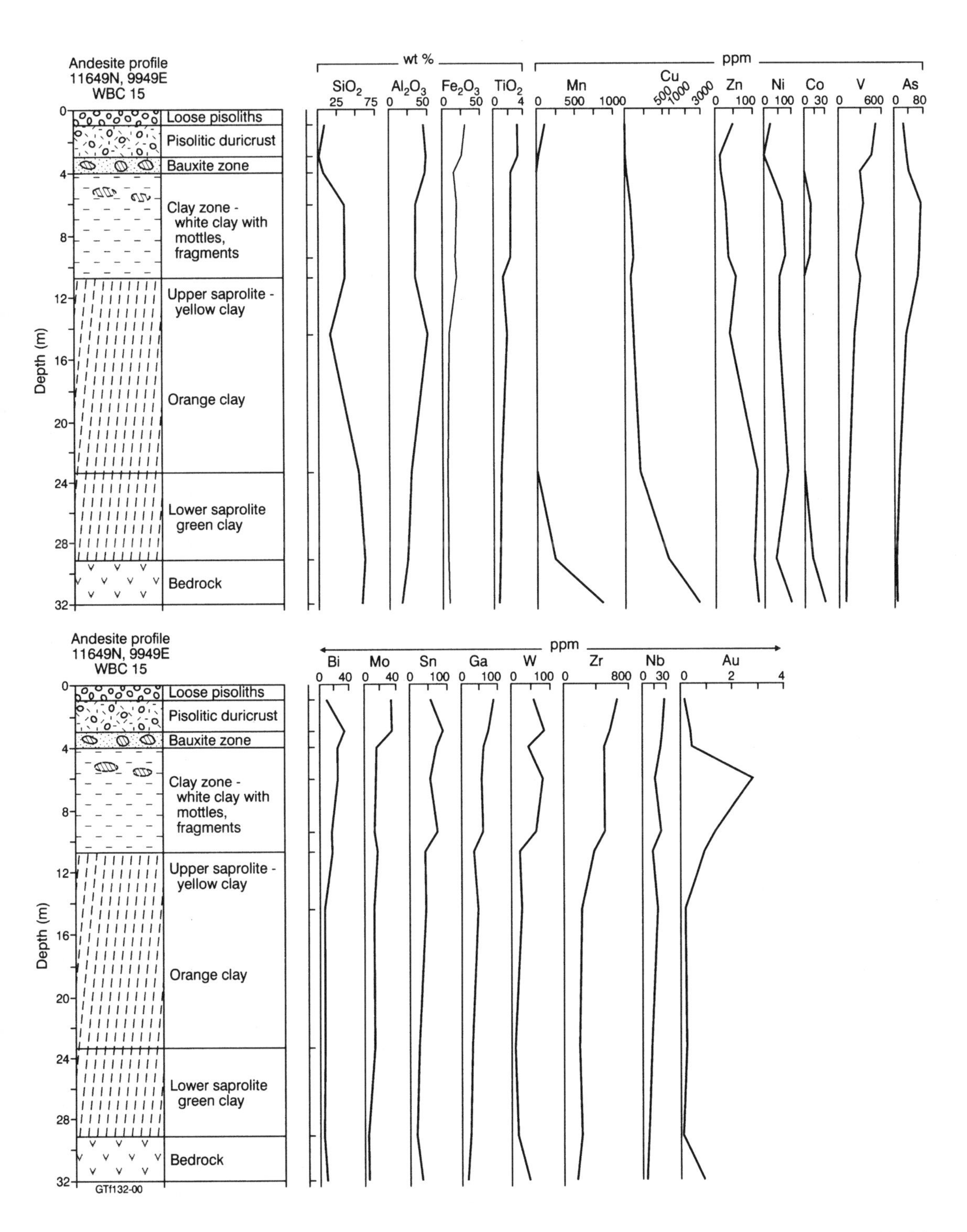

Figure 15.13 *Vertical distribution of major and trace elements in the weathered andesite profile (WBC 15) at Boddington gold mining area (from Anand 1994)*

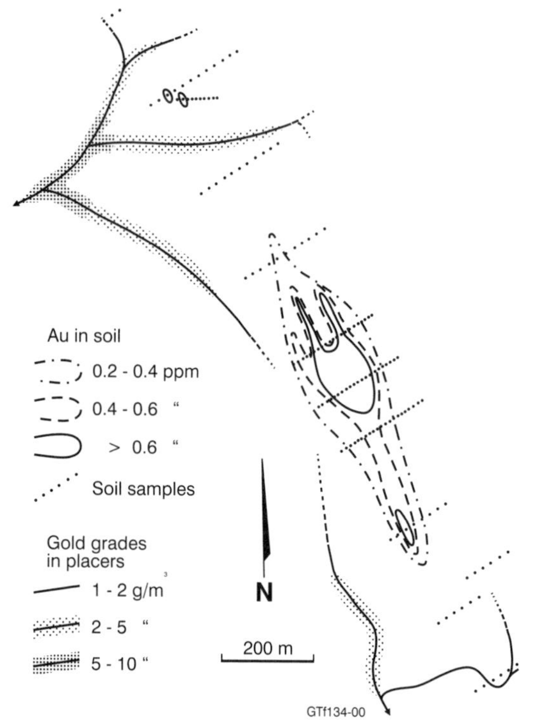

Figure 15.14 *Dondo Mobi Au prospect, Gabon. A mushroom-like Au-dispersion halo in a weathering profile. See text for explanation*

With more detailed regolith mapping and remote investigation it is possible to reconstruct palaeodrainage systems that, unlike their modern counterparts, drain regions shown by old workings to be Au-rich. This increases the potential for discovery of significant placer Au in areas previously thought not to be prospective. The study of palaeodrainage in the West Wyalong district of central New South Wales by Lawrie *et al.* (1999) has identified such a prospect.

Other types of regolithic Au occur fixed on bacteria (*Pedomicrobium*) which reproduce by budding and so may avoid the problems associated with reproducing by division in a chemical environment usually toxic to bacteria (Watterson 1992). Gold grains from Venezuela show tubules of Au radiating from a central region rich in Au (Bischoff *et al.* 1992). They are probably precipitates of Au on the bacterium.

BAUXITE

Bauxite is the ore for Al. Australia is the world's largest producer, mining 37% of the world's ore and 27% of refined products. Guinea, Brazil, Jamaica and India are, in order, the next highest producers.

The main features of 'laterite' and bauxite and its formation are discussed in Chapter 11. The majority of the world's bauxite occurs as 'lateritic' bauxite formed by the weathering of underlying rocks to form upper pisolitic or massive zones rich in the minerals gibbsite, boehmite and/or diaspore. McFarlane (1983) provides figures (Figure 15.15) which show that most bauxite has been formed during the Cainozoic with significant amounts through the Mesozoic and Palaeozoic. She also notes (Figure 15.15) that only a minor proportion is formed by other than 'lateritic' processes during the Cainozoic and late Mesozoic, but that karst bauxite is more abundant in the older deposits (Figure 15.15).

Of the bauxite localities currently worked in Australia most are 'lateritic' bauxite where the upper zone of the weathering profile is composed of massive to moderately pisolitic or nodular gibbsite-rich material (e.g. Darling Ranges, Gove). But at Weipa, the upper bauxitic zone is composed of even-sized pisolites with no matrix. This zone overlies a ferruginous duricrust at the top of a lateritic profile. Although it is Australia's largest bauxite mine, Weipa seems unique in Australia, and perhaps the world. Ollier & Pain (1996) make the point that bauxite deposits are often underlain by kaolin and they contend that three hypotheses have been proposed to explain this phenomenon:

1. the kaolin is resilicified bauxite;
2. the bauxite is desilicified kaolin; and,
3. the kaolin and bauxite are not related to any underlying rock, but are sediments on top.

At Weipa the bauxite has kaolin below the ferricrete and it is in chemical equilibrium with the groundwater, so it is forming (and presumably dissolving) today (Eggleton & Taylor 1999). The bauxite itself, which forms the upper 2–3 m of the deposit, shows some signs of undergoing retrograde weathering, i.e. resilication of the gibbsite, for the pisoliths are coated with a thin kaolinite shell, possibly the result of reaction between aeolian or organically derived silica, and gibbsite. At the top of the bauxite there is evidence of sedimentary reworking by surface processes (Foster 1996). It may be concluded that this deposit is evolving, although it is impossible to say how long it has been developing except in the most general way since it clearly post-dates the mid-Cretaceous at Andoom, where it forms from the Early Cretaceous Rolling Downs Formation, and the Palaeogene at Weipa, where it forms from sediments

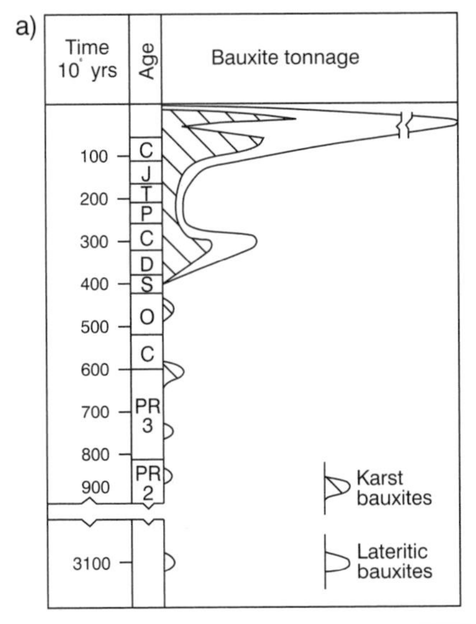

b)

	Karst bauxite percentage of world total
Quaternary - Holocene	1.2
Miocene - Pliocene	24.4
Oligocene	0.6
Palaeocene - Eocene	13.3
Upper Cretaceous	20.0
Lower Cretaceous	8.1
Jurassic	3.2
Triassic	4.0
Permian	3.1
Carboniferous	15.9
Devonian	5.9
Silurian	-
Ordovician	0.1
Cambrian	0.2
	100.0

c)

Duration in millions of years	*Intensity- % karst bauxite formation per million years*
1.5	0.8
35	0.7
15	0.04
20	0.66
36	0.55
36	0.22
60	0.05
30	0.13
55	0.06
65	0.24
50	0.12
40	-
65	-
530	-

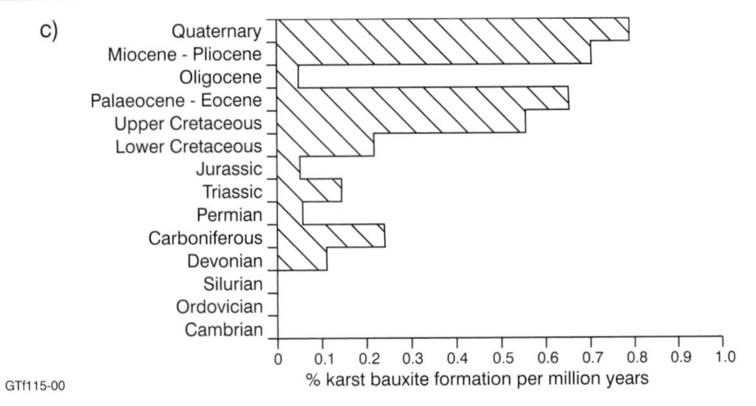

GTf115-00

Figure 15.15 *Karstic and lateritic bauxite formation in the stratigraphic record (after Bardossy 1979, from McFarlane 1983)*

of presumed Palaeogene age. At Weipa all Ollier & Pain's (1996) phenomena (listed above) are occurring simultaneously.

The situation is similar in the Darling Ranges where bauxite clearly developed after the landscape substantially assumed its current shape (Eggleton *et al.* 1998). Further the bauxite grade decreases sympathetically with rainfall, suggesting that the deposits formed since the present climatic regime developed some time during the Late Mesozoic or Palaeogene after the opening of the Indian Ocean and the establishment of the present circum-Antarctic circulation patterns. There are also some data that suggest groundwaters are in chemical equilibrium with the kaolin below the bauxite here. Most other 'lateritic' bauxites are similar to those of the Darling Ranges in many respects.

Some 10% of the world's bauxite occur in karst. At Le Beaux in southern France, the type locality for bauxite, it occurs in karst. Because pure limestone cannot produce aluminous residues karstic bauxite must derive from the solutional residues from impure limestones, or be the result of allochthonous material accumulating in karstic depressions. Bardossy (1982) suggests karst:

- depressions act as traps for the collection of Al-rich material from the surrounding regions;
- depressions provide the good drainage required to form bauxite by desilification of aluminosilicate minerals; and,
- protects bauxites from later erosion.

McFarlane (1983) suggests that karstic bauxite was derived from the erosion of 'lateritic' terrains to their sites of preservation in karst. This implies they are of some provenance and palaeoclimatic significance, but few such data are known to the authors. Figure 15.15 shows karstic bauxites occur from the Precambrian to the Recent.

IRON ORE

Australia has the second largest reserves (16% of global reserves) of Fe-ore after Russia. Most resources are centred in the Hamersley Basin in the Pilbara region of Western Australia where extensive Precambrian banded iron formations (BIFs) provide the source material for the Fe. Similar but minor deposits also occur in South Australia. They are in many ways similar to the BIFs which occur elsewhere in the cratonic regions of the world, except that in Australia, Brazil and Africa, the BIFs have been naturally beneficiated.

Natural beneficiation of the Fe-ores occurs as the BIF (and perhaps the 'shales', actually fine-grained volcaniclastics, of the Hamersley Basin) are weathered over long periods. This causes the silica (and alkalis, Mg, etc.) to be leached, leaving Fe–Al rich 'supergene' enrichments which retain BIF fabrics where they occur over them, and over the 'shales' Fe-rich pisolites occur in massive amounts.

The BIF weathered ironstone has been eroded and deposited as colluvium and alluvium at the base of slopes forming large bodies of naturally beneficiated ore. They occur as thick lenticular bodies along the fronts of existing mountain ranges (Figure 15.16) showing that the landscapes are of some antiquity, at least pre-mid-Miocene (Killick *et al.* 1996). He postulates the Fe-beneficiated material was formed during climates wetter than those now occurring in the Pilbara and began eroding and depositing after that as progressive aridity brought about a change in vegetation. The preservation of the beneficiated deposits is due to bypassing by contemporary erosional regimes. The pisolitic material formed from the volcaniclastic 'shales' are as yet unexplored but hold promise of being a significant resource in the future.

Ollier & Pain (1996) suggest that the Robe River deposits, also in the Hamersley Ranges of Western Australia, are naturally beneficiated by fluvial reworking of earlier weathering profile material. They also point out that these deposits now occur on mesa and plateau surfaces indicating considerable relief inversion in this part of the Hamersleys, in considerable contrast to the areas studied by Killick *et al.* (1996) further south-east.

Bog Fe-ores are formed in lakes and bogs surrounded by Fe-rich rocks. Those formed in lakes generally consist of oolitic to pisolitic Fe-oxide grains cemented together in lens-like masses. They form in shallow water (<1 m deep) around the lake margins. Bog ores form thin sheet-like bodies of earthy to pisolitic Fe-oxides associated with organic-rich sediments. They are formed by the migration of acidic organic-rich groundwater. Iron oxides and oxyhydroxides precipitate where less acidic lake or bog waters enter a more oxidizing environment. Alternatively humic complexes rich in Fe are broken down by bacteria and the Fe-oxides and oxyhydroxides are precipitated.

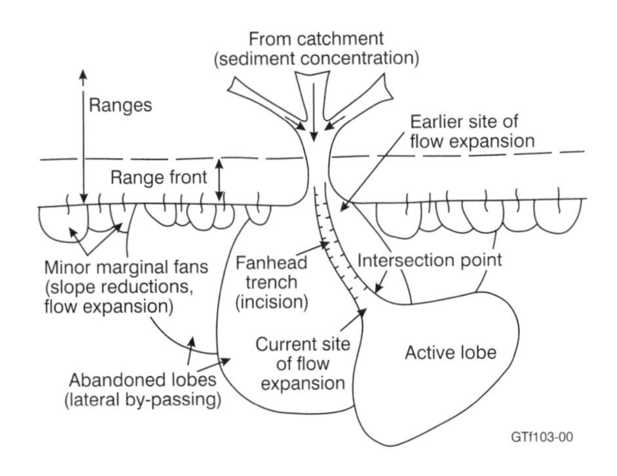

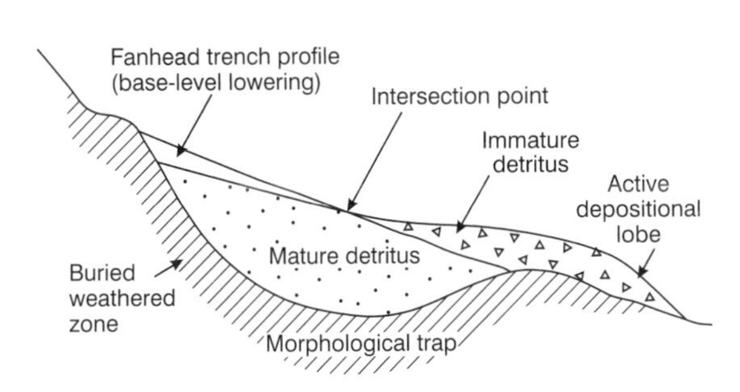

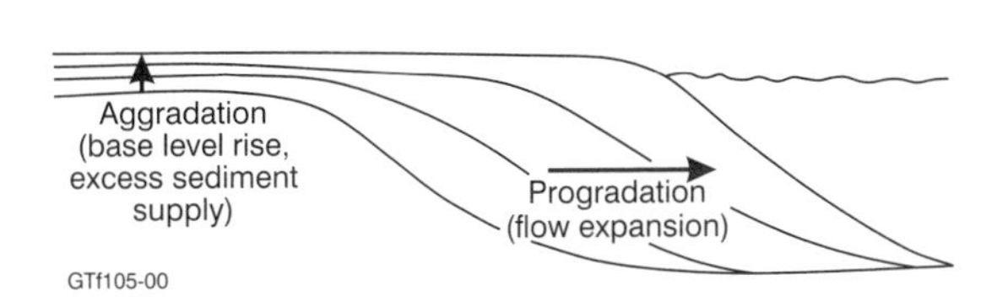

Figure 15.16 *Conceptual model showing sites of flow expansion and gradient reduction associated with weathering beneficiated Fe-ore deposition in the Pilbara (from Killick* et al. *1996)*

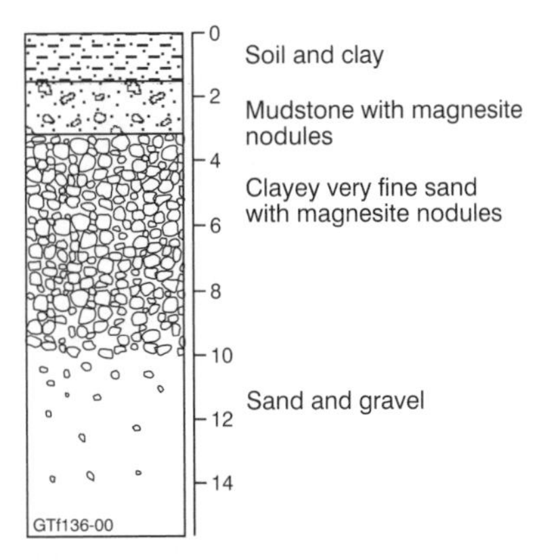

Figure 15.17 *Idealized section and photo of the Kunwarara magnesite mine (photo GT)*

MAGNESITE

Magnesite is a mineral ($MgCO_3$) which occurs in two forms:

- macrocrystalline (spary magnesite) magnesite usually associated with metasomatic or marine environments. This form is mined in Brazil, North Korea, China, Czech Republic, Slovakia and the CIS. No regolith-formed deposits of macrocrystalline magnesite are known; and,
- cryptocrystalline magnesite, commonly associated with serpentinites, occurring as veins, lenses and stockworks. It is also found in sedimentary deposits adjacent to serpentinites. Very dense varieties are called bone magnesite.

Cryptocrystalline magnesite may be primarily associated with hydrothermal activity in serpentinites or formed by the weathering of them. Sedimentary deposits of magnesite form during deposition of the sediments where Mg^{2+} leached from serpentinites during weathering precipitates in alkaline to neutral pH environments as dense nodules of magnesite. Deposits of world significance occur in Turkey, Austria, Greece and Australia. Australia produces only 2% of the world's magnesite (280 000 t) from the largest known resource of cryptocrystalline magnesite (500×10^6 t) at Kunwarara on the central Queensland coastal plain (Figure 15.17).

Figure 15.18 *Photo of deeply weathered serpentinite from Marlborough (photo GT)*

The Kunwarara deposit is in a fluvial system, perhaps with minor lakes along the alluvial tract. The thickness of the deposit varies from 1 to 26 m (Jones 1995). It consists of alluvial sands and gravels containing magnesite nodules in its upper parts overlain by a thin unit (usually <1 m) of more muddy sediments also containing magnesite nodules. This sequence is capped by a clay-rich soil up to 2 m thick (Figure 15.17a). The catchment to the alluvial system hosting the magnesite rises in an area dominated by N–S striking serpentinites that are deeply weathered (Figure 15.18) to form Ni-laterites with chrysoprase veining, and leached of Mg^{2+}. The serpentinites contain significant veins and stockworks of magnesite, but the economic deposits all occur downstream in alluvial and possible lacustrine deposits.

In Turkey at Salda Lake, sediments rich in magnesite have been derived from the erosion of deeply weathered serpentinites and subsequent recrystallization of the magnesite has formed nodules (Schmid 1987). Few other forms of regolith magnesite are known.

MANGANESE

Manganese behaves very similarly to Fe geochemically and the two elements commonly occur together although Fe is many times more abundant than Mn. We have discussed the origins of manganocrete, the regolith-formed variety of Mn-ores, in Chapter 13. Several major regolithic ore deposits are known: in Brazil at Morro da Mina, on Groote Eylandt, Northern Territory, and in the Pilbara of Western Australia. In Brazil the deposits have formed from the weathering of silicate and carbonate rocks which average 30% Mn. The ore is naturally beneficiated by weathering and because of this it forms thicker deposits along joint and dykes in the underlying rocks and along watercourses (Park & MacDiarmid 1975).

At Groote Eylandt the Mn-ores have been naturally beneficiated by deep weathering of mid-Cretaceous calcareous and glauconitic clastic marine sediments which lap on to a Proterozoic quartzite basement. The deep weathering has produced an Mn hardpan or manganocrete at the base of the ore overlain by Mn-pisolites, both loose and, higher in the profile, cemented (Figure 13.29). The Cretaceous sediments contain between 2 and 14% Mn and the ore has an average grade of about 25% Mn. The Mn-ores here are now worked by BHP-Billiton but they have been part of the local Aboriginal people's culture for a very long time. Their paintings, unlike those from other parts of Australia, are based on black from the Mn-minerals with only minor whites (kaolin) and ochres (goethite and haematite). Similar deposits are found in Imini, Morocco and Nikopol in Ukraine (Weber 1997).

Weber (1997) reports Mn-deposits derived from weathering of Precambrian rocks in the Pilbara region of Western Australia. Other regolith-related Mn-deposits occur, but few are of major economic significance as most Mn is mined from Precambrian sequences. She identifies two periods of Mn deposition, one during the Lower to Middle Proterozoic and the Cretaceous to Oligocene deposits mentioned above. These deposits have possibly been enriched by 'lateritic' weathering.

NICKEL LATERITE DEPOSITS

Most of the world's Ni is mined from primary sulfide deposits. In some areas where weathering has been extensive (tropics or regions of low erosion which have been weathering for long periods) nickeliferous weathering products concentrate Ni as silicate or oxide ores. The classical example is New Caledonia, where Ni-laterites were discovered by Garnier (1867). Other Ni-laterites are common in Brazil, the western Pacific and Australia.

Weathering of ultramafic and serpentine rocks, rich in Ni (hosted in olivine or serpentine minerals) may result in the concentration of the Ni in two phases in the weathering profile; as silicates towards the base of the profile and in oxides towards the top (Figure 15.19). These rocks are generally deeply weathered and have profiles similar to laterite profiles. The mineralogical sequence in the profile depends to some

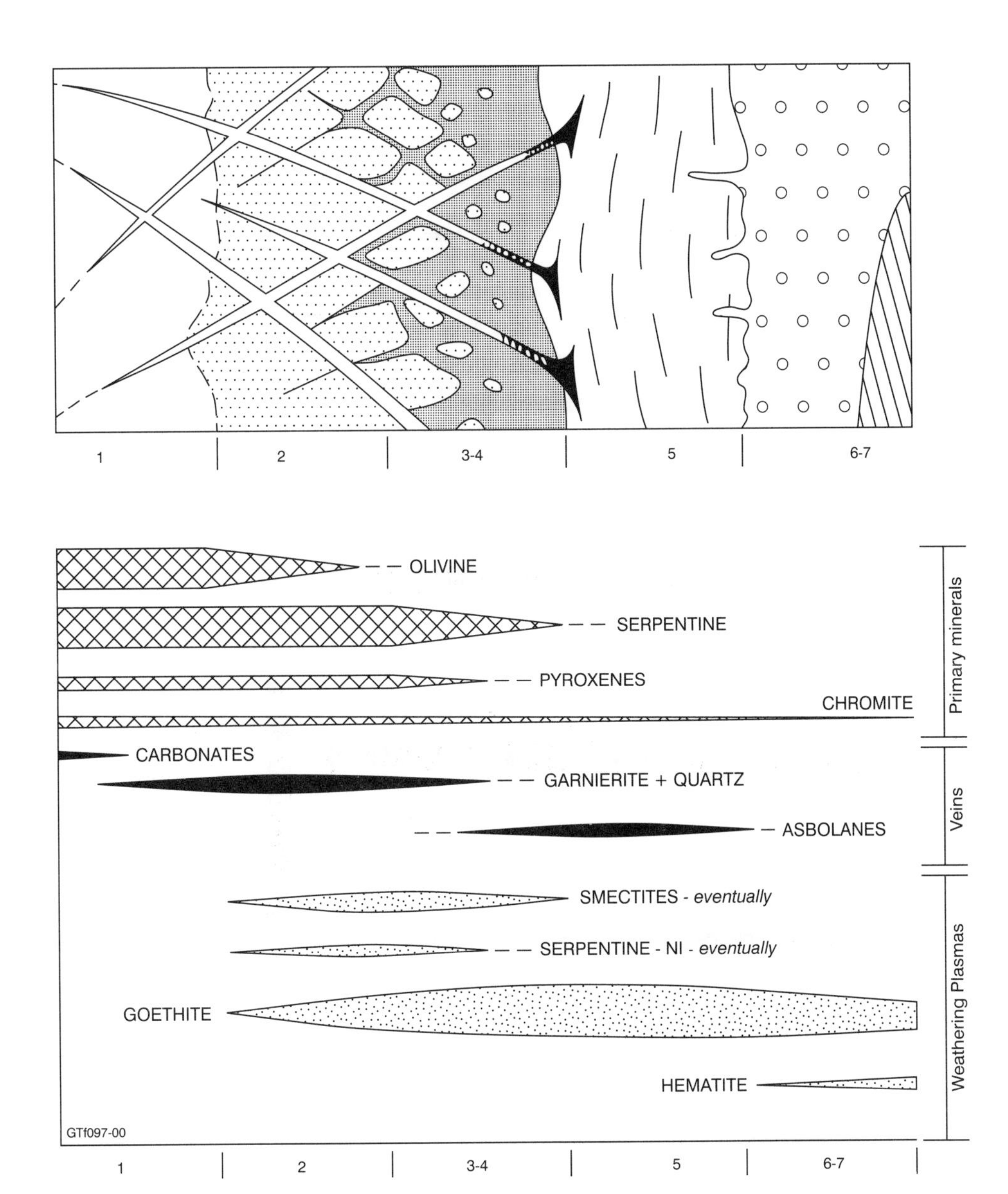

Figure 15.19 *Ni-laterite profile with mineralogical composition of the successive layers. Numbers 1–7 indicate the various layers: (1) fresh bedrock; (2) weakly weathered rock; (3) coarse saprolite (bottom); (4) coarse saprolite (top); (5) ferruginous saprolite; (6) nodular red soil; (7) Fe crust (sometimes present) (from Trescases 1997)*

degree on the type of ultramafic rock weathered. Trescases (1997) provided a chart depicting the progression in mineralogy up a profile, depending on whether it was a serpentine- or olivine-rich parent (Figure 15.20).

The concentration of lateritic Ni depends also on the topography, and the highest-grade ores are generally associated with ridges and spurs where laterally downward transport of Ni is possible. On reaching the base of the profile it precipitates as an Ni-silicate (garnierite). This zone is overlain by a smectite zone which grades upwards into a goethite zone, which may also be Ni-rich. This grades into a goethitic and/or haematitic duricrust.

At Marlborough in central eastern Queensland serpentinite has been deeply weathered and there are

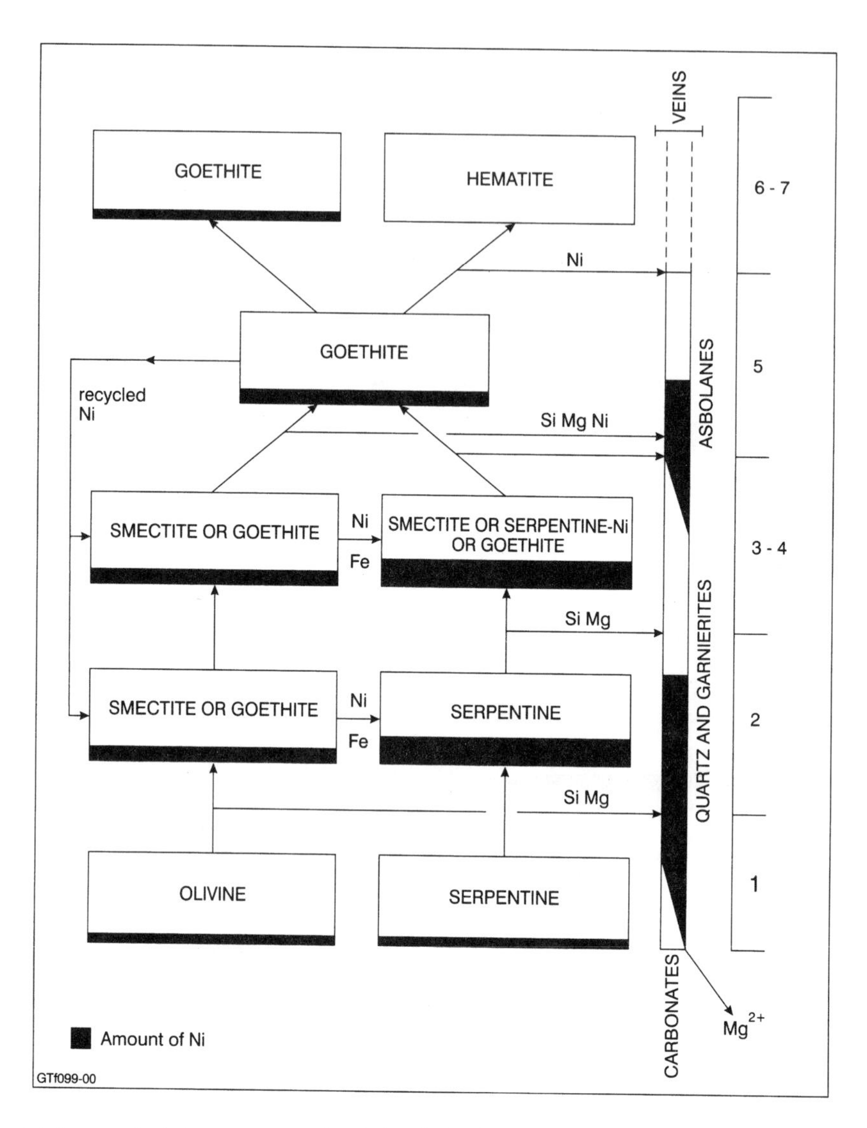

Figure 15.20 *Main mineralogical sequences occurring during the laterite weathering of ultrabasic rocks and chemical transfers (from one mineral to another and from one layer to another). The numbers are as for those in Figure 15.19 (from Trescases 1997)*

significant reserves of lateritic Ni. Here the weathering profile is up to 100+ m deep and consists of (Luke Foster, personal communication 1999):

0–2 m	Soil
0–10 m	'Laterite' or Tertiary sediments which are also frequently 'lateritized'
0–50 m	Silcrete (chalcedonic and chrysoprase) and clay
0–20 m	Brown and green coloured clays
0–20 m	Strongly layered silcrete (chalcedony)
0–20 m	Green clay saprolite
2–35 m	Green saprolite with garnierite and chlorite
? thickness	Fresh ultramafic rocks

Interestingly here, rather than being capped by a ferricrete duricrust many of the ridges are capped by silcrete consisting of chalcedonic quartz and Ni-chalcedony. In fact it seems that it is these Si-rich crusts that maintain the relief over what are thoroughly rotten rocks. They act to preserve topographic relief while the lower areas are not capped and continue to erode. Similar topographic preservation can also be seen associated with the 'greenstones' of the Yilgarn Craton, where the ultramafic rocks, rather than forming topographic lows in an otherwise granitic terrain, as one would expect, are topographically positive, in this case because of ferricrete crusts preserving the relief.

Nickel laterites on the Yilgarn have deep weathering profiles associated with them. An example of such a deposit is at Mt Keith 400 km north-west of Kalgoorlie. Butt & Nickel (1981) described the profile shown in Figure 15.21 and Table 15.3. These data clearly illustrate the massive benefaction of Ni in the oxide zone and the Si-pan zone profile as described from other Ni-laterites. It also shows the platinum group elements (PGEs) are concentrated in the regolith, but in this case sub-economically.

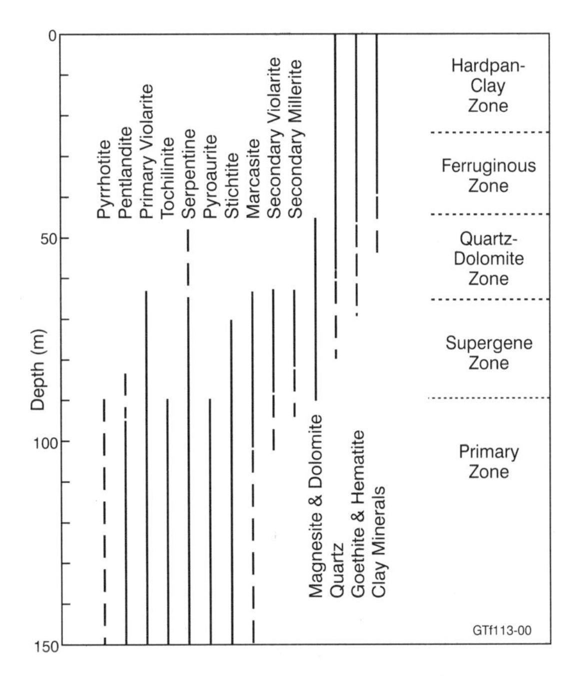

Figure 15.21 *Mineralogy of the regolith over part of the Mt Keith dunite, Western Australia. Full lines indicate the range of maximum abundance; dashed lines indicate lesser abundance (from Butt 1992)*

PLACER DEPOSITS

Placer deposits form important regolith concentrations of Au, diamond, Ti, Sn, Zi and REE minerals as well as some other gemstones. In Chapter 12 we briefly discuss placer concentration mechanisms. Here we point

Table 15.3 *Main weathering horizons formed over mineralized serpentinized dunite at Mt Keith, Western Australia (simplified from Butt 1992)*

Depth (m)	Description	Mean concentrations			
		Pd (ppb)	Pt (ppb)	Cu (ppm)	Ni (ppm)
0–12.3	*Hardpan* – weakly bedded silica cemented red-reddish yellow conglomeratic colluvium with siliceous and ferruginous pebbles in an earthy sand matrix. Fe-pisoliths and accretion dominant below 5 m	0–15	0–22	180	1 030
12.3–25.3	*Clays* – pale green–grey mottled (red–yellow). Pisoliths in upper 2 and lower 2.5 m				
25.3–28.3	*Fe-pisoliths* – usually cemented with up to 30% mottled clay	115	70	995	4 760
28.3–39.0	*Clays* – grey, green and white clays strongly yellow mottled as a matrix to Fe-pisoliths and nodules. ?transported above 35 m				
39.0–42.7	*Massively ferruginized saprolite/lithorelics* – some nodules and strongly mottled green clay and Si-pans				
42.7–45.4	*Mn-horizon* – Mn-oxide veins, nodules and mottles in Fe-saprolite and nodules, green clays and Si-pans	295	125	1505	21 595
45.4–62.5	*Weathered serpentinite* – with abundant brown Si-pans (quartz) to 30 cm thick. Clays (bulk density 0.85–1.5) preserve rock fabric in stichite and Fe-oxides				
62.5–66.1	*Weathered serpentinite* – with occasional Si-pans. Dolomite crystals in voids	95	40	690	14 420
66.1–88.7	*Supergene zone* – partly weathered grey-green serpentinite with secondary sulfides (violarite and pyrite) disseminated and in veins. Si-veins rare. Stichite green at top of zone and pink at bottom	63	27	360	6 020
88.7 +	*Primary zone* – black–green serpentinized dunite with cumulic fabric. Disseminated pentlandite and pyrrhotite, stichite pink	47	19	295	4 495

to some examples of such deposits of economic significance. Gold placers are briefly discussed above.

Mineral sands containing heavy minerals are one of the world's major sources of ilmenite and rutile (Ti-ores) and zircon (Zr-ore). In 1996 Ti-ores from South Africa topped the world production for these sands but Australia has the largest proven resources. Many of these are, however, quarantined because they occur in national parks. Palaeoshoreline deposits (Miocene–Pliocene) around the southern margin of the continent also contain huge mineral sand reserves (Figure 2.43). Beach ridges in the Murray Basin contain extensive resources of rutile and anatase (3.4 Mt in 1988), leucoxene (4.6 Mt), ilmenite (12.5 Mt), zircon (5.1 Mt), monazite (0.6 Mt) and xenotime (0.2 Mt).

Coastal regolith provides important sources of placer minerals. In Australia coastal beach ridge deposits have been, and are being, worked along the west and east coasts for ilmenite, rutile and monazite and they also contain huge zircon reserves.

Diamond alluvial placer deposits are worked from Triassic rocks in Swaziland. Quaternary gravels in Sierra Leone contain diamonds derived from Cretaceous pipes, eroded by about 500 m during the Late Cretaceous and Palaeogene (Thomas 1994). This attests to the longevity of these heavy minerals in the landscape through long periods of weathering, erosion and transport. Modern and Recent diamond placers occur in the Central African Republic (Sutherland 1984). Many other alluvial diamonds have been recorded, but few are economically viable at present. One such example is in New England, New South Wales, where diamonds were reported from alluvial Sn and sapphire workings early this century, but the diamonds have never proved economic, nor has their source been discovered. This is not surprising given the durability of these minerals in the landscape.

Thomas & Thorp (1993), based on studies in north-western Australia, Borneo and West Africa, determined that the important factors in (diamond) placer formation are:

- deeply weathered source rock (e.g. kimberlite diatreme);
- proximity to host or source rocks (<20 km);
- uplifted, preweathered landscape undergoing dynamic erosion (lowering);
- locally adjacent downwarped or faulted basin;
- low gradient stream reaches or colluvial plains on saprolite with fine sediments;
- sediment storage and floodplain formation through the Quaternary;
- coarse clasts in bed-load; and,
- evidence of sediment reworking in buried palaeochannels.

Tin placers off Malaysia have been important sources of Sn, but world prices have curtailed mining.

Other regolith resource materials

OPAL

Opal occurs in weathered rocks either in fractures and voids in the regolith or replacing regolith materials of calcareous/hydroxiapatite fossils. It is also common in diatomite deposits, in some Cainozoic limestones and as vesicle fills in Cainozoic basalts.

There are two forms of opal, common and gem. The difference between the two is not related to composition, but the size and packing of amorphous silica spheres. Gem opal consists of hexagonally packed sheets of microspheres ranging in diameter from 150 to 350 nm (Jones *et al.* 1964). The space between the spheres provides the refractive index change that produces the diffraction effect (colour). The smaller the microspheres the bluer the colour and the larger the redder. Common opal on the other hand is similar to gem opal except the microspheres are not as regular in size nor as well packed.

The majority of opal is common and of no value, but coloured opal is mined from many localities including the famous Lightning Ridge in northern New South Wales. Lightning Ridge produces a variety of opal colours, but the most expensive are the 'black opals' which are dark in colour with large flashes of red. Opal mines extend through northern New South Wales, southern Queensland and central South Australia. The only other significant coloured opals are the 'fire opal' from Mexico. Relatively low value milky opal is found in Europe as well as in the gem opal fields of Australia.

CHRYSOPRASE

Chrysoprase is a green semi-precious stone. It is essentially composed of chalcedony in its fibrous form and cristobalite and tridymite in its massive form. It is found associated with Ni-bearing host rocks where it occurs as veins in deeply weathered serpentinites at

Marlborough, Queensland. It occurs associated with jadite in serpentinized gabbroic and ultramafic rocks in Lower Silesia, Poland and is also mined in the USA, India, Brazil and South Africa. The Marlborough mine is the world's largest and it exports most of its product to south-eastern Asia.

Chrysoprase is green because it contains green nickel-talc (willemsite) as finely disseminated inclusions.

Industrial minerals

A vast range of industrial minerals and materials are garnered from the regolith. Perhaps the most abundantly used are sand and gravel for the construction industry. Most cities and towns have around them quarries and pits from which these resources are excavated. Around our own city, Canberra, sand for concrete and bricklaying is in relatively short supply and is hauled from distances up to about 80 km from deposits along major rivers in the region, as well as from Quaternary beach ridges around Lake George. Gravels are also excavated from the region's rivers and Quaternary beach ridges.

Another very commonly used regolith resource is weathered rock and sediment. It is mainly used for roadbase materials and most roads throughout most countries have what are euphemistically called 'borrow pits' beside or very close to them. These are generally the supply of road base used to construct the nearby road. The actual amount of these materials used is difficult to determine as they are simply located locally by road building authorities, excavated and forgotten. One's impression is that very large volumes of weathered materials are used for this purpose, going by the number of borrow pits one sees. The material extracted must meet some quality constraints. In most cases the presence of expanding clays makes it unacceptable, clay- and sand-sized material must be present in the correct proportions to make the material workable, and it must be close to the site where it is to be used as it is a low-value material.

Clay minerals are one of the major regolith-derived industrial minerals. The most important clay mineral is kaolin, used in a vast array of industrial applications ranging from ceramics to pharmaceuticals. Its first recorded use for ceramics was at Kauling, Jiangxi Province, China, hence both the common term 'china clay' and the anglicized 'kaolin'.

China clay is produced by the washing of granitic saprolite to remove all the coarse-grained particles,

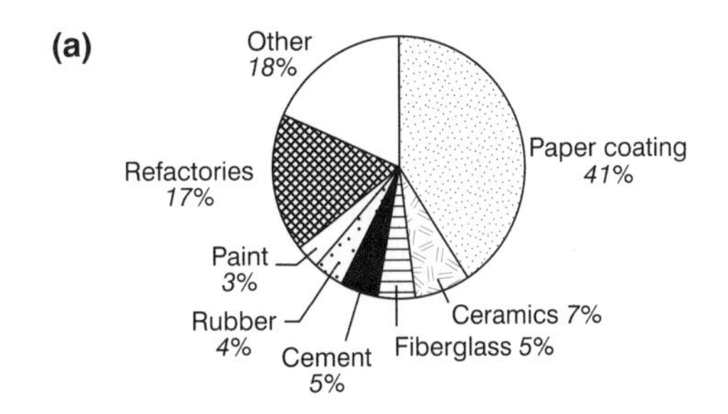

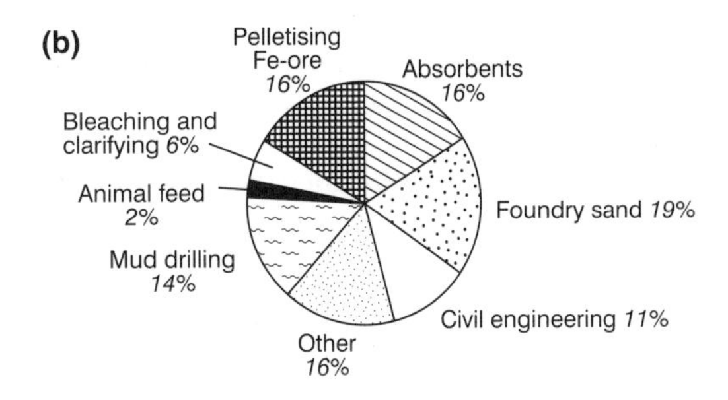

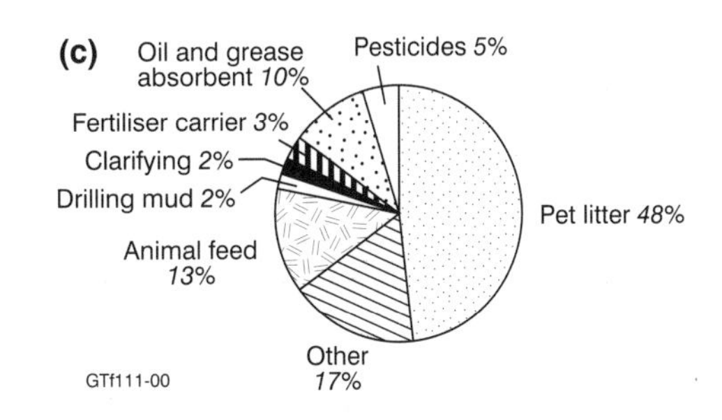

Figure 15.22 *(a) Estimated use of kaolin in Australia for 1994. (b) Estimated uses for bentonite in 1994. (c) Estimated uses for attapulgite during 1994 (from Keeling 1997)*

leaving only the clay minerals. It is used to make fine porcelain or chinaware. Kaolin produced in this way from Cornwall provides the world's major supply of this clay. Figure 15.22a shows the usage of kaolin in Australia. The 'other' category includes plastics, cosmetics, carriers for herbicides and catalysts.

Many kaolin deposits contain halloysite, the tubular or spherical polymorph. Its presence downgrades deposits in respect of some uses, but for ceramic and refractory use it is considered to increase the value of the clay.

Bentonite is the commercial term for the smectite group of expanding clay minerals. Montmorillonite is often the dominant mineral in commercial deposits, but beidellite, hectorite or saponite are also used. The world's major deposits are in Wyoming. They are used for their swelling properties in applications such as viscosity modifiers, bonding agents on foundry sand, pelletizing Fe-ore and in small farm dams as liners to seal the earth wall. In Australia the major uses in 1994 are shown in Figure 15.22b.

Attapulgite is a commercial clay containing a high proportion of the clay minerals palygorskite and/or sepiolite. They are rare fibrous clays having narrow channels that are oriented parallel to the fibre length. This fabric results in a high surface area and porosity and gives them excellent sorption and gelling properties. They occur in alkaline terrestrial or barred marine basins and are formed by the alteration of other clay minerals or by direct precipitation. Most of the world's supply comes from the USA and Spain. Australia produces modest quantities from saline playas. While resources are probably extensive in Australia's arid lake systems, little exploration has occurred. Figure 15.22c shows its uses in 1994.

GLASS SANDS

Window glass composition lies in the range shown in Table 15.4. Silica is the major component, but if glass were made of pure silica it would require fusing at about 1700°C, and be weak. Soda lowers the fusion temperature to about 1200°C, but high soda causes glass to be soluble. Lime, magnesia and alumina add strength. Iron is avoided as it colours the glass.

Most glass sand must be as pure as possible, ideally 100% pure quartz, but such materials do not occur. Iron and alumina contents limit the sand quality as does the presence of refractory minerals like chromite and chrome spinel, zircon, monazite and others. As a result of most sands' composition they must be acid cleaned (mainly to remove Fe_2O_3) and go through a magnetic separator.

Cainozoic 'white sands' from tropical terrains (Thomas 1994) where the sands have been bleached as a result of organo-acid leaching (as for example in podzol formation) provide an, as yet, untapped glass sand resource. Such sands are particularly common in Guyana. High-quality glass sands occur in the Paris Basin, where silification and desilification have yielded very 'clean' sands. In Australia glass sands are garnered from leached coastal dune systems.

Table 15.4 *Typical composition range of window glass*

Oxide	Lower limit (%)	Upper limit (%)
SiO_2	70	74
Na_2O	12	16
CaO	5	11
MgO	1	3
Al_2O_3	1	3
Fe_2O_3	0	0.18

EVAPORITES

Gypsum in many parts of the world is mined from ancient evaporite sequences such as the Triassic of Derbyshire or the Miocene of Tuscany. In countries with more arid climates gypsum is mined from playa lakes. Gypsum is used to produce plaster of Paris by heating it to 107°C when it forms the hemihydrate $2CaSO_4.H_2O$. This form can be readily rehydrated to form gypsum again, but it does it so very quickly that organic setting retardants must be added.

Many other evaporite minerals are mined from regolith materials from arid climates including borax, nitre and sylvite.

DWELLINGS

Despite the commercial value of the regolith deposits discussed above, to mankind the most universal use of regolith is for housing. From the cave dwellings of early man to the castles of Europe to the suburban homes of today, regolith and its material have provided the location, the substrate and the materials for building. Bricks are, for the most part, made from clay won from weathered shales. An ideal brick clay contains kaolin, quartz, small amounts of Fe or Mn-oxyhydroxides and commonly mica or feldspar; the kaolin on firing changes to the fibrous silicate mullite, which bonds to the more inert components to make a hard composite. In Australia annual brick clay manufacture supports a half a billion dollar industry.

The second biggest construction material after bricks is cement, a product made by mixing clay, quartz and limestone. Of these, the clay and quartz are mostly mined from weathered rock or from transported regolith deposits. The glass in the windows, the plaster and paint on the walls, the tiles on the roof, and of course all the timber framing are regolith-derived materials. The regolith is, in every sense, our true home.

References

Abell, R.S. (1985) Geology of the Lake George Basin, N.S.W. *Bureau of Mineral Resources, Geology and Geophysics, Record 1985/4.*

Ab'Saber, A.N. (1982) The paleoclimate and paleoecology of Brazilian Amazonia. In Prance, G.T. (ed.) *Biological Diversification in the Tropics.* Columbia, New York, 1–59.

Acton, G.D. and Kettles, W.A. (1996) Geologic and palaeomagnetic constraints on the formation of weathered profiles near Inverell, eastern Australia. *Palaeogeography, Palaeoclimatology, Palaeoecology*, **126**, 211–225.

Ahnert, F. (1970) Functional relationships between denudation, relief and uplift in large mid-latitude drainage basins. *American Journal of Science*, **268**, 243–263.

Alekin, O.A. and Brazhnikova, L.V. (1966) Removal of solutes from continents by rivers and the relationship of this process to the mechanical erosion of the Earth's surface. In Vinogradov, A.P. (ed.) *Chemistry of the Earth's Crust*, 291–303.

Alekin, O.A. and Brazhnikova, L.V. (1968) Dissolved matter discharge and mechanical and chemical erosion. In Geochemistry, precipitation, evaporation, soil-moisture, hydrometry. *International Association of Science Hydrological Publication*, **78**, 35–41.

Alexander, L.T. and Cady, J.G. (1962) Genesis and hardening of laterite in soils. *Technical Bulletin of the United States Department of Agriculture, 1282.*

Anand, R.R. (1994) Regolith–landform evolution and geochemical dispersion from the Boddington gold deposits, Western Australia. *CSIRO Division of Exploration and Mining Report*, **246R**, 149 pp.

Anand, R.R. (1995) Genesis and clarification of ferruginous regolith materials in the Yilgarn Craton: implications for mineral exploration. In Camuti, K.S. (ed.) *Exploring the Tropics*, Extended Abstracts, 17th International Geochemical Exploration Symposium, Townsville, 1–4.

Anand, R.R. (1996) The value of regolith materials and regolith mapping to exploration. 10 pp. In *Exploration Geochemistry in Regolith Dominated Terrains – Short Course Notes.* CRC LEME, December 1996, Orange, New South Wales.

Anand, R.R. (1998a) An overview of regolith–landform processes of the Yilgarn Craton – implications for exploration. In Scott, K.M., Anand, R.R., Gray, D.J. and Lintern, M.J. (eds) *Sampling Media in Various Australian Regolith Regimes – Short Course Notes.* CRC LEME Report No. 82, Perth.

Anand, R.R. (1998b) Distribution, classification and evolution of ferruginous materials over greenstones on the Yilgarn craton – implications for mineral exploration. In Eggleton, R.A. (ed.) *State of the Regolith.* Geological Society of Australia, Special Publication **20**, 175–193.

Anand, R.R. and Gilkes, R.J. (1984) Weathering of ilmenite in a lateritic pallid zone. *Clays and Clay Minerals*, **32** (5), 363–374.

Anand, R.R. and Smith, R.E. (1993) Regolith distribution, stratigraphy and evolution in the Yilgarn Craton–implications for mineral exploration. In Williams, P.R. and Haldane, J.A. (eds) *Kalgoorlie '93, An International Conference on Crustal Evolution, Metallogeny and Exploration of the Eastern Goldfields, Extended Abstracts. Australian Geological Survey Organisation, Record, 1993/54*, 187–193.

Anand, R.R. and Smith, R.E. (1998) Use of ferruginous materials for geochemical exploration. In Scott, K.M., Anand, R.R., Gray, D.J. and Lintern, M.J. (eds) *Sampling Media in Various Australian Regolith Regimes – Short Course Notes.* CRC LEME Report 82, Perth.

Anand, R.R., Churchward, H.M. and Smith, R.E. (1991) Regolith–landform development and siting and bonding of elements in regolith units, Mt Gibson district, Western Australia. *CSIRO Australia, Division of Exploration Geoscience, Perth. Report 165R/CRCLEME Open File Report 61. CSIRO/AMIRA Project P240: Laterite Geochemistry.* 95 pp.

Anand, R.R., Phang, C., Wildman, J.E. and Lintern, M.J. (1997) Genesis of some calcretes in the southern Yilgarn Craton, Western Australia – implications for mineral exploration. *Australian Journal of Earth Sciences*, **44**, 87–103.

Anand, R.R., Phang, C., Wildman, J. and Lintern, M.J. (1998) Reply: genesis of some calcretes in the Yilgarn Craton, Western Australia: implications for mineral exploration. *Australian Journal of Earth Sciences*, **45**, 179–192.

Anand, R.R., Smith, R.E., Innes, J. and Churchward, H.M. (1989) Exploration geochemistry about Mt Gibson, Western Australia. *CRC LEME Open File Report 35*, Perth, 93 pp.

Anderson, A., Dodds, A.R., McMahon, S. and Street, G.J. (1993) A comparison of airborne and ground electromagnetic techniques for mapping shallow zone resistivity variations. *Exploration Geophysics*, **24**, 323–331.

Anderson, M., Bertsch, P.M. and Miller, W.P. (1990) Beryllium in selected southeastern soils. *Journal of Environmental Quality*, **19**, 347–348.

Andrew, A.S., Carr, G.R., Giblin, A.M. and Whitford, D.J. (1997) Isotope hydrogeochemistry in exploration for buried and blind mineralization, Eyre Peninsula, South Australia. New developments in research for ore deposit exploration. *Geological Society of Australia Abstracts*, **44**, 3.

Angélica, R.S. and da Costa, M.L. (1993) Geochemistry of rare-earth elements in surface lateritic rocks and soils from the Maicuru complex, Para, Brazil. *Journal of Geochemical Exploration*, **47**, 165–182.

Antweiler, R.C. and Drever, J.I. (1983) The weathering of a late Tertiary volcanic ash: importance of organic solutes. *Geochimica et Cosmochimica Acta*, **47**, 623–629.

Arakel, A.V. (1986) Evolution of calcrete in palaeodrainages of the Lake Napperby area, central Australia. *Palaeogeography, Palaeoclimatology, Palaeoecology*, **54**, 283–303.

Ashley, G.M., Shaw, J. and Smith, N.D. (1985) Glacial sedimentary environments. *Society of Economic Paleontologists and Mineralogists, SEPM Short Course No. 16*, Tulsa.

Aspandiar, M.F. (1992) Weathering of a basalt: microsystems and mineralogy. MSc thesis, Australian National University, Canberra, 122 pp.

Aspandiar, M.F. (1998) Regolith and landscape evolution of the Charters Towers Area, North Queensland. PhD thesis, Australian National University, Canberra, 241 pp.

Aspandiar, M.F., Eggleton, R.A., Orr, T., van Eck, M. and Taylor, G. (1997) An understanding of regolith and landscape evolution as an aid to mineral exploration – the Charters Towers experience. *Resourcing the 21st Century*: Australian Institute of Mining and Metallurgy 1997 Annual Conference, Ballarat, 125–129.

Aston, S.R. (ed.) (1983) *Silicon Geochemistry and Biogeochemistry*. Academic Press, London, New York, 248 pp.

Ayliffe, L.K. and Veeh, H.H. (1988) Uranium series dating of speleothems and bones from Victoria Cave, Naracoorte, South Australia. *Chemical Geology (Isotope Section)*, **72**, 211–234.

Baes, C.F. and Mesmer, R.E. (1976) *The Hydrolysis of Cations*. John Wiley, New York, 489 pp.

Bailey, S.W., Brindley, G.W., Johns, W.D., Martin, R.T. and Ross, M. (1971) Summary of national and international recommendations on clay mineral nomenclature. *Clays and Clay Minerals*, **19**, 129–132.

Banfield, J.F. (1985) *The mineralogy and chemistry of granite weathering*. M.Sc. thesis, Australian National University, Canberra, 229 pp.

Banfield, J.F. and Eggleton, R.A. (1988) Transmission electron microscope study of biotite weathering. *Clays and Clay Minerals*, **36** (1), 47–60.

Banfield, J.F. and Eggleton, R.A. (1989) Apatite replacement and rare earth mobilization, fractionation, and fixation during weathering. *Clays and Clay Minerals*, **37**, 113–127.

Banfield, J.F. and Eggleton, R.A. (1990) Analytical transmission electron microscope studies of plagioclase, muscovite and K-feldspar weathering. *Clays and Clay Minerals*, **38**, 77–89.

Banfield, J.F. and Nealson, K.H. (eds) (1997) *Geomicrobiology: Interactions between Microbes and Minerals*. Reviews in Mineralogy, Vol. **35**. Mineralogical Society of America, Washington DC, 448 pp.

Bardossy, G. (1979) The role of tectonism in the formation of bauxite deposits. *Travaux de l'ICSOBA, Zagreb*, **15**, 15–34.

Bardossy, G. (1982) *Karst Bauxites*. Elsevier, Amsterdam; 441 pp.

Bardossy, G. and Aleva, G.J.J. (1990) *Lateritic Bauxites*. Elsevier, Amsterdam, 624 pp.

Barker, W.W., Welch, S.A. and Banfield, J.F. (1997) Biogeochemical weathering of silicate minerals. In Banfield, J.F. and Nealson, K.H. (eds) Review in Mineralogy, Vol. 35. *Geomicrobiology: Interactions between Microbes and Minerals*. Mineralogical Society of America, Washington DC, 391–428.

Barnes, L.C. and Pitt, G.M. (1976) The Mirackina Conglomerate. *Quarterly Geological Notes – Geological Survey of South Australia*, **59**, 2–6.

Bartlett, R.J. and Kimble, J.M. (1976) Behaviour of chromium in soils: I, Trivalent forms. *Journal of Environmental Quality*, **5**, 379–386.

Bauer, F.H. (1959) *The regional geography of Kangaroo Island, South Australia*. Ph.D. thesis, Australian National University, Canberra.

Bayliss, P. and Loughnan, F.C. (1964) Mineralogical transformations accompanying the chemical weathering of clay-slates from New South Wales. *Clay Minerals Bulletin*, **5**, 353–362.

Beattie, J.A. (1970) Peculiar features of soil development in parna deposits in the eastern Riverina, N.S.W. *Australian Journal of Soil Research*, **8**, 145–156.

Beattie, J.A. (1972) Groundsurfaces of the Wagga Wagga region, New South Wales. *CSIRO Soil Publication 28*. CSIRO, Melbourne.

Beavis, F.C. (1985) *Engineering Geology*. Blackwell Scientific, Oxford, 235 pp.

Beavis, F.C. (1960) The Tawonga Fault, north-east Victoria. *Proceedings of the Royal Society of Victoria*, **72**, 95–100.

Becker, G.F. (1895) Gold fields of the southern Appalachians. *Annual Report of the United States Geological Survey, Part III. Mineral Resources of the United States, Metallic Products*, **16**, 251–331.

Beckmann, G.G. (1983) Development of old landscapes and soils. In *Soils, an Australian Viewpoint*. CSIRO, Melbourne, 51–72.

Bell, R.H.V. (1982) The effect of soil nutrient availability on community structure in African ecosystems. In Huntley, B.J. and Walker, B.H. (eds) *Ecology of Tropical Savannas*. Springer-Verlag, Berlin, pp. 193–216.

Benbow, M.C. (1990) Tertiary coastal dunes of the Eucla Basin, Australia. *Geomorphology*, **3**, 9–29.

Bennett, P. and Siegel, D.I. (1987) Increased solubility of quartz in water due to complexing by organic compounds. *Nature*, **326**, 684–686.

Berner, E.K. and Berner, R.A. (1987) *The Global Water Cycle: Geochemistry and Environment*. Prentice-Hall, Englewood Cliffs, NJ, 397 pp.

Berner, R.A. (1978) Rate control of mineral dissolution under Earth surface conditions. *American Journal of Science*, **278**, 1235–1252.

Berner, R.A. (1981) Kinetics of weathering and diagenesis. In Lasaga, A.C. and Kirkpatrick, R.J. (eds) *Kinetics of Geochemical Processes. Reviews in Mineralogy*, **8**, 111–132. Mineralogical Society of America, Washington DC.

Berner, R.A. (1995) Chemical weathering and its effects on atmospheric CO_2 and climate. In White, A.F. and Brantley, S.L. (eds) *Chemical Weathering Rates of Silicate Minerals*. Mineralogical Society of America, Washington DC, 565–583.

Berner, R.A. and Holdren, G.R. (Jr) (1979) Mechanism of

feldspar weathering II: observations of feldspars from soils. *Geochimica et Cosmochimica Acta*, **43**, 1173–1186.

Berner, R.A., Sjöberg, E.L., Velbel, M.A. and Krom, M.D. (1980) Dissolution of pyroxenes and amphiboles during weathering. *Science*, **207**, 1205–1206.

Bettanay, E. and Churchward, H.M. (1974) Morphology and stratigraphic relationships of the Wiluna Hardpan in arid Western Australia. *Journal of the Geological Society of Australia*, **21**, 73–80.

Billett, M.F., Lowe, J.A.H., Black, K.E. and Cresser, M.S. (1997) The influence of parent material on small-scale spatial changes in streamwater chemistry in Scottish upland catchments. *Journal of Hydrology*, **187**, 311–331.

Bird, M.I. and Chivas, A.R. (1988) Oxygen isotope dating of the Australian regolith. *Nature*, **331**, 513–516.

Bird, M.I. and Chivas, A.R. (1989) Stable-isotope geochronology of the Australian regolith. *Geochimica et Cosmochimica Acta*, **53**, 3239–3256.

Bird, M.I. and Chivas, A.R. (1993) Geomorphic and palaeoclimatic implications of an oxygen-isotope chronology for Australian deeply weathered profiles. *Australian Journal of Earth Sciences*, **40**, 345–358.

Bird, M.I., Chivas, A.R. and McDougall, I. (1990) An isotopic study of surficial alunite in Australia. 2. Potassium–argon geochronology. *Chemical Geology*, **80**, 133–145.

Birkeland, P.W. (1984) *Soils and Geomorphology*. Oxford University Press, New York, 372 pp.

Bischoff, G.C.O., Coenraads, R.R. and Lusk, J. (1992) Microbial accumulation of gold: an example from Venezuela. *Neues Jahrbuch für Geologie und Palaeontologie Abhandlungen*, **185**, 131–159.

Bleackley, D. and Khan, E.J.A. (1963) Observations on white sand areas of the Berbice Formation, British Guiana. *Journal of Soil Science*, **14**, 44–51.

Boggs, S. (1987) *Principles of Sedimentology and Stratigraphy*. Merrill Publishing Co., Columbus, Ohio, 784 pp.

Bolton, B.R., Berents, H.W. and Frakes, L.A. (1990) Groote Eylandt manganese deposit. In Hughes, F.E. (ed.) *Geology of the Mineral Deposits of Australia and Papua New Guinea*. Australian Institute of Mining and Metallurgy, Melbourne, 1575–1579.

Bond, G. (1978) Speculations on real sea-level changes and vertical motions of continents at selected times in the Cretaceous and Tertiary Periods. *Geology*, **6**, 247–250.

Bottomley, D.J., Gascoyne, M. and Kamineni, D.C. (1990) The geochemistry, age, and origin of groundwater in a mafic pluton, East Bull Lake, Ontario, Canada. *Geochimica et Cosmochimica Acta*, **54**, 933–1008.

Boulangé, B., Ambrosi, J-P. and Nahon, D. (1997) Laterites and bauxites. In Paquet, H. and Clauer, N. (eds) *Soils and Sediments: Mineralogy and Geochemistry*. Springer, Berlin, 49–65.

Boulet, R. (1975) Toposéquences de sols tropicaux en Haute-Volta; équilibres dynamiques et bioclimats [Toposequences of tropical soils in the Upper Volta; dynamic equilibria and bioclimates]. *Cahiers ORSTOM: Pédologie*, **31** (1), 3–6.

Boulet, R., Lucas, Y., Fritsch, E. and Paquet, H. (1997) Geochemical processes in tropical landscapes: role of the soil covers. In Paquet, H. and Clauer, N. (eds*) Soils and Sediments: Mineralogy and Geochemistry*. Springer, Berlin, 67–96.

Bourman, R.P. (1989) *Investigations of ferricretes and weathered zones in parts of southern and southeastern Australia – a reassessment of the 'laterite' concept*. Ph.D. thesis, University of Adelaide.

Bourman, R.P. (1993) Perennial problems in the study of laterite: a review. *Australian Journal of Earth Sciences*, **40**, 387–401.

Bourman, R.P. (1995) A review of laterite studies in southern South Australia. *Transactions of the Royal Society of South Australia*, **119**, 1–28.

Bowell, R.J., Afren, E.O., Laffoley, N, d'A, Hanssen, E., Abe, S., Yao, R.K. and Pohl, D. (1996) Geochemical exploration for gold in tropical soils – four contrasting case studies from West Africa. *Transactions of the Institute of Mining and Metallurgy, Section B: Applied Earth Science*, **105**, 12–33.

Bowen, H.J.M. (1979) *Environmental Chemistry of the Elements*. Academic Press, London, 316 pp.

Bowler, J.M. (1967) Quaternary chronology of Goulburn Valley sediments and their correlation in southeastern Australia. *Journal of the Geological Society of Australia*, **14**, 287–292.

Bowler, J.M. (1983) Lunettes as indices of hydrologic change; a review of Australian evidence. *Proceedings of the Royal Society of Victoria*, **95**, 147–168.

Bowler, J.M. (1990) The last 500,000 years. In Mackay, N. and Eastburn, D. (eds) *The Murray*. Murray Darling Basin Commission, Canberra, 95–109.

Brand, N.W., Butt, C.R.M. and Elias, M. (1998) Nickel laterites – classification, features and exploration model. *Australian Geological Survey Organization Journal of Australian Geology and Geophysics*, **17**, 81–88.

Brantley, S.L., Crane, S.R., Crerar, D.A., Hellmann, R. and Stallard, R. (1986) Dissolution at dislocation etch pits in quartz. *Geochimica et Cosmochimica Acta*, **50**, 2349–2361.

Braun, J-J., Pagel, M., Herbillon, A. and Rosin, C. (1993) Mobilization and redistribution of REEs and thorium in a syenitic lateritic profile: a mass balance study. *Geochimica et Cosmochimica Acta*, **57**, 4419–4434.

Brewer, R. (1964) *Fabric and Mineral Analysis of Soils*. John Wiley, New York, 470 pp.

Brewer, R. (1968) Clay illuviation as a factor in particle-size differentiation in soil profiles. *Transactions of the 9th International Congress of Soil Science*, Adelaide. Angus and Robertson, **4**, 489–499.

Brimhall, G.H., Lewis, C.J., Ague, J.J., Dietrich, W.E., Hampel, J., Teague, T. and Rix, P. (1988) Metal enrichment in bauxites by deposition of chemically mature aeolian dust. *Nature*, **333**, 819–824.

Brimhall, G.H., Taylor, G., Warin, O., Lewis, G.J., Ford, C. and Bratt, J. (1991) Quantitative geochemical approach to pedogenesis: importance of parent material reduction, volumetric expansion and eolian influx in laterization. *Geoderma*, **52**, 51–91.

Brinkman, R. (1970) Ferrolysis, a hydromorphic soil forming process. *Geoderma*, **3**, 199–206.

Britt, A. (1993) *Regolith, bedrock and geochemistry at Johnson Creek, northwestern Queensland*. B.Sc. Hons. Thesis, Australian National University, Canberra, 190 pp.

Brookins, D.G. (1988) *Eh–pH Diagrams for Geochemistry*. Springer-Verlag, Berlin and New York, 176 pp.

Brooks, R.R. (1987) Serpentine and its vegetation; a multidisciplinary approach. In Dudley, T.R. (ed.) *Ecology,*

Phytogeography and Physiology Series. Dioscorides Press, Portland, 454 pp.

Brotzen, O. (1966) Geochemical ranking of rocks. *Sveriges Geologiska Undersoekning, Serie C, Avhandlingar och Uppsatser*, **2**, 617.

Brown, C.M. (1989) The Ivanhoe Block; its structure, hydrogeology and effect on groundwaters of the Riverine Plain of New South Wales. *Bureau of Mineral Resources Journal of Australian Geology and Geophysics*, **11**, 333–354.

Brown, J.B. (1971) Jarosite–goethite stabilities at 25°C, 1 ATM. *Mineral. Deposita*, **6**, 245–252.

Brown, M.C. and Ollier, C.D. (1975) Geology and scenery of Canberra. *Australian Geographer*, **13**, 97–103.

Browne, W.R. (1914) The geology of the Cooma district, NSW, Part I. *Journal and Proceedings of the Royal Society of New South Wales*, **48**, 172–222.

Browne, W.R. (1972) Greybilly and its associates in Eastern Australia. *Proceedings of the Linnaean Society of New South Wales*, **97**, 98–129.

Bruno, J. and Sellin, P. (1992) Radionuclide solubilities to be used in SKB91. Swedish Nuclear Fuel and Waste Management Co., *Technical Report 92-13*.

Buchanan, D.L. (1988) Platinum-group element exploration. *Developments in Economic Geology 26*, Elsevier, 185 pp.

Buchanan, F. (1807) *A Journey from Madras through the Countries of Mysore, Kanara and Malabar*, 3 volumes. East India Co., London.

Büdel, J. (1957) Die 'doppelten Einebnungsflachen' in den feuchten Tropen. *Zeitschrift für Geomorphologie*, **1**, 201–228.

Büdel, J. (1982) *Climatic Geomorphology*. Princeton University Press, Princeton, 443 pp.

Bureau of Resource Sciences Australia (1997) *Australia's Identified Mineral Resources*, Canberra, 34 pp.

Burnett, A.D., Koirala, N.P. and Hee, A. (1987) Engineering geology and town planning in Hong Kong. *Bulletin Geological Society of Hong Kong*, **3**, 25–42.

Butler, B.E. (1958) Depositional systems of the Riverine Plain of south-eastern Australia. *CSIRO Australia Soil Publication No. 10*.

Butler, B.E. and Churchward, H.M. (1983) Aeolian processes. In *Soils, an Australian viewpoint*. CSIRO, Melbourne, Australia, 91–99.

Butler, B.E. and Hutton, J.T. (1956) Parna in the riverine plain of south-eastern Australia and the soils thereon. *Australian Journal of Agricultural Research*, 7, 536–553.

Butler, B.E., Blackburn, G., Bowler, J.M., Lawrence, C.R., Newell, J.W. and Pels, S. (1973) *A Geomorphic Map of the Riverine Plain of South-eastern Australia*. Australian National University Press, Canberra, 39 pp.

Butt, C.R.M. (1983) Aluminosilicate cementation of saprolites, grits and silcretes in Western Australia. *Journal of the Geological Society of Australia*, **30**, 179–186.

Butt, C.R.M. (1992) Physical weathering and dispersion. In Butt, C.R.M. and Zeegers, H. (eds) *Regolith Exploration Geochemistry in Tropical and Subtropical Terrains*. Elsevier, Amsterdam, 97–113.

Butt, C.R.M. (1998) Exploration in calcrete terrain for commodities other than gold. In Scott, K.M., Anand, R.R., Gray, D.J. and Lintern, M.J. (eds) *Sampling Media in Various Australian Regolith Regimes*, CRC LEME Report 82, Perth.

Butt, C.R.M. and Nickel, E.H. (1981) Mineralogy and geochemistry of the weathering of the disseminated nickel sulphide deposit at Mt. Keith, Western Australia. *Economic Geology*, **76**, 1736–1751.

Butt, C.R.M. and Smith, R.E. (1992) Characteristics of the weathering profile. In Butt, C.R.M. and Zeegers, H. (eds) *Regolith Exploration Geochemistry in Tropical and Subtropical Terrains*. Elsevier, Amsterdam, 299–304.

Butt, C.R.M. and Zeegers, H. (eds) (1992a) *Regolith Exploration Geochemistry in Tropical and Subtropical Terrains*. Elsevier, Amsterdam, 607 pp.

Butt, C.R.M. and Zeegers, H. (1992b) Climate, geomorphological environment and geochemical dispersion models. In Butt, C.R.M. and Zeegers, H. (eds) *Regolith Exploration Geochemistry in Tropical and Subtropical Terrains*. Elsevier, Amsterdam, 3–24.

Butt, C.R.M., Gray, D.J., Lintern, M.J. and Robertson, I.D.M. (1993) Gold and associated elements in the regolith – dispersion processes and implications for exploration. *CSIRO Division of Exploration Geoscience Restricted Report 396R*.

Buurman, P., Meijer, E.L. and van Wijck, J.H. (1988) Weathering of chlorite and vermiculite in ultramafic rocks of Cabo Ortegal, northwestern Spain. *Clays and Clay Minerals*, **36**, 263–269.

Callan, R.A. (1984) Clays of the palygorskite–sepiolite group: depositional environment, age and distribution. In Singer, A. and Galan, E. (eds) *Palygorskite–Sepiolite. Occurrences, Genesis and Uses. Developments in Sedimentology*, **37**, 1–38, Elsevier.

Callen, R.A. (1976) Clay mineralogy and the Neogene climate and geography of the Lake Frome area, South Australia. *Abstracts – International Geological Congress*, **25** (2), 492.

Callender, J.H. (1978) A study of the silcretes near Marulan and Milton, New South Wales. In Langford-Smith, T. (ed.) *Silcrete in Australia*, University of New England, Armidale, 209–221.

Calvo, R.M., Garcia-Rodeja, E. and Macias, F. (1983) Mineralogical variability in weathering microsystems of a granitic outcrop of Galacia (Spain). *Catena*, **10**, 225–236.

Carlisle, D. (1980) *Possible Variation in the Calcrete–Gypcrete Uranium Model*. Open-File Report, United States Department of Energy, GJBX-**53**, 38 pp.

Carlisle, D. (1983) Concentration of uranium and vanadium in calcretes and gypcretes. In Wilson, R.C.L. (ed.) *Residual Deposits: surface related weathering processes and materials. Geol. Soc. London Special Publ. 11*, 185–195, Blackwell.

Carroll, D. (1953) Weatherability of zircon. *Journal of Sedimentary Petrology*, **23** (2), 106–116.

Carson, M.A. and Kirkby, M.J. (1972) *Hillslope Form and Process*. Cambridge University Press, London.

Carvalho, I.G., Mestrinho, S.S.P., Fontes, V.M.S., Goel, O.P. and Souza, F.A. (1991) Geochemical evolution of laterites from two areas of the semi-arid region in Bahia State, Brazil. *Journal of Geochemical Exploration*, **40**, 385–411.

Cas, R.A.F. and Wright, J.V. (1987) *Volcanic Successions, Modern and Ancient: a geological approach to processes, products, and successions*. Allen and Unwin, London, 528 pp.

Chadwick, O.A., Hendricks, D.M. and Nettleton, W.D.

(1989) Silicification of Holocene soils in northern Monitor Valley, Nevada. *Soil Science Society of America Journal*, **53**, 158–164.

Chan, R.A. and Kamprad, J.L. (1995) *Bathurst Regolith Landforms 1:100 000 scale map Sheet 8831*. Australian Geological Survey Organisation, Canberra.

Chan, R.A., Craig, M.A., D'Addario, G.W., Gibson, D.L. and Taylor, G. (1986) The Regolith Terrain Map of Australia 1 : 5 000 000. *Bureau of Mineral Resources, Geology and Geophysics Record, 1986/27*.

Chan, R.A., Craig, M.A., Hazell, M.S. and Ollier, C.D. (1992) Kalgoorlie Regolith Terrain Map and Commentary, Sheet SH51, Western Australia, 1 : 1 000 000 Regolith Series, *Australian Geological Survey Organisation Record, 1992/8*.

Chappell, J. (1974) The geomorphology and evolution of small valleys in dated coral reef terraces, New Guinea. *Journal of Geology*, **82**, 795–812.

Charman, P.E.V. and Murphy, B.W. (1991) *Soils: their properties and management*. Sydney University Press/ Oxford University Press, Melbourne, 363 pp.

Chartres, C.J. (1985) A preliminary investigation of hardpan horizons in north-west New South Wales. *Australian Journal of Soil Research*, **23**, 325–337.

Chartres, C.J. and Walker, P.H. (1988) The effect of aeolian accessions on soil development on granitic rocks in south-eastern Australia. III. Micromorphological and geochemical evidence of weathering and soil development. *Australian Journal of Soil Research*, **26**, 33–53.

Chartres, C.J., Chivas, A.R. and Walker, P.H. (1988) The effect of aeolian accession on soil development on granitic rocks in south-eastern Australia. II. Oxygen-isotope, mineralogical and geochemical evidence for aeolian deposition. *Australian Journal of Soil Research*, **26**, 17–31.

Chen, X.Y. (1994) Pedogenic gypcrete formation in arid central Australia. *Australian Geological Survey Organisation Record, 1994/56*, 17.

Chen, X.Y. (1997) Quaternary sedimentation, parna, landforms, and soils landscapes of the Wagga Wagga 1:100 000 map sheet, south-eastern Australia. *Australian Journal of Soil Research*, **35**, 643–668.

Chen, X.Y., Bowler, J.M. and Magee, J.W. (1991) Aeolian landscapes in central Australia; gypsiferous and quartz dune environments from Lake Amadeus. *Sedimentology*, **38**, 519–538.

Chen, X.Y., Bowler, J.M. and Magee, J.W. (1993) Late Cenozoic stratigraphy and hydrologic history of Lake Amadeus, a central Australian playa. *Australian Journal of Earth Sciences*, **40**, 1–14.

Cherry, J.A., Shaikh, A.U., Tallman, D.E. and Nicholson, R.V. (1979) Arsenic species as an indicator of redox conditions in ground water. *J. Hydrol.*, **43**, 373–392.

Chester, R. and Johnson, L.R. (1971) Trace element geochemistry of North Atlantic aeolian dusts. *Nature*, **231**, 176–178.

Chesworth, W. (1979) The major element geochemistry and the mineralogical evolution of granitic rocks during weathering. In Ahrens, L.H. (ed.) *Origin and Distribution of Elements. Phys. Chem. Earth*. **11**, 305–313.

Chittleborough, D.J. (1991) Indices of weathering for soils and palaeosols formed on silicate rocks. *Australian Journal of Earth Sciences*, **38**, 115–120.

Churchward, H.M. (1961) Soil studies at Swan Hill, Victoria, Australia; I. Soil layering. *Journal of Soil Science*, **12**, 73–86.

Coleman, N.T., LeRoux, F.H. and Cady, J.G. (1963) Biotite–hydrobiotite–vermiculite in soils. *Nature Lond.*, **198**, 409–410.

Colin, F., Alarçon, C. and Vieillard, P. (1993) Zircon: an immobile index in soils? *Chemical Geology*, **107**, 273–276.

Colin, F., Nahon, D., Trescases, J.J. and Melfi, A.J. (1990) Lateritic weathering of pyroxenites at Niquelandia, Goias, Brazil; the supergene behavior of nickel. *Economic Geology*, **85** (5), 1010–1023.

Colman, S.M. and Dethier, D.P. (eds) (1986) *Rates of Chemical Weathering of Rocks and Minerals*. Academic Press, London, 603 pp.

Colman, S.M. and Pierce, K.L. (1981) Weathering rinds on andesitic and basaltic stones as a Quaternary age indicator, western United States. *US Geological Survey Professional Paper 1210*.

Cook, J. (1997) Australia. In Thomas, D.S.G. (ed.) *Arid Zone Geomorphology: process, form and change in drylands*. John Wiley, Chichester, 563–576.

Cooke, R.U. and Doornkamp, J.C. (1974) *Geomorphology in Environmental Management: an introduction*. Clarendon Press, Oxford, 413 pp.

Costin, A.B. and Polach, H.A. (1973) Age and significance of slope deposits, Black Mountain, Canberra. *Australian Journal of Soil Research*, **11**, 13–25.

Coventry, R.J. (1979) The age of a red earth profile in central North Queensland. *Australian Journal of Soil Research*, **17**, 505–510.

Coventry, R.J. (1982) The distribution of red, yellow and grey earths in the Torrens Creek Area, north Queensland. *Australian Journal of Soil Research*, **20**, 1–14.

Coventry, R.J. and Walker, P.H. (1977) Geomorphological significance of late Quaternary deposits of the Lake George area, N.S.W. *Australian Geographer*, **13**, 370–376.

Craft, F.A. (1931a) The physiography of the Shoalhaven River Valley. I. Tallong–Bungonia. *Proceedings of the Linnean Society of New South Wales*, **56**, 99–132.

Craft, F.A. (1931b) The physiography of the Shoalhaven River Valley. II. Nerrimunga Creek. *Proceedings of the Linnean Society of New South Wales*, **56**, 242–260.

Craft, FA (1931c) The physiography of the Shoalhaven River Valley. III. Bulee Ridge. *Proceedings of the Linnean Society of New South Wales*, **56**, 261–265.

Craft, F.A. (1931d) The physiography of the Shoalhaven River Valley. IV. Nerriga. *Proceedings of the Linnean Society of New South Wales*, **56**, 412–430.

Craft, F.A. (1933) The surface history of the Monaro. *Proceedings of the Linnean Society of New South Wales*, **58**, 229–244.

Craig, M.A. (1993) Hardpan as a dating medium for regolith: Yilgarn examples. In Klootwijk, C. (compiler) *Palaeomagnetism in Australasia: Dating, Tectonic, and environmental applications. Abstracts*. Australian Geological Survey Organisation Record 1993/20, 84.

Craig, M.A. and Brown, M.C. (1984) Permian glacial pavements and ice movement near Moyhu, north-east Victoria. *Australian Journal of Earth Sciences*, **31**, 439–444.

Creasey, J., Edwards, A.C., Reid, J.M., MacLeod, D.A. and Cresser, M.S. (1986) The use of catchment studies for assessing chemical weathering rates in two contrasting upland areas in Northeast Scotland. In Colman, S.M. and

Dethier, D.P. (eds) *Rates of Chemical Weathering of Rocks and Minerals*. Academic Press, Orlando, 467–502.

Croke, J., Magee, J. and Price, D. (1996) Major episodes of Quaternary activity in the lower Neales River, northwest of Lake Eyre, Central Australia. *Palaeogeography, Palaeoclimatology, Palaeoecology*, **124**, 1–15.

Crook, K.A.W. (1968) Weathering and roundness of quartz sand grains. *Sedimentology*, **11**, 171–182.

CSIRO (1983) *Soils: an Australian viewpoint*. CSIRO, Melbourne, 928 pp.

Currey, D.T. (1977) The role of applied geomorphology in irrigation and groundwater studies. In Hails, J.R. (ed.) *Applied Geomorphology: a perspective of the contribution of geomorphology to interdisciplinary studies and environmental management*. Elsevier, Amsterdam, 51–83.

da Costa, M.L. (1993) Gold distribution in lateritic profiles in South America, Africa, and Australia: applications to geochemical exploration in tropical regions. *Journal of Geochemical Exploration*, **47**, 143–163.

da Costa, M.L. and Araújo, E.S. (1996) Application of multi-element geochemistry in Au-phosphate-bearing lateritic crusts for identification of their parent rocks. *Journal of Geochemical Exploration*, **57**, 257–272.

Dalrymple, J.B., Blong, R.J. and Conacher, A.J. (1968) 'A hypothetical nine-unit landsurface model'. *Zeitschrift für Geomorphologie*, **12**, 60–76.

Dammer, D., Chivas, A.R. and McDougall, I. (1996) Isotopic dating of supergene manganese oxides from the Groote Eylandt deposit, Northern Territory, Australia. *Economic Geology*, **91**, 386–401.

Dana, J.D. (1985) *Manual of Mineralogy*. Klein, C. and Hurlbut, C.S. Jr. (eds) 20th edition. John Wiley, New York, 596 pp.

Dangic, A. (1983) Kaolinization of bauxite: a study in the Vlasenica bauxite area Yugoslavia. I. Alteration of matrix. *Clays and Clay Minerals*, **33** (6), 517–524.

Davis, J.A. and Kent, D.B. (1990) Surface complexation modeling in aqueous geochemistry. In Hochella, M.F. Jr and White, A.F. (eds) *Mineral Water Interface Geochemistry. Reviews in Mineralogy*, **23**, 177–260. Mineralogical Society of America.

Davis, W.M. (1889) Methods and models in geographic teaching. *The American Naturalist*. University of Chicago Press, Chicago, 566–583.

Davy, R. (1979) A study of laterite profiles in relation to bed rock in the Darling Range near Perth, W.A. *Geological Survey of Western Australia, Report* 8.

Delvigne, J., Bisdom, E.B.A., Sleeman, J. and Stoops, G. (1979) Olivines, their pseudomorphs and secondary products. *Pédologie*, **XXIX** (3), 247–309.

Delvigne, J.E. (1998) *Atlas of Micromorphology of Mineral Alteration and Weathering*. Canadian Mineralogist Special Publication 3. Mineralogical Association of Canada, 494 pp.

Dennen, W.H. and Moore, B.R. (1986) *Geology and Engineering*. Wm. C. Brown Publishers, Dubuque, Ind., 423 pp.

De Ploey, J. (1964) Stone-lines and clayey-sandy mantles in Lower Congo: their formation and the effect of termites. In Boullion, A. (ed.) *Etudes sur les Termites Africains*, Université de Louvanium, Leopoldville, 399–414.

De Ploey, J. (1965) Position géomorphologique, génèse et chronologie de certains dépôts superficiels au Congo occidentale. *Quaternaria*, **7**, 131–154.

Dixon, J.B. and Weed, S.B. (eds) (1989) *Minerals in Soil Environments*. Soil Science Society of America, Madison, Wis.

Dixon, J.C. and Young, R.W. (1981) Character and origin of deep arenaceous weathering mantles on the Bega Batholith, southeastern Australia. *Catena*, **8**, 97–109.

d'Orey, F.L.C. (1975) Contribution of termite mounds to locating hidden copper deposits. *Transactions of the Institute of Mining and Metallurgy, Section B, Applied Earth Science*, **84**, 150–151.

Douglas, I. (1968) The effects of precipitation chemistry and catchment area lithology on the quality of river water in selected catchments in eastern Australia. *Earth Science Journal*, **2**, 126–144.

Drever, J.I. (1988) *The Geochemistry of Natural Waters*. Prentice-Hall, Englewood Cliffs, New Jersey, 437 pp.

Drever, J.I. and Clow, D.W. (1995) Weathering rates in catchments. In White, A.F. and Brantley, S.L. (eds) *Chemical Weathering Rates of Silicate Minerals. Reviews in Mineralogy 31*. Mineralogical Society of America, 463–484.

Drever, L., Fontes, J.Ch. and Riche, G. (1987) Isotopic approach to calcite dissolution and precipitation in soils under semiarid conditions. *Chemical Geology (Isotope Sections)*, **66**, 307–314.

Drexel, J.F. (1988) *Mineral Resources Review – South Australia Department of Mines and Energy*, **156**. South Australia, Department of Mines and Energy, Adelaide, 129 pp.

Drexel, J.F. and Preiss, W.V. (eds) (1995) *The Geology of South Australia*. Volume 2, *The Phanerozoic*. South Australian Geological Survey Bulletin, 54, 347 pp.

Duparc, L., Molly, E. and Borloz, A. (1927) *Sur la Birbirite une roche nouvelle*. Compte Rendu des Séances de la Société de Physique et d'Histoire Naturelle de Genève, **44**, 137–139.

Dury, G.H. (1968) An introduction to the geomorphology of Australia. In Dury G.H. and Logan, M.I. (eds) *Studies in Australian Geography*. Heinemann Educational, Melbourne, 1–36.

Dury, G.H. and Habermann, G.M. (1978) Australian silcretes and northern-hemisphere correlatives. In Langford-Smith, T. (ed.) *Silcrete in Australia*, University of New England, Armidale, 223–259.

Du Toit, A.L. (1954) *The Geology of South Africa*. Oliver and Boyd, Edinburgh, 611 pp.

Edenharter, A., Guernot, F. Oberhänsli, R. and Stalder, H.A. (1980) Niob-Anatas aus dem Binntal. *Schweizer Strahler*, **5**, 261–267.

Edmunds, W.M. and Trafford, J.M. (1993) Beryllium in river baseflow, shallow groundwaters and major aquifers of the U.K. *Applied Geochemistry*, Supplementary Issue No. 2: 223–233.

Edwards, C.A. and Lofty, J.R. (1977) *Biology of Earthworms*. Chapman & Hall, London, 333 pp.

Eggleton, A., Aspandiar, M., le Gleuher, M., Morgan, M., Taylor, G. and Tilley, D. (1998) The laterite profile: origins and variations. In Britt, A.F. and Bettenay, L. (eds) *Regolith '98: New Approaches to an Old Continent – Program and Abstracts*, CRC LEME, Perth, 8.

Eggleton, R.A. (1975) Nontronite topotaxial after hedenbergite. *American Mineralogist*, **60**, 1063–1068.

Eggleton, R.A. (1984) Formation of iddingsite rims on olivine: a TEM study. *Clays and Clay Minerals*, **32**, 1–11.

Eggleton, R.A. (2001) *The Regolith Glossary*. CRC LEME, Perth.

Eggleton, R.A. and Boland, J.N. (1982) Weathering of enstatite to talc through a series of transitional phases. *Clays and Clay Minerals*, **30**, 11–20.

Eggleton, R.A. and Taylor, G. (1999) Selected thoughts on 'laterites'. In Taylor, G. and Pain, C.F. (eds) *New Approaches to an Old Continent*. CRC LEME, Perth, 209–226.

Eggleton, R.A. and Tilley, D.B. (1998) Hisingerite: a ferric kaolin mineral with curved morphology. *Clays and Clay Minerals*, **46**, 400–413.

Eggleton, R.A., Varkevisser, D. and Foudoulis, C. (1987) The weathering of basalt: changes in bulk chemistry and mineralogy. *Clays and Clay Minerals*, **35**, 161–169.

Elderfield, H., Upstill-Goddard, R. and Sholkovitz, E.R. (1990) The rare earth elements in rivers, estuaries, and coastal seas and their significance to the composition of ocean waters. *Geochimica et Cosmochimica Acta*, **54**, 971–991.

Evans, A.M. (1995) *Introduction to Mineral Exploration*. Blackwell Scientific, Oxford, 396 pp.

Evans, W.R., Williams, R.M., Barnett, S.R. and Hoxley, G. (1991) Groundwater resource and salinity management by hydrogeological mapping in the Murray Basin. In Ventriss, H. (ed.) *Proceedings of the International Conference on Groundwater in Large Sedimentary Basins*. Australian Water Resources Council Series, **20**, 531–537.

Fanning, D.S., Rabenhorst, M.C. and Bigham, J.M. (1993) Colors of acid sulfate soils. In Bigham, J.M. and Ciolkosz, E.J. (eds) *Soil Color*. Soil Science Society of America, Special Publication 31, 91–108.

FAO (1985) *Soil Map of the World 1 : 5 000 000*. Food and Agriculture Organization of the United Nations, Paris.

Fee, J.A., Gaudette, H.E., Lyons, W.B. and Long, D.T. (1992) Rare-earth element distribution in Lake Tyrrell groundwaters, Victoria, Australia. *Chemical Geology*, **96**, 67–93.

Firman, J.B. (1981) Regional stratigraphy of the regolith on the southwest margin of the Great Australian Basin Province, South Australia. *South Australian Department of Mines and Energy Unpublished Report 81/40*, 144 pp.

Firman, J.B. (1994) Paleosols in laterite and silcrete profiles: evidence from the southeastern margin of the Australian Precambrian Shield. *Earth Science Reviews*, **36**, 149–179.

Fitzpatrick, R.W. and Schwertmann, U. (1982) Al-substituted geothite – an indicator of pedogenic and other weathering environments in South Africa. *Geoderma*, **27**, 335–347.

Flach, K.W., Nettleton, W.D. and Nelson, R.E. (1974) The micromorphology of silica-cemented soil horizons in western North America. In Rutherford G.K. (ed.) *Soil Microscopy*. The Limestone Press, Kingston, 714–729.

Fleming, A., Summerfield, M.A., Stone, J.O., Fifield, L.K. and Cresswell, R.G. (1999) Denudation rates for the southern Drakensberg escarpment, SE Africa, derived from *in-situ* produced cosmogenic ^{36}Cl: initial results. *Journal of the Geological Society London*, **156**, 209–212.

Flicoteaux, R. and Lucas, J. (1984) Weathering of phosphate minerals. In Nriagu, J.O. and Moore, P.B. (eds) *Phosphate Minerals*. Springer-Verlag, Berlin and New York, 292–317.

Fookes, P.G. (1997) Geology for engineers: the geological model, prediction and performance. *Quarterly Journal of Engineering Geology*, **30**, 293–424.

Foster, K.A., Shirtliff, G.J. and Hill, S.M. (2000) *Balaclava Regolith-Landforms Sheet 7133-1-5*. Australian Geological Survey Organization, Canberra.

Foster, K.A., Shirtliff, G. and Hill, S.M. (2000) Balaclava 1:25 000 Regolith Landform Map Sheet, CRC LEME, Perth.

Foster, L.D. (1996) *Sedimentary reworking across the Weipa bauxite deposit*. B.Sc. (Hons) thesis, Australian National University, Canberra, 127 pp.

Fournier, F. (1960) *Climat et Erosion: la relation entre l'érosion du sol par l'eau et les précipitations atmosphériques*. Presses Universitaires de France, Paris.

Fox, C.S. (1936) Buchanan's laterite of Malabar and Kanara. *Records of the Geological Survey of India*, **69**, 389–422.

Frakes, L.A. and Bolton, B.R. (1984) Origin of manganese giants: sea level changes and anoxic–oxic history. *Geology*, **12**, 83–86.

Francis, G. and Walker, G.T. (1978) Silcretes of subaerial origin in southern New England. *Search*, **9**, 321–323.

Freyssinet, Ph. (1994) Gold mass balance in lateritic profiles from savanna and rainforest zones. *Catena*, **21**, 159–172.

Frick, C. (1985) A study of the soil geochemistry of the Platreef in the Bushveld complex, South Africa. *Journal of Geochemical Exploration*, **24**, 51–80.

Frick, C. (1986) The behaviour of uranium, thorium and tin during leaching from a coarse-grained porphyritic granite in an arid environment. *Journal of Geochemical Exploration*, **25**, 261–282.

Fritz, S.J. (1988) A comparative study of gabbro and granite weathering. *Chemical Geology*, **68**, 275–290.

Frondel, C. (1962) *The System of Mineralogy of James Dwight Dana and Edward Salisbury Dana, Yale University 1837–1892*, 7th edn. Vol. 3 *Silica Minerals*. John Wiley, New York, 334 pp.

Frost, R.R. and Griffin, R.A. (1977) Effect of pH on adsorption of arsenic and selenium from landfill leachate by clay minerals. *Soil Science Society of America Journal*, **41**, 53–57.

Fuchs, W.A. and Rose, A.W. (1974) The geochemical behavior of platinum and palladium in the weathering cycle in the Stillwater Complex, Montana. *Economic Geology*, **69**, 332–346.

Furrer, G. and Stumm, W. (1986) The coordination chemistry of weathering: I. Dissolution kinetics of δ-Al_2O_3 and BeO. *Geochimica et Cosmochimica Acta*, **50**, 1847–1860.

Gallagher, K., Hawkesworth, C.J. and Mantovani, M.S.M. (1994) The denudation history of the onshore continental margin of SE Brazil inferred from apatite fission track data. *Journal of Geophysical Research, B: Solid Earth and Planets*, **99**, 117–118.

Gardner, L.R., Kheoruenromne, I. and Chen, H.S. (1978) Isovolumetric geochemical investigation of a buried granite saprolite near Columbia, SC, USA. *Geochimica et Cosmochimica Acta*, **42**, 417–424.

Garnett, D.L., Rea, W.J. and Fuge, R. (1982) Geochemical exploration techniques applicable to calcrete-covered areas. In Glenn, H.W. (ed.) *Proceedings of the 12th*

Commonwealth Mining and Metallurgical Institute Congress, Johannesburg, Geological Society of South Africa, 945–955.

Garnier, J. (1867) Essai sur la géologie et les ressources minérales de la Nouvelle-Calédonie. *Ann. Mines, Paris*, **6**, 1–92.

Garrels, R.M. and Christ, C.L. (1965) *Solutions, Minerals and Equilibria*. Freeman, Cooper, San Francisco, 450 pp.

Garrels, R.M. and Mackenzie, F.T. (1971) *Evolution of Sedimentary Rocks*. W.W. Norton and Company, New York, 397 pp.

Gatehouse, R. and Greene, R. (eds) (1998) *Aeolian Dust: implications for Australian mineral exploration and environmental management*. CRC LEME Report 102, 33 pp.

Geidans, L. (1973) Bauxitic laterites of the south western part of Western Australia. Australian Institute of Mining and Metallurgy Western Australia Conference, Parkville, Victoria, 173–182.

Geological Society (1990) Engineering Group Working Party Report: tropical residual soils. *Quarterly Journal of Engineering Geology*, **23**, 1–101.

Gerth, J., Davey, G.B. and Cockayne, D.J.H. (1985) Incorporation of trace metals and radionuclides into Goethite. *Proceedings, 4th Australian Conference on Nuclear Techniques of Analysis*, AINSE, 6–8 November 1985, 108–110.

Gibbs, R.J. (1967) The geochemistry of the Amazon River system, Part I. The factors that control the salinity and composition of the suspended solids. *Geological Society of America Bulletin*, **78**, 1203–1232.

Gibson, D.L. and Wilford, J.R. (1995) *Broken Hill Regolith-Landforms Map, 1:500 000*. CRC LEME, Perth.

Gieseking, J.E. (ed.) (1975) *Soil Components* Vol. 2. *Inorganic Components*. Springer-Verlag, Berlin, 684 pp.

Giral, S., Savin, S.M., Girad, J-P. and Nahon, D.B. (1993) The oxygen isotope geochemistry of kaolinites from lateritic profiles: implications for pedology and paleoclimatology. *Chemical Geology*, **107**, 237–240.

Glasby, G.P. (1971) The influence of aeolian transport of dust particles on marine sedimentation in the south-west Pacific. *Journal of the Royal Society of New Zealand*, **1**, 285–300.

Glassford, D.K. and Semeniuk, V. (1995) Desert-aeolian origin of late Cenozoic regolith in arid and semi-arid southwestern Australia. *Palaeogeography, Palaeoclimatology, Palaeoecology*, **114**, 131–166.

Glenn, R.C., Jackson, M.L., Hole, F.D. and Lee, G.B. (1960) Chemical weathering of layer silicate clays in loess-derived Tama silt-loam of southwestern Wisconsin. *Proceedings of the National Conference, Clays and Clay Minerals*, **8**, 63–83.

Glover, P.E., Glover, E.C., Trump, E.C. and Wateridge, L.E.D. (1964) The lava caves of Mount Suswa, Kenya, with particular reference to their ecological role. *Studies in Speleology*, **1**, 51–66.

Goldberg, E.D. (1978) *Biogeochemistry of Estuarine Sediments*. UNESCO, Paris, 293 pp.

Gomes, C.S.F. and Massa, M.E. (1992) Allophane and spherulitic halloysite, weathering products of trachytic pumice fall-outs in the Caldeira Velha area (S. Miguel-Azores). *Mineralogica et Petrographica Acta*, **XXXV**(A), 283–288.

Gordon, M. Jr and Tracey, J.I. Jr (1952) Origin of the Arkansas bauxite deposits. In *Problems of Clay and Laterite Genesis – Symposium*, American Institute of Mining and Metallurgical Engineers, New York, 12–34.

Götze, J. and Plötze, M. (1997) Investigation of trace-element distribution in detrital quartz by electron paramagnetic resonance (EPR). *European Journal of Mineralogy*, **9**, 529–537.

Goudie, A.S. (1973) *Duricrusts in Tropical and Subtropical Landscapes*. Clarendon Press, Oxford, 174 pp.

Goudie, A.S. and Pye, K. (1983) *Chemical Sediments and Geomorphology*. Academic Press, London, 439 pp.

Graham, R.C., Weed, S.B., Bowen, L.H. and Buol, S.W. (1989) Weathering of iron-bearing minerals in soils and saprolite on the North Carolina Blue Ridge Front: I. Sand-sized primary minerals. *Clays and Clay Minerals*, **37**, 19–28.

Grana-Gomez, M.J., Barral-Silva, M.T. and Seoane Labineira, S. (1990) Cobalt fractionation in surface horizons of soils from the Province of Lugo (Spain). *Chemical Geology*, **84**, 68–69.

Grandstaff, D.E. (1986) The dissolution rate of forsteritic olivine from Hawaiian beach sand. In Colman, S.M. and Dethier, D.E. (eds) *Rates of Chemical Weathering of Rocks and Minerals*. Academic Press, Orlando, 41–59.

Gray, D.J. (1998) Selective extraction techniques for the recognition of buried mineralisation, Curara Well, Western Australia. In Britt, A.F. and Bettenay, L. (eds) *Regolith '98: New Approaches to an Old Continent – Program and Abstracts*, CRC LEME, Perth, 31.

Gray, D.J., Schorin, K.H. and Butt, C.R.M. (1996) Mineral associations of platinum and palladium in lateritic regolith, Ora Banda Sill, Western Australia. *Journal of Geochemical Exploration*, **57**, 245–255.

Gray, D.J., Butt, C.R.M. and Lawrance, L.M. (1992) The geochemistry of gold in lateritic terrain. In Butt, C.R.M. and Zeegers, H. (eds) *Regolith Exploration Geochemistry in Tropical and Subtropical Terrains*. Elsevier, Amsterdam, 461–482.

Greffie, C., Parron, C., Benedetti, M., Amouric, M., Marion, P. and Colin, F. (1993) Experimental study of gold precipitation with synthetic iron hydroxides: HRTM-AEM and Mössbauer spectroscopy investigation. *Chemical Geology including Isotope Geology*, **107**, 297–300.

Grey, I.E. and Reid, A.F. (1975) The structure of pseudorutile and its role in the natural alteration of ilmenite. *American Mineralogist*, **60**, 898–906.

Grimes, K.G. (1979) The stratigraphic sequence of old land surfaces in North Queensland. *Bureau of Mineral Resources Journal of Australian Geology and Geophysics*, **4**, 33–46.

Grimes, K.G. (1980) The Tertiary geology of North Queensland. In Henderson, R.A. and Stephenson, P.J. (eds) *The Geology and Geophysics of North-eastern Australia*. Geological Society of Australia, Queensland Division, Brisbane, 329–347.

Gunn, R.H. (1974) Soils of the Balonne–Maranoa area. In *Lands of the Balonne–Maranoa Area, Queensland*. CSIRO Australian Land Use Research Series 34, 148–179.

Gunn, R.H. (1985) Shallow groundwaters in weathered volcanic, granite and sedimentary rocks in relation to dryland salinity in southern New South Wales. *Australian Journal of Soil Research*, **23**, 355–371.

Gunn, R.H. and Galloway, R.W. (1978) Silcretes in south-central Queensland. In Langford-Smith, T. (ed.) *Silcrete in*

Australia. Department of Geography, University of New England, Armidale, 51–71.

Gupta, A. (1987) Quaternary geology in Singapore: a review. In Wezel, F. and Rau, J.L. (eds) *Proceedings of the CCOP Symposium on Developments in Quaternary Geological Research in East and Southeast Asia during the Last Decade*. United Nations Economic and Social Commission for Asia and the Pacific, Bangkok, 263–289.

Gurnell, A.M. and Clark, D.J. (eds) (1987) *Glacio-fluvial Sediment Transfer: an alpine perspective*. John Wiley, Chichester, 524 pp.

Güven, N. and Hower, W.F. (1979) A vanadium smectite. *Clay Minerals*, **14**, 241–245.

Habermehl, M.A. (1980) The Great Artesian Basin. *Bureau of Mineral Resources Journal of Australian Geology and Geophysics*, **5**, 9–38.

Harder, E.C. (1952) Examples of bauxite deposits illustrating variations in origin. In *AIME, Problems of Clay and Laterite Genesis – Symposium*, Feb. 1951, 35–64.

Harnois, Luc (1988) The CIW index; a new chemical index of weathering. *Sedimentary Geology*. **55** (3–4), 319–322.

Harvey, A.M. (1997) The role of alluvial fans in arid zone fluvial systems. In Thomas, D.S.G. (ed.) *Arid Zone Geomorphology: process, form and change in drylands*. John Wiley, Chichester, 231–259.

Hay, R.L. and Reeder, R.J. (1991) Calcretes of Olduvai Gorge and the Ndolanya Beds of northern Tanzania. *Sedimentology*, **25**, 649–672.

Hays, J. (1967) Land surfaces and laterites in the north of the Northern Territory. In Jennings, J.N. and Mabbutt, J.A. (eds) *Landform Studies from Australia and New Guinea*. Australian National University Press, Canberra, 182–210.

Helios-Rybicka, E., Calmano, W. and Breeger, A. (1995) Heavy metal sorption/desorption in competing clay minerals; an experimental study. *Applied Clay Science*, **9**(5), 369–381.

Henderson, R.A. (1996) Tertiary units and landscape development in the Townsville–Mackay hinterland, North Queensland. *Geological Society of Australia Abstracts*, **41**, 190.

Herbillon, A.J. and Stone, W.E.E. (1985) Du rôle de l'hydrolyse des cations ferriques induite par des interfaces dans des phénomènes de laterisation [The role of hydrolysis of ferric cations induced by interfaces in laterization processes]. In Alexandre, J. and Symoens, J-J. (eds) *Les processus de lateritisation; journee d'étude – Het laterisatieproces; studiedag* [Processes of lateritization; study day]. Academie Royale des Sciences d'outre-Mer, Brussels, Belgium, 63–72.

Herczeg, A.L. and Chapman, A. (1991) Uranium-series dating of lake and dune deposits in southeastern Australia: a reconnaissance. *Palaeogeography, Palaeoclimatology, Palaeoecology*, **84**, 285–298.

Hesse, P.P. (1994) The record of continental dust from Australia in Tasman Sea sediments. *Quaternary Science Reviews*, **13**, 257–272.

Hesse, P.P. (1997) Mineral magnetic 'tracing' of aeolian dust in southwest Pacific sediments. *Palaeogeography, Palaeoclimatology, Palaeoecology*, **131**, 327–353.

Heyligers, P.C. (1963) Vegetation and soil of a white-sand savanna in Suriname. *Verhandelingen der koninkluke Nederlandse*, Akademie van Wetenschappen, AFD Natuurkunde. N.V. Noord-Hollandsche Uitgevers Maatschappij, Amsterdam, 148 pp.

Hieronymus, B., Boulegue, J. and Kotschoubey, B. (1990) Gallium behaviour in some intertropical environment alterations. In Noack, Y. and Nahon, D. (eds) *Geochemistry of the Earth's Surface and of Mineral Formation. Chemical Geology*, **84** (1–4), 78–82.

Hill, R.J. and Milnes, A.R. (1974) Phosphate minerals from Reaphook Hill, Flinders Ranges, South Australia. *Mineralogical Magazine*, **39**, 684–695.

Hill, S.M. (1995) Biological applications of regolith mapping. In McQueen, K.G. and Craig, M.A. (eds) *Developments and New Approaches in Regolith Mapping*. Centre for Australian Regolith Studies Occasional Publication No. 3, 69–72.

Hill, S.M. (1996) Mesozoic regolith and palaeo-landscape features in southeastern Australia. *Geological Society of Australia – Abstracts*, **43**, 246–250.

Hill, S.M. and Kohn, B.P. (1998) Morphotectonic evolution of the Mundi Mundi Range Front, Broken Hill Block, western NSW. In Taylor, G. and Pain, C. (eds) *Regolith '98, New Approaches to an Old Continent, Proceedings*. CRC LEME, Perth, 319–334.

Hill, S.M., McQueen, K.G. and Foster, K. (1998) Regolith carbonate accumulations in western and central NSW: Characteristics and potential as an exploration sampling medium. In Taylor, G. and Pain, C. (eds) *Regolith '98: New Approaches to an Old Continent, Proceedings*. CRC LEME, Perth, 191–208.

Hills, J.M. (1975) *Exploration from the Mountains to the Basin*. El Paso Geological Society, El Paso, 190 pp.

Hine, R., Williams, I.S., Chappell, B.W. and White, A.J.R. (1978) Contrasts between I- and S-type granitoids of the Kosciusko Batholith. *Journal of the Geological Society of Australia*, **25**, 219–234.

Hochella, M.F. Jr and Banfield, J.F. (1995) Chemical weathering of silicates in nature: a microscopic perspective with theoretical considerations. In White, A.F. and Brantley, S.L. (eds) *Chemical Weathering Rates of Silicate Minerals*. Mineralogical Society of America, Washington, DC, 353–406.

Holdren, G.R. Jr and Berner, R.A. (1979) Mechanism of feldspar weathering. I. Experimental studies. *Geochimica et Cosmochimica Acta*, **43**, 1161–1171.

Holeman, J.N. (1968) The sediment yield of the major rivers of the world. *Water Resources Research*, **4**, 737–747.

Holt, J.A. and Greenslade, P.J.M. (1979) Ants (Hymenoptera: Formicidae) in mounds of *Amitermes laurensis* (Isoptera: Termitidae). *Journal of the Australian Entomological Society*, **18**, 349–361.

Hough, D.J. (1983) *The influence of parent rocks on the nature of surficial materials in the Eden region*. M.Sc. thesis, Australian National University, Canberra.

Huang, W.H. and Keller, W.D. (1971) Dissolution of clay minerals in dilute organic acids at room temperature. *American Mineralogist*, **56**, 1082–1095.

Hubble, G.D., Isbell, R.F. and Northcote, K.H. (1983) Features of Australian soils. In *Soils, an Australian Viewpoint*. CSIRO, Melbourne, 17–47.

Huggett, R.J. (1991) *Climate, Earth Processes and Earth History*. Springer-Verlag, Berlin, 281 pp.

Hunt, C.B. (1967) *Physiography of the United States*. United States Geological Survey, San Francisco, 480 pp.

Hunt, C.B. (1986) *Surficial Deposits of the United States.* Van Nostrand Reinhold, New York, 189 pp.

Hutton, J.T., Twidale, C.R. and Milnes, A.R. (1978) Characteristics and origin of some Australian silcretes. In Langford-Smith, T. (ed.) *Silcrete in Australia.* University of New England Press, Armidale, 19–39.

Idnurm, M. and Senior, B.R. (1978) Palaeomagnetic ages of Late Cretaceous and Tertiary weathered profiles in the Eromanga Basin, Queensland. *Palaeogeography, Palaeoclimatology, Palaeoecology*, **24**, 263–277.

Ihnat, M. (ed.) (1989) *Occurrence and Distribution of Selenium.* CRC Press, Boca Raton, Fla.

Ildefonse, P. (1978) *Mécanismes de l'altération d'une roche gabbroïque du Massif du Pallet (Loire Atlantique).* Thèse, l'Université de Poitiers, 142 pp.

Isbell, R.F. (1996) *The Australian Soil Classification. Australian Soil and Land Survey Handbook*, Volume 4. CSIRO, Melbourne, 143 pp.

Jackson, J.A. (1997) *Glossary of Geology*, 4th edn. American Geology Institute, Alexandria, Va, 769 pp.

Jacobs, L.W. (ed.) (1989) Selenium in agriculture and the environment. *Soil Science Society of America Special Publication no. 23.* SSSA, Madison, Wis, 232 pp.

Jansen, J.M.L. and Painter, R.B. (1974) Predicting sediment yield from climate and topography. *Journal of Hydrology*, **21**, 371–380.

Jeong, G.Y. (1992) *Mineralogy and genesis of kaolin in the Sancheong district, Korea.* Ph.D. thesis, Seoul National University, 325 pp.

Jeong, G.Y. (1998) Formation of vermicular kaolinite from halloysite aggregates in the weathering of plagioclase. *Clays and Clay Minerals*, **46**, 270–279.

Jessup, R.W. and Norris, R.M. (1971) Cainozoic stratigraphy of the Lake Eyre Basin and part of the arid region lying to the south. *Journal of the Geological Society of Australia*, **18**, 303–331.

Johnson, N.M., Driscoll, C.T., Eaton, J.S., Likens, G.E. and McDowell, W.H. (1981) 'Acid rain', dissolved aluminum and chemical weathering at the Hubbard Brook Experimental Forest, New Hampshire. *Geochimica et Cosmochimica Acta*, **45**, 1421–1437.

Jones, B.F., Kennedy, V.C. and Zellweger, G.W. (1974) Comparison of observed and calculated concentrations of dissolved Al and Fe in stream water. *Water Resource Res.*, **10**, 791–793.

Jones, H.A. (1965) Ferruginous oolites and pisolites. *Journal of Sedimentary Petrology*, **35**, 838–845.

Jones, J.B., Sanders, J.V. and Segnit, E.R. (1964) Structure of opal. *Nature*, **204**, 990–991.

Jones, M.R. (1995) Magnesite in review. *Queensland Government Mining Journal*, **96**, 11–20.

Joshi, V.U. and Kale, V.S. (1997) Colluvial deposits in northwest Deccan, India – their significance in the interpretation of late Quaternary history. *Journal of Quaternary Science*, **12**, 391–403.

Jutson, J.T. (1914) An outline of the physiographical geology (physiography) of Western Australia. *Geological Survey of Western Australia Bulletin*, **61**, 240 pp.

Jutson, J.T. (1934) The physiography (geomorphology) of Western Australia. *Geological Survey of Western Australia Bulletin*, **95**, 366 pp.

Jutson, J.T. (1950) The physiography of Western Australia. 3rd edn. *Geological Survey of Western Australia Bulletin*, **95**, 366 pp.

Kabata-Pendias, A. and Pendias, H. (1984) *Trace Elements in Soils and Plants.* CRC Press, Boca Raton, Fla.

Kalkowsky, E. (1901) Die Verkieselung der Gesteine in der nordlichen Kalahari. *Abh. Naturwiss Ges. Isis, Dresden*, 55–107 (in German).

Kauranne, K., Salminen, R. and Eriksson, K. (1992) *Regolith Exploration Geochemistry in Arctic and Temperate Terrains.* Elsevier, Amsterdam, 443 pp.

Keeling, J.L. (1997). Industrial clays: meeting the challenge of increased technical demands and shifting markets. *Resourcing the 21st Century*: AusIMM 1997 Annual Conference, Ballarat, AusIMM Publication Series, 1/97, 239–246.

Keeling, J.L. and Self, P.G. (1996) Garford paleochannel palygorskite. *MESA Journal*, **1**, 20–23.

Kerr, M.H. (1955) On the occurrence of silcretes in Southern England. *Leeds Philosophical and Literary Society, Scientific Section*, **6**, 328–337.

Keywood, M.D., Chivas, A.R., Fifield, L.K., Cresswell, R.G. and Ayres, G.P. (1997) The accession of chloride to the western half of the Australian continent. *Australian Journal of Soil Research*, **35**, 1177–1189.

Kiefert, L. and McTainsh, G.H. (1996) Oxygen isotope abundance in the quartz fraction of aeolian dust: implications for soil and ocean sediment formation in the Australasian region. *Australian Journal of Soil Research*, **34**, 467–473.

Kiernan, K. (1990) Weathering as an indicator of the age of Quaternary glacial deposits in Tasmania. *Australian Geographer*, **21**, 1–17.

Killick, M.F., Churchward, H.M. and Anand, R.R. (1996) Regolith terrain analysis for iron ore exploration in the Hammersley Province, Western Australia. CRC LEME Restricted Report, **7R**, 94 pp.

King, L.C. (1945) Geomorphology of the Natal Drakensberg. *Transactions of the Geological Society of South Africa*, **47**, 255–282.

King, L.C. (1951) The geomorphology of the eastern and southern districts of Southern Rhodesia. *Transactions of the Geological Society of South Africa*, **54**, 33–64.

King, L.C. (1956) Drakensberg scarp of South Africa; a clarification. *Geological Society of America Bulletin*, **67**, 121–122.

King, L.C. (1957) Uniformitarian nature of hillslopes. *Transactions of the Edinburgh Geological Society*, **17**, 81–102.

Kirkham, J.H. (1975) Clay mineralogy of some tephra beds of Rotorua area, North Island, New Zealand. *Clay Minerals*, **10**, 437–450.

Knighton, D. and Nanson, G. (1997) Distinctiveness, diversity and uniqueness in arid zone river systems. In Thomas, D.S.G. (ed.) *Arid Zone Geomorphology: process, form and change in drylands.* John Wiley, Chichester, 185–203.

Kraus, M.J. (1999) Paleosols in clastic sedimentary rocks: their geological applications. *Earth Science Reviews*, **47**, 41–70.

Kump, L.R., Brantley, S.L. and Arthur, M.A. (2000) Chemical Weathering, Atmospheric CO_2 and Climate. *Annual Review of Earth and Planetary Sciences*, **28**, 611–667.

Lai, K.W. and Taylor, B.W. (1984) The classification of colluvium in Hong Kong. In Yim, W.W.S. (ed.) *Geology of Surficial Deposits in Hong Kong*. Geological Society of Hong Kong, 75–85.

Lambeck, K. and Stephenson, R. (1986) The post-Palaeozoic uplift history of south-eastern Australia. *Australian Journal of Earth Sciences*, **33**, 253–270.

Lamplugh, G.W. (1902) Calcrete. *Geological Magazine*, **9**, 575.

Lamplugh, G.W. (1907) The Geology of the Zambezi Basin around the Batoka Gorge. *Quarterly Journal of the Geological Society of London*, **63**, 162–216.

Langford-Smith, T. (1978) *Silcrete in Australia*. Department of Geography, University of New England, Armidale, 304 pp.

Lasaga, A.C. (1995) Fundamental approaches in describing mineral dissolution and precipitation rates. In White A.F. and Brantley, S.L. (eds) *Chemical Weathering Rates of Silicate Minerals. Reviews in Mineralogy*, **31**, 23–86. Min. Soc. America.

Law, K.R., Nesbitt, H.W. and Longstaffe, F.J. (1991) Weathering of granitic tills and the genesis of a podzol. *American Journal of Science*, **291**, 940–955.

Lawrence, C.R. (1976) Murray Basin. In Douglas, J.G. and Ferguson, J.A. (eds) *Geology of Victoria*. Geological Society of Australia Special Publication, **5**, 276–288.

Lawrie, K.C., Chan, R.A., Gibson, D.L. and de Souza Kovacs, N. (1999) Alluvial gold potential in buried palaeochannels in the Wyalong district, Lachlan Fold Belt, New South Wales. *AGSO Research Newsletter*, **30**, 1–5.

Leah, P.A. (1996) Relict lateritic weathering profiles in the Cobar District, NSW. In Cook, W.G., Ford, A.J.H., McDermott, J.J., Standish, P.N., Stegman, C.L. and Stegman, T.M. (eds) *The Cobar Mineral Field – a 1996 Perspective*. Australian Institute of Mining and Metallurgy, Melbourne, 157–177.

Lecomte, P. and Zeegers, H. (1992) Humid tropical terrains (rainforests). In Butt, C.R.M. and Zeegers, H. (eds) *Regolith Exploration Geochemistry in Tropical and Subtropical Terrains*. Elsevier, Amsterdam, 241–294.

Lee, K.E. and Wood, T.G. (1971) *Termites and Soils*. Academic Press, London, 251 pp.

Lees, B.G., Hayne, M. and Price, D.M. (1993) Marine transgression and dune initiation on western Cape York, northern Australia. *Marine Geology*, **114**, 81–89.

Lees, B.G., Lu, Y. and Price, D.M. (1992) Thermoluminescence dating of dunes at Cape St. Lambert, East Kimberleys, northwestern Australia. *Marine Geology*, **106**, 131–139.

Lees, B.G., Stanner, J. and Price, D.M. (1995) Thermoluminescence dating of dune podzols at Cape Arnhem, north Australia. *Marine Geology*, **129**, 63–75.

Lelong, F., Tardy, Y., Grandin, G., Trescases, J.J. and Boulangé, B. (1976) Pedogenesis, chemical weathering, and processes of formation of some supergene ore deposits. In Wolf, K.H. (ed.) *Supergene and Surficial Ore Deposits; Textures and Fabrics*. Handbook of Stratabound and Stratiform Ore Deposits, 3. Elsevier, Amsterdam, 93–173.

Levitski, A., Filanovski, B., Bourenko, T., Tannenbaum, E., Bar-Am, G. and Dukhanin, A. (1995) A preliminary study of the diffusion sampling technique in locating buried mineralisation and an oil field in southern Israel. *Journal of Geochemical Exploration*, **54**, 73–86.

Lidmar-Bergstrom, K. (1989) Exhumed Cretaceous landforms in south Sweden. *Zeitschrift für Geomorphologie*, **72**, 21–40.

Lidmar-Bergstrom, K. (1995) Relief and saprolites through time on the Baltic Shield. *Geomorphology*, **12**, 45–61.

Lindsay, W.L. (1988) Solubility and redox equilibria of iron compounds in soils. In Stucki, J.W., Goodman, B.A. and Schwertmann, U. (eds) *Iron in Soils and Clay Minerals*. NATO ASI Series, 37–62.

Lindsay, W.L. (1979) *Chemical Equilibria in Soils*. John Wiley, New York, 449 pp.

Lintern, M.J. and Butt, C.R.M. (1998) Gold exploration using pedogenic carbonate (calcrete): In Eggleton, R.A. (ed.) *The State of the Regolith*; Second Australian Conference on Landscape Evolution and Mineral Exploration, Brisbane, November 1996. Geological Society of Australia Special Publication 20, 200–208.

Lintern, M.J. and Scott, K.M. (1990) The distribution of gold and other elements in soils and vegetation at Panglo, Western Australia. *CSIRO Australia, Exploration Geoscience Restricted Report*, **129R**, 96 pp.

Lintern, M.J. and Sheard, M.J. (1998) Silcrete – a potential new exploration sample medium: a case study from the Challenger gold deposit. *Quarterly Earth Resources Journal of Primary Industries and Resources South Australia*, **11**, 16–20.

Litchfield, W.H. (1969) Soil surfaces and sedimentary history near the MacDonnell Ranges, N.T. *CSIRO Australia Soil Publication No. 25*.

Litchfield, W.H. and Mabbutt, J.A. (1962) Hardpan in soils of semi-arid Western Australia. *Journal of Soil Science*, **13**, 148–159.

Livingstone, D.A. (1963) Chemical compositions of rivers and lakes. *US Geological Survey Professional Paper 440-G*.

Loganathan, P., Burau, R.G. and Fuerstenau, D.W. (1977) Influence of pH on the sorption of Co^{2+}, Zn^{2+} and Ca^{2+} by a hydrous manganese oxide. *Soil Science Society of America Journal*, **41**, 57–62.

Lopatin, G.V. (1952) Nanosy rek SSSR (obrazovanie i perenos). *Izvestiya Vaesoyznogo Geograficheskogo Obshohestva*, **14**. St Petersburg Nauka, 366 pp.

Loughnan, F.C. (1960) *The Origin, Mineralogy, and Some Physical Properties of the Commercial Clays of New South Wales*. University of New South Wales Geological Series No. 2, 348 pp.

Loughnan, F.C. (1969) *Chemical Weathering of the Silicate Minerals*. Elsevier, Amsterdam, 154 pp.

Lucas, Y., Boulet, R. and Chauvel, A. (1988) Intervention simultanée des phénomènes d'enforcement vertical et de transformation latérale dans la mise en place de systèmes sols ferrallitiques-podzols de l'Amazonie Brésilienne. *C.R. Académie des Sciences Paris, Série II*, **306**, 1395–1400.

Lucas, Y., Boulet R., Chauvel, A. and Veillon, L. (1987) Systèmes sols ferrallitiques-podzols en région amazonienne. In Righi, D. and Chauvel, A. (eds) *Podzols et Podzolisation*. Comptes Rendus de la Table Ronde Internationale, 1986, Association Française pour l'Étude du Sol, INRA/ORSTOM, Poitiers, 53–65.

Lumb, P. (1975) Slope failures in Hong Kong. *Quarterly Journal of Engineering Geology*, **8**, 31–65.

Ma Chi (1996) *The ultra-structure of kaolin*. Ph.D. thesis, Australian National University, Canberra.

Mabbutt, J.A. (1984) Landforms of the Australian deserts. In El Baz, F. (ed.) *Deserts and Arid Lands*. Nijhoff, The Hague, 79–94.

Macaire, J.J., Cocirta, C., Karrat, L. and Perruchot, A. (1994) Basalt weathering and fluvial sedimentary particles; comparison of two watersheds in the middle Atlas Mountains, Morocco. *Journal of Sedimentary Research, Section A: Sedimentary Petrology and Processes*, **64**, 490–499.

McArthur, J.M., Turner, J.V., Lyons, W.B., Osborn, A.O. and Thirlwall, M.F. (1991) Hydrochemistry on the Yilgarn Block, Western Australia: ferrolysis and mineralisation in acidic brines. *Geochimica et Cosmochimica Acta*, **55**, 1273–1288.

McArthur, W.H. and Bettenay, E. (1960) The development and distribution of the soils of the Swan Coastal Plain Western Australia. *CSIRO Australian Soils Publication No. 16.*

McDonald, R.C., Isbell, R.F., Speight, J.G., Walker, J. and Hopkins, M.S. (1990) *Australian Soil and Land Survey Field Handbook*, 2nd edn. Melbourne, Inkata Press, Melbourne, 198 pp.

McEwan-Mason, J.R.C. (1991) The late Cainozoic magnetostratigraphy and preliminary palynology of Lake George, New South Wales. In Williams, M.A.J., De Deckker, P. and Kershaw, A.P. (eds) *The Cainozoic in Australia: a reappraisal of the evidence*. Geological Society of Australia Special Publication No. 18, 195–209.

McFarlane, M.J. (1976) *Laterite and Landscape*. Academic Press, London, 151 pp.

McFarlane, M.J. (1983) Laterites. In Goudie, A.S. and Pye, K. (eds) *Chemical Sediments and Geomorphology*. Academic Press, London, 7–58.

McFarlane, M.J. and Bowden, D.J. (1992) Mobilization of aluminium in the weathering profiles of the African surface in Malawi. *Earth Surface Processes and Landforms*, **17**, 789–805.

McFarlane, M.J., Bowden, D.J. and Giusti, L. (1994) The behaviour of chromium in weathering profiles associated with the African surface in parts of Malawi. In Robinson, D.A. and Williams, R.B.G. (eds) *Rock Weathering and Landform Evolution*. John Wiley, Chichester, 321–338.

McFarlane, M.J., Ringrose, S., Giusti, L. and Shaw, P.A. (1995) The origin and age of karstic depressions in the Darwin–Koolpinyah area of the Northern Territory of Australia. In Brown, A.G. (ed.) *Geomorphology and Groundwater*. John Wiley, Chichester, 93–120.

McGowran, B. (1978) The Tertiary of Australia; stratigraphic sequences and episodic geohistory. *Proceedings of the Regional Congress on Geology, Mineral and Energy Resources of Southeast Asia (GEOSEA)*, **3**, 73–80.

Machette, M.N. (1985) Calcic soils of the southwestern United States. In Weide, D.L. and Faber, M.L. (eds) *Soils and Quaternary geology of the southwestern United States. Geological Society of America Special Paper*, **203**, 1–21.

McKenzie, R.M. (1980) The adsorption of lead and other heavy metals on oxides of manganese and iron. *Australian Journal of Soil Resources*, **18**, 61–73.

Mackey, T., Lawrie, K., Wilkes, P., Munday, T., de Souza Kovacs N., Chan, R., Gibson, D., Chartres, C. and Evans, R. (2000) Paleochannels near West Wyalong, New South Wales: a case study in delineation and modelling using aeromagnetics. *Exploration Geophysics*, **31**, 1–7.

McNally, G.H. (1977) Coober Pedy town water supply – groundwater prospects and completion report, Stuart Range Bores 7–10. South Australian Department of Mines Unpublished Report No. 77/77.

McNally, G.H. (1992). Engineering geology of duricrusts. In Branagan, D.F. and Williams, K.F. (eds) *Papers of the Fifth Edgeworth David Day Symposium*, September 1992. The Edgeworth David Society, Sydney. University of Sydney, 89–96.

McNally, G.H. and Wilson, I.R. (1995) Silcretes of the Mirackina Palaeochannel, Arckaringa, South Australia. *Australian Geological Survey Organisation Journal of Australian Geology and Geophysics*, **16**, 295–301.

McSaveney, M.J. (1992) Weathering-rind dating for grey sandstones of the Torlesse Supergroup; the version of 1992. *Geological Society of New Zealand Miscellaneous Publication*, **63A**, 106.

McTainsh, G. and Broughton, W.C. (1993) *Land Degradation Processes in Australia*. Longman Cheshire, Melbourne, 389 pp.

McTainsh, G.H., Nickling, W.G. and Lynch, A.W. (1997) Dust deposition and particle size in Mali, West Africa. *Catena*, **29**, 307–322.

Macumber, P.G. (1978) Permian glacial deposits, tectonism, and the evolution of the Loddon Valley. *Mining Geology and Energy Journal of Victoria*, **7**, 34–36.

Macumber, P.G. (1991) *Interaction between Groundwater and Surface Systems in Northern Victoria*. Dept of Conservation and Environment, 345 pp.

Magee, J.W., Bowler, J.M., Miller, G.H. and Williams, D.L.G. (1995) Stratigraphy, sedimentology, chronology and palaeohydrology of Quaternary lacustrine deposits at Madigan Gulf, Lake Eyre, South Australia. In Chivas, A.R. (ed.) *Arid Zone Palaeoenvironments. Palaeogeography, Palaeoclimatology, Palaeoecology*. **113**, 3–42.

Maignien, R. (1966) *Review of Research on Laterites*. Natural Resources Research IV, UNESCO, Paris, 148 pp.

Maignien, R. (1958) Le cuirassement des sols en Guinée, Afrique Occidentale. *Mémoire du Service de la Carte Géologique d'Alsace et de Lorraine*, **16**, Strasbourg, 239 pp.

Manceau, A. and Charlet, L. (1992) X-ray absorption spectroscopic study of the sorption of Cr(III) at the oxide/water interface. I. Molecular mechanism of Cr(III) oxidation on Mn oxides. *Journal of Colloid and Interface Science*, **148**, 425–442.

Manceau, A., Llorca, S. and Calas, G. (1987) Crystal chemistry of cobalt and nickel in lithiophorite and asbolane from New Caledonia. *Geochimica et Cosmochimica Acta*, **51**, 105–113.

Mann, A.W. (1998) Oxidised gold deposits: relationships between oxidation and relative position of the water-table. *Australian Journal of Earth Sciences*, **45**, 97–108.

Mann, A.W. and Deutscher, R.L. (1980) Solution geochemistry of lead and zinc in water containing carbonate, sulphate and chloride ions. *Chemical Geology*, **29**, 293–311.

Mann, A.W. and Ollier, C.D. (1985) Chemical diffusion and

ferricrete formation. In Jungerius, P.D. (ed.) *Soils and Geomorphology. Catena Supplement*, **6**, 151–157.

Mann, A.W., Birrell, W.D., Lawrance, L.M., Mann, A.T. and Gardner, K.R. (1993) The new Mobile Metal Ion approach to the detection of buried mineralisation. In Williams, P. R. and Haldane, J. A. (eds) *Kalgoorlie '93, an International Conference on Crustal Evolution, Metallogeny and Exploration of the Eastern Goldfields*, Australian Geological Survey Organisation, Record 1993/54, 223–225.

Marker, A. and De Oliveira, J.J. (1990) The formation of rare earth element scavenger minerals in weathering products derived from alkaline rocks of SE-Bahia, Brazil. *Chemical Geology*, **84**, 373–374.

Markovics, G. (1977) *Chemistry of weathering of the Toorongo Granodiorite, Mt. Baw Baw, Vic.* B.Sc. with Hons. thesis, Latrobe University.

Martini, I.P. and Chesworth, W. (1992) *Weathering, Soils and Paleosols*. Elsevier, Amsterdam, 618 pp.

Marzo, M. and Puigdefabregas, C. (1993) *Alluvial Sedimentation*. Blackwell Scientific, Oxford, 586 pp.

Mason, M.G. (1975) Groundwater near Giddi Giddinna Creek, northeast of Coober Pedy, South Australia. *Quarterly Geological Notes – South Australian Geological Survey*, **56**, 2–6.

Mathieu, D., Bernat, M. and Nahon, D. (1995) Short-lived U and Th isotope distribution in a tropical laterite derived from granite (Pitinga river basin, Amozonia, Brazil): applications to assessment of weathering rate. *Earth and Planetary Science Letters*, **136**, 703–714.

May, H.M., Helmke, P.A. and Jackson, M.L. (1979) Gibbsite solubility and thermodynamic properties of hydroxy-aluminum ions in aqueous solution at 25°C. *Geochimica et Cosmochimica Acta*, **43**, 861–868.

Mazzucchelli, R.H., Chapple, B.E.E. and Lynch, J.E. (1980) Northern Yorke Peninsular Cu, Gawler Block, SA. *Journal of Geochemical Exploration*, **12**, 203–207.

Measures, C.I. and Edmond, J.M. (1983) Beryllium in the Orinoco and Amazon Basins. *EOS Transactions, AGU*, **64** (45), 698.

Merrill, G.P. (1897) *A Treatise on Rocks, Rock Weathering and Soils*. Macmillan, New York, 411 pp.

Meunier, A. and Velde, B. (1979) Weathering mineral facies in altered granites: the importance of local small-scale equilibria. *Mineralogical Magazine*, **43**, 261–268.

Meybeck, M. (1976) Total mineral dissolved transport by world major rivers. *Hydrological Sciences Bulletin*, **21**, 265–284.

Meybeck, M. (1979) Concentrations des eaux fluviales éléments majeurs et apports en solution aux océans. *Revue de Géologie Dynamique et de Géographie Physique*, **21**, 215–246.

Miall, A.D. (1990) *Principles of Sedimentary Basin Analysis*. Springer-Verlag, New York, 668 pp.

Miall, A.D. (1992) Alluvial deposits. In Walker, R.G. and James, N.P. (eds) *Facies models; response to sea level change*. Geological Association of Canada, St Johns, 119–142.

Miall, A.D. (1996) *The Geology of Fluvial Deposits: sedimentary facies, basin analysis, and petroleum geology*. Springer, Berlin, 582 pp.

Middleburg, J.J., van der Weijden, C.H. and Woittiez, J.R.W. (1988) Chemical processes affecting the mobility of major, minor and trace elements during weathering of granitic rocks. *Chemical Geology*, **68**, 253–273.

Middleton, N.J. (1997) Desert dust. In Thomas, D.S.G. (ed.) *Arid Zone Geomorphology: process, form and change in drylands*. John Wiley, Chichester, 413–436.

Milburn, D. and Wilcock, S. (1994) The Kunwarara magnesite deposit, central Queensland. In Holcombe, R.J., Stephens, C.J. and Fielding, C.R. (eds) *Field Conference, Capricorn Region, Central Queensland Coast*. Geological Society of Australia Inc., 99–107.

Miller, D.S. and Duddy, I.R. (1989) Early Cretaceous uplift and erosion of the northern Appalachian Basin, New York, based on apatite fission track analysis. *Earth and Planetary Science Letters*, **93**, 35–49.

Miller, W.R. and Drever, J.I. (1977) Chemical weathering and related controls on surface water chemistry in the Absaroka Mountains, Wyoming. *Geochimica et Cosmochimica Acta*, **41**, 1693–1702.

Milliman, J.D. and Meade, R.H. (1983) World-wide delivery of river sediment to the oceans. *Journal of Geology*, **91**, 1–21.

Millot, G. (1970) *Geology of Clays*. Springer-Verlag, New York, 429 pp.

Millot, R., Gaillardet, J., Dupré, B. and Allegre, C.T. (1999) Global silicate weathering rates in the Mackenzie River basin, Northwest Territories, Canada: influence on dissolved organic matter. *European Union of Geosciences Journal of Conference Abstracts*, **4**. Cambridge Publications, UK.

Milne, G. (1935) Some suggested units of classification and mapping, particularly for east African soils. *Soil Research*, **4**, 183–198.

Milne, G. (1947) A soil reconnaissance journey through parts of Tanganyika Territory, December 1935 to February 1936. *Journal of Ecology*, **35**, 192–265.

Milnes, A.R. (1983) Silicification in Cainozoic landscapes of arid central Australia. *Abstracts – Geological Society of Australia*, **9**, 214–215.

Milnes, A.R. and Fitzpatrick, R.W. (1989) Titanium and zirconium minerals. In Dixon, J.B. and Weed, S.B. (eds) *Minerals in Soil Environments*, 2nd edn. Soil Science Society of America, Madison, Wis., 1131–1205.

Milnes, A.R. and Hutton, J.T. (1974) The nature of microcryptocrystalline titania in 'silcrete' skins from the Beda Hill area of South Australia. *Search*, **5**, 153–154.

Milnes, A.R. and Hutton, J.T. (1983) Calcretes in Australia. In *Soils: an Australian viewpoint*. CSIRO, Melbourne, 119–162.

Milnes, A.R. and Twidale, C.R. (1983) An overview of silicification in Cainozoic landscapes of arid central and southern Australia. *Australian Journal of Soil Research*, **21**, 387–410.

Milnes, A.R., Bourman, R.P. and Northcote, K.H. (1985) Field relationships of ferricretes and weathered zones in southern South Australia; a contribution to 'laterite' studies in Australia. *Australian Journal of Soil Research*, **23** (4), 441–465.

Mineral Industry Quarterly, South Australia (1988) Heavy mineral sand exploration in South Australia. South Australia Department of Mines and Energy, Adelaide, **51**, 8–9.

Mongelli, G. and Moresi, M. (1990) Biotite-kaolinite transformation in a granitic saprolite of the serre (Calabria,

Southern Italy). *Mineralogica et Petrographica Acta*, **XXXIII**, 273–281.

Mongelli, G. (1993) REE and other trace elements in a granitic weathering profile from 'Serre', southern Italy. *Chemical Geology*, **103**, 17–25.

Moore, C.L. (1996a) Evaluation of regolith development and element mobility during weathering using the isocon technique. *Geological Society of Australia Special Publication*, **20**, 141–147.

Moore, C.L. (1996b) *Process of chemical weathering of selected Cainozoic eastern Australian basalts*. Ph.D. Thesis, Australian National University, Canberra.

Moore, M.E., Gleadow, A.J. and Lovering, J.F. (1986) Thermal evolution of rifted continental margins: new evidence from fission tracks in basement apatites from southeastern Australia. *Earth and Planetary Science Letters*, **78**, 255–270.

Morris, D.A. and Johnson, A.I. (1967) Summary of hydrologic and physical properties of rock and soil materials, as analyzed by the hydrologic laboratory of the U.S. Geological Survey. *United States Geological Survey Water-Supply Paper 1839–D*.

Moss, A.J. (1972) Initial fluviatile fragmentation of granitic quartz. *Journal of Sedimentary Petrology*, **42**, 905–916.

Moss, A.J. and Green, P. (1975) Sand and silt grains; predetermination of their formation and properties by microfractures in quartz. *Journal of the Geological Society of Australia*, **22**, 485–495.

Moss, A.J., Walker, P.H. and Hutka, J. (1973) Fragmentation of granitic quartz in water. *Sedimentology*, **20**, 489–511.

Mosser, C. and Zeegers, H. (1988) The mineralogy and geochemistry of two copper-rich weathering profiles in Burkina Faso, West Africa. *Journal of Geochemical Exploration*, **30**, 145–166.

Mosser, C., Petit, S., Parisot, J-C., Decarreau, A. and Mestdagh, M. (1990). Evidence of Cu in octahedral layers of natural and synthetic kaolinites. *Chemical Geology*, **84**, 281–282.

Moulton, K.L. and Berner, R.A. (1999) Quantitative effects of plants on weathering: forest sinks for carbon. In Armannsson, H. (ed.) *Geochemistry of the Earth's Surface*. Balkema, Rotterdam, 57–60.

Mulcahy, M.J. (1967) Landscapes, laterites and soils in southwestern Australia. In Jennings, J.N. and Mabbutt, J.A. (eds) *Landform Studies from Australia and New Guinea*. Australian National University Press, Canberra, 211–230.

Nahon, D. and Merino, E. (1996) Pseudomorphic replacement versus dilation in laterites: petrographic evidence, mechanisms, and consequences for modelling. *Journal of Geochemical Exploration*, **57**, 217–225.

Nahon, D. and Tardy, Y. (1992) The ferruginous laterites. In Butt, C.R.M. and Zeegers, H. (eds) *Regolith Exploration Geochemistry in Tropical and Subtropical Terrains*. Elsevier, Amsterdam, 41–55.

Nahon, D.B. (1991) *Introduction to the Petrology of Soils and Chemical Weathering*. John Wiley, New York, 313 pp.

Nanson, G.C., Chen, X.Y. and Price, D.M. (1995) Aeolian and fluvial evidence of changing climate and wind patterns during the past 100 ka in the Western Simpson Desert. *Palaeogeography, Palaeoclimatology, Palaeoecology*, **113**, 87–102.

Nanson, G.C., Rust, B.R. and Taylor, G. (1986) Coexistent mud braids and anastomosing channels in an arid-zone river; Cooper Creek, central Australia. *Geology*, **14**, 175–178.

Nash, D.J., Thomas, D.S.G. and Shaw, P.A. (1994) Siliceous duricrusts as palaeoclimatic indicators: evidence from the Kalahari Desert of Botswana. *Palaeogeography, Palaeoclimatology, Palaeoecology*, **112**, 279–295.

Naylor, H., Turner, P., Vaughan, D.J. and Fallick, A.E. (1989) The Cherty Rock, Elgin: a petrographic and isotopic study of a Permo-Triassic calcrete. *Geological Journal*, **24**, 205–221.

Nesbitt, H.W. (1979) Mobility and fractionation of rare earth elements during weathering of granodiorite. *Nature*, **279**, 206–210.

Nesbitt, H.W. and Markovics, G. (1997) Weathering of granodioritic crust, long-term storage of elements in weathering profiles, and petrogenesis of siliciclastic sediments. *Geochimica et Cosmochimica Acta*, **61** (8), 1653–1670.

Nesbitt, H.W. and Young, G.M. (1982) Early Proterozoic climates and plate motions inferred from major element chemistry of lutites. *Nature*, **299**, 715–717.

Nesbitt, H.W. and Young, G.M. (1989) Formation and diagenesis of weathering profiles. *Journal of Geology*, **97**(2), 129–147.

Netterberg, F. (1978) Dating and correlation of calcretes and other pedocretes. *Transactions of the Geological Society of South Africa*, **81**, 379–391.

Netterberg, F. (1980) Geology of southern African calcretes; 1. Terminology, description, macrofeatures, and classification. *Transactions of the Geological Society of South Africa*, **83**, 255–283.

Netterberg, F. (1985) Pedocretes. In Brink, A.B.A. (ed.) *Engineering Geology of Southern Africa*. **4**, Building Publications (CSIR Reprint RR 430), Silverton. 286–307.

Nicholas, D.J.D. and Egan, A.R. (eds) (1975) *Trace Elements in Soil–Plant–Animal Systems*. Academic Press, New York, 417 pp.

Nicholson, K. (1992) Contrasting mineralogical–geochemical signatures of manganese oxides: guides to metallogenesis. *Economic Geology*, **87**, 1253–1264.

Nickel, E.H. and Daniels, J.L. (1985) Gossans. In Wolf K.H. (ed.) *Handbook of Strata-bound and Stratiform Ore Deposits*, **13**, Elsevier, 261–340.

Nordstrom, D.K. (1982) The effect of sulfate on aluminum concentration in natural waters: some stability relations in the system Al_2O_3–SO_3–H_2O at 298K. *Geochimica et Cosmochimica Acta*, **46**, 681–692.

Nordstrom, D.K. and Southam, G. (1997) Geomicrobiology of sulfide mineral oxidation. In Banfield, J.F. and Nealson, K.H. (eds) *Geomicrobiology: interactions between microbes and minerals. Reviews in Mineralogy*, **35**, 361–390. Mineralogical Society of America.

Norrish, K. (1972) Factors in the weathering of mica to vermiculite. *Proc. 1972 International Clay Conference*, 417–432.

Norrish, K. (1975) The geochemistry and mineralogy of trace elements. In Nicholas, D.J.D. and Egan, A.R. (eds) *Trace Elements in Soil–Plant–Animal Systems*. Academic Press, New York, 55–81.

Norrish, K. and Pickering, J.G. (1983) Clay minerals. In *Soils, an Australian viewpoint*. CSIRO, Melbourne, 281–308.

Norrish, K. and Rosser, H. (1983) Mineral phosphate. In *Soils: an Australian viewpoint*, CSIRO, Melbourne, 335–361.

Nott, J. (1995) Long-term landscape evolution on Groote Eylandt, Northern Territory. *Australian Geological Survey Organisation Journal of Australian Geology and Geophysics*, **16**, 303–307.

Nott, J., Young, R. and McDougall, I. (1996) Wearing down, wearing back, and gorge extension in the long-term denudation of a highland mass; quantitative evidence from the Shoalhaven Catchment, Southeast Australia. *Journal of Geology*, **104**, 224–232.

Nye, P.H. (1955) Some soil-forming processes in the humid tropics. Part II. The development of the upper slope member of the catena. *Journal of Soil Science*, **6**, 51–62.

Okumura, S. (1986) Weathering process of Nabari gabbroic body. 2. Congruent dissolution of plagioclase. *Journal of the Japanese Association of Mineralogists, Petrologists and Economic Geologists*, **81**(3), 116–128.

Okumura, S. (1988) Geochemistry and mineralogy of weathering process of Nabari Gabbroic Body, Southwest Japan. *Journal of Geosciences Osaka City University*, **31**, 123–171.

Oliveira, S.M.B. and Campos, E.G. (1991) Gold-bearing iron duricrust in central Brazil. *Journal of Geochemical Exploration*, **41**, 309–323.

Ollier, C. and Pain, C. (1996) *Regolith, Soils and Landforms*. John Wiley, Chichester, 306 pp.

Ollier, C.D. (1965) Some features of granite weathering in Australia. *Zeitschrift für Geomorphologie*, **4**, 43–52.

Ollier, C.D. (1969) *Weathering*, 1st edn. Longman, London, 304 pp.

Ollier, C.D. (1984) *Weathering*, 2nd edn. Longman Inc., London, 270 pp.

Ollier, C.D. (1991a) *Ancient Landforms*. Belhaven Press, London, 233 pp.

Ollier, C.D. (1991b) An hypothesis about antecedent and reversed drainage. *Geografia Fisica e Dinamica Quaternaria*, **14**, 243–246.

Ollier, C.D. (1995) An unreliable history of regolith mapping. In McQueen, K.G. and Craig, M.A. (eds) *Developments and New Approaches in Regolith Mapping*. Centre for Australian Regolith Studies, Occasional Publication 3, 1–16.

Ollier, C.D. and Galloway, R.W. (1990) The laterite profile, ferricrete and unconformity. *Catena*, **17**, 97–109.

Ollier, C.D. and Pain, C.F. (1995) Landscape evolution and tectonics in southeastern Australia; reply. *Australian Geological Survey Organisation Journal of Australian Geology and Geophysics*, **16**, 325–331.

Ollier, C.D. and Rajaguru, S.N. (1989) Laterite of Kerala (India). *Geografia Fisica e Dinamica Quaternaria*, **12**, 27–33.

Ollier, C.D., Chan, R.A., Craig, M.A. and Gibson, D.L. (1988) Aspects of landscape history and regolith in the Kalgoorlie region, Western Australia. *Bureau of Mineral Resources Journal of Australian Geology and Geophysics*, **10**, 309–321.

Östhols, E., Bruno, J. and Grenthe, I. (1994) On the influence of carbonate on mineral dissolution: III. The solubility of microcrystalline ThO_2 in CO_2–H_2O media. *Geochimica et Cosmochimica Acta*, **58**, 613–623.

Ostwald, J. (1992) Genesis and paragenesis of the tetravalent manganese oxides of the Australian continent. *Economic Geology*, **87**, 1237–1252.

O'Sullivan, P.B., Brown, R.W., Kohn, B.P. and Gleadow, A.J.W. (1998) The palaeoplain model for passive margins – does it really work for the SE Australian margin? In Britt, A.F. and Bettenay, L. (eds) *Regolith '98, New Approaches to an Old Continent, Program and Abstracts*. CRC LEME, Perth, 5.

Pain, C., Chan, R., Craig, M., Hazell, M., Kamprad, J. and Wilford, J. (1991) *RTMAP – BMR Regolith Database Field Handbook*. Minerals and Landuse Program, Series 1. Bureau of Mineral Resources, Geology and Geophysics, Record 1991/29.

Pain, C.F. (1986) Scarp retreat and slope development near Picton, New South Wales, Australia. *Catena*, **13**, 227–239.

Pain, C.F. and Ollier, C.D. (1995a) Inversion of relief; a component of landscape evolution. *Geomorphology*, **12**, 151–165.

Pain, C.F. and Ollier, C.D. (1995b) Regolith stratigraphy: principles and problems. *Australian Geological Survey Organisation Journal of Australian Geology and Geophysics*, **16**, 197–202.

Pain, C.F. and Wilford, J. (1994) *Ebagoola Regolith-Landforms Map*. Australia 1:250 000 Regolith-Landform Series. Sheet SD 54-12. Australian Geological Survery Organization, Canberra.

Pain, C.F., Wilford, J.R. and Dohrenwend, J.C. (1994) Regolith-landforms of the Ebagoola 1 : 250 000 sheet area (SD 54-12), North Queensland. *Australian Geological Survey Organisation, Record, 1994/7*.

Palmer, M.R. and Edmond, J.M. (1993) Uranium in river water. *Geochimica et Cosmochimica Acta*, **57**, 4947–4955.

Palmer, M.R. and Swihart, G.H. (1996) Boron isotope geochemistry: an overview. In Grew, E.S. and Anovitz, L.M. (eds) *Boron: Mineralogy, Petrology and Geochemistry. Reviews in Mineralogy* **33**. Mineralogical Society of America, 709–744.

Paquet, H., Colin, F., Duplay, J., Nahon, D. and Millot, G. (1987) Ni, Mn, Zn, Cr-smectites, early and effective traps for transition elements in supergene ore deposits. In Rodrigez-Clemente R, and Tardy, Y. (eds) *Geochemistry and Mineral Formation in the Earth Surface*, 221–229. Consejo Superior de Investigaciones Cientificas, Centre National de la Recherche Scientifique, 893 pp.

Parc, S., Nahon, D., Tardy, Y. and Viellard, P. (1989) Estimated solubility products and fields of stability for cryptomelane, nsutite, birnessite and lithiophorite based on natural lateritic weathering sequences. *American Mineralogist*, **74**, 466–475.

Parfitt, R.L. (1978) Anion adsorption by soils and soil materials. *Advances in Agronomy*, **30**, 1–50.

Parfitt, R.L. (1990) Allophane in New Zealand – a review. *Australian Journal of Soil Research*, **28**, 343–60.

Parfitt, R.L., Russell, M. and Orbell, G.E. (1983) Weathering sequence of soils from volcanic ash involving allophane and halloysite, New Zealand. *Geoderma*, **29**, 41–57.

Parfitt, R.L., Saigusa, M. and Cowie, J.D. (1984) Allophane and halloysite formation in a volcanic ash bed under different moisture conditions. *Soil Science*, **138**, 360–364.

Parham, W.E. (1969) Formation of halloysite from feldspar: low temperature artificial weathering vs natural weathering. *Clays and Clay Minerals*, **17**, 13–22.

Parianos, J.M. (1994) Geology of the Brolga Ni–Co laterite deposit central Queensland. In Holcombe, R.J., Stephens, C.J. and Fielding, C.R. (eds) *Field Conference, Capricorn Region, Central Queensland Coast*. Geological Society of Australia, Inc., 108–117.

Park, C.F. Jr and MacDiarmid, R.A. (1975) *Ore Deposits*. W.H. Freeman and Co., San Francisco, 529 pp.

Paton, T.R. and Williams, M.A.J. (1972) The concept of laterite. *Annals of the Association of American Geographers*, **62**, 42–56.

Paton, T.R., Humphreys, G.S. and Mitchell, P.B. (1995) *Soils: a new global view*. University College London Press, London, 213 pp.

Pauling, L. (1930) The structure of micas and related minerals. *Proceedings of the National Academy of Sciences*, **16**, 123–129.

Pavich, M.J. (1985) Appalachian piedmont morphogenesis: weathering, erosion and Cenozoic uplift. In Morisawa, M. and Hack, J.T. (eds) *Tectonic Geomorphology*. Allen and Unwin, London, 299–319.

Pédro, G. (1966) Essai sur la caractérisation géochimique des différents processus zonaux résultant de l'altération des roches superficielles (cycle aluminosilicique). *Comptes Rendus Hebdomadaires des Séances de l'Académie des Sciences, Série D: Sciences Naturelles*, **262** (17), 1828–1831.

Pédro, G. (1968) Distribution des principaux types d'altération chimique à la surface du globe. *Revue de Géographie Physique et de Géologie Dynamique*, **19**, 457–470.

Pédro, G. and Delmas, A.B. (1970) Principes géochimiques de la distribution des éléments-traces dans les sols. *Annales Agronomiques, Paris*, **21**, 483–518.

Pell, S.D. and Chivas, A.R. (1995) Surface features of sand grains from the Australian Continental Dunefield. *Palaeogeography, Palaeoclimatology, Palaeoecology*, **113**, 119–132.

Pell, S.D., Chivas, A.R. and Williams, I.S. (1999) Great Victoria Desert: development and sand provenance. *Australian Journal of Earth Sciences*, **46**, 289–299.

Pell, S.D., Chivas, A.S. and Williams, I.S. (2000) The Simpson, Strzelecki and Tirari Deserts: development and sand provenance. *Sedimentary Geology*, **130**, 107–130.

Pell, S.D., Williams, I.S. and Chivas, A.R. (1997) The use of protolith zircon-age fingerprints in determining the protosource areas for some Australian dune sands. *Sedimentary Geology*, **109**, 233–260.

Pels, S. (1971) River systems and climatic changes in southeastern Australia. In Mulvaney, D.J. and Golson, J. (eds) *Aboriginal Man and Environment in Australia*. Australian National University Press, Canberra, 38–46.

Penck, W. (1924) *Die Morphologische Analyse: Ein Kapitel der Physikalischen Geologie*. Geographische Abhandlungungen, **2**, Reihe, Heft, 2, Stuttgart.

Pereira, L.C.J., Waerenborgh, J.C., Figueiredo, M.O., Prudêncio, M.I., Gouveia, M.A., Silva, T.P., Morgado, I. and Lopes, A. (1993) A comparative study of biotite weathering from two different granitic rocks. *Chemical Geology*, **107**, 301–306.

Pettijohn, F.J. (1963) Chemical composition of sandstones; excluding carbonate and volcanic sands. Chapter S, in *Data of Geochemistry*, 6th edn, US Geological Survey Professional Paper 440-5, S1–S21.

Phillips, F.M., Zreda, M.G., Grosse, J.C., Klein, J., Evenson, E.B., Hall, R.D., Chadwick, O.A. and Sharma, P. (1997) Cosmogenic ^{36}Cl and ^{10}Be ages of Quaternary glacial and fluvial deposits of the Wind River Range, Wyoming. *Bulletin of the Geological Society of America*, **109**, 1453–1463.

Phillips, G.N. and Hughes, M.J. (1996) The geology and gold deposits of the Victorian Gold Province. *Ore Geology Reviews*, **11**, 255–302.

Pillans, B. (1987) Lake shadows; aeolian clay sheets associated with ephemeral lakes in basalt terrain, southern New South Wales. *Search*, **18**, 313–315.

Pillans, B. (1997) Soil development at a snail's pace: evidence from a 6 Ma soil chronosequence on basalt in north Queensland, Australia. *Geoderma*, **80**, 117–128.

Pillans, B. (1998) *Regolith Dating Methods: a guide to numerical dating techniques*. CRC LEME, Perth, 30 pp.

Pillans, B., Tonui, E. and Idnurm, M. (1999) Palaeomagnetic dating of weathered regolith at Northparkes Mine, NSW. In Taylor, G. and Pain, C. (eds) *Regolith '98, New Approaches to an Old Continent, Proceedings*. CRC LEME, Perth, 237–242.

Pillans, B., Tonui, E. and Idnurm, M. (1999) Palaeomagnetic dating of weathered regolith. In Taylor, G. and Pain, C.F. (eds) *New Approaches to an Old Continent*, Proceedings of Regolith '98 Conference, Kalboorlie. CRC LEME, Perth, 237–242.

Plafker, G. and Ericksen, G.E. (1978) Nevados Huascaran avalanches, Peru. In Voight, B. (ed.) *Rockslides and Avalanches*, Vol. 1. Elsevier, Amsterdam, 277–314.

Playford, P.E., Cockbain, A.E. and Low, G.H. (1976) Geology of the Perth Basin, Western Australia. *Bulletin of the Geological Survey of Western Australia* **124**.

Pollack, H.N. and Chapman, D.S. (1993) Underground records of changing climate. *Scientific American*, June, 16–22.

Pons, L.J., van Breeman, N. and Driessen, P.M. (1982) Physiography of coastal sediments and development of potential soil acidity. In Kittrick, J.A., Fanning, D.S. and Hossner, L.R. (eds) *Acid Sulfate Weathering*. Soil Science Society of America Special Publication, **10**, 1–18.

Pracejus, B. and Bolton, B.R. (1992) Geochemistry of supergene manganese oxide deposits, Groote Eylandt, Australia. *Economic Geology*, **87**, 1310–1335.

Pracejus, B., Bolton, B.R. and Frakes, L.A. (1988) Nature and development of supergene manganese deposits, Groote Eylandt, Northern Territory, Australia: a model for modern and ancient weathering environments. *Ore Geology Reviews*, **4**, 71–98.

Prescott, J.A. and Pendleton, R.L. (1952) Laterite and lateritic soils. *Commonwealth Bureau of Soil Science Technical Communication*, **47**.

Price, G.D., Valdes, P.J. and Sellwood, B.W. (1997) Prediction of modern bauxite occurrence: implications for climate reconstruction. *Palaeogeography, Palaeoclimatology, Palaeoecology*, **131**, 1–13.

Price, R.C., Gray, C.M., Wilson, R.E., Frey, F.A. and Taylor, S.R. (1991) The effects of weathering on rare-earth element, Y and Ba abundances in Tertiary basalts from southeastern Australia. *Chemical Geology*, **93**, 245–265.

Probst, A., Dambrine, E., Viville, D. and Fritz, B. (1990) Influence of acid atmospheric inputs on surface water chemistry and mineral fluxes in a declining spruce stand within a small granitic catchment (Vosges Massif, France). *Journal of Hydrology*, **116**, 101–124.

Proust, D. (1982) Supergene alteration of metamorphic chlorite in an amphibolite from Massif Central, France. *Clay Minerals*, **17**, 159–173.

Proust, D. and Meunier, A. (1989) Phase equilibria in weathering processes. In *Weathering: its products and deposits*. Volume I, *Processes*. Theophrastus Publications, Greece, 121–145.

Proust, D. and Velde, B. (1978) Beidellite crystallization from plagioclase and amphibole precursors: local and long-range equilibrium during weathering. *Clay Minerals*, **13**, 199–209.

Pye, K. (1987) *Aeolian Dust and Dust Deposits*. Academic Press, London, 334 pp.

Pye, K. (1993) *The Dynamics and Environmental Context of Aeolian Sedimentary Systems*. Geological Society, London, 332 pp.

Pye, K. (1994) *Sediment Transport and Depositional Processes*. Blackwell Scientific, Oxford, 397 pp.

Pye, K. and Mazzullo, J. (1994) Effects of tropical weathering on quartz grain shape; an example from northeastern Australia. *Journal of Sedimentary Research, Section A: Sedimentary Petrology and Processes*, **64**, 500–507.

Radke, U. and Brückner, H. (1991) Investigation on age and genesis of silcretes in Queensland (Australia) – preliminary results. *Earth Surface Processes and Landforms*, **16**, 547–554.

Raymahashay, B.C. (1968) A geochemical study of rock alteration by hot springs in the Paint Pot Hill area, Yellowstone Park. *Geochimica et Cosmochimica Acta*, **32**, 499–522.

Reading, H.G. (1996) *Sedimentary Environments: processes, facies, and stratigraphy*. Blackwell Scientific, Cambridge, Mass., 688 pp.

Reedman, J.H. (1979) *Techniques in Mineral Exploration*. Applied Science Publishers, London, 533 pp.

Reiche, P. (1943) Graphic representation of chemical weathering. *Journal of Sedimentary Petrology*, **13** (2), 58–68.

Reiche, P. (1950) A survey of weathering processes and products. *University of New Mexico Publications in Geology*, No. 3, 95 pp.

Retallack, G.J. (1990) *Soils of the Past: an introduction to paleopedology*. Unwin Hyman, Boston, 520 pp.

Retallack, G.J. (1997) Palaeosols in the Upper Narrabeen Group of New South Wales as evidence of early Triassic palaeoenvironments without exact analogues. *Australian Journal of Earth Sciences*, **44**, 185–201.

Ride, W.D.L., Taylor, G., Walker, P.H. and Davis, A.C. (1989) Zoological history of the Australian Alps – the mammal fossil-bearing deposits of the Monaro. In Good, R. (ed.) *The Scientific Significance of the Australian Alps*. Australian Alps Liaison Committee and Australian Academy of Science, 79–110.

Rivers, C.J., Eggleton, R.A. and Beams, S.D. (1995) Ferricretes and deep weathering profiles of the Puzzler Walls, Charters Towers, north Queensland. *Australian Geological Survey Organisation Journal of Australian Geology and Geophysics*, **16**, 203–211.

Robertson, I.D.M. and Eggleton, R.A. (1991) Weathering of granitic muscovite to kaolinite and halloysite and of plagioclase-derived kaolinite to halloysite. *Clays and Clay Minerals*, **39**, 113–126.

Robertson, I.D.M., Butt, C.R.M. and Chaffee, M.A. (1998) Fabric and chemical composition: from parent lithology to regolith. *CRC LEME Report*, **72**, 157–174.

Rodriguez-Navarro, C., Doehne, E. and Sebastian, E. (1999) Origins of honeycomb weathering; the role of salts and wind. *Geological Society of America Bulletin*, **111** (8), 1250–1255.

Rogers, J. (1977) Sedimentation on the continental margin off the Orange River and the Namib Desert. *Bulletin – South African National Committee for Oceanographic Research, Marine Geology Programme*, 7. University of Cape Town, Department of Geology, Cape Town, South Africa, 162 pp.

Roquin, C., Freyssinet, Ph., Zeegers, H. and Tardy, Y. (1990) Element distribution patterns in laterites of southern Mali: consequence for geochemical prospecting and mineral exploration. *Applied Geochemistry*, **5**, 303–315.

Rose, A.W. and Bianchi-Mosquera, G.C. (1993) Adsorption of Cu, Pb, Zn, Co, Ni and Ag on goethite and hematite: a control on metal mobilization from red beds into stratiform copper deposits. *Economic Geology*, **88**, 1226–1236.

Ross, G.J., Wang, C. and Protz, R. (1983) Soil mineralogical evidence as an indicator of Post- and Pre-Wisconsinian weathering in Canada. In Mahaney, W.C. (ed.) *Correlation of Quaternary Chronologies*. Geobooks, Toronto, 191–202.

Ruxton, B.P. (1968) Measures of the degree of chemical weathering of rocks. *Journal of Geology*, **76**, 518–527.

Ruxton, B.P. and Berry, L. (1957) Weathering of granite and associated erosional features in Hong Kong. *Geological Society of America Bulletin*, **68**, 1263–1291.

Ruxton, B.P. and Taylor, G. (1982) The Cainozoic geology of the Middle Shoalhaven Plain. *Journal of the Geological Society of Australia*, **29**, 239–246.

Ryan, J.L. and Rai, D. (1984) Thorium(IV) hydrous oxide solubility. *Inorganic Chemistry*, **26**, 4140–4142.

Salomons, W. and Förstner, U. (1984) *Metals in the Hydrocycle*. Springer-Verlag, Berlin, 349 pp.

Salonen, V.P. (1986) Length of boulder transport in Finland. In Phillips, W.J. (chair), *Prospecting in Areas of Glaciated Terrain 1986*. Institute of Mining and Metallurgy, London, 211–215.

Sánchez, C. and Galán, E. (1995) An approach to the genesis of palygorskite in a Neogene–Quaternary continental basin using principal factor analysis. *Clay Minerals*, **30**, 225–238.

Santosh, M. and Omana, P.K. (1991) Very high purity gold from lateritic weathering profiles of Nilambur, southern India. *Geology*, **19**, 746–749.

Savigear, R.A.G. (1953) Some observations on slope development in south Wales. *Institute of British Geographers, Publication No.* **18**, 31–51.

Schafer, B.M. and McGarity, J.W. (1980) Genesis of red and dark brown soils on basaltic parent materials near Armidale, N.S.W., Australia. *Geoderma*, **23**, 31–47.

Schellmann, W. (1981) Considerations on the definition and classification of laterites. In Chowdhury, M.K. Roy, Radhakrishna, B.P., Vaidyanadhan, R., Banerjee, P.K. and Ranganathan, K. (eds) *Lateritisation Processes*. A.A. Balkema, Rotterdam, 1–10.

Schellmann, W. (1983) Geochemical principles of lateritic

nickel ore formation. In Melfi, A.J. and Carvalho, A. (eds) *Lateritization Processes. Proc 2nd Int. Sem. on Lateritization Processes*, São Paulo, Brazil, 119–135.

Schindler, P.W. (1990) Co-adsorption of metal ions and organic ligands: formation of ternary surface complexes. In Hochella, M.F. Jr. and White, A.F. (eds) *Mineral-water Interface Geochemistry. Reviews in Mineralogy*, **23**. Mineralogical Society of America, 281–307.

Schmid, H. (1987) Turkey's Salda Lake; a genetic model for Australia's newly discovered magnesite deposits. *Industrial Minerals*, **239**, 19–31.

Schmidt, P.W., Prasad, V. and Ramam, P.K. (1983) Magnetic ages of some Indian laterites. *Palaeogeography, Palaeoclimatology, Palaeoecology*, **44**, 185–202.

Schmitt, J.M. (1983) Albitization in relation to the formation of uranium deposits in the Rouergue area (Massif Centrale, France). *Sciences Géologiques. Mémoires. (Strasbourg)*, **73**, 185–194.

Schnoor, J.L. and Stumm, W. (1985) Acidification of aquatic and terrestrial systems. In Stumm, W. (ed.) *Chemical Processes in Lakes*. John Wiley, New York, 311–338.

Schulze, D.G. (1984) The influence of aluminium on iron oxides: VIII. Unit cell dimensions of Al-substituted goethites and estimation of Al from them. *Clays and Clay Minerals*, **32**, 36–44.

Schumann, A. (1993) Changes in mineralogy and geochemistry of a nepheline syenite with increasing bauxitization, Poços de Caldas, Brazil. *Chemical Geology*, **107**, 327–331.

Schumm, S.A. (1963) The disparity between present rates of denudation and orogeny. *US Geological Survey Professional Paper 454–H.*

Schumm, S.A. (1975) Rates of denudation in selected small catchments in eastern Australia. *Earth-Science Reviews*, **11**, 88–89.

Schwertmann, U. (1985) Occurrence and formation of iron oxides in various pedogenic environments. In Stucki, J.W., Goodman, B.A. and Schwertmann, U. (eds) *Iron in Soils and Clay Minerals*. NATO ASI, 267–308.

Schwertmann, U., Gasser, U. and Sticher, H. (1989) Chromium-for-iron substitution in synthetic geothites. *Geochimica et Cosmochimica Acta*, **53**, 1293–1297.

Scott, K.M. and Dotter, L.E. (1990) *The Mineralogical and Geochemical Effects of Weathering on Shales at the Panglo Deposit, Eastern Goldfields, WA*. CSIRO Division of Exploration Geoscience Restricted Report 171R.

Selby, M.J. (1982) *Hillslope Materials and Processes*, 1st edn. Oxford University Press, Oxford, 264 pp.

Selby, M.J. (1993) *Hillslope Materials and Processes*, 2nd edn. Oxford University Press, Oxford, 273 pp.

Selley, R.C. (1988) *Applied Sedimentology*. Academic Press, London, 446 pp.

Senior, B.R. (1975) Notes on the Cooper Basin in Queensland. *Queensland Government Mining Journal*, **76**, 260–265.

Senior, B.R. (1978) Silcrete and chemically weathered sediments in southwest Queensland. In Langford-Smith, T. (ed.) *Silcrete in Australia*. Department of Geography, University of New England, Armidale, 41–50.

Senior, B.R. (1979) Mineralogy and chemistry of weathered and parent sedimentary rocks in Southwest Queensland. *Bureau of Mineral Resources Journal of Australian Geology and Geophysics*, **4**, 111–124.

Senior, B.R. and Senior, D.A. (1972) Silcrete in southwest Queensland. *Bureau of Mineral Resources, Australia, Bulletin*, **125**, 25–31.

Shannon, R.D. (1976) Revised effective ionic radii and systematic studies of interatomic distances in halides and chalcogenides. *Acta Crystallographica*, **A32**, 751–767.

Short, S.A., Lowson, R.T., Ellis, J. and Price, D.M. (1989) Thorium–uranium disequilibrium dating of Late Quaternary ferruginous concretions and rinds. *Geochemica et Chosmochemica Acta*, **53**, 1379–1389.

Sikora, W.S. and Budek, L. (1994) Sorption of Cu, Zn, Cd and Pb by basalt-derived clays Krzeniów (Lower Silesia). *Mineralogia Polonica*, **25**, 99–104.

Silva, P.G., Harvey, A.M., Zazo, C. and Goy, J.L. (1992) Geomorphology, depositional style and morphotectonic relationships of Quaternary alluvial fans in the Guadalentin depression (Murcia, southeast Spain). *Zeitschrift für Geomorphologie NF*, **36**, 325–341.

Singh, B. and Gilkes, R.J. (1991) Weathering of a chromian muscovite to kaolinite. *Clays and Clay Minerals*, **39**, 571–579.

Singh, B. and Gilkes, R.J. (1992a) An electron optical investigation of the alteration of kaolinite to halloysite. *Clays and Clay Minerals*, **40**(2), 212–229.

Singh, B. and Gilkes, R.J. (1992b) Properties and distribution of iron oxides and their association with minor elements in the soils of south-western Australia. *Journal of Soil Science*, **43**, 77–98.

Singh, B. and Gilkes, R.J. (1995) The natural occurrence of χ-alumina in lateritic pisoliths. *Clay Minerals*, **30**, 39–44.

Singh, B. and Mackinnon, I.D.R. (1996) Experimental transformation of kaolinite to halloysite. *Clays and Clay Minerals*, **44**, 825–834.

Singh, B., Gilkes, R.J. and Butt, C.R.M. (1992) An electron optical investigation of aluminosilicate cements in silcretes. *Clays and Clay Minerals*, **40**, 707–721.

Singh, G., Opdyke, N.D. and Bowler, J.M. (1981) Late Cainozoic stratigraphy, palaeomagnetic chronology and vegetational history from Lake George, New South Wales. *Journal of the Geological Society of Australia*, **28**, 435–452.

Sircombe, K.N. (1999) Tracing provenance through the isotope ages of littoral and sedimentary detrital zircon, eastern Australia. *Sedimentary Geology*, **124**, 47–67.

Sivarajasingham, S., Alexander, L.T., Cady, J.G. and Cline, M.G. (1962) Laterite. *Advances in Agronomy*, **14**, 1–60.

Skinner, B.J. and Porter, S.C. (1987) *Physical Geology*. John Wiley, New York, 750 pp.

Smale, R.J. (1973) Silcretes and associated silica diagenesis in southern Africa and Australia. *Journal of Sedimentary Petrology*, **43**, 1077–1089.

Smith, B.A. (1996) *Calcretes: their nature and association with gold and other metals*. B.Sc. (Honours) thesis, Australian National University, Canberra.

Smith, G.I. and Medrano, M.D. (1996) Continental borate deposits of Cenozoic age. In Grew, E.S. and Anovitz, L.M. (eds) *Boron: Mineralogy, Petrology and Geochemistry. Reviews in Mineralogy*, **33**. Mineralogical Society of America, 263–298.

Smith, I.E.M., White, A.J.R., Chappell, B.W. and Eggleton, R.A. (1988) Fractionation in a zoned monzonite pluton: Mt Dromedary, southeastern Australia. *Geological Magazine*, **125**, 273–284.

Smith, K.L., Milnes, A.R. and Eggleton, R.A. (1987)

Weathering of basalt: Formation of iddingsite. *Clays and Clay Minerals*, **35**, 418–428.
Smith, R.E. (1983) Generalised model for secondary dispersion haloes in transported overburden in weathered terrain. In Smith, R.E. (ed.) *Geochemical Exploration in Deeply Weathered Terrain*. CSIRO Division of Mineralogy, Floreat Park, 153–154.
Smith, R.E. (1989) Using lateritic surfaces to advantage in exploration. In Garland, G.D. (ed.) *Proceedings of Exploration '87, Third Decennial Conference on Geophysical and Geochemical Exploration for Minerals and Groundwater*. Ontario Geological Survey, Special Volume 3, 312–322.
Soil Survey Staff (1975) *Soil Taxonomy: a basic system of soil classification for making and interpreting soil surveys*. USDA Soil Conservation Service, Washington DC.
Sollmann, T. (1957) *A Manual of Pharmacology and its Applications to Therapeutics and Toxicology*, 8th edn. W.B. Saunders, Philadelphia and London.
Soubies, F., Melfi, A.J. and Autefage, F. (1990) Geochemical behaviour of rare earth elements in alterites of phosphate and titanium ore deposits in Tapira, Minas Gerais, Brazil: the importance of the phosphates. *Chemical Geology*, **84**, 377.
Sposito, G. (1984) *The Surface Chemistry of Soils*. Clarendon Press, Oxford 234 pp.
Sposito, G. (1989) *The Chemistry of Soils*. Oxford University Press, New York, 277 pp.
Stace, H.C.T., Hubble, G.D., Brewer, R., Northcote, K.H., Sleeman, J.R., Mulcahy, M.J. and Hallsworth, E.G. (1968) *A Handbook of Australian Soils*. Rellim Tech. Publs, Glenside, SA, 429 pp.
Stallard, R.F. (1985) River chemistry geology, geomorphology and soils in the Amazon and Orinoco Basins. In Drever, J.I. (ed.) *The Chemistry of Weathering*. Reidel Publishing, Dordrecht, 293–316.
Stallard, R.F. and Edmond, J.M. (1981) Geochemistry of the Amazon: 1. Precipitation chemistry and the marine contribution to the dissolved load at the time of peak discharge. *Journal of Geophysical Research*, **86** (C10), 9844–9858.
Stallard, R.F. and Edmond, J.M. (1983) Geochemistry of the Amazon: 2. The influence of geology and weathering environment on the dissolved load. *Journal of Geophysical Research*, **88** (C14), 9671–9688.
Stallard, R.F. and Edmond, J.M. (1987) Geochemistry of the Amazon: 3. Weathering chemistry and limits to dissolved inputs. *Journal of Geophysical Research*, **92** (C8), 8293–8302.
Stanjek, H. and Schwertmann, U. (1992) The influence of aluminum on iron oxides. Part XVI: Hydroxyl and aluminum substitution in synthetic hematites. *Clays and Clay Minerals*, **40**, 347–354.
Stauffer, R.E. and Wittchen, B.D. (1991) Effects of silicate weathering on water chemistry in forested, upland, felsic terrane of the USA. *Geochimica et Cosmochimica Acta*, **55**, 3253–3271.
Stefansson, A., Gislason, S.R. and Arnorsson, S. (2001) Dissolution of primary minerals in natural waters. II. Mineral saturation state. *Chemical Geology*, **172**, 251–276.
Stephens, C.G. (1971) Laterite and silcrete in Australia: a study of the genetic relationships of laterite and silcrete and their companion materials, and their collective significance in the formation of the weathered mantle, soils, relief and drainage of the Australian continent. *Geoderma*, **5**, 5–52.
Stirling, C.H., Esat, T.M., McCullock, M.T. and Lambeck, K. (1995) High-precision U-series dating of corals from Western Australia and implications for the timing and duration of the Last Interglacial. *Earth and Planetary Science Letters*, **135**, 115–130.
Stone, A.T. (1997) Reactions of extracellular organic ligands with dissolved metal ions and mineral sources. In Banfield, J.F. and Nealson, K.H. (eds) *Geomicrobiology: interactions between microbes and minerals. Reviews in Mineralogy*, **35**, 309–344. Mineralogical Society of America.
Stortz, M. (1928) Die sekundaren authigenen Kieselsaure in ihrer petrogenetischgeologischen Bedeutung. *Monogr. Geol. Paleontol.*, **II**, 481 pp.
Strakhov, N.M. (1967) *Principles of Lithogenesis*. Oliver and Boyd, Edinburgh.
Strusz, D.L. (1971) *Canberra, Australian Capital Territory and New South Wales; 1:250,000 geological series, sheet SI/55-16, international index, explanatory notes*. Australian Bureau of Mineral Resources Geology and Geophysics, 47 pp.
Stumm, W. and Morgan, J.J. (1970) *Aquatic Chemistry: an introduction emphasizing chemical equilibria in natural waters*, 1st edn. Wiley-Interscience, New York, London, Sydney, Toronto, 583 pp.
Stumm, W. and Morgan, J.J. (1981) *Aquatic Chemistry: an introduction emphasizing chemical equilibria in natural waters*, 2nd edn. John Wiley, New York.
Stumm, W. and Morgan, J.J. (1996) *Aquatic Chemistry: an introduction emphasizing chemical equilibria in natural waters*, 3rd edn. Wiley Interscience, 1022 pp.
Suchecki, R.K., Hubert, J.F. and Birney-de-Wet, C.C. (1988) Isotopic imprint of climate and hydrogeochemistry on terrestrial strata of the Triassic–Jurassic Hartford and Fundy rift basins. *Journal of Sedimentary Petrology*, **58**, 801–811.
Sueoka, T. (1988) Identification and classification of granitic residual soils using chemical weathering index. In *Geomechanics of Tropical Soils, Proceedings of the Second International Conference on Geomechanics in Tropical Soils*, Singapore, A.A. Balkema, Rotterdam, 55–61.
Sugden, D.E. (1978) Glacial geomorphology. *Progress in Physical Geography*, **2**, 309–320.
Summerfield, M.A. (1983) Geochemistry of weathering profile silcretes, southern Cape Province, South Africa. In Wilson, R.C.L. (ed.) *Residual Deposits. Special Publication Geological Society London*, **11**, 167–178.
Summerfield, M.A. (1991) *Global Geomorphology; an introduction to the study of landforms*. John Wiley, New York, 537 pp.
Sutherland, D.G. (1984) Geomorphology and mineral exploration: some examples from exploration for diamondiferous placer deposits. *Zeitschrift für Geomorphologie*, **51**, 95–108.
Sverdrup, H. and Warfvinge, P. (1995) Estimating field weathering rates using laboratory kinetics. In White, A.F. and Brantley, S.L. (eds) *Chemical Weathering Rates of Silicate Minerals. Mineralogical Society of America, Reviews in Mineralogy*, **31**, 485–541.
Tapley, I.J. (1998a) *Landform and Regolith Mapping in the North-eastern Goldfields Region, Western Australia, and north Drummond Basin, Queensland, Using AIRSAR Radar Data.*

Volumes I and II. CRC LEME Restricted Report 55R, 64 pp. and unpaginated.

Tapley, I.J. (1998b) *Final Report: landform, regolith and geological mapping in Australia using polarimetric and interferometric radar sites.* CRC LEME Restricted Report 90R, 18 pp.

Tardy, Y. (1969) *Géochimie des altérations. Etude des arènes et des eaux de quelques massifs cristallins d'Europe et d'Afrique.* Thèse Doc. Fac. Sci., Strasb., *Mém. Serv. Carte Géol. Als. Lorr.*, **31**, 199 pp.

Tardy, Y. (1971) Characterization of the principal weathering types by the geochemistry of waters from some European and African crystalline massifs. *Chemical Geology*, **7**, 253–271.

Tardy, Y. (1992) Diversity and terminology of lateritic profiles. In Martini, I.P. and Chesworth, W. (eds) *Weathering, Soils and Paleosols.* Elsevier, Amsterdam, 379–405.

Tardy, Y. (1993) *Pétrologie des laterites et des sols tropicaux.* Masson, Paris, 459 pp.

Tardy, Y. and Nahon, D. (1985) Geochemistry of laterites, stability of Al-goethite, Al-hematite, and Fe^{3+}-kaolinite in bauxites and ferricretes: an approach to the mechanism of concretion formation. *American Journal of Science*, **285**, 865–903.

Tardy, Y. and Roquin, C. (1992) Geochemistry and evolution of lateritic landscapes. In Martini, I.P. and Chesworth, W. (eds) *Weathering, Soils and Paleosols.* Elsevier, Amsterdam, 407–443.

Tardy, Y., Bocquier, G., Paquet, H. and Millot, G. (1973) Formation of clay from granite and its distribution in relation to climate and topography. *Geoderma*, **10**, 271–284.

Tardy, Y., Boeglin, J-L., Novikoff, A. and Roquin, C. (1995) Petrological and geochemical classification of laterites. *Proceedings of the 10th International Clay Conference*, Adelaide, 1993, 481–486.

Taylor, G. (1978) Silcretes in the Walgett–Cumborah region of New South Wales. In Langford-Smith, T. (ed.) *Silcrete in Australia.* Department of Geography, University of New England, Armidale, 187–193.

Taylor, G. (1994) Landscapes of Australia: their nature and evolution. In Hill, R.S. (ed.) *History of the Australian Vegetation: Cretaceous to Recent.* Cambridge University Press, Cambridge, 60–80.

Taylor, G. and Butt, C.R.M. (1998) The Australian regolith and mineral exploration. *Australian Geological Survey Organisation Journal of Geology and Geophysics*, **17**, 55–67.

Taylor, G. and McNally, G.H. (1992) *Engineering and Environmental Geology of the Regolith.* Key Centre for Mines, University of New South Wales, Sydney.

Taylor, G. and Ruxton, B.P. (1987) A duricrust catena in south-east Australia. *Zeitschrift für Geomorphologie*, **31**, 385–410.

Taylor, G. and Smith, I.E. (1975) The genesis of sub-basaltic silcretes from the Monaro, New South Wales. *Journal of the Geological Society of Australia*, **22**, 377–385.

Taylor, G. and Walker, P.H. (1986a) Tertiary Lake Bunyan, northern Monaro, NSW; Part I. Geological setting and landscape history. *Australian Journal of Earth Sciences*, **33**, 219–229.

Taylor, G. and Walker, P.H. (1986b) Tertiary Lake Bunyan, northern Monaro, NSW; Part II. Facies analysis and palaeoenvironmental implications. *Australian Journal of Earth Sciences*, **33**, 231–251.

Taylor, G. and Woodyer, K.D. (1978) Bank deposition in suspended-load streams. In Miall, A.D. (ed.) *Fluvial Sedimentology. Canadian Society of Petroleum Geologists, Memoir* **5**, 257–275.

Taylor, G., Eggleton, R.A., Holzhauer, C.C., Maconachie, L.A., Gordon, M., Brown, M.C. and McQueen, K.G. (1992) Cool climate lateritic and bauxitic weathering. *Journal of Geology*, **100**, 669–677.

Taylor, G., Truswell, E.M., Eggleton, R.A. and Musgrave, R. (1990a) Cool climate bauxites. *Chemical Geology*, **84**, 183–184.

Taylor, G., Truswell, E.M., McQueen, K.G. and Brown, M.C. (1990b) The early Tertiary palaeogeography, landform evolution and palaeoclimates of the southern Monaro, New South Wales, Australia. *Palaeogeography, Palaeoclimatology, Palaeoecology*, **78**, 109–134.

Taylor, G.F. and Thornber, M.R. (1992) Gossan and ironstone surveys. In Butt, C.R.M. and Zeegers, H. (eds) *Handbook of Exploration Geochemistry*, Vol. 4. Elsevier, 139–202.

Taylor, G.F., Wilmshurst, J.R., Butt, C.R.M. and Smith, R.E. (1980) Ironstones and gossans. *Journal of Geochemical Exploration*, **12**, 118–122.

Taylor, R.M. (1988) Proposed mechanism for the formation of soluble Si–Al and Fe(III)–Al hydroxy complexes in soils. *Geoderma*, **42**, 65–77.

Taylor, R.M. and McKenzie, R.M. (1966) The association of trace elements with manganese minerals in Australian soils. *Australian Journal of Soil Research*, **4**, 29–39.

Tazaki, K. and Fyfe, W.S. (1987) Primitive clay precursors formed on feldspar. *Canadian Journal of Earth Sciences*, **24**(3), 506–527.

Teale, G.S., Flood, R.H. and Shaw, S.E. (1995) A Devonian cordierite-bearing volcanic pebble from the Cretaceous Bulldog Shale, South Australia. *Geological Survey of South Australia Quarterly Notes*, **128**, 17–23.

Tejan-Kella, M.S., Chittleborough, D.J. and Fitzpatrick, R.W. (1991) Weathering assessment of heavy minerals in age sequences of Australian sandy soils. *Soil Science Society of America Journal*, **55**, 427–438.

Théveniaut, H. and Freyssinet, Ph. (1999) Paleomagnetism applied to lateritic profiles to assess saprolite and duricrust formation processes: the example of Mont Baduel (French Guiana). *Palaeogeography, Palaeoclimatology, Palaeoecology*, **148**, 209–231.

Thiry, M. (1981) Sédimentation continentale et altérations associées: calcitisations, ferruginisations et silicification, les agriles plastiques du Sparnacien du Bassin de Paris. *Sciences Géologiques. Mémoires (Strasbourg)*, **64**, 173 pp.

Thiry, M. (1997) Continental silicifications: a review. In Paquet, H. and Clauer, N. (eds) *Soils and Sediments: mineralogy and geochemistry.* Springer, Berlin, 191–221.

Thiry, M. and Milnes, A.R. (1991) Pedogenic and groundwater silcretes at Stuart Creek Opal Field, South Australia. *Journal of Sedimentary Petrology*, **61**, 111–127.

Thiry, M. and Simon-Coinçon, R. (1996) Tertiary paleoweathering and silcretes in the southern Paris Basin. *Catena*, **26**, 1–26.

Thiry, M. and Simon-Coinçon, R. (1999) *Palaeoweathering, Palaeosurfaces and Related Continental Deposits.* International Association of Sedimentologists, Special Publication Number **27**, 406 pp.

Thiry, M. and Turland, M. (1985) Paleotoposequences of ferruginous soils and siliceous crust in the Siderolithic of the northern Central Massif, Montlugon–Domerat Basin. In *Symposium on Paleoalteration and Associated Landscapes. Géologie de la France: 1985*, **2**, 175–192. Bureau de Recherches Géologiques et Minières, Orleans.

Thomas, M.F. (1966) Some geomorphological implications of deep weathering patterns in crystalline rocks in Nigeria. *Transactions of the Institute of British Geographers*, **40**, 173–193.

Thomas, M.F. (1994) *Geomorphology in the Tropics: a study of weathering and denudation in low latitudes*. John Wiley, New York, 460 pp.

Thomas, M.F. and Thorp, M.B. (1993) The geomorphology of some Quaternary placer deposits. *Zeitschrift für Geomorphologie*, **87**, 183–194.

Thomas, M.F. and Thorp, M.B. (1995) Geomorphic response to rapid climatic and hydrologic change during the Late Pleistocene and Early Holocene in the humid and sub-humid tropics. *Quaternary Science Reviews*, **14**, 193–207.

Thompson, C.H. (1981) Podzol chronosequences on coastal dunes of eastern Australia. *Nature*, **291**, 59–61.

Thompson, J.B. Jr. (1978) Biopyriboles and polysomatic series. *American Mineralogist*, **63**, 239–249.

Thorarinsson, S. (1967) The eruptions of Hekla in historical times. The eruption of Hekla 1947–1948. *Icelandica Science Society*, **1**, 1–170.

Thornber, M.R. (1985) Supergene alteration of sulphides. vii. Distribution of elements during the gossan-forming process. *Chemical Geology*, **53**, 279–301.

Thornber, M.R. (1992) The chemical mobility and transport of elements in the weathering environment. In Butt, C.R.M. and Zeegers, H. (eds) *Handbook of Exploration Geochemistry*, Vol. 4. Elsevier, 79–96.

Thornber, M.R. and Taylor, G.F. (1992) The mechanisms of sulphide oxidation and gossan formation. In Butt, C.R.M. and Zeegers, H. (eds) *Handbook of Exploration Geochemistry*, Vol. 4. *Regolith Exploration Geochemistry in Tropical and Subtropical Terrains*, 119–138.

Thornber, M.R. and Wildman, J.E. (1984) Supergene alteration of sulphides. VI The binding of Cu, Ni, Zn, Co and Pb with gossan (iron-bearing) minerals. *Chemical Geology*, **44**, 399–434

Thorp, M.B., Thomas, M.F., Martin, T. and Whalley, W.B. (1990) Late Pleistocene sedimentation and land-form development in western Kalimantan (Indonesian Borneo). *Geologie en Mijnbouw*, **69**, 133–150.

Tilley, D.B. (1994) *The evolution of bauxitic pisoliths*. Ph.D. thesis, Australian National University, Canberra, 214 pp.

Tilley, D.B. (1998) The evolution of bauxitic pisoliths at Weipa in northern Queensland. In Eggleton, R.A. (ed.) *The state of the regolith*. Proceedings of the Second Australian Conference on Landscape Evolution and Mineral Exploration. Geological Society of Australia Special Publication, **20**, 148–156.

Tilley, D.B. and Eggleton, R.A. (1994) Tohdite ($5Al_2O_3.H_2O$) in bauxites from Northern Australia. *Clays and Clay Minerals*, **42**, 485–488.

Tilley, D.B. and Eggleton, R.A. (1996) The natural occurrence of eta-alumina (η-Al_2O_3) in bauxite. *Clays and Clay Minerals*, **44**, 658–664.

Tilley, D.B., Morgen, C.M. and Eggleton, R.A. (1994) The evolution of bauxitic pisoliths from Weipa, North Queensland. In Pain, C.F., Craig, M.A. and Campbell, I.D. (eds) *Australian Regolith Conference '94 Abstracts, Broken Hill. Australian Geological Survey Organisation Record 1994/56*, 58.

Torrent, J., Schwertmann, U. and Barrón, V. (1992) Fast and slow phosphate sorption by goethite-rich natural materials. *Clays and Clay Minerals*, **40**, 14–21.

Trendall, A.F. (1962) The formation of 'apparent peneplains' by a process of combined lateritisation and surface wash. *Zeitschrift für Geomorphologie*, (NF), **6**, 183–197.

Trescases, J.J. (1997) The lateritic nickel-ore deposits. In Paquet, H. and Clauer, N. (eds) *Soils and Sediments: mineralogy and geochemistry*. Springer, Berlin, 125–138.

Trescases, J.J., Dino, R. and Oliveira, S. (1987a) Un gisement de nickel supergene en zone semiaride; Sao Joao do Piaui (Bresil) [A supergene nickel mineralization in the semi-arid zone; Sao Joao do Piaui, Brazil]. In Rodriguez-Clemente, R. and Tardy, Y. (eds) *Geochemistry and Mineral Formation in the Earth Surface*. Consejo Superior de Investigaciones, Cientificas Centre National de la Recherche Scientifique, 273–288.

Trescases, J.J., Fortin, P., Melfi, A. and Nahon, D. (1987b) Rare earths elements accumulation in lateritic weathering of Pliocene sediments, Curitiba Basin, Brazil. In Rodrigez-Clemente, R. and Tardy, Y. (eds) *Geochemistry and Mineral Formation in the Earth Surface*. Consejo Superior de Investigaciones, Cientificas Centre National de la Recherche Scientifique, 259–272.

Tricart, J. and Cailleux, A. (1972) *Introduction to Climatic Geomorphology*. Longman, London, 295 pp.

Twidale, C.R. (1983) Australian laterites and silcretes: ages and significance. *Revue de Géologie Dynamique et de Géographie Physique*, **24**, 35–45.

Twidale, C.R. and Campbell, E.M. (1995) Pre-Quaternary landforms in the low latitude context: the example of Australia. *Geomorphology*, **12**, 17–35.

Ullyott, J.S., Nash, D.J. and Shaw, P.A. (1998) Recent advances in silcrete research and their implications for the origin and palaeoenvironmental significance of sarsens. *Proceedings of the Geologists' Association*, **109**, 255–270.

Ure, A.M. and Berrow, M.L. (1982) The chemical constituents of soils. In Bowen, H.J.M. (ed.) *Environmental Chemistry*, Vol. 2. R. Soc. Chem., Burlington House, London, 94–202.

Valeton, I. (1972) *Bauxites*. Elsevier, Amsterdam, 226 pp.

Van Baalen, M.R. (1993) Titanium mobility in metamorphic systems: a review. *Chemical Geology*, **110**, 233–249.

Van Breeman, N. (1982) Genesis, morphology, and classification of acid sulfate soils in coastal plains. In Kittrick, J.A., Fanning, D.S. and Hossner, L.R. (eds) *Acid Sulfate Weathering*. Soil Science Society of America, 95–104.

Van Breeman, N. (1988) Long-term chemical, mineralogical, and morphological effects of iron-redox processes in periodically flooded soils. In Stucki, J.W., Goodman, B.A. and Schwertmann, V. (eds) *Iron in Soils and Clay Minerals*. NATO ASI Series, 811–823.

Van Dijk, D.C. and Beckmann, G.G. (1978) The Yuleba Hardpan and its relationship to soil geomorphic history in the Yuleba–Tara region Southeast Queensland. In Langford-Smith, T. (ed.) *Silcrete in Australia*, University of New England, Armidale, 73–91.

Van Olphen, H. (1977) *An Introduction to Clay Colloid Chemistry: for clay technologists, geologists, and soil scientists*, 2nd edn. John Wiley, New York, 318 pp.

Vasconcelos, P. (1996) Geochronological evidence for the preservation of Cretaceous weathering profiles in northwestern Queensland. *Geological Society of Australia, Abstracts*, **43**, 543–544.

Vasconcelos, P.M., Renne, P.R., Brimhall, G.H. and Becker, T.A. (1994) Direct dating of weathering phenomena by $^{40}Ar/^{39}Ar$ and K–Ar analysis of supergene K–Mn oxides. *Geochimica et Cosmochimica Acta*, **58**, 1635–1665.

Veblen, D.R. and Burnham, C.W. (1978) New biopyriboles from Chester, Vermont: I. Descriptive mineralogy. *American Mineralogist*, **63**, 1000–1009.

Velbel, M.A. (1983) A dissolution–reprecipitation mechanism for the pseudomorphous replacement of plagioclase feldspar by clay minerals during weathering. *Sci. Géol., Mém.*, **71**, 139–147.

Velbel, M.A. (1985) Hydrogeochemical constraints on mass balances in forested watersheds of the southern Appalachians. In Drever, J.I. (ed.) *The Chemistry of Weathering*. Reidel, Dordrecht, 231–247.

Velbel, M.A. (1989) Weathering of hornblende to ferruginous products by a dissolution–reprecipitation mechanism: petrography and stoichiometry. *Clays and Clay Minerals*, **37**, 515–524.

Velbel, M.A. (1993) Formation of protective surface layers during silicate-mineral weathering under well-leached, oxidizing conditions. *American Mineralogist*, **78**, 405–414.

Velbel, M.A. (1996) The natural weathering of staurolite: crystal-surface textures, relative stability, and the rate-determining step. *American Journal of Science*, **296**, 453–472.

Vepraskas, M.J., Jongmans, A.G., Hoover, M.T. and Bouma, J. (1991) Hydraulic conductivity of saprolite as determined by channels and porous groundmass. *Journal of the Soil Science Society of America*, **55**, 932–938.

Vlasov, K.A. (1968) Geochemistry and mineralogy of rare elements and genetic types of their deposits. Translated from Russian by Z. Lerman, Jerusalem, Israel Program for Scientific Translation, 1966–68, Academy of Sciences of the USSR. State Geological Committee of the USSR. Institute of Mineralogy, Geochemistry and Crystal Chemistry of Rare Elements. Translation of *Geokhimiia, mineralogiia i geneticheskie tipy mestorozhdenii redkikh elementov*, Moscow, 1964.

Vlek, P.L.G. and Lindsay, W.L. (1977) Thermodynamic stability and solubility of molybdenum minerals in soils. *Journal of the Soil Science Society of America*, **41**, 42–46.

Vogel, D.E. (1975) Precambrian weathering in acid metavolcanic rocks from the Superior Province, Villebon township, south-central Quebec. *Canadian Journal of Earth Science*, **12**, 2080–2085.

von Gunten, H.R., Roessler, E., Lowson, R.T., Reid, P.D. and Short, S.A. (1999) Distribution of uranium and thorium series radionuclides in mineral phases of a weathered lateritic transect of a uranium orebody. *Chemical Geology*, **160**, 225–240.

Wada, K. (1989) Allophane and imogolite. In Dixon, J.B. and Weed, S.B. (eds) *Minerals in Soil Environments*. Soil Science Society of America, 1051–1087.

Wagh, A.S. and Pinnock, W.R. (1987) Occurrence of scandium and rare earth elements in Jamaica bauxite waste. *Economic Geology*, **82**, 757–761.

Walden, J., White, K. and Drake, N.A. (1996) Controls on dune colour in the Namib sand sea; preliminary results. *Journal of African Earth Sciences*, **22**, 349–353.

Walker, P.H. (1962) Terrace chronology and soil formation on the South Coast of NSW. *Journal of Soil Science*, **13**, 178–186.

Walker, P.H. (1966) Postglacial environments in relation to landscape and soils on the Cary drift, Iowa. *Iowa State University, Agriculture and Home Economy Experiment Station Research Bulletin*, **549**, 835–875.

Walker, P.H. and Butler, B.E. (1983) Fluvial processes. In *Soils, an Australian Viewpoint*. CSIRO, Melbourne, 83–90.

Walker, P.H. and Gillespie, R. (1978) Notes on five radiocarbon dates from sites near Canberra. *CSIRO Australia Division of Soils Technical Memorandum No. 5/78*.

Walker, P.H. and Green, P. (1976) Soil trends in two valley fill sequences. *Australian Journal of Soil Research*, **14**, 291–303.

Walker, P.H., Chartres, C.J. and Hutka, J. (1988) The effect of aeolian accessions on soil development on granitic rocks in south-eastern Australia: I. Soil morphology and particle size distribution. *Australian Journal of Soil Research*, **26**, 1–16.

Walker, R.G. (ed.) (1984) *Facies Models*. Geological Association of Canada, Toronto, 317 pp.

Walling, D.E. and Webb, B.W. (1983) Patterns of sediment yield. In Gregory, K.L. (ed.) *Background to Palaeohydrology; a perspective*. John Wiley, Chichester, 69–100.

Walther, J. (1916) Das geologische Alter und die Bildung des Laterits. *Petermanns Geographischer Mitteilungen*, **62**, 1–7, 46–53.

Wang, Q-M. (1988) *Mineralogical aspects of monzonite alteration: an investigation by electron microscopy and chemistry*. Ph.D. thesis, Australian National University, Canberra, 192 pp.

Ward, W.T. (1977) Geomorphology and soils of the Stratford–Bairnsdale area, East Gippsland, Victoria. *CSIRO Australia Soils and Land Use Series No. 57*.

Wasklewicz, T., Dorn, R.I., Clark, S., Hetrick, J., Pope, G., Liu, T., Krinsley, D.H., Dixon, J., Moore, R.B. and Clark, J. (1993) Olivine does not necessarily weather first. *Singapore Journal of Tropical Geography*, **14**, 72–79.

Wasson, R.J. (1983) Dune sediment types, sand colour, sediment provenance and hydrology in the Strzelecki–Simpson dunefield, Australia. In Brookfield, M.E. and Ahlbrandt T.S. (eds) *Eolian Sediments and Processes*. Elsevier, Amsterdam, 165–195.

Wasson, R.J. (1986) Climate, rates and timing of Quaternary continental dune accumulation in Australia. In *Sediments Down-Under; 12th International Sedimentological Congress, Abstracts*. Bureau of Mineral Resources, Geology and Geophysics, Canberra, 322.

Watson, J.A.L. and Gay, F.J. (1970) The role of grass eating termites in the degradation of a mulga ecology. *Search*, **1**, 43.

Watson, J.P. (1970) Contribution of termites to development of zinc anomalies in Kalahari sand. *Transactions of the*

Institute of Mining and Metallurgy, Section B, Applied Earth Science, **79**, 53–59.
Watterson, J.R. (1992) Preliminary evidence for the involvement of budding bacteria in the origin of Alaskan placer gold. *Geology*, **20**, 315–318.
Watts, S.H. (1978) The nature and occurrence of silcrete in the Tibooburra area of northwestern New South Wales. In Langford-Smith, T. (ed.) *Silcrete in Australia*. Department of Geography, University of New England, Armidale, 167–185.
Wayland, E.J. (1934) Peneplains and other erosional platforms. In *Annual Report, Geological Survey of Uganda, 1933*, 77–78.
Weaver, C.E. (1984) Origin and geologic implications of the palygorskite deposits of S.E. United States. In Singer, A. and Galan, E. (eds) *Palygorskite-Sepiolite. Occurrences, Genesis and Uses. Developments in Sedimentology*, **37**, 39–58, Elsevier.
Weaver, C.E. and Beck, K.C. (1977) Miocene of the S.E. United States: a model for chemical sedimentation in a peri-marine environment. *Developments in Sedimentology*, **22**. Elsevier, Amsterdam and New York, 234 pp.
Webb, J.A. and Golding, S.D. (1998) Geochemical mass-balance and oxygen isotope constraints on silcrete formation and its palaeoclimatic implications in southern Australia. *Journal of Sedimentary Research*, **68**, 981–993.
Weber, F. (1997) Evolution of lateritic manganese deposits. In Paquet, H. and Clauer, N. (eds) *Soils and Sediments: mineralogy and geochemistry*. Springer, Berlin, 97–124.
Webster, J.G. (1986) The solubility of gold and silver in the system Au–Ag–S–O_2–H_2O at 25°C and 1 atm. *Geochimica et Cosmochimica Acta*, **50**, 1837–1845.
Wedepohl, K.H. (ed.) (1969–78) *Handbook of Geochemistry*. Springer, Berlin.
Wells, M.A. and Gilkes, R.J. (1999) Synthetic Ni geothite and hematite: reproducing hosts for nickel mineralization in Ni-laterites. In Taylor, G. and Pain, C. (eds) *New Approaches to an Old Continent. Proceedings of Regolith '98*. Co-operative Research Centre for Landscape Evolution and Mineral Exploration, Perth, 299–310.
Wernicke, R.S. and Lippolt, H.J. (1993) Botryoidal hematite from the Schwarzwald (Germany): heterogeneous uranium distributions and their bearing on the helium dating method. *Earth and Planetary Science Letters*, **114**, 287–300.
Wheeler, W.M. (1910) *Ants: their structure, development and behaviour*. Columbia University Press, New York.
White, A.J.R., Williams, I.S. and Chappell, B.W. (1976) The Jindabyne Thrust and its tectonic, physiographic and petrogenetic significance. *Journal of the Geological Society of Australia*, **23**, 105–112.
White, A.J.R., Williams, I.S. and Chappell, B.W. (1977) *Geology of the Berridale 1:100 000 Sheet*. Geological Survey of New South Wales, 138 pp. plus map.
White, D.E., Hem, J.D. and Waring, G.A. (1963) *Chapter F.* Chemical Composition of Subsurface Waters. In Fleischer, M. (ed.) *Data of Geochemistry*, 6th edn. Geological Survey Professional Paper 440–F, United States Government Printing Office, Washington.
Whitney, J.W. and Harrington, C.D. (1993) Relict colluvial boulder deposits as paleoclimatic indicators in the Yucca Mountain region, southern Nevada. *Bulletin of the Geological Society of America*, **105**, 1008–1018.
Widdowson, M. (1997) Tertiary palaeosurfaces of the SW Deccan, western India; implications for passive margin uplift. In Widdowson, M. (ed.) *Palaeosurfaces; recognition, reconstruction and palaeoenvironmental interpretation*. Geological Society Special Publication, **120**, 221–248.
Wiewióra, A., Dubi'nska, E. and Iwasi'nska, I. (1982) Mixed layering in Ni-containing talc-like minerals from Szklary, Lower Silesia, Poland. In van Olphen, H. and Veniale, F (eds) *Proceedings of the International Clay Conference 1981, Developments in Sedimentology 35*, Elsevier, 111–125.
Williams, G.E. (1969) Glacial age of piedmont alluvial deposits in the Adelaide area, South Australia. *Australian Journal of Soil Science*, **32**, 257.
Williams, L.A., Parks, G.A. and Crerar, D.A. (1985) Silica diagenesis; I. Solubility controls. *Journal of Sedimentary Petrology*, **55**, 301–311.
Williams, M.A.J. (1968) Termites and soil development near Brock's Creek, Northern Territory. *Australian Journal of Science*, **31**, 153–154.
Williams, P.A. (ed.) (1990) *Oxide Zone Geochemistry*. Ellis Horwood, New York, 286 pp.
Williamson, W.O. (1957) Silicified sedimentary rocks in Australia. *American Journal of Science*, **255**, 23–42.
Willis, S.M. (1999) *Regolith geology and landscapes of the Thackaringa–West area, Broken Hill, NSW*. Honours thesis, University of Canberra.
Winkler, E.M. and Wilhelm, E.J. (1970) Salt burst by hydration pressures in architectural stone in urban atmosphere. *GSA Bulletin*, **81**, 567–572.
Windom, H.L. (1970) Contribution of atmospherically transported trace metals to South Pacific sediments. *Geochimica et Cosmochimica Acta*, **34**, 509–514.
Wollast, R. and Chou, L. (1985) Kinetic study of the dissolution of albite with a continuous flow-through fluidized bed reactor. In Drever, J.J. (ed.) *The Chemistry of Weathering*. Reidel, Dordrecht, The Netherlands, 75–96.
Wood, S.A. (1990) The aqueous geochemistry of the rare-earth elements and yttrium. *Chemical Geology*, **82**, 159–186.
Woodall, R. (1981) The science and art of mineral exploration. An address to the CSIRO Executive Seminar 'Assisting the mineral industry – present and future.'
Woolnough, W.G. (1927) Presidential Address: Part I. The chemical criteria of peneplanation. Part II. The duricrust of Australia. *Journal and Proceedings of the Royal Society of New South Wales*, **61**, 1–53.
Woolnough, W.G. (1930) The influence of climate and topography in the formation and distribution of products of weathering. *Geological Magazine*, **67**, 123–132.
Wopfner, H. (1978) Silcretes of northern South Australia and adjacent regions. In Langford-Smith, T. (ed.) *Silcrete in Australia*. Department of Geography, University of New England, Armidale, 93–141.
Wopfner, H. (1983) Environment of silcrete formation: a comparison of examples from Australia and the Cologne Embayment. In Wilson, R.C.L. (ed.) *Residual Deposits: surface related weathering processes and materials*, Blackwell, Oxford, 151–158.
Wright, V.P. (1990) A micromorphological classification of fossil and Recent calcic and petrocalcic microstructures. In Douglas, L.A. (ed.) *Soil Micromorphology: a basic and applied science*. Proceedings of the International Working Meeting on Soil Micromorphology, 8, Sub-commission of

Soil Micromorphology of the International Society of Soil Science, Paris, 401–407.
Wright, V.P. and Marriott, S.B. (1993) The sequence stratigraphy of fluvial depositional systems: the role of floodplain storage. *Sedimentary Geology*, **86**, 203–210.
Wright, V.P. and Tucker, M.E. (1991) Calcretes: an introduction. In Wright, V.P. and Tucker, M.E. (eds) *Calcretes*. International Association of Sedimentologists Reprint Series 2, 1–22.
Wyatt, D.H. and Webb, A.W. (1970) Potassium–argon ages of some northern Queensland basalts and an interpretation of Late Cenozoic history. *Journal of the Geological Society of Australia*, **17**, 39–51.
Yaalon, D.H. (1988) Calcic horizon and calcrete in acidic soils and paleosols: progress in the last 20 years. *Soil Science Society of America, Agronomy Abstracts*.
Yatsu, Eiju (1988) *The Nature of Weathering: an introduction*. Sozosha, Tokyo, 624 pp.
Yeliseyeva, O.P. and Omel'yanenko, B.I. (1988) Behavior of uranium during weathering-crust formation at the Alekseyevka kaolinite deposit, north Kazakhstan. *Geochemistry International*, **25**(1), 83–89.
Yim, W.W.S. (1985) Chrome-bearing spinels and silcrete formation in North West Bay, southern Tasmania. *Search*, **16**, 166–167.
You, C-F., Lee, T. and Li, Y-H. (1989) The partitioning of Be between soil and water. *Chemical Geology*, **77**, 105–118.
Young, R.W. (1978) Silcrete in a humid landscape: the Shoalhaven Valley and adjacent coastal plains of southern New South Wales. In Langford-Smith, T. (ed.) *Silcrete in Australia*. Department of Geography, University of New England, Armidale, 195–207.
Young, R.W. (1981) Denudational history of the South-Central uplands of New South Wales. *Australian Geographer*, **15**, 77–88.
Young, R.W. (1985) Silcrete distribution in eastern Australia. *Zeitschrift für Geomorphologie*, **29**, 21–36.
Young, R.W. (1986) Tower karst in sandstone: Bungle Bungle massif, northwestern Australia. *Zeitschrift für Geomorphologie*, **30**, 189–202.
Young, R.W. and McDougall, I. (1982) Basalts and silcretes on the coast near Ulladulla, southern New South Wales. *Journal of the Geological Society of Australia*, **29**, 425–430.
Young, R.W., Nanson, G.C. and Jones, B.G. (1987) Weathering of Late Pleistocene alluvium under a humid temperate climate: Cranebrook terrace, southeastern Australia. *Catena*, **14**, 469–484.
Zeegers, H. and Lecomte, P. (1982) Seasonally humid tropical terrains (savannas). In Butt, C.R.M. and Zeegers, H. (eds) *Regolith Exploration in Tropical and Subtropical Terrains*. Elsevier, Amsterdam, 203–240.
Zhou, H., Murray, H.H. and Harvey, C.C. (1999) Formation of the Guanshan palygorskite, Anhui Province, P.R. China. In Kodama, H. (ed.) *Clays for our Future. Proceedings of the 11th International clay conference*, 89–96. ICC97 Organizing Committee, Ottawa, Canada.
Zoltai, T. and Stout, J.H. (1984) *Mineralogy: concepts and principles*. Burgess, Minneapolis, Minn., 505 pp.

Author Index

Subject Index